Sponsoring Editor: Anne Kelly
Developmental Editor: Rebecca Strehlow
Project Editor: Carol Zombo
Design Administrator: Jess Schaal
Text and Cover Design: Terri Ellerbach
Cover Illustration: Painting © Helen Hardin 1976, *Prayers of Harmonious Chorus.* Acrylic Painting. Photo © Cradoc Bagshaw 1994.
Photo Researcher: Carol Parden
Production Administrator: Randee Wire
Compositor: Interactive Composition Corporation
Printer and Binder: R.R. Donnelley & Sons Company
Cover Printer: R.R. Donnelley & Sons Company

Faces of Mathematics, Third Edition

Library of Congress Cataloging-in-Publication Data

Roberts, A. Wayne (Arthur Wayne)
Faces of mathematics / A. Wayne Roberts. -- 3rd ed.
p. cm.
Includes index.
ISBN 0-06-501069-8
1. Mathematics. I. Title.
QA39.2.R6 1994
510--dc20
94-2399
CIP

94 95 96 97 9 8 7 6 5 4 3 2 1

A. Wayne Roberts
Macalester College

HarperCollins*CollegePublishers*

Contents

Preface

It has long been held that anyone who aspires to be educated must study mathematics, and *Faces of Mathematics* is intended to be a source book for those who want to see what mathematics can contribute to a liberal education. In particular, it is addressed to college students who plan to take just one or two semesters of mathematics. Perhaps they want to satisfy a curriculum requirement, or perhaps they are prospective elementary school teachers who need a broad perspective on the field, together with some depth in understanding the underlying concepts of elementary mathematics.

A number of books have addressed this audience. Some try to survey the content of mathematics, offering a smorgasbord from which users may choose according to their tastes. Others emphasize the methodology rather than the content of mathematics. The former books are often intriguing but superficial; the latter are impressive but often too difficult for the audience we have in mind.

Those who believe, as I believe, that educated people should study mathematics know that mathematics can help us learn something about thinking itself: how to state our problems clearly, sort out the relevant from the irrelevant, argue coherently, and abstract some common properties from many individual situations. I wish to move toward these goals by steering a middle course among the available texts. Insofar as it is consistent with maintaining a light, readable style that will appeal to our audience, I have selected topics that can be presented in some depth. Moreover, I have continually addressed the larger contention that mathematics is the ideal arena in which to develop skill in the areas of organizing information, analyzing a problem, and presenting an argument.

A Word about the Title

The original title, *Faces of Mathematics*, was chosen for two reasons. The first was to emphasize the fact that mathematics was developed by human beings, real people with real faces. True, they may have had special talents, but on the whole they lived their lives subject to the same constraints as anyone else. Results in mathematics do not arise through divine revelation; they represent the hard work of individual men and women. The faces and brief biographies of many of the most significant contributors to this field appear in this book.

The second goal was to suggest that mathematics is like a finely cut diamond; it must be seen from several sides to be fully appreciated. Each view exposes a new face with its own distinctive features. Four of these faces—solving problems, finding order, reasoning and modeling, and abstracting from the familiar—reflect activities characteristic of mathematicians. The text is organized around these four faces.

An Emphasis on Involvement

A truism frequently seen in mathematics departments proclaims, "Mathematics is not a spectator sport," and I have tried in various ways to get readers involved in the stuff of the subject. To a large extent, I have done this by trying to engage the reader in solving problems.

So what's new about problems in a mathematics course? Plenty, because my purpose is not to develop particular computational skills, but to give insight into how one approaches problems. This enables me to replace the long lists of similar exercises designed to develop skills with puzzle type problems that serve my purposes just as well, and are much more engaging to work on. I believe, moreover, that the principles of problem solving I describe can be carried over into many of the areas of human activity where all of us are called upon to solve problems.

Part I of this book consists, therefore, of three chapters on problem solving. I seek to develop the heuristics that provide a framework for getting started, suggestions for methods that might be tried, and encouragement to think carefully about a proposed solution. These same themes are emphasized throughout the rest of the book, and reminders of problem-solving skills prompt students from the margins of most chapters. I also use a problem to open every section through the rest of the book, a problem chosen to be memorable (the kind that intrigues you, that you tell a friend about during lunch), as well as to be a kind of "hook" for wanting to read the material in that section.

Students should find in the lists at the end of each section in Part I a few problems that engage them in a sustained effort, that provoke discussion, and that require some careful written expression if their solution is to be understood. This kind of involvement is encouraged in Parts II, III, and IV of the book by including after the problem set in each section an outline of a project under the title *For Research and Discussion*.

An Emphasis on the Human Element

Consistent with some of the purposes described in the explanation of why this book is called *Faces*, the biographies of mathematicians scattered throughout the book focus not so much on their work as on other aspects of their lives. I have tried in my choices not so much to include the most important contributors (though that certainly was a consideration) as to emphasize that contributions to mathematics have been international in character, that interest in some problems has spanned generations, that women have made contributions and encountered resistance to their work, that the ugly specter of nationalism has affected the way the subject has grown, and so on. In short, I have tried to emphasize mathematics as human activity.

There are numerous things that most mathematicians would like their neighbors, friends, and even family members to understand about their subject. Many of them are explained in the essays that begin each chapter. Taken together, these fourteen essays give insight into how mathematicians perceive themselves and the work that they do.

Changes in the Third Edition

Those familiar with previous editions of this book might appreciate a summary of what is different in this edition.

The role of problem solving, always cited by users of the book, has been strengthened in three ways. First, Part I, the part that introduces problem solving heuristics, has been enlarged both in coverage and with new problems. It contains Chapter 1, "Getting Started," which focuses on what to do when you don't know what to do; Chapter 2, "Methods of Solution," intent on developing a mental checklist; and Chapter 3, "Reflecting on Solutions," reminding readers that there's more to do than check. Second, the ideas introduced in Part I are emphasized in exposition throughout the rest of the book. Finally, every section of the book now opens with a titled problem that I hope readers will find memorable in the ways explained above.

This edition reflects the author's agreement with that increasing number of mathematics teachers who have come to believe that students should be asked to do some research, writing, and discussion in mathematics just as they are in other classes. Thus, except for Part I where students have long lists of problems that provide for the activities just mentioned, every section of the book ends with a suggestion *For Research and Discussion.*

The essays mentioned above, found at the beginning of each chapter, are set forth with great hope that teachers will delight in assigning them, that students will greatly enjoy reading them, and that discussion of them will help students to better understand not only mathematics, but the people who work at mathematics.

The text has been brought up to date in a variety of ways. Biographic material now includes some living mathematicians; discussion of some recent advances (Fermat's last theorem, the four color problem) emphasizes that mathematics continues to grow as a discipline; outdated material (notably the chapter on computers) has been dropped; and the identification of problems where a calculator might be useful has been replaced with the idea that every student has and will make use of a calculator wherever that makes sense.

Advice to Teachers

This text can be used in a variety of ways. The book contains sufficient material for a full-year (two-semester) course. It is also easy to make selections for the typical semester course offered at many colleges. One-term courses that emphasize problem solving can be built around Chapters 1, 2, 3, 7, 8, and 12, for example. One-term courses that are more philosophical, with particular attention to clear thinking and precise writing, may use most of Chapters 1, 2, 3, 9, 10, 11, 12, 13, and 14. There are many other possibilities. The Dependence Chart that follows the preface will help you design a course to your liking. It illustrates the way chapters build on each other and cluster so that you may rearrange them in logical order.

Supplement Package

The new edition of *Faces of Mathematics* is accompanied by a supplements package intended to meet a range of teaching and learning needs.

Student Solutions Manual, prepared by the author, with complete, worked-out solutions to most odd-numbered section exercises.

Instructor's Manual has three components. In the first the author describes the way in which he handles classroom discussion and presentation of material on a section-by-section basis. This includes discussion-provoking questions and is helpful with difficult material that has been found to be effective. The second section includes two hour exams, each exam based on 100 points, exactly in a form that might be used in a class, for each chapter in the book. Finally there is a section giving answers to even-numbered exercises in the text.

HarperCollins Test Generator/Editor for Mathematics with Quizmaster is available in IBM and Macintosh versions and is fully networkable. The test generator enables instructors to select questions by objective, section, or chapter, or to use a ready-made test for each chapter. The editor enables instructors to edit any preexisting data or to easily create their own questions. The software is algorithm driven, allowing the instructor to regenerate constants while maintaining problem type, providing a nearly unlimited number of available test or quiz items. Instructors may generate tests in multiple-choice or open-response formats, scramble the order of questions while printing, and produce up to

25 versions of each test. The system features printed graphics and accurate mathematical symbols. It also features a preview option that allows instructors to view questions before printing and to replace or skip questions if desired. *Quizmaster* enables instructors to create tests and quizzes using the Test Generator/Editor and save them to disk so that students can take the test or quiz on a stand-alone computer or network. *Quizmaster* then grades the test or quiz and allows the instructor to create reports on individual students or classes.

GraphExplorer With this sophisticated software, available in IBM and Macintosh versions, students can graph rectangular, conic, polar, and parametric equations; zoom; transform functions; and experiment with families of equations quickly and easily.

GeoExplorer Available in IBM and Macintosh versions, this software package enables students to draw, measure, modify, and transform geometric shapes on the screen.

StatExplorer This software package for IBM and Macintosh computers helps students enhance their understanding of statistics by exploring a wide range of statistical representations including graphs, centers and spreads, and transformations.

Acknowledgments

It will be evident to anyone familiar with the first two editions of this book that my greatest debt is to Dale Varberg, my esteemed co-author, who has retired and chose not to participate in the substantial rewriting we felt necessary to bring this book up to date. We are more than co-authors; we are friends, and our wives are friends. Though Dale has not actively participated in the project this time around, it is certainly to him that I have turned for advice on all the questions that occur along the way. The essays introduced in this edition certainly reflect the way the two of us think about our discipline, and discussing the ideas they embody has been one of many joys we have shared together.

Many reviewers lent invaluable support to this revision by reviewing the second edition, the proposed revisions to the text, and the manuscript for this edition. Gratitude is extended to:

Sharon Abramson, Nassau Community College
Charles L. Adie, Northern Essex Community College
Elizabeth A. Calog, St. Louis Community College at Florissant Valley
Ann M. Collier, George Mason University
Duane E. Deal, Ball State University
Normand A. Dion, Franklin Pierce College
George Duchossois, University of South Dakota
Joe Fisher, University of Cincinnati
William M. Fitzgerald, Michigan State University

Art Fruhling, Yuba College
Malcolm Goldman, New York University
Richard Jerrard, University of Illinois
Charles Jones, Ball State University
Susan Knights, Casper College
Warren S. Loud, University of Minnesota
Charles W. Nelson, University of Florida
Stewart M. Robinson, Cleveland State University
Ben Roth, University of Wyoming
Sandra Savage, Orange Coast College
C. Donald Smith, Louisiana State University in Shreveport
Todd M. Swanson, Hope College
Stephen J. Willson, Iowa State University

Many problems carried over from previous editions were initially published in one of several excellent collections, including *Amusements in Mathematics* and *536 Puzzles and Curious Problems* by H. E. Dudeney, *Puzzles for Pleasure* by E. R. Emmet, *Puzzles in Math & Logic* by A. J. Friedland, *Mathematical Puzzles* by Martin Gardner, *Litton's Problematical Recreations* by J. F. Hurley, *The Moscow Puzzles* by B. A. Kordemsky, and *Puzzles, Patterns, and Pastimes* by C. F. Linn.

A Special Word to Students

Many years of teaching have convinced me that most students who fall within this book's intended audience approach mathematics with fear and trembling. I have made every effort to ease this anxiety by using simple examples, clear explanations, and a limited technical vocabulary. My aim is to demonstrate that mathematics is interesting, relevant, and learnable.

I believe that problem solving is the heart of mathematics, and that you must try problems—many problems—if you are to understand the subject. Every section begins with a problem; every section ends with a host of problems for you to try. They are carefully arranged in order of increasing difficulty, the most challenging being identified with an asterisk. Be sure to work at the problems; it is the only way to learn mathematics. It is also the activity most likely to help you in later life.

A. Wayne Roberts

Dependence Chart

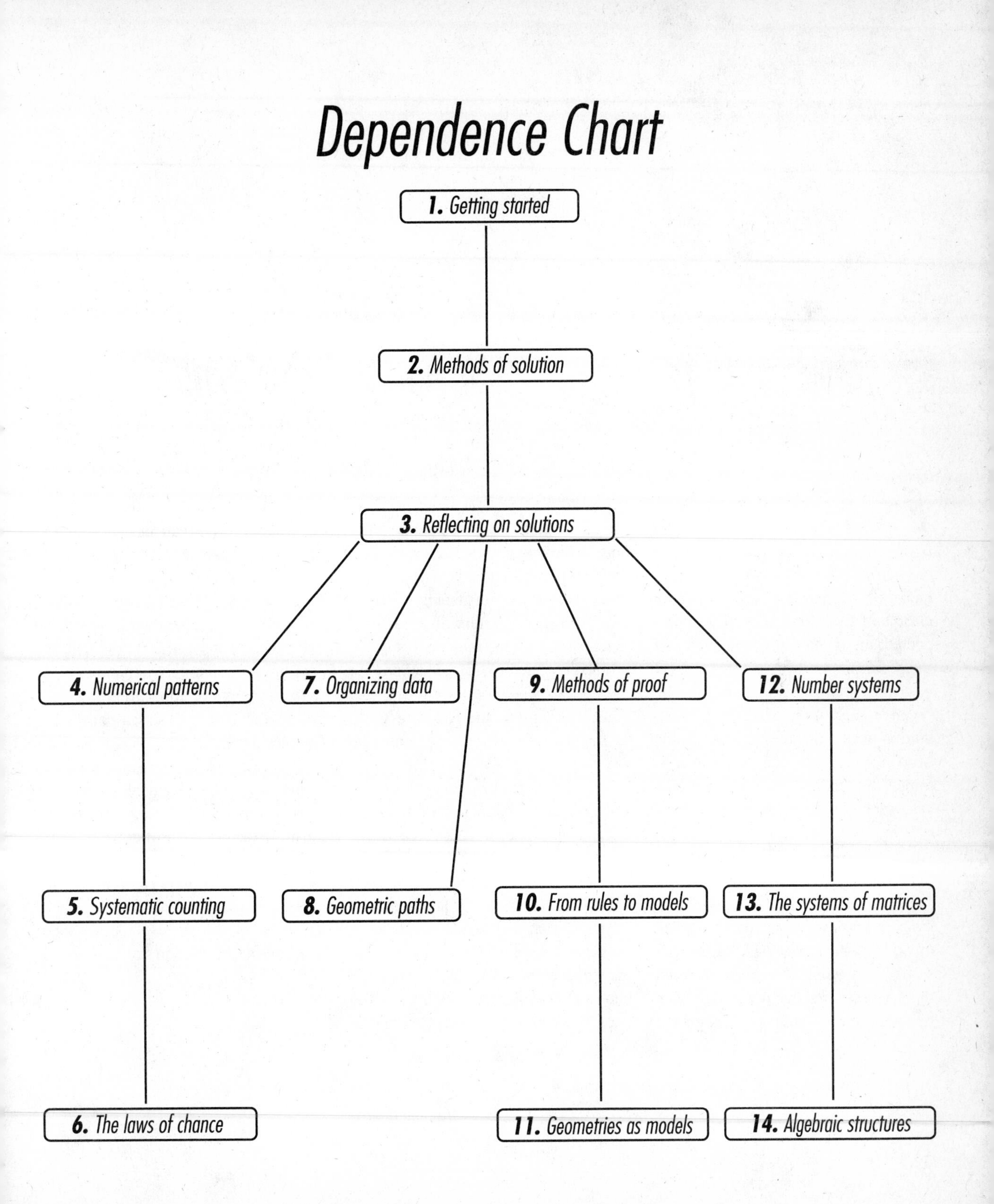

part I
Solving Problems

Every subject has its problems. In contrast, however, to many useful areas of human inquiry (medicine, psychology, economics, . . .) where a clear and final answer is seldom expected, mathematical problems admit the possibility of incontestably correct answers. They therefore afford us an excellent medium in which to focus attention not on the answers, but on the thinking process by which they are obtained. In Part I, we undertake such a study, always looking for principles that may be applied to any setting in which we are called upon to solve a problem.

It is quite common to hear someone say, when reflecting on a finished piece of work, "The hardest part was getting started." A small child, told to clean the playroom, may be reduced to tears by the hopeless feeling of not knowing where to start. English courses offer units on how to get started on an essay, and still professional writers talk about the horror of facing a blank page. Chapter 1 offers an outline of practical steps, illustrated with specific problems, that provide guidelines for those times when we have the feeling that, "I haven't a clue how to begin."

In the same way we contend that one can develop a mental framework to sort through almost automatically whenever a problem is to be solved. What committee or task force is not helped greatly by a member who has a knack for trying a small example first, who is willing to float some trial balloons so that reactions will quickly show what might be wrong with them, or who sees how to partition the job into manageable parts that can be worked on one at a time. Chapter 2 suggests that one can indeed adopt a mind-set for problem solving.

Finally, we suggest that much can be learned from reflecting on solutions, comparing them with our original expectations, subjecting them to credibility checks, or examining them to see if they have application to other problems. That is the emphasis of Chapter 3.

It is not essential for our purposes to consider only practical problems. What is essential is that our problems illustrate the principles that we have in mind, that they pose a challenge—yet seem enough within grasp to be tantalizing—and that they draw out from our imagination creative ideas about which we are pleased to say, "I thought of that."

A great discovery solves a great problem but there is a grain of discovery in the solution of any problem. Your problem may be modest; but if it brings into play your inventive faculties, and if you solve it by your own means, you may experience the tension and enjoy the triumph of discovery.

GEORGE POLYA

George Polya was known for his ability to ask fruitful questions, for his ability to penetrate classical fields of mathematics so deeply as to notice remarkable relationships that had escaped previous investigators, and for the breadth of his work. He was born in Hungary, making him one of the disproportionately large number of famous mathematicians to come from that small country. After teaching for 26 years at the Swiss Federal Institute of Technology in Zürich, he went to Stanford where he wrote his 250th mathematical research paper at the age of 97.

He attributed his ability to contribute to so many areas of mathematics to the fact that his own training did not start with mathematics, but moved progressively from law ("This I could stand just for one semester.") to languages and literature (He was certified in Hungary to teach Latin.) to philosophy, then physics, and finally mathematics. He is often quoted as saying, "I thought I am not good enough for physics and I am too good for philosophy. Mathematics is in between."

Polya's research in mathematical analysis, probability, combinatorics, geometry, algebra, and number theory assured him a place of honor among the world's leading mathematicians of the 20th century. It is all the more remarkable then, that he also took a lively interest in how best to teach the art of problem solving.

His books, *How to Solve It,* 2nd ed. (Garden City, N.Y.: Doubleday, 1959), *Mathematics and Plausible Reasoning* (2 vols.) (Princeton, N.J.: Princeton University Press, 1954), and *Mathematical Discovery* (2 vols.) (New York: Wiley, 1962), exhibit mathematical writing at its best, and have led teachers around the world to emulate his "Guess and Test" style of classroom teaching. His ideas have certainly influenced the approach in Part I of this text.

George Polya
(1887–1985)

chapter 1

Getting Started

Success in the solution of a problem generally depends in a great measure on the selection of the most appropriate method of approaching it.

W. A. WHITWORTH

That's a Great Problem!

Everyone associates mathematics with problems. Unfortunately, the ones most people think of are the drill problems that came at them in lists of 30 or so in their school courses. Such people may be excused for regarding as a bit peculiar anyone who could refer to a problem in mathematics as great.

Yet, mathematicians do delight in what they might call a great problem. Moreover, though there will be variations in taste, we can be rather specific about the characteristics of a great problem. To illustrate them, let us consider a problem that every mathematician would call great. It was first posed by Pierre de Fermat in 1637.

> The equation $x^2 + y^2 = z^2$ is satisfied by many sets of integers: $x = 3$, $y = 4$, $z = 5$, and $x = 5$, $y = 12$, $z = 13$ are but two sets of three integers that work. Can you find, for any $n > 2$, a set of three integers that satisfy $x^n + y^n = z^n$?

Fermat, who enjoyed working puzzles, wrote in the margin of a book that he had a truly wonderful proof that the answer to the puzzle was no, but that it was unfortunately too long to fit into the margin.

What makes this such a great problem? First, it is easy to state to virtually anyone. Long technical explanations, or technical definitions are unnecessary.

Secondly, there is an element of surprise. Since it is so easy to find many different sets of integers that satisfy $x^2 + y^2 = z^2$, why would it not be relatively easy to find a set of integers for which $x^3 + y^3 = z^3$? or $x^4 + y^4 = z^4$? Ah! Try it.

The problem also seems tractable; that is, there seem to be ways to "get a handle on it." Since $x^2 + y^2 = z^2$ may be related to the sides of a right triangle, perhaps $x^3 + y^3 = z^3$ can be related to some geometric figure. People strong in algebra will recall that $x^3 + y^3 = (x + y)(x^2 - xy + y^2)$, and may try to make something of that.

Let us not minimize that the problem was posed by someone who had mathematical taste. Fermat was perhaps the best amateur mathematician who ever lived, and it was known that he had a knack for posing problems that, if worked on, might lead one to a lot of interesting discoveries.

Not only was the proposer known to be a great mathematician, but other great mathematicians are known to have tried it; and tried it; and tried it. For over 350 years they have been trying it. It has staying power.

Easy to state in nontechnical terms, having an element of surprise, tractable, connected to other interesting ideas, capable of exciting the interest of others, not bounded by place or time; these are the characteristics that make a great problem. Though the problems we have chosen to open each section of this book are much simpler than Fermat's problem, we have chosen them with these characteristics in mind, hoping that all readers will find themselves saying at least once in a while, "That's a great problem."

1.1 Clarify the Problem

Here's the Pitch

Suppose that in a regulation nine-inning baseball game, a major league pitcher gets credit for a complete game that is not shortened by rain or for any other reason. Suppose further that this pitcher accomplishes this feat by throwing the minimum number of pitches possible. How many?

The goal of this chapter is to help you get started when confronting a problem. We use mathematical problems to give us something of substance to think about, but our thesis is that the guidelines we provide here can give you a helpful way to get started on any problem that comes your way. In this section we provide some initial steps you might take in order to clarify in your own mind just what the problem is that you are being asked to solve.

How Did This Problem Come Up?

Homer Slideby, who normally avoids the Mathematics Department, stopped in one day to ask Professor Branebom how to solve

$$3s + 6g = 310$$

She pointed out that there were many solutions: $s = \frac{10}{3}$, $g = 50$, or $s = 100$, $g = \frac{5}{3}$, or . . . When Homer interrupted to say that he expected an answer of about $s = 50$ and $g = 25$, Professor Branebom asked where this problem had come from.

Homer explained that he had been in charge of ticket sales for a production by theater students, and that he was now trying to figure out how many student tickets at \$3 per ticket and how many general admission tickets at \$6 had been sold.

When Professor Branebom understood that s and g represented the number of student tickets and the number of general admission tickets sold, it was clear to her that there would be no applause for Homer's performance. She reasoned that if s and g were whole numbers (positive integers in her language), then $3s + 6g$ would also be an integer. Moreover, since $3s + 6g = 3(s + 2g)$, any integer choices for s and g would give an integer that was a multiple of 3. Since 310 is not a multiple of 3, Homer's problem as stated had no solution. A mistake had been made in making change at the ticket booth, or in counting the money; 310 was not a possible total if everything had been done correctly.

For Professor Branebom to come to the right conclusion, she needed first to see the problem in its natural context. Problem solving does bear similarities to a good play. We have a better chance of making sense of it at the end if we don't try to come in during the middle of the act.

All the world's a stage,
And all the men and women merely players:
Shakespeare

Here is another problem that underscores the same point. Consider Amy's plight when the battery in her watch went dead. She had to choose one morning between two old "wind-up" watches in her drawer. One, she knew, gained 8 minutes a day. The other, she had observed while doing some serious clock watching in one of her classes, lost 20 seconds in an hour. Which watch, she wondered, was best?

We first observe, since $20(24) = 480$ seconds $= 8$ minutes, that both watches, 24 hours after having been set, differ from the correct time by 8 minutes. To answer the question, we need to know more about the criteria for determining "best."

If Amy has a part-time job driving the campus bus, the watch that has her running early will leave her riders in hot pursuit of her as well as her bus; but if her main concern is to get to lunch on time, then the fast watch really is ahead in the running. Rather than asking which watch is "best," it might be better to ask which watch will best serve Amy's needs.

This problem, like the one before it, underscores our first suggestion for anyone attempting to solve a problem. Try to understand the context in which the problem arose.

Try Some Small Examples First

If asked a question that involves big numbers, try to ask a very similar question that involves smaller numbers. If a problem suggests that some property holds for all integers n, try writing the property out for small cases. It is never a bad idea to begin by accumulating some data.

Consider the problem of finding the sum

$$S = 1 + 3 + 5 + \cdots + 99.$$

Rephrased, this asks for the sum of the odd numbers from 1 to 99. Why not get some experience by finding the sum of the odd numbers from 1 to 3? from 1 to 5? from 1 to 7?

$$1 + 3 = 4$$
$$1 + 3 + 5 = 9$$
$$1 + 3 + 5 + 7 = 16$$

Do you notice anything about the sums? (If you don't notice anything, go do something else for a while.) The three statements could be written this way.

The sum of the first 2 odd integers is 2^2.

The sum of the first 3 odd integers is 3^2.

The sum of the first 4 odd integers is 4^2.

Which odd number is 99? Note that 3, the second odd integer, is $2(2) - 1$; 5, the third odd integer, is $2(3) - 1$. In the same way, $99 = 2(50) - 1$ is seen to be the 50th odd integer, and a reasonable guess is that

The sum of the first 50 odd integers is 50^2.

Now try this one.

A magician calls for a volunteer from the audience and instructs the Poor Nut to pick a number between 1 and 20. The Poor Nut does so, without revealing the number to the magician, of course. The magician then gives the following instructions:

(a) multiply the number by 2
(b) subtract 1
(c) multiply by 3
(d) add 1
(e) divide by 2

The Poor Nut does all of these things, and then responds to the magician's question by telling him that the resulting number is 32. What number does the magician announce as the number chosen by the Poor Nut?

Let's put ourselves in the place of the Poor Nut, choose some numbers, and do what we're told.

	Step a	*Step b*	*Step c*	*Step d*	*Step e*
Pick 5	10	9	27	28	14
Pick 10	20	19	57	58	29
Pick 15	30	29	87	88	44

There are now two ways to proceed.

From the evidence accumulated so far, we might just guess. We came close to 32 with a pick of 10, and bigger choices of initial picks seem to give bigger results. Go ahead; make a guess. Then check it.

We might also try to follow what happens to a first pick of n. We get

	Step a	*Step b*	*Step c*	*Step d*	*Step e*
Pick n	$2n$	$2n - 1$	$3(2n - 1)$ $= 6n - 3$	$6n - 2$	$\frac{1}{2}(6n - 2)$ $= 3n - 1$

Notice that this formula works for each n we tried. Now ask what value of n gives $3n - 1 = 32$.

Don't Impose Conditions That Aren't There

There was a blind beggar who had a brother, but this brother had no brothers. What was the relationship between the two?

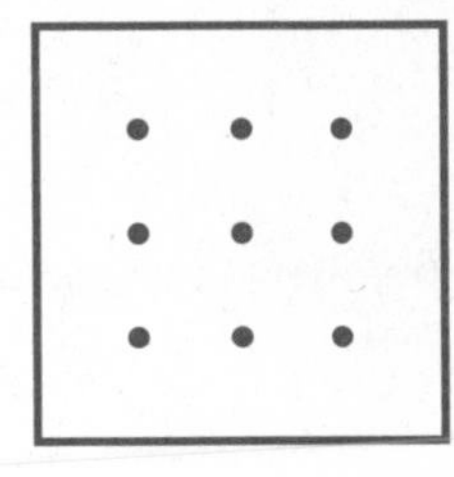

If you have trouble answering this question, it is because you are imposing a condition that isn't there. Try again.

Here is another one to try. Without taking your pencil off the paper, draw four straight line segments that will pass through the nine dots in the margin. Again, if you have trouble, it is because you are unconsciously assuming something that is not stated in the problem.

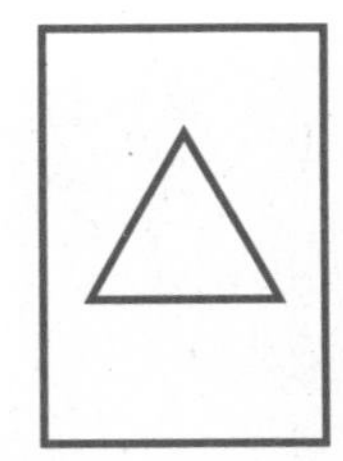

Three points can be arranged so that the three distances they determine are equal. Can four points be arranged so that the six distances they determine are equal?

It is quite common for us to make unconscious assumptions that then lead us to work on a problem quite different from the one given to us. The problems above, trivial though they be, are interesting because people often assume, in turn, that all blind beggars are male, that line segments cannot extend beyond the square containing the given points, and that the four points all have to be in the same plane.

You can see if you're getting the idea by trying this.

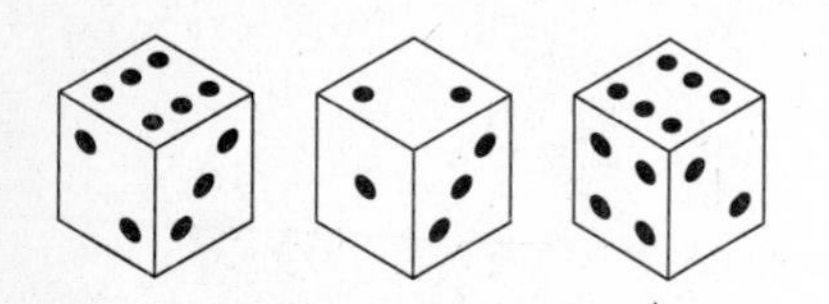

In the figure at the left, you are given three views of the same cube. How many dots are there on the bottom face (opposite the six) in the left-most view?

Friedland, *Puzzles in Math and Logic*

Our opening problem, **Here's the pitch,** fits into a discussion of making assumptions unconsciously, but people usually need a little longer to catch this one. Some get hooked right away by the notion that this would be a great pitching performance, and so get drawn into this line of reasoning. This is great pitching, and the supreme accomplishment of a pitcher is to strike out a batter on three pitches. The answer is therefore found by multiplying 9 (the pitches necessary to strike out 3 batters in an inning) times 9 (the number of innings) to get 81. That is way off base.

The idea that this is a great pitching performance is one imposed by the problem solver. The side may have been retired on three towering fly balls, each caught by a desperate outfielder showing his backside to the pitcher. That could happen on three pitches, and it's the kind of thinking that leads to the most common answer of $3 \times 9 = 27$ which is much closer, but still wrong. The hint to Problem 26 at the end of this section will help you if you ultimately need a reliever on this one.

Our second maxim for problem solvers is, therefore, that they be careful not to place restrictions on themselves not imposed by the problem itself. By the way, in the problem showing three views of a cube, did you assume that the cube pictured is a standard die?

Strip the Problem of Irrelevant Details

We have already considered the problem in which the magician asks the Poor Nut to choose a number between 1 and 20. Was it essential (or even relevant) to the "trick" that the number be between 1 and 20? Not at all; that condition is what magicians call a distracter, intended to get the audience thinking about something irrelevant to what is really important.

Problems, especially those from real life, often come to us cluttered by details having nothing to do with finding a solution. They are usually not intended to hinder us, but they do serve as distracters, and our first job is to pare them away so as to expose the core of the matter.

A newspaper account tells of an unlikely and tragic accident in which a man leaves his home in New York City at noon on a fine sunny day to surprise his family by joining them on a vacation they are having in the mountains. After making just 35 miles in the first hour and a half because of traffic congestion, he is able to increase his speed to 55 mph. Unbeknown to him, his wife leaves the cabin at 1:15 that same afternoon, hurrying into the city at 65 mph to bring a sick child home. Rain develops along the way, and in a highly improbable event, his car skids on a curve and hits the one being driven by his wife. Both are killed. "Who'" wonders Grandpa Uphill as he reads about it, "was closer to home when it happened?"

There is hardly a piece of information in the whole paragraph that has anything to do with the question asked. Stripped of the details that make the story newsworthy, the question is this: if two objects (cars) are at the same spot, which of them is farther from another spot (a home in New York City)? So stated, it is obvious that they are equally far from home.

That was really too easy. See if this one mixes you up.

Consider two cylindrical glass containers, one holding 2 liters of water, the other 1 liter of red wine. Ten milliliters of water are transferred to the container of wine and mixed in thoroughly. Then ten milliliters of the mixture is transferred back to the container of water. Is there then more wine in the water or water in the wine?

Everyone sees that the redness of the wine and the shape of the containers is irrelevant. It generally takes a little longer to realize that the amounts of water and wine we start, with the amounts transferred (provided they are equal), and the thoroughness of the mixing are also irrelevant. The only thing that really counts is that we end up with as much liquid in each container as we had in the beginning. Then it is clear that the water removed from the container of water has been replaced by wine, and the wine removed from the container of wine has been replaced by water. Thus, there is exactly as much wine in the water as there is water in the wine.

Now ask yourself the same question if 100 two-step transfers were made.

Rózsa Péter
1905–1977

Rózsa Péter

Rózsa Péter started out to study chemistry, but was drawn to mathematics by what she believed to be its intrinsic beauty. She graduated in 1927, hoping to teach secondary mathematics in Hungary, but did not succeed in finding a full-time position until 1945. She started her mathematical research in number theory, but was much discouraged when some of what she discovered turned out to be a rediscovery of results already published by others. In the 1930s, captivated by the beauty of some very theoretic and seemingly esoteric results, she saw how to use the same methods to develop the new field of recursive functions. Years later, distinguished as the first Hungarian female mathematician to become an Academic Doctor of Mathematics and honored with numerous prizes, she said

> "I would not have dreamed that this theory could also be applied practically. And today? My book on recursive functions was the second Hungarian mathematical book to be published in the Soviet Union, and precisely on the practical grounds that its subject matter has become indispensable to the theory of computers." [Rózsa Péter, "Mathematics is Beautiful," *Mathematical Intelligencer,* 12, No. 1, 1990, pp 58–64]

Interested in mathematics education, she once wrote, "It is a great misconception that the mathematician mainly calculates. I have picked out mathematically minded students with the following riddle." The riddle she used is the one posed in the text about wine in the water.

This paper is a translation, with excerpting and editing by the Intelligencer, of an address that Rózsa Péter delivered to high school teachers and students in 1963 in Rostock, German Democratic Republic. The original version was published in Mathematik in der Schule 2 (1964), 81–90.

Summary

Most of us have served on ad hoc committees set up to deal with a specific problem. We can recall that the committee did or could have profited from someone being persistent in determining just what problem the committee was expected to address. The gathering of some data or specific information is also to be recommended as an early step, and it is a rare committee indeed that does not have to be drawn back from discussion that is irrelevant to the problem at hand.

A medical doctor meeting for the first time with a new patient, or roommates trying to reorganize the space in their room might similarly benefit from the outline we are proposing.

A. How did this problem come up? What are we trying to achieve?
B. Can we try a few small tests here before we launch into something massive?
C. Let's try to identify and ignore things irrelevant to solving the problem before us.

Can we give advice to the person who gets stuck on almost everything? The best advice is probably this: don't try to solve a problem too quickly. Force yourself to go through the steps, simple as they seem, that are listed above.

If, after a real effort, you are still stumped, note that hints are given for selected problems at the very end of the section. Since it is hard to give hints that do not immediately ruin the problem, you are encouraged to refer to the hints only as a desperation measure. Remember that this is quite different from books you may have used in which you learn a technique and then practice it on a list of very similar problems. Rather than using the hints to solve as many problems as you can, you should, in the spirit of our opening comments, preserve for yourself in as many cases as possible the right to say of a solution, "I thought of that."

Problem Set 1.1

1. If there are 12 one-cent stamps in a dozen, how many two-cent stamps are there in a dozen?

2. Homer had seven apples and ate all but three. How many were left?

3. A chemist discovered that a certain reaction took 1 hour and 20 minutes when he wore a blue tie, but only 80 minutes when he wore a red tie. Why the difference?

4. A patient called to say that he had to cancel his appointment and wanted to reschedule it. The receptionist entered the cancellation in the computer, and then called upon the search routine to find the next available appointment. What did she do wrong?

5. Five apples are in a basket. How can you divide them among five girls so that each gets an apple but one apple remains in the basket?

6. Which is worth more: 1 pound of $10 gold pieces or $\frac{1}{2}$ pound of $20 gold pieces?

7. A ship stands anchored offshore with a rope ladder hanging over its side. The ladder has 12 rungs, and the distance between the rungs is 12 inches. At 4:00 P.M. the ocean is calm, and the bottom rung just touches the water. But then the tide begins to come in, raising the water level 3 inches every hour. When will the water just cover the third rung from the bottom?

8. An apartment house has six stories, each the same height. How many times as long will it take me to run up the stairs to the sixth floor as it does to run up to the third floor? Assume that I start at the first floor.

9. Homer always buys the same style of socks, choosing either green or blue. He knows that there are 28 pairs of socks in his drawer, but in the dark he cannot tell colors. How many socks should he take out to the light to be sure he has a match?

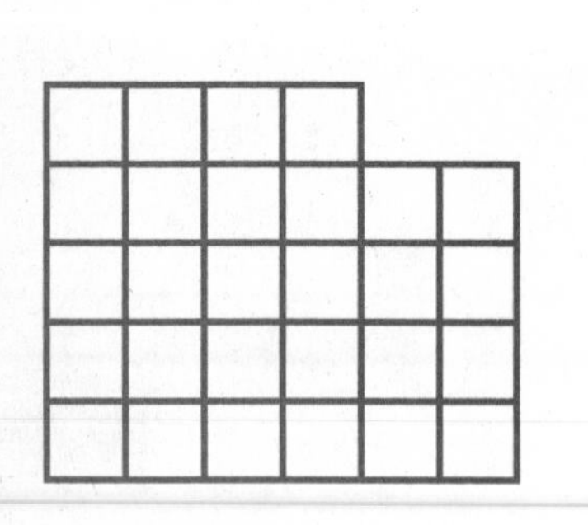

10. Three kinds of apples are mixed in a basket. How many apples must you take out to be sure of getting at least two of one kind? At least three of one kind?

11. If 3 hens lay 3 eggs in 3 days, how many eggs will 300 hens lay in 300 days?

12. If 3 cats catch 3 rats in 3 minutes, how many cats will catch 100 rats in 100 minutes?

13. Grandma Uphill makes tasty French toast in a pan that holds only two slices. After browning one side of a slice, she turns it over. Each side takes 30 seconds. How can she brown both sides of three slices in $1\frac{1}{2}$ instead of 2 minutes?

14. Trace and then cut the top figure in the top margin to make two identical pieces. Make your cuts along the lines.

15. Without raising your pencil from the paper, draw five straight line segments connecting all 12 of the displayed dots and make sure you end where you started.

16. Without raising your pencil from the paper, draw six line segments connecting the 16 dots shown in the margin. Can you do this in such a way that you end where you started?

17. In a rectangular room, how can you place 10 small tables so there are 3 on each wall?

18.* Zippo is an intercity messenger service which provides 1-day delivery for any box, the maximum dimension (length, width, or height) of which does not exceed 15 inches. The architectural firm of Woff and Toppel wishes to send an acetate drawing which cannot be folded but can be rolled. The drawing is 40 inches by 24 inches. How can they use the Zippo service?

19.* A man walks 1 mile south, 1 mile west, and then 1 mile north, ending where he began. Where did he start his journey? Is there more than one answer?

20.* Three pennies can all be placed in contact as indicated in the diagram. Four pennies can be placed so that each touches all the others by laying the fourth on top of the other three. How can you place five pennies so that each penney touches the other four?

1 2 3 4 5 6
7 8 9 10 11 12
13 14 15

21.* Homer has five pieces of chain which he wants to join into a long chain (see diagram below). He could open ring 3 (first operation), link it to ring 4 (second operation), then open ring 6 and link it to 7, etc. In this way he would complete the job in eight operations, but it can be done in six. How?

22.* Hugo has six pieces of chain, each with four links. It costs one cent to cut a link and two cents to weld one closed. How can the six pieces be fashioned into one chain for as little money as possible?

23. Joan admires her Grandmother's gold chain, and so it is agreed that Joan will weed Grandma's garden this summer on 23 days, one for each link in the chain. Joan insists on being paid daily, so that she has n links after the nth day of work. Grandma agrees to this, but points out that they should be willing to exchange cut links or pieces of chain so as to minimize the number of links to be cut. They agree to this. What links (they are numbered for convenience in the figure) should be cut to achieve this goal?

1
3
5
7
9
11
13
15
17
19
21
23

24. At age 24, when he had just $1000, Quigley P. Sharp discovered a way to double his savings every three years. At age 39, he realized that he had already saved half as many years as would be necessary to reach his retirement goal. At what age did he have half as much money as he wanted for retirement?

25. If you have not done so already, now is the time to get to the bottom of the first (left) view of the cube depicted in this section.

26. What is the minimum number of pitches that a pitcher must throw if he pitches an entire game not cut short for any reason?

Hints to Selected Problems

2. Four is wrong. Read the question again—carefully.
5. Don't add conditions not stated.
7. Ships float.
9. How many socks can he take out *without* getting a match?
10. How many apples can you take out that are all different?
11. How many eggs do three hens lay in 300 days?
12. How many rats can three cats catch in 1 minute?
15. Start drawing horizontally from a point 2 units to the left of the upper left dot.
17. Can a table be on two walls?
18. This is a cockeyed problem.
19. If you thought of the North Pole, fine. But don't stop with that. There are many other answers.
21. It's as easy as 1, 2, 3.
25. In the left view, draw a line through the two dots on one side and another through the three dots on the other to form a V. Note that this V opens toward a side showing six dots. Now draw the V determined by the two dots and three dots in the middle view.
26. If the home team is ahead after the visitors bat in the first half of the ninth inning, the last of the ninth is not played; but don't jump at 24.

1.2 Organize the Information

The Handshake Problem

Dick and Jane (leaving Spot at home) went to a party at which there were three other couples. A great deal of handshaking went on. Jane observed that no one shook hands with his or her partner, and that if she excluded herself, everyone at the party shook hands with a different number of people. Jane herself also shook some hands. With how many people did Dick shake hands?

Would you agree that our thinking about the problem of wine in the water (page 12) was somehow helped by the picture of the two glass containers? The picture, in the end, had very little to do with the solution, but it got us started.

It seems always to help if we can find some new way to represent the information in a problem. We might do no more than to break a complex paragraph into a list of short statements, each statement declaring just one bit of information. But we can also use pictures, proportional line segments to represent propotional quantities, diagrams, tables, or clever representations that bring out symmetries not previously noticed. Good problem solvers look for several different ways to organize the same information

Draw a Picture

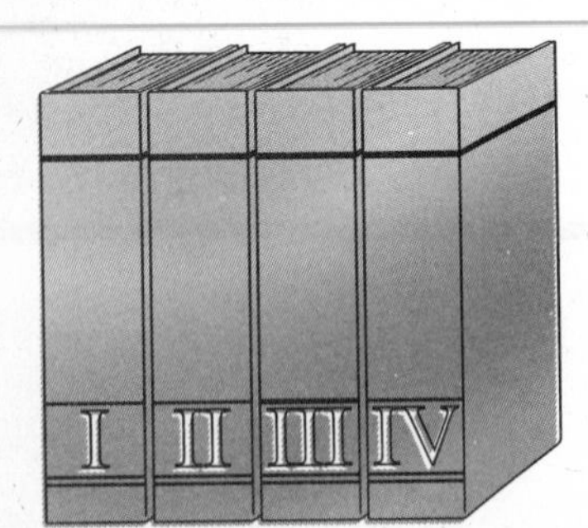

Hugo Hardback, the librarian, has in his collection of rare books a Greek lexicon which was published in four volumes, each having exactly the same number of pages. Unfortunately, the volumes have stood lo these many years, side by side, volumes I, II, III, and IV, without ever having been used. In fact, a bookworm has bored a hole straight through the set, from the first page of volume I to the last page of volume IV, taking care, however, not to betray its presence by boring through the outside covers of the first and last volumes. The pages in each volume measure 2 inches from first to last, and each cover is $\frac{1}{8}$ inch. Assuming that the bookworm has accomplished its feat with the shortest tunnel possible, how long is the tunnel?

Most people, especially if they hear this problem orally, take it as a challenge to add the given measurements in their head. Those who are successful in this mental exercise usually respond with $8\frac{3}{4}$ inches, opening themselves up to the observation that they don't spend enough time in the library.

Would you have drawn a picture for yourself if one had not been provided in the margin? This problem is tricky, and without a picture, you may be tempted by Problem 24 at the end of this section to dig in too far.

Make a Diagram

Each day at noon a ship leaves New York for LeHavre, France, and another ship leaves LeHavre for New York. The trip across the Atlantic takes six full days. How many of the ships that left LeHavre will a ship leaving New York meet on the high seas (as opposed to meeting in the harbor) during the journey to LeHavre?

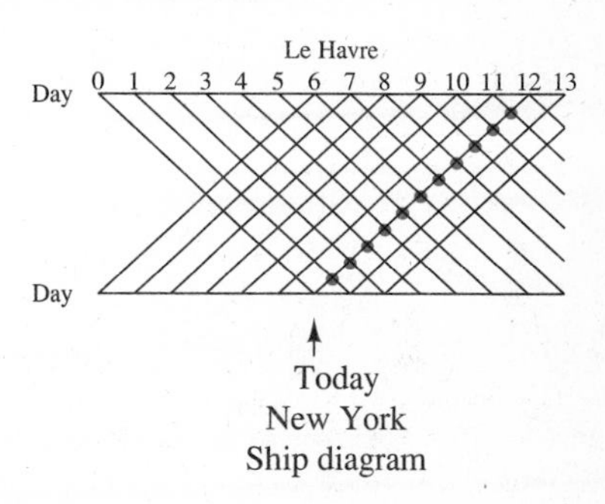

Ship diagram

If you answer 6, you are forgetting about the ships already enroute. The diagram at the right clarifies matters. A ship leaving New York today will encounter 11 ships on the high seas. If we include the ships it meets in each of the harbors, the answer is 13.

A well chosen diagram helps greatly with **The Handshake Problem** at the top of this section. Think of labeling each of the partygoers, excluding Jane, with the number of hands they have shaken. Since no one shook hands with a partner, the largest label we could possibly have is 6, and since there are seven people to be labeled, each with a different integer, we shall need all of the possible labels, 0, 1, 2, 3, 4, 5, 6. Picture the partygoers standing in a circle, wearing their labels.

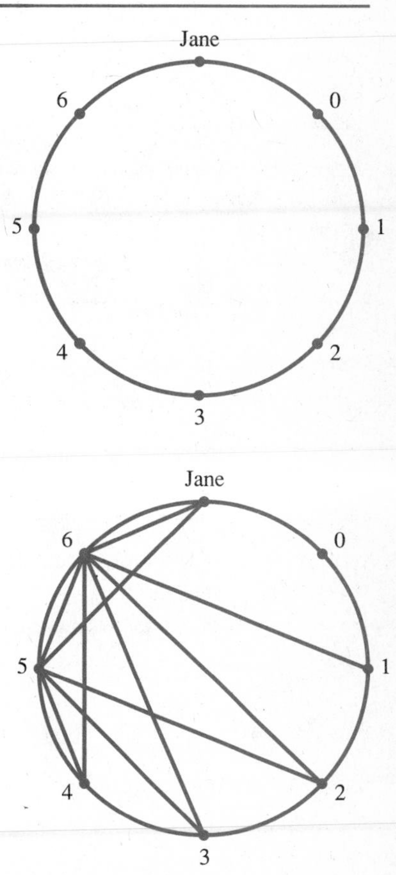

Now we shall connect with a line all those who shook hands. Begin with No. 6, since it is obvious that that person shook hands with everyone not labeled 0. Note that No. 6 must have No. 0 as a partner; the most gregarious is paired with the one most shy (or aloof—we cannot tell from the information given; but see how easily we stray to what is irrelevant.)

Move on to No. 5, remembering that No. 1 is no longer available to shake any hands. Who is the partner of No. 5? Continue until you have, as they say, come full circle. How is Dick labeled?

Make a Table

Besides providing a succinct summary of information, the advantages of a table in problem solving are that it gets you started, it encourages you to proceed step by step as you see new ways to deduce bits of information, and it focuses your attention on what remains to be done. We can illustrate all of these ideas and a few more by solving the following puzzle.

Allen, Bill, Clint, and Dan live in Allentown, Billings, Clinton, and Danville, but none of the four live in the town associated with their name. Danville is not the town where Allen lives. The hometown of Bill has the name associated with the man who lives in the town that has the name of the man who lives in Clinton. Where does Dan live?

Before setting up a table to organize what we know and keep track of our progress, let's agree on some notation, and then illustrate the idea of breaking a complex paragraph into a list of declarative sentences. We shall designate the four men as a, b, c, d and the towns with the associated names as A, B, C, D.

(1) a does not live in D.

Adopt notation that is suggestive of its meaning.

Since we do not know the hometown of Bill nor the name of the man who lives in Clinton, designate them respectively by X and y. The confusing sentence now says

(2) b lives in X.
(3) x lives in Y.
(4) y lives in C.

Method consists entirely in properly ordering and arranging things to which we should pay attention.

René Descartes

Observe immediately from (2) that

(5) The man called x is not Bill; i.e. $x \neq b$ and from (4) that
(6) The man called y is not Clint; $y \neq c$

Putting what we know so far in a table creates a pretty sparse table, but it establishes the format we intend to follow.

Towns	*A*	*B*	*C*	*D*
Residents				not a

Let's start at the very beginning, a very good place to start.

Rodgers and Hammerstein

Might x = a? Then (1), together with (3) would imply that $y \neq$ d, leaving y = b as the only possibility. Our table would be

Towns	*A*	*B*	*C*	*D*
Residents	b by (2)	a by (3)	b by (4)	not a

This is impossible: $x \neq$ a.

Since (5) tells us that $x \neq b$, we move on to consider the possibility that $x = c$. That would allow y to be a, b, or d. We explore each possibility separately.

	A	B	C	D	Objection
$y = a$	c by (3)		b by (2) a by (4)	not a	Two men in one town
$y = b$		c by (3)	b by (2)	not a	No place for a
$y = d$			b by (2) d by (4)		Two men in one town

The conclusion is that $x \neq c$. We are down to only one possibility: $x = d$. Our table takes on a little more permanence.

Towns	A	B	C	D
Residents				b by (2)

We said that one advantage of using a table is that it provides step-by-step encouragement as you see it getting filled in. We hope you are now encouraged enough to complete this as Problem 26.

This may be a good place to emphasize again that the way we have organized the information for this problem is not the only way, or perhaps even the best way, to do it. You may choose, for instance, to put the towns around the edge of a circle, and to think of trying to get the four men in the center of the circle each back to his own town. Organize in whatever way appeals to you, and if you get hung up on your first effort, try an entirely different scheme.

Keep a Systematic Record

In how many ways can you add eight odd numbers together to get 20? Numbers may be repeated, but changes in order are not to be counted as new solutions.

This type of problem demands a system. If one looks for solutions in a random way, he or she is almost bound to miss some.

As in packing a suitcase, it is wise to start with the largest items and work down. Now, we can't use 19, 17, or 15 in the solution. They use up too much space—it's impossible to add seven more odd numbers and still get 20. But 13 works:

$$13 + 1 + 1 + 1 + 1 + 1 + 1 + 1 = 20$$

Next we try 11, then 9, etc.:

$$11 + 3 + 1 + 1 + 1 + 1 + 1 + 1 = 20$$
$$9 + 5 + 1 + 1 + 1 + 1 + 1 + 1 = 20$$
$$9 + 3 + 3 + 1 + 1 + 1 + 1 + 1 = 20$$
$$7 + 7 + 1 + 1 + 1 + 1 + 1 + 1 = 20$$
$$7 + 5 + 3 + 1 + 1 + 1 + 1 + 1 = 20$$
$$7 + 3 + 3 + 3 + 1 + 1 + 1 + 1 = 20$$
$$5 + 5 + 5 + 1 + 1 + 1 + 1 + 1 = 20$$
$$5 + 5 + 3 + 3 + 1 + 1 + 1 + 1 = 20$$
$$5 + 3 + 3 + 3 + 3 + 1 + 1 + 1 = 20$$
$$3 + 3 + 3 + 3 + 3 + 3 + 1 + 1 = 20$$

Note that we have placed the bigger numbers to the left. It gives us a systematic way of writing down all 11 solutions.

Here's a slightly greater challenge.

The digits of 261, for example, add up to 9; so do those of 522. How many of the thousand integers 1, 2, 3, . . . , 1000 have digits that add up to 9?

If you were being paid by the hour, you could of course list all one thousand integers, add the digits in each case, and then count those that summed to 9; but if I were the one paying you by the hour, I'd look for someone else to hire for the next job I wanted done. I'd look for someone who could organize the attack a bit.

Here is one way to organize the job. Look for such integers that have a zero in the units column, then look among those that have a 1 in the units column, etc.

Units Col.	*0*	*1*	*2*	*3*	*4*	*5*	*6*	*7*	*8*	*9*
	90	81	72						18	9
Listed	180	171	162						108	
integers	270	261	252							
are those										
with digits										
that sum	630	621	612							
to 9.	720	711	702							
	810	801								
	900									
Total in Col.	10	9	8	7	6	5	4	3	2	1

It only remains to add the totals in each column:

$$10 + 9 + 8 + 7 + 6 + 5 + 4 + 3 + 2 + 1 = 55$$

Notice that we have not filled in the entire table. An advantage to getting work organized is that you often see a pattern, enabling you to say "how it would go" without actually doing it. We have already mentioned, and we will prove in Chapter 4, that the sum $10 + 9 + \cdots + 2 + 1$ can also be obtained quickly without actually adding it. People who work this way are the ones we like to pay by the hour.

Create and Exploit Symmetry

Find the sum of all the digits you will write down if you make a list of all the integers from 1 to 1000 inclusive.

Clarify the question.

Try a small example first.

First let's make sure the problem is understood. We are not being asked for the sum of all the integers. If we had been asked for the sum of all the digits we would write down in making a list of all the integers from 10 to 15 inclusive, the answer would not be

$$10 + 11 + 12 + 13 + 14 + 15 = 75$$

It would be

$$(1 + 0) + (1 + 1) + (1 + 2) + (1 + 3) + (1 + 4) + (1 + 5) = 21$$

While looking at the small example, we might also ask ourselves if we could get the answer following a pattern that could be exploited for the larger problem that was posed. Consider listing the numbers this way.

10	15
11	14
12	13

The goal is to add together each of the digits written down. Note that as the integers are displayed, the digits on each horizontal line sum to 7;

$$1 + 0 + 1 + 5 = 7$$
$$1 + 1 + 1 + 4 = 7$$
$$1 + 2 + 1 + 3 = 7$$

There are three lines; the grand total must be $3 \times 7 = 21$.

This suggests a way to take advantage of symmetry in solving the original problem, stated later as Problem 27. Try

0	999
1	998
2	997
⋮	⋮
9	990
10	989
⋮	⋮
99	900
100	899
⋮	⋮
498	501
499	500

Watch out though; 1000 is missing from the list.

Summary

When a problem has many details, it is hard to keep track of them in our minds. A list, chart, or diagram may bring these details into clearer focus and help us perceive relations between them. And if a picture can be drawn that somehow represents the problem, draw it. This is a maxim that great problem solvers follow religiously.

Problem Set 1.2

1. Fourteen clothespins are placed on a line at 7-foot intervals. How far is it from the first to the last?
2. At six o'clock the wall clock struck six times. Checking my watch, I noticed that the time between the first and last strokes was 30 seconds. How long will the clock take to strike twelve midnight?
3. Each hour, on the hour, a bus leaves Dallas for Houston and another bus leaves Houston for Dallas. The trip takes 5 hours. How many buses will a bus leaving Dallas at 10:00 A.M. meet on its way to Houston?
4. After hearing a gloomy Sunday sermon on the danger of smoking, Hugo promised himself to cut back 2 cigarettes a day, smoking 2 less on Monday than he had on Sunday, etc. He kept his promise for a week, that is, through Saturday. During the whole week he smoked 63 cigarettes. How many did he smoke the day of the sermon?
5. A lazy and rather careless frog fell into a cistern 21 feet deep. It was the morning of July 4, and it may have been a firecracker that did it. After thinking over its plight, the frog started to climb and made 3 feet by nightfall. Next morning it discovered that it had slid back 1 foot

during its sleep. Satisfied with this pattern (3 feet upward progress during the day, 1 foot downward slide at night), the frog worked its way up the side of the cistern. On what date did it get out?

6. I live in a place where the temperature goes up sharply during the day and down at night. This affects my watch. I notice that it gains $\frac{1}{2}$ minute during the day but then loses $\frac{1}{3}$ minute during the night; thus it is $\frac{1}{6}$ minute fast for a 24-hour period. One morning—July 1—my watch showed the right time. On what date did if first show 5 minutes fast?

7. How many triangles are there in the picture at the right?

8. Amy plans to invite six of her friends (call them A, B, C, D, E, and F) to lunch on five different days. She has three small tables at which only two people can sit. How can she arrange it so that no two people ever sit together twice? You are to assume that Amy never sits down.

9. A detachment of 10 soldiers must cross a river. The bridge is down, and the river is deep. What should they do? Suddenly the sergeant in charge spots two boys playing in a little rowboat near shore. The boat is so tiny that it can hold only two boys or one soldier. Still the soldiers manage to cross the river. How?

10. A man has to take a wolf, a goat, and some cabbage across a river. His boat has room enough for the man (who must row) plus either the wolf or the goat or the cabbage. If he takes the cabbage with him, the wolf will eat the goat. If he takes the wolf, the goat will eat the cabbage. Only when the man is present are the goat and the cabbage safe from their enemies. Nevertheless, the man carries the wolf, goat, and cabbage across the river. How?

11. A chemist has two identical test tubes, each able to hold 50 milliliters of liquid. In a beaker she has 48 milliliters of solution, and she wishes to measure off exactly 42 milliliters. How should she proceed?

12. In how many ways can you add six even numbers (bigger than zero) to get 20? Changes in order are not counted as new solutions.

13. In how many ways can you add five positive whole numbers to get a sum of 11? Changes in order are not counted as new solutions.

14. If there are nine people in a room and every person shakes hands exactly once with each of the other people, how many handshakes will there be?

15. A railroad line runs straight from Posthole to Podunk. Center City is on this line, exactly halfway in between. Klondike is just as far from Posthole as it is from Center City, and Center City is as far from Klondike as it is from Podunk. All cities are connected by direct railroad lines. If it is 6 miles from Posthole to Klondike, how far is it from Klondike to Podunk?

16. I have already covered one-third of the distance from Podunk to Boondocks, and after I walk another kilometer I'll be halfway there. How far is it from Podunk to Boondocks?

17. The new pastor at First Lutheran Church plans to post the page numbers for three hymns at each Sunday service. To do it, he must buy

plastic cards each with one large digit on it, but obviously he wishes to minimize the number of cards to be bought. The hymnal has hymns numbered from 1 to 632. How many cards must he buy to make sure that any selection of the three hymns is possible?

18. We expected you to assume that 6's and 9's required different cards in Problem 17. Suppose 6's can be turned upside down to form 9's. Then how many cards are required?

19. What is plausible basis for the order in which the following 10 digits are arranged?

$$8 - 5 - 4 - 9 - 1 - 7 - 6 - 3 - 2 - 0$$

20.* What is the sum of all the counting numbers from 1 to 100? What is the sum of the *digits* in all the counting numbers from 1 to 100?

21.* Of the members of three athletic teams at a certain school, 21 are on the basketball team, 26 on the baseball team, and 29 on the football team. Fourteen play baseball and basketball, 15 play baseball and football, and 12 play football and basketball. Eight are on all three teams. How many team members are there altogether?

22.* A pet store offered a baby monkey for sale at \$1.25. The monkey grew, and the next week it was offered at \$1.89. Not one to monkey around, the shopkeeper subsequently raised the price to \$5.13, then to \$5.94, and next to \$9.18. Finally, during the sixth week, an organ grinder bought the monkey at \$12.42. How were the prices figured?

23.* Andrew Algaard was born on April 1, 1863, and died on May 25, 1950. How many days did he live? (Don't forget leap years.)

24. How long is the tunnel bored by the bookworm described in this section?

25. Refering to **The Handshake Problem,** how many hands did Dick shake? What about Jane?

26. Finish the problem of determining where Allen, Bill, Clint, and Dan live.

27. Find the sum of all the digits you will write down if you make a list of all the digits from 1 to 1000 inclusive.

28. Consider the arrangement of coins shown in the margin. We wish to interchange the position of the penny and the nickel by moving one coin at a time adjoining open space. Can it be done, and if so how?

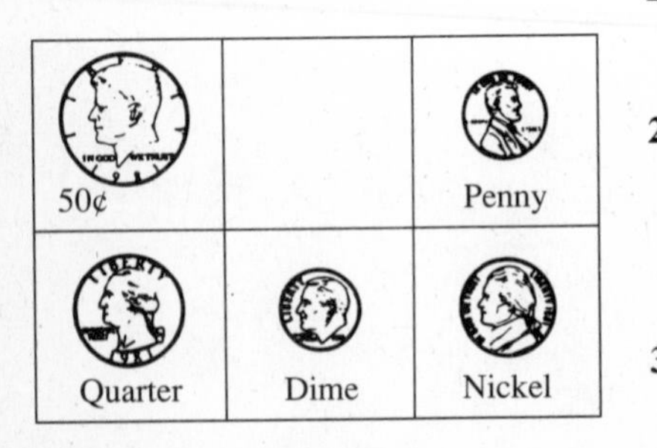

29.* A coin collector has 12 coins that look alike, but one is counterfeit and has a weight different from that of the other 11. Her scale is not sensitive enough to detect the slight difference, but she does have an excellent two-pan balance. Show how she can find the counterfeit coin in the least number of weighings.

30.* Do Problem 29 assuming there are 13 coins, one of which is counterfeit.

31.* We have 10 piles of 10 quarters each. One pile consists entirely of counterfeits, and each coin in this pile weighs 4.6 grams. A good quarter weighs 5.6 grams. You have a single-pan scale which is accurate to hundredths of a gram. What is the minimum number of weighings needed to identify the bad pile?

Hints to Selected Problems

1. Draw a picture; try the problem first with just three clothespins.

3. See the LeHavre-New York ship problem in the text.

5. Make a diagram showing the position of the frog each morning.

7. You will have to find a systematic way of counting. How many triangles are not subdivided into smaller triangles? How many are subdivided into exactly two smaller triangles, etc.?

9. A picture may help you get started.

11. As a first step, divide the liquid equally between the two test tubes (most test tubes are transparent, so this is usually possible). Keep track of each step.

	Beaker	*Tube A*	*Tube B*
Start	48	0	0
Step 1	0	24	24
Step 2			

14. Label the people A, B, C, . . . , I. Let A shake hands with B, C, . . . , I. Then let B shake hands with C, D, . . . , I, and so on.

15. Draw a good picture.

17. First figure out how many 0's are needed, then how many 1's, then 2's, and so on.

19. Eight, five, four, etc.

21. Draw three overlapping circles, each circle representing the players from one of the teams as in the margin. Put appropriate numbers in each of the seven regions.

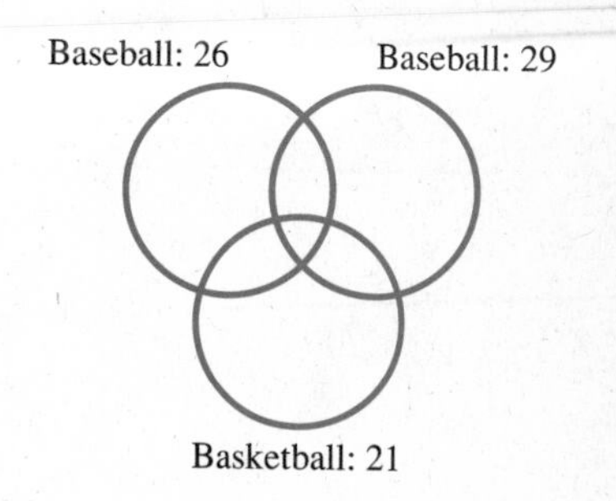

22. Count pennies; the increases were 64, 324, 81, 324, and 324. Make a chart showing prices and increases.

24. Look at the picture. Where is page 1 of Volume I? the last page of Volume IV?

25. From work done in the text, we see that 6 and 0 are partners, and that 5 and 1 are partners. Find the partner for 4; who is left for Jane?

29. You don't have to put all your coins on the balance the first time.

1.3 Anticipate the Solution

Planting Trees, Envisioning a Forest

How many trees can be planted on a square plot of land that is 100 feet on each edge if the seedlings are to be no closer than 10 feet to each other? Trees may be planted on the boundaries of the field.

Friedland, *Puzzles in Math and Logic*

A student told of an experience on a summer job in which he was to drive a delivery truck to a site in a new housing development where street signs had not yet been erected. How was he to find it? Envisioning what the site would look like with numerous trucks moving in and out, he hit upon the successful plan of driving to the general area, and then looking for muddy tracks on the street to follow. There are many situations in which the best first steps involve anticipating what the solution will look like.

Establish Boundaries

Looking for boundaries works on more than just jigsaw puzzles. Here is an example.

What are the possible areas for a rectangle if it is known that the lengths of the edges are integers, and that the area in square units is equal to the length of the perimeter?

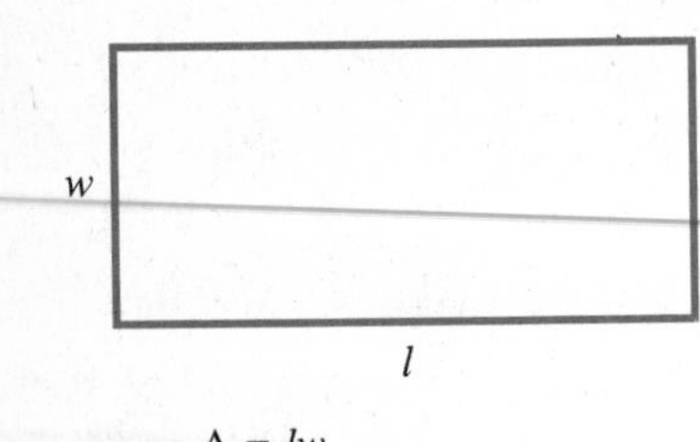

$A = lw$
$P = 2l + 2w$

We immediately picture a rectangle of length l and width w. For such a rectangle, the area $A = lw$ and the perimeter P is given by $P = 2l + 2w$. We could just start guessing values for w and l, hoping to hit on a trend that would lead to equal values for the area A and the perimeter P. We have, after all, argued in the last section for looking at small values and accumulating evidence, so we don't want to knock the idea now.

Two considerations suggest, however, that we do a little analysis before we start guessing. First, we have no clue yet as to possible sizes of the rectangle, or of its general shape (close to a square, or very long and thin)? Secondly, there may be many choices of w and l that work. Finding one example gives no clue as to how many more there might be.

We therefore begin with a little analysis, noting that we seek w and l for which area = perimeter, that is

$$lw = 2l + 2w$$
$$lw - 2l = 2w$$
$$l(w - 2) = 2w$$
$$l = \frac{2w}{w - 2}$$

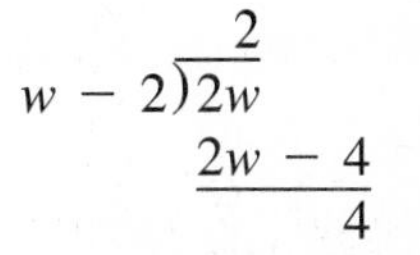

Since $l > 0$, it must be that $w > 2$; and since w is an integer, $w \geq 3$. Rewrite our expression for l by carrying out the division of $\frac{2w}{w-2}$.

$$l = 2 + \frac{4}{w - 2}$$

The smallest possible value for w, $w = 3$, gives $l = 2 + 4 = 6$; larger values of w give smaller values of l. Since $w \leq l$, we have now restricted our search substantially.

$$3 \leq w \leq l \leq 6$$

It's time to make a table of possible values

w	$l = \frac{2w}{w-2}$	*Area = Perimeter*
3	6	18
4	4	16
5	10/3	(doesn't count; l is not an integer
6	3	18

We need go no further. The possible areas are 16 and 18.

We move on to the problem of **Planting Trees, Envisioning a Forest.** If we impose a grid of parallel lines at intervals of 10 feet on the plot, we can plant a tree at each of the $(11)(11) = 121$ intersections, so we know we can do at least that well. The question is, can we do better?

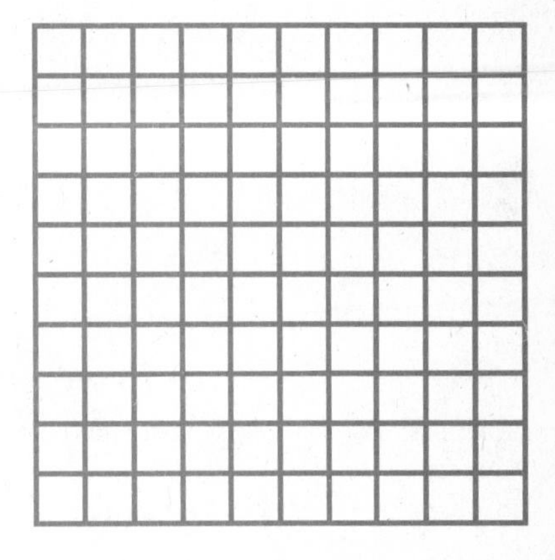

We certainly cannot do better than 11 trees per row, and every tree stands at the center of a circle of radius 10 in which no other tree can be planted. We might consider, however, moving the next row up by staggering its trees. Might this enable us to plant more trees? More than what? More than the known lower bound of 121. We leave it as a question for you to investigate later in Problem 12.

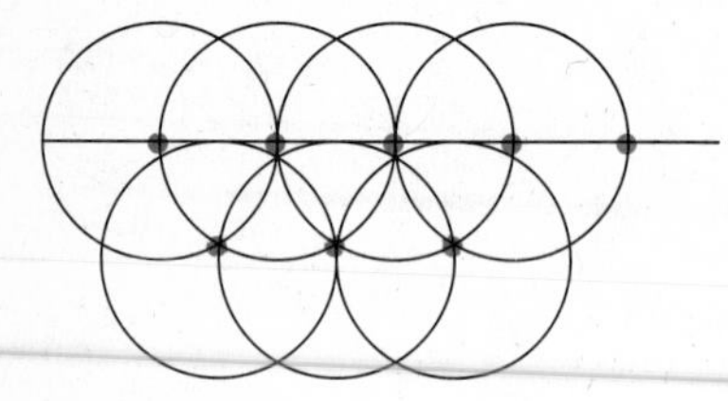

Because problems have such great variety, only the most general advice can be given for establishing boundaries. If your problem involves positive integers n, explore what happens for $n = 1$ and $n = 2$. Ask what happens as n gets very large. Similarly, if time t is involved, ask what happens when $t = 0$; when t gets very large. Do a lot of preliminary estimating to get "ballpark" figures. (This is especially important if you use a calculator, where pushing a single wrong key can have dramatic effects on the answer.)

Work Backwards

The rules for a game between two players are as follows. Place six pennies on the table. Player 1 takes either 1 or 2 pennies. Then Player 2 takes 1 or 2 pennies, and the game continues in this way until the pennies are gone. The winner is the one who takes the last penny. If you were playing, would you want to go first or second?

To win the game, you must make the last move, say move L. Work backwards, the next to last move being L-1, etc.

Move	*Player*	*Pennies You'd Like to See on the Table*
L	you	1 or 2 (so you could take them all)
L-1	opponent	3 (so your opponent will leave 1 or 2)
L-2	you	4 or 5 (so you can leave 3)

Therefore, be polite and let your opponent go first, which will be move L-3.

The point is not that you (to say nothing of your hapless opponent) will find this game a challenge or even a lot of fun. It does illustrate very nicely, however, a situation in which a complete analysis is achieved by looking at the position you want to be in at the end, and then working backwards.

Summary

Truth emerges more readily from error than from confusion.

Francis Bacon

One way for the Chair to proceed at a first meeting of a committee, provided that the members understand the purpose and are not disposed to feel "railroaded," is to present an outline of what their final report might look like. This will serve to quickly focus the thinking of both those who like it and those who think it is terrible.

It is almost always true that we will make better progress if we start out with some idea of where we want to go, and what should be true when we get there. Where numbers are involved, we will be greatly aided in assessing the correctness of our work in the end if we have some estimate of the size of our likely answer in the beginning.

Problem Set 1.3

1. As Homer approaches the check-out line at the grocery store, he wonders if he has enough money with him to cover his purchases. Items he has in his basket are marked as follows: $1.31, 2.45, 3 for 1.79 (he has two), 1.16, 2.41, 3.18, 2 for 1.73 (he has three), 6.52, 4.07, 3.21, 11.52, .43, .82, .21, 1.35. Make your own quick estimate; then find the actual sum with your calculator.

2. Amy checks her bank balance once each month when the statement comes from the bank. Her register looks like this.

Payments	*Deposits*	*Balance*
		346.15
36.50		
18.72		
43.21		
185.00		
19.54		
	143.50	
11.21		
19.40		
115.82		
12.00		
	250.00	
9.12		

Estimate her new balance. Then calculate it.

3. Estimate each of the following. Then use your calculator to find the exact value. Compare

(a) $\dfrac{\sqrt{(41)^2 + (16)^2}}{8.15 + (7.3)^2}$ **(b)** $\dfrac{413 + (9.7)^2}{7 + \sqrt{17.6}}$

4. Follow the instructions for Problem 3.

(a) $400(0.93)^{10}$ (b) $\dfrac{2.7^3 + (8.1)^2}{(0.1)^2}$

5. If it is known that $s > 4$ and that both r and s are integers, what are the possible values for each, given that

$$r = 2 + \frac{3s}{s - 3}?$$

6. If it is known that $x \geq 0$, what can you say about the values of

$$y = \frac{3x - 2}{x + 1}?$$

7. A square is said to be contained in a polygon if no portion of the square lies outside the polygon. Draw a square contained in a 6×7 rectangle R and color it. Then draw a square in R that does not overlap any colored area (except possibly for a common border), and color it. Continue in this way until you have colored 5 nonoverlapping squares contained in R. What is the maximum area you can color following these rules?

8. A drawer contains 8 black socks, 18 green socks, 12 blue socks, and 11 brown socks (always one sock lost). If the drawer is in a darkened room, how many socks must you grab if your only concern is to have for a two-day trip
   (a) four matching socks?
   (b) two pair of matching socks?

9. Three drivers happen to leave Plainview for the monotonous freeway drive of 96 miles to Westflats. All set their cruise control and left it at a constant speed for the entire trip. When the first driver arrived, she was 6 miles ahead of the next car and 10 miles ahead of the third. How far behind was the third car when the second car arrived? (*Altantic and Pacific,* 1978)

10. It was stormy the night of the murder, and the electricity went off, leaving me without lights and stopping the clock at 7:46. I immediately lighted two candles that I keep in the cabin for such emergencies. Though the candles were new and the same height, I knew from experience that the red one would last for 4 hours, the white one for only 3 hours. I then read and otherwise occupied myself for what seemed like several hours until I heard a shot outside. I immediately blew out my candles and spent a fearful night behind locked doors. In the morning I noted that the red candle was exactly three times as long as the white one. What time did the shooting occur?

11. Analyze the game of picking up pennies for a game that starts with 7 pennies; then try with 8 pennies.

12. Complete **Planting Trees, Envisioning a Forest.**

Hints for Selected Problems

6. Note that x and y are not restricted to integer values. You might try drawing a graph.

12. If the trees are staggered in alternate rows, the rows need be only h feet apart, where $h^2 + 5^2 = 10^2$. The offset rows will not, however, accommodate 11 trees, so you want to use as many rows of 11 as possible. If you start at the north edge of the plot, bear this in mind as you approach the south edge.

chapter 2

Methods of Solution

Solving a problem is similar to building a house. We must collect the right material, but collecting the material is not enough; a heap of stones is not yet a house. To construct the house or the solution, we must put together the parts and organize them into a purposeful whole.

GEORGE POLYA

It Takes Time

"You just go and think about it."

Think about it! What did she think I had been doing? I couldn't get the problem; that's why I had come for help in the first place. And what help did I get? ". . . just go and think about it."

I no longer remember the details. It was a problem about mixing nuts, so many at one price, so many at another to get a mix at still another; and I was the one going nuts. Ms. Ledbetter had looked at my work, asked a question or two, and no doubt concluded that I had all the technical skills needed; concluded that if I struggled with it, I'd get it. She was, in retrospect, one of my best teachers.

All too often, a mathematics class consists of learning a method for doing a certain kind of problems; and it is followed by the assignment of 20 of the same kind to be done as homework. Such assignments teach two unintended lessons. The first is that for every problem, there is a method which, if properly understood, will solve the problem presto. The second is that if you can't solve a problem presto, it should be skipped until someone shows you again how to solve that kind of problem.

It's just the opposite that should be taught. Learning a particular mathematical technique pales in importance next to the more important objective of learning that success on a mathematical problem is, to an extent greater than most people realize, the reward of a persistent effort. You think hard, you think long, you carry it with you to ponder while engaged in some repetitious activity, or while waiting for the bus. You skim through some books looking for a related idea, and you talk to your friends about it (if you've chosen the right kind of friends).

This process is what mathematicians call living with a problem. At their desks' they concentrate on it intensely. In the coffee room (the mathematics laboratory), they discuss it with colleagues. They go to hear another mathematician lecture on another problem, hoping that serendipity will strike, that somehow an idea picked up at the lecture will give them a new approach to their problem. Strangely, it sometimes works.

The role of the subconscious is not well understood in this process, but some mathematicians claim that once a problem is well-formulated, sleep or semisleep often plays an almost mystical role in bringing them to a solution.

> "On being very abruptly awakened by an external noise, a solution long searched for appeared to me at once without the slightest instant of reflection on my part."
>
> J. Hadamard

There are, then, some good arguments for spending a lot of time in bed. (There are also some poor ones.) This much seems clear: those who hit on fruitful ideas while asleep have always thought very deeply about the problem at hand while wide awake.

This also seems clear: many people consider themselves poor at mathematical problems because solutions do not come to them as quickly as they (falsely) believe they come to those who "get math." They could experience the satisfaction of solving a challenging problem if they would simply take the time. Think about it!

2.1 Solve a Simplified Version First

Pictures Perfect

A company prints up 8 distinct picture postcards. They are to be sold individually, in packages of 2, packages of 3, etc., but the company wishes to abide by two rules in packaging. The cards in any package shall be distinct, and any two packages shall always have at least one card in common–so that a customer who buys two packages, perhaps in different stores, will recognize upon examining them that they are produced by the same company. How many different packages can be formed?

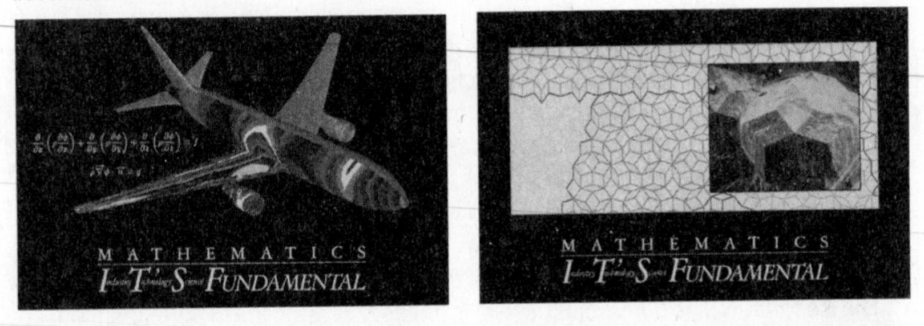

If you seek advice about a teacher before taking a class, the comment that, "Old Fogbottom can't teach," is really not too helpful. It may be that your source is a young Fogtop who can't think. If, however, you are told that, "Old Fogbottom always lectures and never plans any time for questions," you have some specific information that you can then check out, perhaps directly with Fogbottom. You will also be more aware as you check out a specific complaint that it is just one aspect of what goes into teaching, enabling you to weigh it in an overall evaluation of whether this class fits your style of learning.

In this section, you will be advised to look carefully at special cases of a problem. Two features of this approach should be underscored. Be specific, and work out the details carefully. Then remember that one example does not tell the whole story. You do not solve a mathematical problem by offering one example, or even by offering several examples. You only get ideas of how a thing may work out in general. That, however, is exactly what we are trying to do–get an idea.

$$5! = 5 \cdot 4 \cdot 3 \cdot 2 \cdot 1 = 120$$

$$7! = 7 \cdot 6 \cdot 5 \cdot 4 \cdot 3 \cdot 2 \cdot 1 = 7 \cdot 6 \cdot 5! = 5040$$

Try Small Cases

The symbol $n!$ stands for the product of all the positive integers from 1 to n inclusive. The following question appeared on the 1991 Minnesota High School Mathematics League competition.

The sum $S = 1(1!) + 2(2!) + 3(3!) + \cdots + 1991(1991!)$ can be written in the form $K! - 1$. Find K.

A participant who had our principle in mind might have noted the pattern and tried it for much smaller cases.

$$1(1!) + 2(2!) = 1 + 4 = 5 = 6 - 1 = 3! - 1$$
$$1(1!) + 2(2!) + 3(3!) = 1 + 4 + 18 = 23 = 24 - 1 = 4! - 1$$

A reasonable guess for the given problem would be 1992. It remains, of course, to prove that, but proofs are much easier if we have some notion of what it is that we're trying to prove. (Problem 12)

Here is another problem from the same competition (originally from *Mathematics Magazine,* 64, No. 3, June 1991, p. 188).

Let K and C be the circumscribed and the inscribed circles, respectively, of a regular polygon having 1992 edges of length 1. What is Area(K) − Area(C)?

In case you have forgotten what a regular polygon is, or what is meant by the circumscribed and inscribed circles, examples are provided in the margin for regular polygons of 3 sides, 4 sides, and 6 sides. The principle of this section says that even if you have not forgotten terms from elementary geometry, you should still look at these small cases before (way before) you try to draw one with 1992 sides.

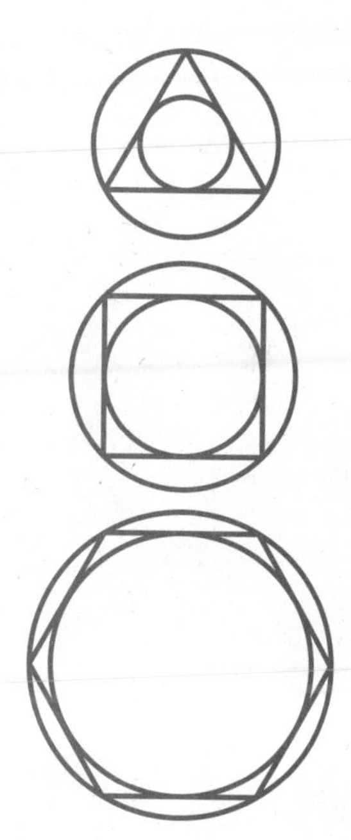

Let the circumscribed circle have radius R, the inscribed circle have radius r. Notice that in every case, those drawn as well as for any regular polygon with sides of length 1 that might be drawn, we obtain a right triangle that has one side of length 1/2. From the theorem of Pythagoras,

$$\left(\frac{1}{2}\right)^2 + r^2 = R^2.$$

so

$$R^2 - r^2 = \left(\frac{1}{2}\right)^2$$

Then

$$\text{Area}(K) - \text{Area}(C) = \pi R^2 - \pi r^2 = \pi(R^2 - r^2) = \pi\left(\frac{1}{2}\right)^2.$$

The answer is always $\frac{\pi}{4}$, independent of the number of sides of the polygon. Looking at a few small cases led this time to a solution that is proved.

Supply Your Own Trial Data

Here is a variation of a problem that was widely discussed in mathematical literature for several years because it was found that in spite of its apparent simplicity, it was invariably missed by a distressingly large number of students (reportedly by as many as 40%) when put on a *college* entrance exam.

Using p to represent the number of professors and s to represent the number of students, express algebraically the statement that there are ten times as many students as professors at this college.

It is reasonable to assume that anyone who tried this problem wrote down either

$$p = 10s \qquad \text{or} \qquad s = 10p.$$

Some probably wrote down both and pondered them before making a choice.

How should one proceed? Writing down the two candidates is not a bad start. Then supply some data for yourself. If there were just 1 professor ($p = 1$), we should surely have $s = 10$. In which expression do these numbers work? $s = 10p$. Presto! Let's try another.

Center City is right at the mouth of Swift River where it flows into Quiet Lake. Upstream a few miles is the thriving town of Klondike, while across the lake and the same distance away is the village of Tubville. Horace proposes a race with his good friend Hugo. Horace will row upstream to Klondike and return. Hugo will row across Quiet Lake to Tubville and return. Assuming that they are equally good oarsmen and that they start from Center City at the same time, who will win?

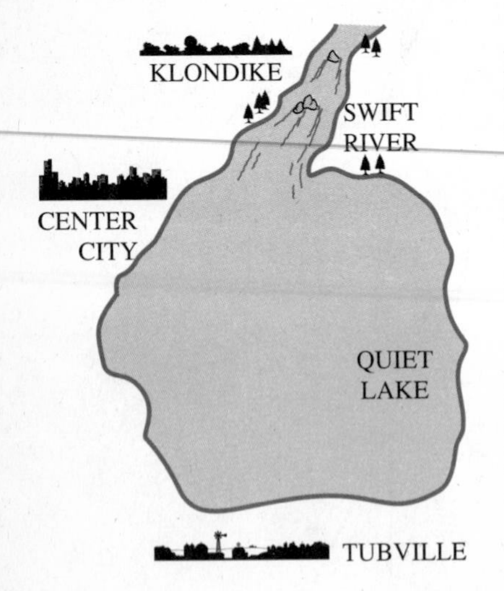

Since Horace will be helped by the current in one direction and hindered in the other, it is tempting to guess that he and Hugo will arrive back at Center City in a dead tie. That's a good guess, but it happens to be wrong.

Our problem is notable for its lack of information. We don't know how far it is between the cities, how fast the men can row, or the speed of the current in Swift River. Can it be that none of these things make any difference?

Let's begin with a concrete, simple case. Suppose that it is 12 miles from Center City to Klondike, that the men can row 6 miles per hour (mph), and that Swift River flows at 2 mph. Then Horace will make $6 - 2 = 4$ mph upstream, and he will reach Klondike in 3 hours. Downstream he will make $6 + 2 = 8$ mph. It will take him $1\frac{1}{2}$ hours, or a total of $4\frac{1}{2}$ hours for the roundtrip. In the meantime, Hugo, rowing at a steady 6 mph in Quiet Lake, will make the 24-mile roundtrip in 4 hours flat.

This is enough to eliminate our original guess. Is it enough to declare Hugo the winner? No. Let's try a different set of data. Suppose that the two oarsmen can row at only 4 mph and that the other information is as before. Now Horace can go only 2 mph upstream. It will take him 6 hours to reach Klondike. On the return he will make 6 mph, and it will take 2 hours. This gives a total of 8 hours altogether. Meanwhile, Hugo will need only $\frac{24}{4} = 6$ hours to make the whole trip. He will be back at Center City by the time Horace gets to Klondike.

Now we begin to see what happens in general. It will certainly take Horace longer to row upstream than downstream. Thus the current will hinder him for a longer time than it will help him. He can't possibly compensate for the time lost going upstream on the return trip, and so he will always lose.

Of course, there is a nice algebraic demonstration of this fact. But that is the subject Section 2.5.

Summary

If a problem involves a large number that occurs in an expression with a definite pattern to it, explore the same pattern for much smaller numbers. If a statement is made, or if you are inclined as part of the solving process to make a statement (in the English language, or as an algebraic formula) intended to relate two quantities, try it for some simple data of your own choosing. Does it say what you intend for it to say?

Problem Set 2.1

1. Assume the earth to be a perfect sphere, and that a rope has been tightly tied around the equator. Snip the rope, insert a 6-foot section of extra rope, and take up the slack by pulling the now lengthened rope into a circle concentric with (but slightly larger than) the equator. Is the extra room so obtained enough for an ant to slip through? A mouse? A cat? How about you, good reader, could you slip through? (The circumference of the earth is approximately 25,000 miles.)

2. A ball of yarn of a certain size will make one mitten. If its diameter were doubled, how many mittens would it make?

3. Homer and Hugo started together to go from Pitstop to Posthole. Homer ran half of the time and walked half of the time. Hugo ran half of the distance and walked half of the distance. Assuming they ran and walked at the same rate, who got to Posthole first?

4. Rodney Roller drives 20 miles to work each morning. The first 10 miles are through heavy traffic and he averages 20 mph. The second 10 miles are further out of the city where he averages 40 mph. What is his average speed for the whole trip?

5. Show that, if every student at a certain college has at least one friend there, at least two students have the same number of friends.

6. There are 1 million dots on a sheet of paper. Can a straight line be drawn, no matter how the dots are situated, so that no point is on the line and exactly one-half of the dots are on each side of the line?

7. Find the sum of

$$\frac{1}{1 \cdot 2} + \frac{1}{2 \cdot 3} + \frac{1}{3 \cdot 4} + \cdots + \frac{1}{99 \cdot 100}.$$

8. Seven bags of marbles contain 15, 18, 19, 22, 24, 25, and 26 marbles, respectively. One bag contains chipped marbles only. The other six bags contain no chipped marbles. Jane takes four of the bags and George takes two of those that are left. Only the bag of chipped marbles remains. If Jane gets twice as many marbles as George, how many chipped marbles are there?

9. How many triples of integers (x, y, z), $1 \le x < y < z \le 100$, can you find having the property that the sum of any two of the integers is divisible by the third?

10. How many triples of integers (x, y, z), $1 \le x \le y \le z \le 100$, can you find having the property that the sum of any two of the integers is divisible by the third?

11. Solve the lead problem of this section, **Pictures perfect.**

12. Prove that $S = 1(1!) + 2(2!) + 3(3!) + \cdots + 1991(1991!) = 1992! - 1$.

Hints to Selected Problems

1. $C = 2\pi r$, where π is a little more than 3. What happens to C when r is increased by 1?

2. $V = 4\pi r^3/3$, where V is the volume and r is the radius.

3. Did Hugo run half of the time?

4. Thirty is wrong. What does average speed mean?

6. Try the same question with 10 dots, noting that the 10 dots determine $(10)(9)/2 = 45$ slopes. Use a line with a slope different from these 45.

7. First find $\frac{1}{1 \cdot 2} + \frac{1}{2 \cdot 3}$, then $\frac{1}{1 \cdot 2} + \frac{1}{2 \cdot 3} + \frac{1}{3 \cdot 4}$. Continue to look at simplified versions of

$$\frac{1}{1 \cdot 2} + \frac{1}{2 \cdot 3} + \cdots + \frac{1}{(n-1)n}$$

11. With no constraints at all, how many different packages can be formed? Now note that because any two packages must contain a common card, for every package that is used (say a package containing three cards), there is another package (in the example, the other five cards) that cannot be used. Try the problem using only four or five cards to start with.

2.2 Experiment, Guess, Demonstrate

Track It Down

Teams A, B, and C participated in a track and field meet in which each team entered one contestant in each event. Points were awarded in each event for a third-place finish, more points for a second-place finish, and still more for finishing first. At least three events were held. A won the meet with a total of 14 points, while B and C tied with 7 points each. B won first place in the high jump. Who won the pole vault, assuming that no ties occurred in any event?

The first hurdle to get over in **Track it Down** is the shock of being expected to answer such a question.

There Is a Role for Guessing

To get off the starting block, suppose that there are exactly three events, and that first, second, and third places are awarded 3, 2, and 1 points, respectively.

Now let's take the advice of Section 1.2 and organize the meager information that we have.

	Team		
Event	***A***	***B***	***C***
High jump		3	
Pole vault			
Mystery event			
TOTAL	14	7	7

We notice almost immediately that our guess must be wrong, since there is no way that team A could have accumulated 14 points; the most possible would have been 8.

We could just guess again—trying four events or five events; or we could assign more points in each event. And if nothing else occurred to us, this is the course of action we would probably follow.

But actually, there is more to observe from our table than the mere fact that we are wrong. The teams earned a total of $14 + 7 + 7 = 28$ points. We asked, how many points were awarded after each event?

Our first guess was $3 + 2 + 1 = 6$. That would mean there were $\frac{28}{6} = 4\frac{2}{3}$ events—which is ridiculous. So we are now conscious of something that should have occurred to us in the first place.

$$\begin{pmatrix}\text{Number of points awarded} \\ \text{after each event}\end{pmatrix} \text{ times } \begin{pmatrix}\text{number of} \\ \text{events}\end{pmatrix} \text{is } 28$$

If we had noticed this at the beginning, we never would have tried guesses of 6 and 3.

We know that a certain number of points was awarded for third place, more for second, and still more for first. The total therefore must be *at least* $1 + 2 + 3 = 6$, but we know that 6 does not work. Let's try 7, which would mean there were four events. Moreover, a little experimenting with groups of three positive integers that add up to 7 quickly convinces us that $7 = 4 + 2 + 1$ is the only possibility; 4 points were awarded for first place, 2 for second, and 1 for third. Our table now takes a slightly different form.

	Team		
Event	***A***	***B***	***C***
High jump		4	
Pole vault			
Event 3			
Event 4			
TOTAL	14	7	7

Again we ask if there is any way that A could have obtained 14 points. This time there is—just one. Team A would have had to take first in all events save the high jump, and there it would have had to take second. But that hurdle we didn't have to jump. Team A won the pole vault.

Since we have reached our conclusion on the basis of an experiment, two questions naturally occur. Is our present guess consistent with the rest of the information (can we make assignments of the points available so that B and C each have 7 points)? Second, is there any other set of trial values that is consistent with the given information? When you have checked these out, you will most certainly agree that team A won the pole vault.

Eliminate Wrong Answers

If the number of possible answers is reasonably small, one can often find the correct one by a systematic process of eliminating the wrong ones. (This procedure is a favorite of students taking multiple-choice exams.) Here is a well-known puzzle in which this process ultimately yields the solution:

A farmer's wife drove to town to sell a basket of eggs. To her first customer she sold half her eggs and half an egg. To the second she sold half of what she had left plus half an egg. And to a third she sold half of what she then had and half an egg. Three eggs remained. How many did she start with? She did not break any eggs.

Now, two things are clear. The answer is an integer considerably greater than 3, and yet it can't be very large (baskets hold only so many eggs). We could start by trying some reasonable number, say 17, and then 18, 19, 20, etc., eliminating wrong answers along the way. Eventually, we reach 31 and, lo, it works, To her first customer she sold $\frac{31}{2} + \frac{1}{2} = 16$, to her second she sold $\frac{15}{2} + \frac{1}{2} = 8$, and to her third $\frac{7}{2} + \frac{1}{2} = 4$, leaving 3 eggs unsold.

Summary

When faced with a problem that has many potential solutions, experiment a little. If you work in a systematic way, you may find the answer rather quickly. At the very least, you may observe a pattern that suggests the correct solution. If so, make a conjecture. State it boldly. But don't forget to check it carefully. Mathematicians are not satisfied until they have demonstrated that a solution is correct.

Problem Set 2.2

1. In a certain family, each boy has as many sisters as brothers, but each sister has only half as many sisters as brothers. How many brothers and sisters are there in the family?
2. On a 24-item true-false test, Professor Witquick gave 5 points for each correct response but took off 7 points for each wrong one. Homer answered all the questions and came up with a big fat zero for his score. How many did he get right?
3. Place the numbers 5 through 9 in the vacant squares at the right so that every row, horizontal, vertical, or diagonal, has the same sum.

	1	
3		
4		2

4. With a pile of about 1000 identical cubical bricks, a cubical monument is to be built, and it is to stand on top of a square plaza constructed from the same pile of bricks. There are to be equal numbers of bricks in the monument and in the plaza. What is the largest monument that can be built?
5. Move only three coins from the first arrangement to produce the second.

6. The odometer of the family car showed 15,951 kilometers. The driver noticed that this number was palindromic; that is, it read the same backward as forward. "Curious," the driver said to herself. "I imagine it will be a long time before that happens again." But 2 hours later, the odometer showed a new palindromic number. How far had the car gone during the 2 hours?

7. Notice that $(1 + 2)(3 + 4 + 5 + 6 + 7 + 8 + 9) = 126$. Place mathematical signs between all the following digits to make the equation correct:

$$1\ 2\ 3\ 4\ 5\ 6\ 7\ 8\ 9 = 100$$

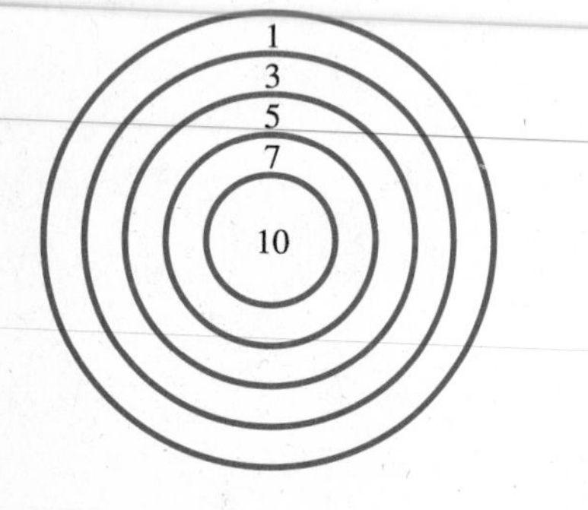

8. Debra hit the dartboard pictured in the margin with all five darts that she threw. One of the following numbers is her score: 12, 30, 46. Which one?

9. An electronics firm had a large bin containing connectors that go on the ends of wires. The connectors come in six colors, and the same color must go on each end of a wire. Suppose you are to prepare 10 wires. You need 20 connectors (10 sets). If the bin is in a dimly lit area where you cannot see the colors, what minimum number of connectors must you take in order to be certain of having 10 matching sets?

10. I asked the election judge how the candidates were ranked. "Guess," she responded. Having no idea, I responded with an alphabetical listing: *A*llan, *B*etty, *C*arl, *D*ennis, *E*thel (A B C D E). "Way off," she told me. "You not only got each person in the wrong position, but no one in your ranking follows the correct immediate predecessor. With at least that much information, I guessed again: DAECB. "An improvement," she told me. "You have two in their proper position, and you have two following correctly their immediate predecessor. I then found that I had enough information to be certain of the correct ranking of the candidates. What was it?

11. Allison, Betty, and Claire were seated at a round table having lunch. They were discussing their part-time jobs as a waitress, a drugstore clerk, and a bus driver—though not necessarily in that order. Each woman shared a part of her lunch with the woman on her right. Betty gave some carrot sticks to the waitress. Allison gave a brownie to the woman who gave the potato chips to the bus driver. Match the women to their jobs.

12. Gary was sent to the bakery to buy some fancy cupcakes for his little party, and reported to his mother upon his return that he had spent \$6.00 to buy three different kinds, costing 36 cents, 42 cents, and 51 cents, respectively. His mother asked how many he had purchased altogether, and he told her. She seemed satisfied, and continued working on her crossword puzzle, but soon was doing some figuring in the margin of the newspaper. "I'm trying to figure out how many you got of each kind," she said. Did you buy only one of one kind? Gary answered her with only one word, but it was enough to enable his mother to solve the little problem she had posed for herself. How many had he bought at each price? [Hunter and Maduchy, *Math Diversions* (a variation).]

13. Three hockey teams, A, B, and C, play a three-game tournament. We have the following information:
(a) Each team played two games.
(b) A won two games; B tied one game.
(c) B scored a total of two goals; C scored three goals,
(d) One goal was scored against A, four against B, and seven against C. Find the scores of all three games.

14. For a series of games involving five hockey teams, we have the following partial information. Fill in the rest of the table, figure out who played whom, and determine what the score was for each game.

	Played	*Won*	*Lost*	*Tied*	*Goals for*	*Goals against*
A	3	2	0		7	0
B	2	2	0		4	1
C	3	0	1		2	4
D	3	1	1		4	4
E	3	0			0	

15. By drawing straight lines across the face of a clock, can you divide it into three regions so that the sums of the numbers in the three regions are equal? Four regions? Six regions?

16. Write the numbers 1 through 19 in the circles shown so that the numbers in any three circles on a straight line total 30.

17. A Hamline graduate attended a recent class reunion and sent in the following account. "I met 15 former classmates. More than half were nurses, the rest were lawyers. Of the nurses, most were females and there were still more female lawyers. Both of the latter facts were true even if you included me. My friend, a noted lawyer, left his wife at home. What conclusion can you draw about me?"

18. The firm that designed the streetcar line from Podunk to East Podunk forgot to allow for expansion of the rails due to the heat of the summer sun. When laid in February, the rails were perfectly straight and exactly 1 mile (5280 feet) long. By mid-July, they had expanded in length by 2 feet, causing them to buckle up in the middle, thus forming the shape of an isosceles triangle. How high above the earth did the middle point rise? First guess, and then figure it out mathematically.

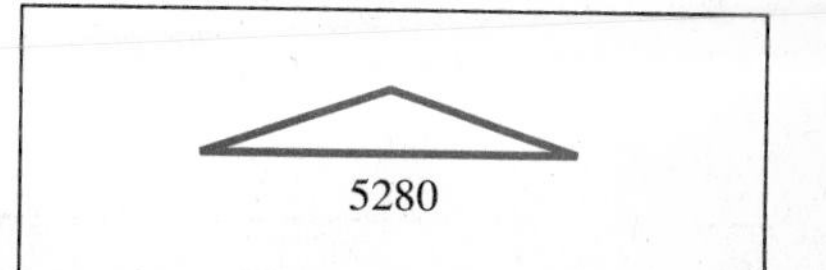

19. If a two digit number is multiplied by the product of its digits, a three digit number is formed, each digit of which is the unit digit of the original number. Find the original number.

Hints to Selected Problems

1. There must be one more boy than girl in the family.

2. Try various possibilities.

4. If the cube is $4 \times 4 \times 4$, then the plaza must be 8×8 (since they are to use the same number of bricks). We cannot use a cube of $5 \times 5 \times 5$, since 125 is not a perfect square. Is there any number larger than 4 that works?

6. 16,961 is a new palindromic number, but it's not the first one after 15,951. Find the first one.

11. Remember the table we used in Section 1.2 to identify the hometowns of Allen, Bill, Clint, and Dan.

14. Do things in the order mentioned. To fill in the table, first ask yourself how many games were played, then how many ties there were, and then which teams played to ties.

2.3 Divide and Conquer

```
F O R T Y
    T E N
    T E N
---------
S I X T Y
```

It All Adds Up

Each letter in the addition problem at the right stands for a digit, and all are distinct. Find Y. (The O is an "oh," as in, "oh, oh," rather than the numeral zero.)

We come now to a method commonly associated with large problems that can be handled by identifying subgoals to be worked on somewhat independently, perhaps even by different people, so long as there is good coordination of the effort. It is an approach commonly used in engineering and computer science. Problems that illustrate the method are naturally more complex than the ones we have been looking at so far, but this is also a part of mathematical thinking that we believe has lessons for all types of problem solving. A problem need not be solved all at once.

Focus on One Part At a Time

Here is one aspect of the method that can be illustrated with a short little puzzle.

If 4 cats eat 4 cans of cat food in 4 days, how much food should be left to feed 2 cats for 6 days?

The trick is to not think about numbers of cats, cans, and days all at once. Focus. Consider the following sequence of steps in which, as we move from one line to the next, one item always remains fixed.

	Cats	*Cans*	*Days*
Given	4	4	4
How much for 1 cat for 4 days?	1	1	4
How much for 1 cat for 6 days?	1	3/2	6
How much for 2 cats for 6 days?	2	3	6

Set Subgoals

Even when you are working alone, it sometimes helps to think about how you might parcel out the work if you had a team of people to work. You can then play the role of the individual members, focusing your attention in this way on just one piece of the work at a time, but being assured that there is excellent coordination of the overall project.

We illustrate the idea with an attack on **It All Adds Up.** Suppose we number the columns, starting from the units column, and imagine that we can assign one person to carefully analyze each column. Their reports might come back something like this.

```
5 4 3 2 1
F O R T Y
    T E N
    T E N
---------
S I X T Y
```

(1) Since N + N ends in a zero, either N = 0 or N = 5. The carry to (2) is accordingly either 0 or 1. There are no constraints on Y here, and it does not appear anywhere else, so it will no doubt be determined at the end by elimination as we match up the ten letters E, F, I N, Q, R, S, T, X, Y and the ten digits 0, 1, 2, 3, 4, 5, 6, 7, 8, 9.

(2) Since E + E ends in zero, either E = 0 or E = 5. There can be no incoming carry to this column, and the carry forward will be either 0 or 1.

(3) Looking ahead to (4), we see that there must be a carry from this column of at least 1. Note that the carry forward cannot exceed 2 since the sum of three digits R + T + T added to a carry of no more than 1 cannot be as much 30.

(4) Either a 1 or 2 is carried in, and looking ahead to (5), we see that there must be a carry forward. Since the digit represented by O ("oh") added to a carry of 1 or 2 cannot be as much as 20, the carry out must be a 1. If $I \neq 0$ then $O > 8$, and a 2 must have been carried in, in which case O = 9 and I = 1.

(5) There is a carry in of 1; F + 1 = S, meaning that F and S are successive integers.

As these reports are received and compared, several conclusions leap out. Reports (1) and (2) surely imply that N = 0 and E = 5. Then since 0 is assigned, (4) implies that O = 9, I = 1, and that a 2 must have been carried into column 4.

Answer Partial Questions

We can summarize progress thus far in two ways, writing both the addition itself with known digits filled in, and writing down the correspondence of digits to letters as far as we know them.

```
 F 9 R T Y
     T 5 0        0 1 2 3 4 5 6 7 8 9
     T 5 0        N I       E       O
 ---------
 S 1 X T Y
```

We also know that S and F must be fitted adjacent to each other in the correspondence chart, and that since the carry into column 4 is a 2 and $X \neq 0$, we must have $R + T + T > 20$. The fact that 9 is already assigned means that $R \leq 8$, so in turn, $2T > 12$, $T > 6$. The choices for T are exactly two; either $T = 7$ or $T = 8$.

Can $T = 7$? It is now clear that there is a carry of 1 to column 3, so we could with $T = 7$ have column (3) as follows:

```
1        (carried forward)
R
7
7
―
X + 20 (because the carry into column 4 is 2)
```

Thus, $R + 15 = X + 20$. $R = X + 5$. Consider in this situation our correspondence of digits to letters

```
0 1 2 3 4 5 6 7 8 9
N I       E   T   O
```

We cannot set $X = 2$ because 7 cannot be assigned to R, and for similar reasons, we cannot set $X = 4$. The only possibility is to set $X = 3$, $R = 8$. Then where can S and F, the two adjacent digits, be placed? We conclude that $T \neq 7$.

If you have faith that there is a solution, you are now sure that $T = 8$. It is hoped that you will, with $T = 8$, carry out an analysis similar to the last paragraph to see that with this choice, the puzzle can be completed. (Problem 8)

Summary

Again we have used puzzles to illustrate a problem-solving technique that is important, certainly in mathematics, engineering, and computer science, but we think in a much wider context as well. The young child is overwhelmed by the command to pick up the area in which he or she has been playing for several hours, and needs to be shown that by concentrating on one thing at a time, the job can be done. The same lesson applies over and over to jobs that in the beginning seems to have no starting-point. Conceive of the job as a series of tasks; focus on one at a time, until they begin to fit together as a whole. Divide and conquer.

Problem Set 2.3

1. If a chicken and one half lays an egg and one half in a day and one half, how many eggs will 30 chickens lay in 30 days?
2. An urn contains 100 red chips, 100 white chips, and 100 blue chips. What is the minimum number of chips that must be drawn to be assured that
   (a) you will have 30 chips of one color?
   (b) there will be at least 20 of one color and at least 10 of another?
   (c) there will be 10 of each color?
   (d) either (a) or (b) or (c).
3. Allowing any combination whatever of United States coins and bills less than $20, what is the largest amount of money you could have without being able to give exact change to someone wanting to break a $20 bill?
4. Five positive integers, when added four at a time, give the sums 119, 130, 134, 149, and 152. What is the smallest of the five integers?
5. From a set {a, b, c, d} of four numbers, six sums of two numbers can be formed: a + b, a + c, a + d, b + c, b + d, c + d. Partition {1, 2, 3, 4, 5, 6, 7, 8} into two sets A and B of four numbers each so that the six sums formed by the numbers in set A are exactly the same sums (perhaps in different order) formed by the numbers in set B.
6. Five back packs were ready to go. Bob Morebrawn, contemplating the possibility of being asked to carry two packs, weighed all ten possible combinations. The weights were 110, 112, 113, 114, 115, 116, 117, 118, 120, 121. If the packs are designated in order of increasing weight by A (lightest), B, C, D, E (heaviest), how much does B weigh?
7. Three hunting companions agree to meet at a rented cabin. When they arrive, they find that there is no wood for the fireplace. The first man has three commercially produced pressed logs in the trunk of his car which he contributes to the cause, and the second man has 5 such logs which he contributes. The third man, having no logs, contributes $4. If the cost of the fire is being borne equally by all three men, how much of the $4 should the first man get?
8. Complete **It All Adds Up.**

Problems 9, 10 and 11 come from A. J. Friedland, Puzzles in Math and Logic, *Dover, 1970*

9. With the same rules as were in force for **It All Adds Up,** solve the two alphametrics in which each letter represents the same numeral in both additions.

```
  O N E          O N E
+ O N E      + F O U R
-------      ---------
  T W O        F I V E
```

10. Follow the instructions for Exercise 9, knowing that this time two distinct solutions are possible.

```
A B C        A D G
D E F        B E H
-----        -----
G H I        C F I
```

11. Each $*$ represents a different numeral in the following multiplication, and all are different from 1. Find the integers being multiplied.

```
    * * *
      * *
  -------
  * * * * 1
```

12. In the following multiplication, digits are represented by letters and asterisks. Identical letters stand for identical digits, different letters stand for different digits, and an asterisk stands for any digit.

```
      A B C
      B A C
    -------
    * * * *
    * * A
* * * B
-----------
* * * * * *
```

[Kordemsky, Boris A., *The Moscow Puzzles* (New York: C. Scribner's Sons, 1972)]

Hints to Selected Problems

2. It helps to ask in each case how many chips can be drawn without achieving the goal.

3. Focus on one denomination of bill or coin at a time, starting with the largest.

4. Let the integers be $n_1 \leq n_2 \leq n_3 \leq n_4 \leq n_5$. You can find $4(n_1 + \cdots + n_5)$.

5. Since there is only one way to get a sum of 3, 1 and 2 cannot be in the same set. What about 1 and 3? Focus on one pair at a time.
6. See the hint to Problem 4.
7. What is the cost of the fire?

2.4 Transform the Problem

Run, Trot, Run

It is exactly 100 miles from Posthole to Pitstop. At 8 A.M. Horace sets out from Posthole to visit his friend Hilda in Pitstop. Unbeknown to Horace, Hilda sets out at exactly the same time to visit Horace. Both ride at 10 mph. As Hilda leaves, her tireless dog Trot runs on ahead at 30 mph. When the dog meets Horace, he immediately turns tail and heads back to Hilda, and when he meets her, he turns again and races back to Horace. Trot persists in this behavior until the two friends meet midway between the two towns. How far did Trot run?

Often a problem that sounds very complicated becomes extremely simple when looked at the right way. Just as one turns a jewel to examine every facet, so one should look at a problem from every angle. Often an elegant solution will appear when the problem is seen in the right light.

Take the problem above, for example. If one begins with the first idea that occurs to him, he will probably try to find the length of the first leg of Trot's journey. Then he will add on the length of the return trip, which is of course somewhat shorter. Continuing this way, he will have to compute and add together the lengths of all of Trot's trips. But how many back-and-forth trips does Trot make? And how long is each trip? Just thinking about Trot's trips tires most of us.

But wait. This problem is one of those jewels that ought to be seen from another angle. Ask yourself how long Trot ran. As long as Hilda (or Horace) rode. And since Hilda rode 50 miles at 10 mph, that is exactly 5 hours. Now, Trot ran all this time at 30 mph, and so he covered 150 miles. There it is—a simple, elegant solution to what appeared to be a hard problem.

> You turn the problem over and over in your mind; try to turn it so that it appears simpler. The aspect of the problem you are facing at this moment may not be the most favorable: Is the problem as simply, as clearly, as suggestively expressed as possible? Could you restate the problem?
>
> *George Polya*

Note how we obtained this solution. We changed the original question, How far did Trot run? to one that is much easier to answer, How long did Trot run? The answer to the second question allowed us to answer the first.

Reformulate the Question

One hundred seventeen tennis players enter a park district tournament. It is an elimination tournament in which the loser of a match is out. In the first round, 1 player draws a bye and 58 matches are played, so that 59 players go into the second round. How many matches will have to be played in order to determine the champion?

To get a feel for the problem, let's start with a simple case, say 5 players, and work out the results of a typical tournament as shown below. Note that four matches are played. But is this the way to approach a problem with 117 players? And what if there were 1017 players? Surely none of us would want to draw a diagram in that case.

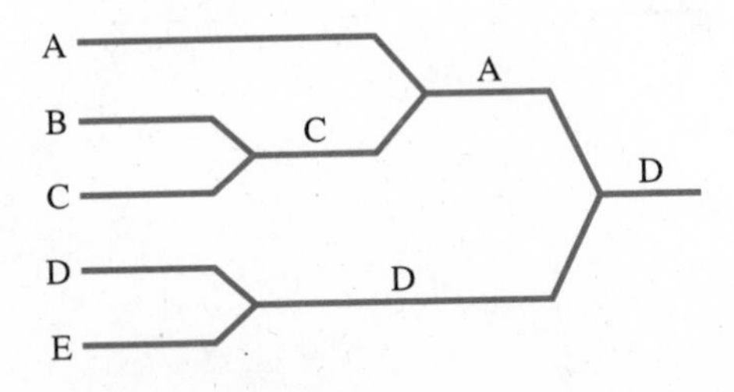

Let's try to reformulate the question. What happens in a tennis match? Well, one person wins and one person loses. But if a person loses, she is out. So every match has a loser, and there are exactly as many matches as there are losers. So our question becomes, How many losers are there? Answer: All but one of the players are losers. So, how many matches will be played if we start with 117 players? It must be 116; and if we start with 1017 players, there will be 1016 matches.

Suddenly the whole problem has become a triviality. It was simply a matter of restating the question. By the way, each match also has a winner. Why doesn't it help to look at the problem from this viewpoint?

Represent the Information in New Ways

Problem 19 of Section 1.2 asked what plausible basis there might be for the order in the listing of the ten digits as

8 5 4 9 1 7 6 3 3 2 0

Writing the sequence in the form

eight, five, four, nine, one, seven, six, three, two, zero

gives us the idea. They are arranged alphabetically.

Having looked back, let us now look ahead, to Problem 2 of Section 5.5 where we are asked how many solutions there are in non-negative integers to

$$x_1 + x_2 + x_3 = 10.$$

One such solution is $x_1 = 5, x_2 = 3, x_3 = 2$. A clever way to represent this solution is with the picture

$$***** \mid *** \mid **$$

The solution $x_1 = 7, x_2 = 0, x_3 = 3$ would be represented by

$$******* \mid \mid ***$$

and any solution could be represented using ten asterisks and two vertical line segments. As you will see in Section 5.2, it turns out to be very easy to answer the question, "In how many ways can we arrange 10 identical asterisks and 2 identical segments in a row?"

Find a New Problem Equivalent to the Old One

A Tibetan monk, starting at daybreak and pausing often to rest, slowly made his way up the winding trail to the peak of Mt. Arazza, arriving just at sunset. There at a pagoda he meditated all night. The next morning he started back down over the same trail, arriving at the base exactly at noon. Demonstrate that he was at some point along the trail at exactly the same time on both days.

To do this, let us imagine (mathematicians must be able to pretend) a second monk who has exactly the same habits as the first one. Let him start from the base of the mountain on the second day and follow exactly the same pattern the first monk followed on the first day. That is, let him pause for rest, hobble along, and do everything in exactly the same way and at the same speed. This second monk must meet the first monk coming down, and their meeting point is the solution to the original problem.

Summary

Sometimes it is possible, for a given question, to ask a different question which, if answered, will yield the answer to the original one. What is it that you want to find? Is there something else directly related to it that would be easier to find? Can the whole problem be reformulated? These ideas are worth thinking about even if you can clearly see your way through the problem. A chisel that cuts a stone cleanly, exposing its colored layers, is much to be preferred over a sledge hammer that smashes it beyond recognition.

Problem Set 2.4

1. How many matches will be required to determine the champion in a tennis tournament that starts with 89 players?

2. In a certain state, the champion basketball team is determined as follows. All 508 teams enter district tournaments, the winners go to regional tournaments, and finally the regional winners go to the state tournament. How many games must be scheduled to determine the champion team?

3. Thirteen teams enter a double-elimination baseball tournament (in which a team must lose twice to be eliminated). How many games must be played to determine a champion if the winner never loses? If the winner loses one game?

4. Homer has a pile of old lead shot which he plans to melt and pour into a form requiring 1 quart of lead. How can he tell if he has enough? All he has to measure with is a 1-gallon container which happens to be about half full of water.

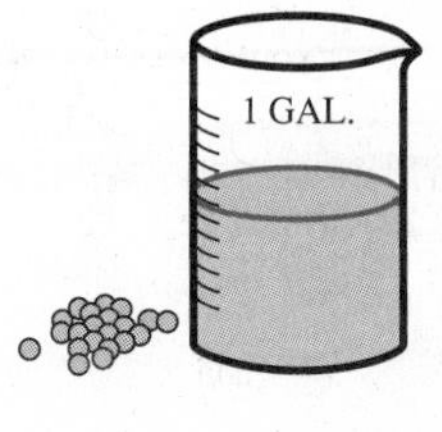

5. Henry has a 12-inch ruler marked only in inches. He wants to divide a $7\frac{1}{4}$-inch-wide board into six strips of equal width. How should he mark the board for sawing?

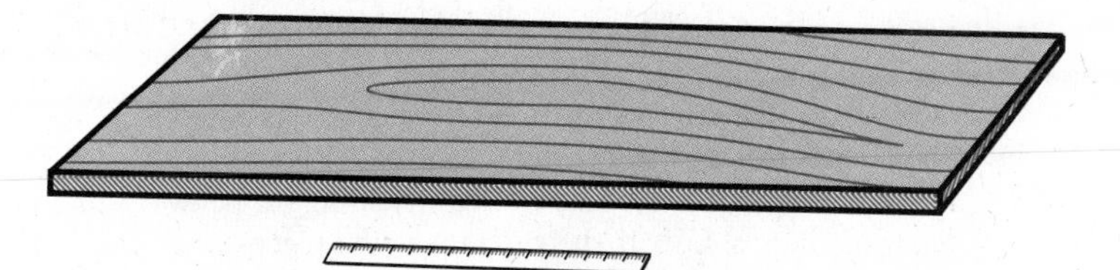

6. Amy and Susan, looking into a cylindrical barrel which appears to be about half full of liquid, argue about whether it is in fact a little more or a little less than half full. How can they settle the argument without a measuring device?

7. The distance from Chicago to New York is 800 miles. A nonstop train leaves Chicago for New York at 60 mph. Another train leaves New York for Chicago at 40 mph. How far apart are the trains 1 hour before they meet?

8. Two college students were traveling from Evanston to Chicago on an electric train. "I notice," said one of them, "that trains coming in the opposite direction pass us every 10 minutes. What do you think—how many trains from Chicago arrive in Evanston every hour, given equal speeds in both directions?" "Six, of course," the other answered, "because 60 divided by 10 is 6." The first student did not agree. What is your answer?

9. In the opening problem about Trot and the two bicyclists, suppose Horace races along at 15 mph while Hilda pedals at a leisurely 10 mph. If Trot races back and forth at 30 mph, how far will he run before the cyclists meet?

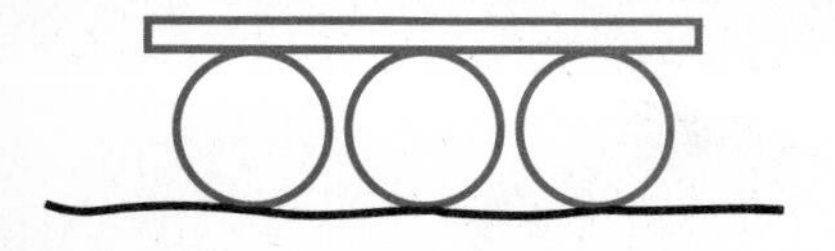

10. The rollers in the diagram at the left have circumferences of 1 foot. How far does the slab move each time the rollers go through a complete revolution?

11. Farmer Brown has followed the same routine for years now. Each morning he milks the cow at the barn, carries the milk can to the river to cool, and then walks on to the house for breakfast. Long ago he figured out exactly where he should hit the river in order to minimize the length of his path. How did he do it?

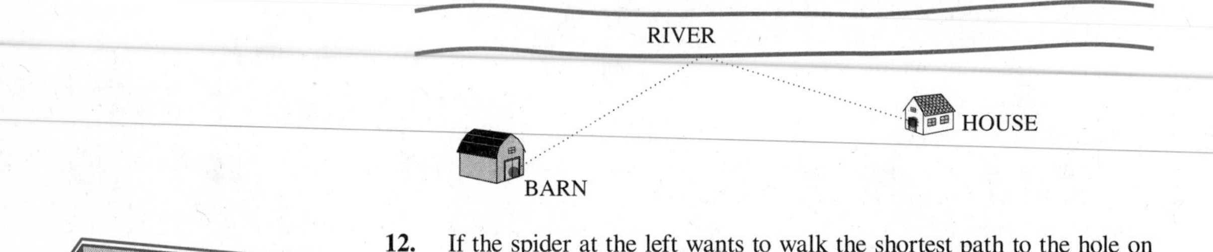

12. If the spider at the left wants to walk the shortest path to the hole on the opposite side, tell it how this path can be found.

13. A farmer's wife with a carving knife has a cube of cheese 3 inches on a side. She wishes to slice the cube into 27 small cubes 1 inch on a side. She can do this with the six cuts shown in the figure below. Can she reduce the number of cuts if she rearranges the pieces after each cut?

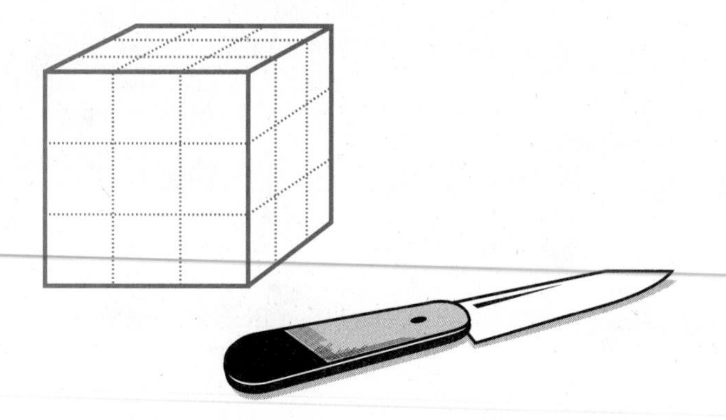

14. With a large power mower, it takes 3 hours to mow a lawn. Using a smaller mower, it takes 6 hours. If Amy and Homer work together using the two mowers, how long will it take?

15. Two opposite corners have been removed from a standard chess-board. Can this mutilated chessboard be covered with 31 dominoes, each of which can cover two squares?

16.* If the squares removed in Problem 15 are of opposite colors (but anywhere on the board), can the remaining squares be covered with 31 dominoes?

17.* My only timepiece is a wall clock. One day I forgot to wind it, and it stopped for I don't know how long. Anyway, I wound it up again and went to visit a friend across town whose clock is always correct. I stayed awhile and then returned home. Then I made a simple calculation and set my clock correctly. How did I do it?

18.* Sadie entered life at 7 pounds and died exactly 100 years later weighing 79 pounds. Show that, at some time in her life, her age was equal to her weight.

19.* A rubber string is laid out along a line. Then it is stretched along this line by pulling both ends, but not necessarily the same amount. Show that at least one point remains stationary.

Hints to Selected Problems

5. Laying the ruler squarely across the board won't help.

6. Tip the barrel, but how far?

7. How fast are they closing the distance between them?

10. Think of the motion of the slab as being due to two factors: (a) the movement of the centers of the rollers, and (b) the rotation of the rollers about their centers.

11. Try a reflection and then remember that a line provides the shortest distance between two points.

13. Study the center cube.

15. One domino always covers two squares—but of what kind?

2.5 The Use of an Unknown

A Balancing Act

Using the theme of an old-fashioned store, Mr. Gladgrind set up in a mall a shop that sold, among other things, a selection of coffee beans that he would grind as requested. He used an old-fashioned pan balance in which he would put weights on one side, coffee on the other. Because it was a genuine antique that had gotten damaged, one pan was six inches from the balance point, the other six and one half inches. He compensated for this by weighing half the order using the right-hand pan for coffee, the left for weights, and then reversing sides for the other half of the order. Thus, if you ordered two pounds of coffee, he would put a pound weight in the left pan and balance it with coffee in the right pan. Then he would put his weight in the right pan and balance it with coffee in the left. Does this work? Exactly how much coffee would you get if you ordered two pounds. (If you need to know, the pans weigh one pound each.)

We have left until last the methods of problem solving traditionally associated with mathematics. This was done not only to impress you with the fact that there are other ways to approach problems, but because intelligent use of unknowns utilizes the other methods as well. For many students, x marks the spot where arithmetic leaves off and real mathematics begins. Its appearance no doubt disappoints some of our readers who think it means a return to what they found to be both boring and difficult in high school. Others are delighted, feeling we have finally gotten to something significant.

Both groups may be helped by this advice. Don't confuse mathematics with x-itus. The aimless manipulation of x's is not mathematics any more than a random collection of words is an essay. Mathematics is the process of reasoning from a clearly stated question or problem to an irrefutable answer or solution. For us, the use of a letter, usually x, to stand for an unknown number is a device, a strategy in solving problems. The correct manipulation of x's, like the correct spelling of words, is not the chief end; it is a means to something more important.

> I hope that I shall shock a few people in asserting that the most important single task of mathematical instruction in the secondary schools is to teach the setting up of equations to solve word problems.
>
> *George Polya*

Getting the Equation

For most people, the difficult part of using a letter to represent an unknown quantity in a problem is not the algebraic solution of the equation, but the setting up of the equation in the first place. And it is for this part of the process that our suggestions in Chapter 1 for getting started can be most helpful.

It is still of utmost importance to clarify the question. Pictures, diagrams, charts and tables will still help to organize the information. It is especially wise to anticipate what will be a reasonable answer.

To these bits of sound advice, we must emphasize two more specific steps in the process if you plan to use a letter (and it doesn't have to be x) to represent an unknown quantity.

(1) Be very specific in writing down exactly what it is that the letter represents. Don't say that t represents time. Say that it represents hours (or minutes, or whatever) since an incident of interest occurred. Sometimes you will want to let the letter represent whatever you seek to find, but sometimes it is wise to make another choice. If you seek John's age, and if John is three times as old as Sally, you may choose to use a for Sally's age, $3a$ for John's age, rather than to introduce fractions by letting John's age be a, making Sally's age $a/3$.

(2) When you believe that you see how to set up an equation, state it in words first. If you cannot express your idea in a crisp, grammatically correct English sentence, the chances are excellent that it won't make sense as an algebraic expression either.

When you can clearly define what the letter represents and can say in clear English just how the facts of the problem can be related in a statement of equality, the translation into algebra is usually very simple. So also is the actual solution of the algebraic equation.

Finally, we remind you to check your answer against your anticipated solution. Avoid the foolish error of completing your algebraic manipulation and announcing (without reviewing how you got started in the first place) that $x = 175$. If x represents Grandma Uphill's age, or someone's dollar expenditure at McDonald's, a check of your work is in order.

Now it's time to illustrate these principles.

An Age Old Problem

When a father was 42, his son was 8. Now he is three times as old as his son. How old is the son?

Be Definite

Do not say x = age.

Do not say x = son's age.

Do say x = son's age now.

We begin by specifying the unknown. *Let* x *denote the son's age right now*. Next, we draw a picture that describes the problem.

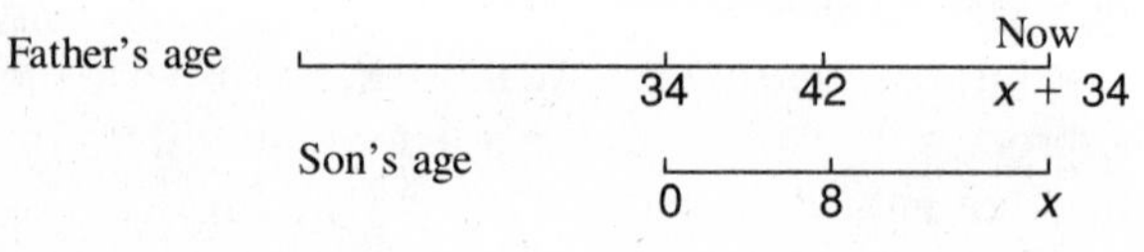

Now read the first sentence of the problem (and look at the picture). It says that the father was (and therefore always will be) 34 years older than his son. So his age now is $x + 34$. And the second sentence says

(the father's age now) is 3 times (the son's age now).

This is easily transformed into an algebraic equation

$$(x + 34) \qquad = 3 \qquad (x)$$

Now solve.

$$x + 34 = 3x$$

Add $-x$: $$34 = 2x$$

Multiply by $\frac{1}{2}$: $$17 = x$$

Thus, the son is 17 and the father is $17 + 34 = 51$.

Do these answers make sense? Yes. If we had gotten 170 and 204, we would have immediately gone back to look for a mistake. In any case, it is easy to check our answers. We note that 51 is three times 17, and that, under these circumstances, when the son was 8, the father was 42.

A Mixture Problem

How many liters of pure sulfuric acid should be added to 5 liters of 15 percent sulfuric acid to obtain a solution that is 50 percent sulfuric acid?

We'll begin with a picture that displays all the given information in a visual way. Note that we have *let* x *represent the number of liters of pure acid to be added.* But how do we get an equation? We look at the amount of *pure* acid in the three containers in the figure.

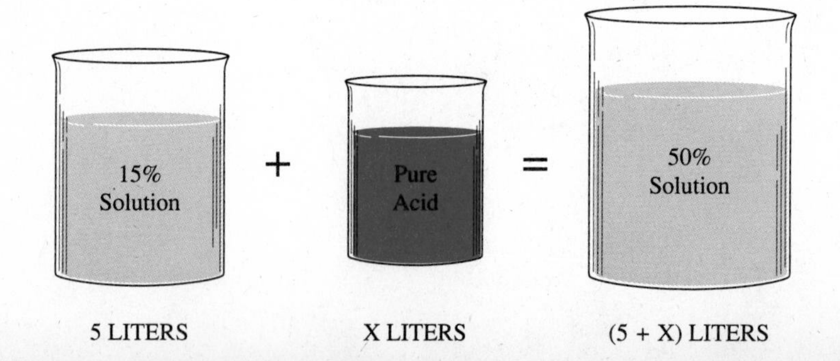

$$\begin{pmatrix}\text{The pure acid}\\ \text{we start with}\end{pmatrix} \text{ plus } \begin{pmatrix}\text{the pure acid}\\ \text{we add}\end{pmatrix} \text{ is } \begin{pmatrix}\text{the pure acid}\\ \text{we end with}\end{pmatrix}$$

$$(0.15)(5) \quad + \quad x \quad = (0.50)(5 + x)$$

To solve the equation, we might use the following steps.

Remove parentheses:	$0.75 + x = 2.50 + 0.50x$
Multiply by 100:	$75 + 100x = 250 + 50x$
Add -75:	$100x = 175 + 50x$
Add $-50x$:	$50x = 175$
Divide by 50:	$x = \dfrac{175}{50} = 3.5$

The answer of 3.5 liters of pure acid to be added seems reasonable. You can check that it is exactly right.

A Gas Mileage Problem

Rodney Roller bought a compact car that goes 15 miles farther on a gallon of gas than did his old desert yacht. He finds that he can now get to Grandma Uphill's house on 4 gallons of gas whereas it used to require 9 gallons. How far does she live from Rodney?

We will surely need the fact that

$$\text{miles per gallon} = \frac{\text{distance traveled}}{\text{gallons used}}$$

or equivalently

$$\text{distance traveled} = (\text{gallons used})(\text{miles per gallon})$$

Since the problem asks how far Grandma Uphill lives from Rodney, it seems sensible to let x represent that distance. Thus,

$$x = (\text{gallons used}) \cdot (\text{miles per gallon})$$

Depending on whether we are referring to the old car or the new car this gives

$$x = 9 \cdot (\text{miles per gallon with old car})$$
$$x = 4 \cdot (\text{miles per gallon with new car})$$

This almost asks that we equate the two expressions for x. When we do, we get

$$9(\text{miles per gallon with old car}) = 4(\text{miles per gallon with new car})$$

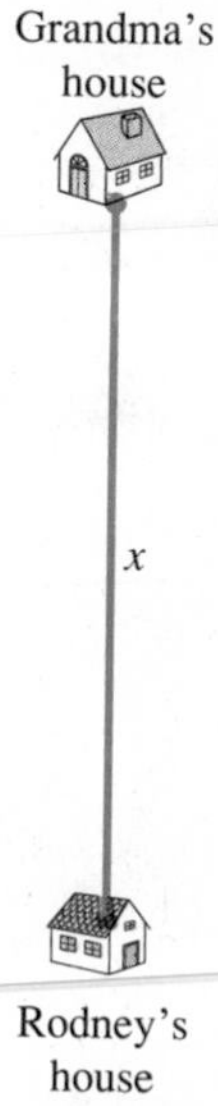

It may seem unfortunate that our unknown x has now disappeared but this apparent misfortune forces us to realize there are other un-

Grandma's house

x

Rodney's house

knowns in this problem. We don't know how many miles per gallon either car gets, though we do know how they are related.

This turns out to be one of those problems which yields to the strategy of asking a related question: Can we find the miles per gallon the old car got? *Let* m *be that number*. Then, $m + 15$ is the number of miles per gallon with the new car. And our equation becomes

$$9m = 4(m + 15)$$

This equation is easily solved for m.

$$9m = 4m + 60$$
$$5m = 60$$
$$m = 12$$

If m is 12, then $x = 9(12) = 108$. It is 108 miles to Grandma Uphill's house. You can check that this answer is correct.

Run, Trot, Run—Again

Suppose that in the problem, **Run, Trot, Run,** we were asked for the time when Trot returns to Hilda for the first time. We begin by clarifying the problem, restating it as a series of one-fact sentences.

- It is exactly 100 miles from Posthole to Pitstop.
- At 8:00 A.M. Horace leaves Posthole, Hilda leaves Pitstop.
- They ride toward each other at 10 mph.
- Trot leaves Pitstop at 8:00 A.M., running ahead of Hilda at 30 mph.
- When Trot meets Horace, he immediately turns and runs back to Hilda.
- What time is it when Trot meets Hilda?

Divide and conquer.
Stage 1

We will treat this problem in two stages, finding first the time for Trot to meet Horace, then finding the time it takes Trot to return to Hilda. Begin with a picture of the situation when Trot meets Horace.

Draw a picture or diagram.

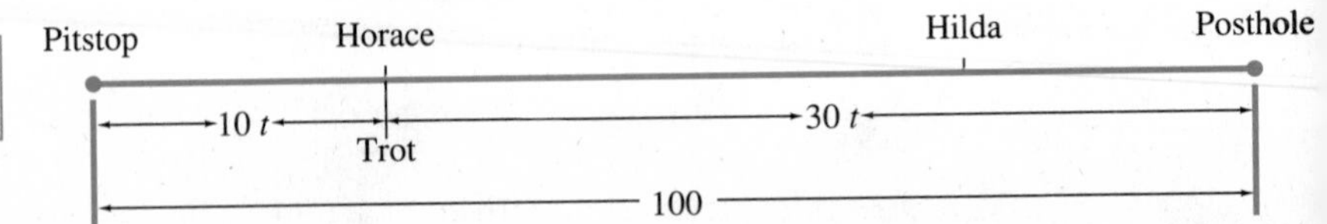

d = rt

Let t be the hours that have elapsed since 8:00 A.M. to the moment when Trot meets Horace. Horace will have traveled $10t$ miles; Trot will have run $30t$.

(Horace's distance)	plus	(Trot's distance)	is 100
$10t$	$+$	$30t$	$= 100$

Notice that the algebraic equation we have written is a direct translation of the English sentence above it. The solution of this equation presents no problem.

$$40t = 10$$

$$t = \frac{100}{40} = \frac{5}{2}$$

Observe that when Trot meets Horace, Horace and Hilda are separated by 50 miles.

Now let's draw a second picture that shows the situation when Trot gets back to Hilda.

Divide and conquer.
Stage 2

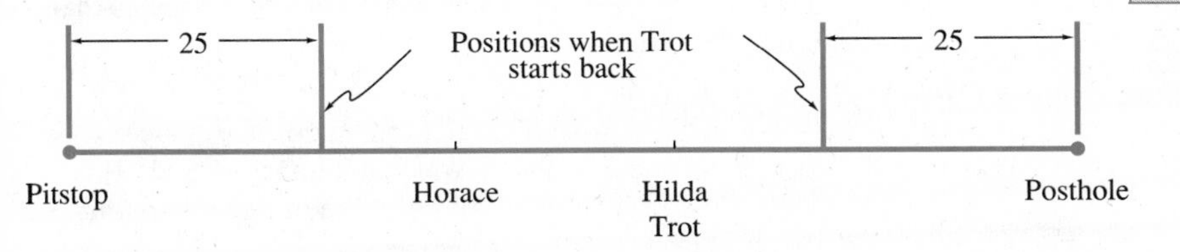

Let T be the hours that have elapsed while Trot ran from where he left Horace back to where he met Hilda. Hilda will in that time travel $10T$ and Trot will travel $30T$

(Hilda's distance)	plus	(Trot's distance)	is 50
$10T$	$+$	$30T$	$= 50$

$$40T = 50$$

$$T = \frac{50}{40} = \frac{5}{4}$$

The total elapsed time is $\frac{5}{2} + \frac{5}{4} = \frac{15}{4} = 3$ hours, 45 minutes. The time is 11:45 when Trot returns.

Summary

The use of a letter to represent an unknown quantity is one more useful method for solving a problem. When it is used, one should

- still use the techniques previously discussed for getting started.
- write a very clear, specific definition of what the letter represents.
- write in clear English prose the statement to be expressed as an algebraic equation.

It is especially important, after algebraic manipulation determines the value of the unknown, to check this value against anticipated values.

Problem Set 2.5

1. Write an algebraic expression using x that represents each English phrase. For example, a number plus one-half its square could be represented by $x + \frac{1}{2}x^2$.
(a) One-half the sum of a number and its square.
(b) The sum of 3 and twice a number.
(c) The sum of three consecutive even numbers if the smallest is x.
(d) The perimeter (distance around) a square of side x.
(e) The area (length times width) of a rectangle of width x if the length is 3 less than twice the width.
(f) The volume of a box (length times width times height) with square base of x meters and height 3 more than one-half the width.
(g) The value in cents of the coins in Ellen's purse if it contains x nickels, three times as many dimes as nickels, 4 fewer pennies than dimes, and no other coins.

2. Follow the directions of Problem 1.
(a) One-third the product of a number and four times its square.
(b) Fifty-two percent of the amount by which a number exceeds 10.
(c) The total cost of the tickets for x children, three more adults than children, if adult tickets are \$3.50 each and children's tickets are \$1.25 each.
(d) The number of miles traveled if I drove 55 miles per hour for x hours.
(e) The time it took to drive 400 miles if Susan drove at x miles per hour.
(f) The cost of gas at \$1.55 per gallon for a 600-mile trip if my car gets x miles per gallon.
(g) Amy's weekly pay if she worked 40 hours at \$6.80 per hour and x additional hours at time and one-half.

3. The sum of three consecutive positive integers is 636.
(a) Let x be the smallest of the three. What are the next two integers?
(b) What expression represents the sum of the three integers written in part (a)? Set this expression equal to 636.
(c) Find the three integers.

4. The sum of three consecutive even numbers is 192. Find them, following the procedure outlined in Problem 3.

5. The sum of three consecutive integers is 63 more than the smallest one.
(a) Let s represent the smallest integer. What expressions represent the next two integers?
(b) Write an equation, using the fact that the sum of the three integers represented in part (a), decreased by 63, must equal s.
(c) Find the three integers.

6. A number plus a third of that number and a fifth of that number is equal to 46.

(a) Let n be that number. What expressions represent a third and a fifth of that number?
(b) Translate the statement of the problem into a mathematical equation.
(c) Find the number.

7. A rectangle is 4 feet longer than twice its width; its perimeter is 200 feet.
(a) Let w equal the width. What is the length?
(b) Using the expressions from part (a), what is the perimeter?
(c) Find the dimensions of the rectangle.

8. The width of a rectangle is one-third its length and its perimeter is 300 centimeters. Find its width, following the procedure outlined in Problem 7.

9. A father is three times as old as his son, but 15 years from now he will be only twice as old as his son.
(a) Let x designate the son's age now. What is the father's age now?
(b) What will be the age of the son in 15 years? The father? Express the fact that the father is (in 15 years) twice as old as the son.
(c) How old is the son now?

10. John is four times as old as Amy is now, but in 16 years he'll only be twice as old. How old is John now? Follow the procedure outlined in Problem 9.

11. A tank contains 10 gallons of 5 percent salt brine. An 8 percent salt brine is to be obtained by evaporating some of the water in the original mixture.
(a) Note that the original mixture contains 10(.05) gallons of salt.
(b) Let x represent the number of gallons of water to be evaporated. What expression represents the number of gallons of salt in the tank after evaporation?
(c) How does the salt in the tank before evaporation compare with the salt in the tank after evaporation?
(d) How much water should be evaporated?

12. The cooling system in Rodney Roller's car has a capacity of 14 quarts. At the present time, 25 percent of the coolant is antifreeze. To protect his car for a Minnesota winter, he ought to have a 50 percent mixture.
(a) Suppose he drains off x quarts of his present mixture. Before adding any pure antifreeze, how much antifreeze will he have in his tank?
(b) What algebraic expression will represent the quarts of antifreeze in the tank after refilling?
(c) How many quarts of antifreeze does he want in his tank?
(d) How many quarts should he drain off?

13. How many gallons of a 60 percent solution of nitric acid should be added to 10 gallons of a 30 percent solution to obtain a 50 percent solution of the acid?

14. How much commercial bleach, which contains 0.51 pound of sodium hypochlorite per gallon, should be added to 1 gallon of water to obtain a solution that contains 0.06 pound of sodium hypochlorite per gallon?

15. The Joneses plan to put in a concrete drive from their garage to the street. The drive is to be 36 feet long, and they have been advised to make it 4 inches thick. Since there is an extra delivery charge for less than 4 cubic yards of ready-mixed concrete, the Joneses decide to make the drive just wide enough to use 4 cubic yards. How wide should they make it? Hint: Put everything in the same units. Volume is length times width times height. And there are 27 cubic feet in a cubic yard.

16. Not content merely to keep up with the Joneses, the Smiths decide to put in a concrete drive 48 feet long and 6 inches thick. How wide should it be to use up all concrete in a 9 cubic yard load?

17. When you rent a boat to fish by trolling in a river, you are told that it will take you upstream at 3 miles per hour and bring you back at 7 miles per hour. You set out at 6:00 A.M. going upstream, having rented the boat until noon.
    (a) Let t represent the time spent going upstream. How much time will be spent going downstream?
    (b) Express in algebraic form the fact that the distance traveled upstream equals the distance traveled downstream.
    (c) At what time should you turn around?

18. Going at a certain speed it took Mary 8 hours to get to Center City. If she could have driven 10 miles an hour faster, it would only have taken $6\frac{2}{3}$ hours. How far did she drive?

19. Trains traveling on parallel tracks start at noon from cities 800 miles apart. One train travels at 40 miles per hour, the other at 75 miles per hour. At what time will they meet?

20. A group of students plans to charter a flight to take them to a ski resort. They are told that they will have to pay so much each if 75 students sign, but can save $65 each if they can find another 25 who want to go. What is the price if only 75 go? Assume the total price stays the same.

21. Susan Sharp, manager of a shoe department, must sell a certain pair of shoes for $18.60 to make her usual profit. She knows the store is planning a storewide sale in which everything will be sold at a 20 percent discount. How should she price the shoes to maintain her profit?

22. Tammy and Pat were engaged to paint the parking meters on Main Street. Tammy arrived early and had painted 3 meters on the south side of the street when Pat turned up and pointed out that Tammy's contract was for the north side. So Tammy started afresh on the north side while Pat continued on the south. When Pat finished, she went across and, starting with the last meter, worked backwards toward Tammy, painting six meters for Tammy, by the time they met. There were an equal number of meters on each side. Which girl painted more meters, and how many more did she paint?

23. A factory received an order for widgets that would normally have taken 12 days to produce. The customer was in a hurry, however, so steps were taken to increase productivity by 25%. As a result, in just 10 days the factory produced 42 more widgets than were required to fill the order. What is the normal output of widgets per day in this factory?

24. Solve **A Balancing Act,** the problem that opened this section.

25.* A classic problem is the courier problem. If a column of soldiers 3 miles long is marching at 5 miles per hour, how long will it take a courier on a motorcycle traveling at 25 miles per hour to deliver a message from the end of the column to the front and then return?

26.* How long after 4:00 P.M. will the minute hand on a clock overtake the hour hand?

27.* What is the first time after twelve o'clock noon when the hands of the clock will be together again?

28.* At what time between four and five o'clock do the hands form a straight line?

29. Jack can do a certain job in 4 days, and Jill can do it in 3 days. How long will it take them working together?

30.* A can do a piece of work in one-third the time B can; B can do it in three-fourths the time C can; all together they can do it in 12 days. How long would it take A to do it alone?

31.* The combined age of a ship and its boiler is 48 years. The ship is twice as old as the boiler was when the ship was half as old as the boiler will be when the boiler is three times as old as the ship was when the ship was three times as old as the boiler. How old is the ship?

Hints to Selected Problems

14. After g gallons of the commercial bleach has been added, there is to be $(1 + g)(.06)$ pounds of sodium hypochlorite in the solution.

19. The sum of the distances they travel must be 800 miles.

22. Tammy painted 3+[meters on a side − 6]. How many did Pam paint?

23. If the factory normally produce w widgets per day, then it produces $w + \frac{w}{4}$ when productivity is increased.

25. Divide and conquer. Let t be the time it will take until the courier reaches the front of the column; find t. Then as step 2, find the time to return.

26. The two hands both make the same angle measured from their noontime positions. The hour makes an angle of $120 + t(\frac{30}{60})$ at t minutes after 4:00 P.M.

27. See the hint for Exercise 26.

28. When the hands are straight, the angle the minute hand makes with its noontime position is 180° more than the angle the hour hand makes with its noontime position.

29. Jack does $\frac{1}{4}$ in a day while Jill does $\frac{1}{3}$ of the job.

2.6 The Use of Several Unknowns

PAT vs. BUN

Two political parties vie for seats in the General Assembly of the Imagin nation. In the last election, 20 percent of the seats were regarded as marginal, the others being regarded as safe for the incumbents. If the Bread for the Underdeveloped Needy (BUN) party had won all of the marginal seats, it would have had 86 more seats than the Patriots Against Titillation (PAT). If PAT had won all the marginal seats, then it would have had 22 more seats than the BUNs. Alas, both parties won exactly half of the marginal seats while every seat regarded as safe was won by the party expected to win it. How many seats do the BUNs have in the General Assembly?

Some problems are most easily solved by using more than one letter to represent unknowns. The guiding principles do not change. Take full advantage of all that we have learned about getting started. Write in full detail what each unknown represents. State your ideas in good English before trying to translate into algebraic equations.

A general (and, at this stage, deliberately imprecise) guideline is that if you use n letters to represent unknowns, you will need n equations to determine their values. Most of the examples here use $n = 2$; a few use $n = 3$. Since the same values for the unknowns must work in each equation, it is said that we have a system of simultaneous equations.

Though our focus here is on setting up the equations, the part that always causes the most trouble, the solutions provided below include a brief refresher on how such systems are solved.

Nickel and Dime Stuff

Homer has 50 coins in his pocket, all in nickels and dimes, and altogether he has \$3.50. How many nickels and how many dimes has he?

Here is a simple way to set up this problem. *Let* x *be the number of nickels and* y *the number of dimes.*

English	*Algebra*
(The number of nickels) plus (the number of dimes) is 50	$x + y = 50$
(The value of the nickels in cents) plus (the value of the dimes in cents) is 350	$5x + 10y = 350$

And here is the way to solve the two equations:

Multiply the first equation by -5:	$-5x - 5y = -250$
Write down the second equation:	$5x + 10y = 350$
Add the two equations:	$5y = 100$
Multiply by $\frac{1}{5}$:	$y = 20$
Substitute $y = 20$ into one of the original equations:	$x + 20 = 50$
Add -20:	$x = 30$

Thus, Homer has 30 nickels and 20 dimes. That is certainly 50 coins. The nickels are worth \$1.50, the dimes \$2.00, a total of \$3.50.

A Two-unknown Mixture Problem

Suppose that you are applying for a job at Harold's Nut Shop. To test your ability, Harold points to two bins of nuts, walnuts costing \$6.00 per pound and cashews costing \$8.40 per pound. He tells you to prepare 25 pounds of a nut mix to be sold at \$6.80 per pound. How many pounds of each kind should you use?

On an empty paper sack, you should write something like the following.

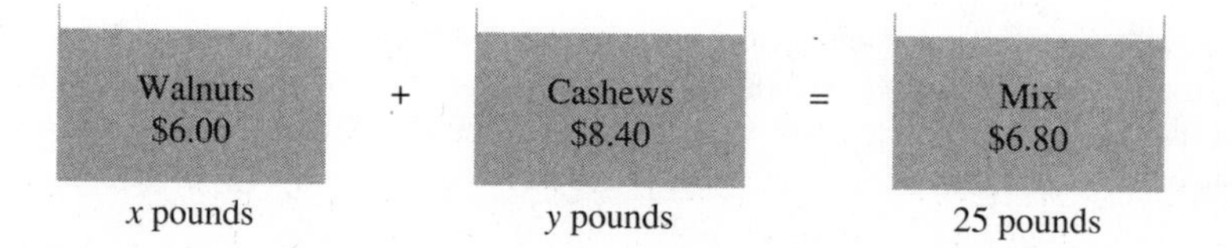

Setting it up:

$$x = \text{number of pounds of walnuts}$$
$$y = \text{number of pounds of cashews}$$
$$6.00x + 8.40y = 6.80(25) \qquad \text{(looking at values)}$$
$$x + y = 25 \qquad \text{(looking at pounds)}$$

To solve, multiply the second equation by -6 and then add the two equations.

$$\begin{array}{rcl} 6.0x + 8.4y & = & 6.8(25) \\ -6.0x - 6.0y & = & -6.0(25) \\ \hline 2.4y & = & .8(25) \end{array}$$

$$y = \frac{25}{3}$$

Now substitute $y = \frac{25}{3}$ into the second of the original equations.

$$x + \frac{25}{3} = 25$$

$$x = \frac{50}{3}$$

You should mix $16\frac{2}{3}$ pounds of walnuts with $8\frac{1}{3}$ pounds of cashews.

Isaac Newton

Isaac Newton (1642–1727) is universally acclaimed as one of the most brilliant and influential scientists ever to have lived. His contributions to mathematics extend from his use of the binomial theorem to the invention of calculus. Some of the basic principles he set down (to every action, there is an equal and opposite reaction, for instance) are familiar to people who have only the most superficial understanding of science, and those who are best versed in science are necessarily familiar with a great deal more that he did. Every schoolchild has some notion that Newton saw in a falling apple the core his gravitational theory, and one of the most famous pictures of Newton shows him using a prism to study the properties of light.

In spite of it all, few modern readers can appreciate the way in which Newton's work fundamentally changed the way everyone understood the world around them. Scholars may (and do) debate whether Alexander Pope was writing with sincerity, or with a touch of weariness over the public adulation of Newton even in his own time, but his lines do in fact convey a realistic assessment of the impact Newton's work had on a modern "world view."

Nature and Nature's laws lay hid in the night:
God said "Let Newton be!" and all was light.

A Problem From Newton

Here is a problem attributed to Isaac Newton who is said to have let no grass grow under his feet, but who assumed for this problem that grass would grow at a constant rate under the feet of grazing oxen.

Three pastures with grass of identical height, density, growth rate, color, and nutritional value are 10/3 acres, 10 acres, and 24 acres. If the first pasture can feed 12 oxen for 4 weeks, and the second can feed 21 oxen for 9 weeks, how many oxen can be fed on the third pasture for 18 weeks?

There are three unknown quantities in this problem to which we find it convenient to give names.

s = the height, measured in convenient units, at the time the oxen are admitted to the pastures.

g = the additional length, measured in the same units, that the grass would grow in a week (if left alone by the oxen.)

n = the number of oxen that can be fed in the third pasture.

We are assuming that the grass grows g units per week and provides the same total nutrition whether it is periodically "mowed down" by the oxen, or left to grow for the week, and then eaten all at once. The volume of grass necesary to feed k oxen for a week will be called k oxen-weeks. Thus, the first field is to provide for $12 \times 4 = 48$ oxen-weeks. In the first field,

$$\begin{pmatrix}\text{vol. of begin-}\\ \text{ning grass}\end{pmatrix} + \begin{pmatrix}\text{vol. of grass that}\\ \text{grows in 4 wks.}\end{pmatrix} = \begin{pmatrix}\text{vol. of grass for}\\ \text{48 oxen-weeks}\end{pmatrix}$$

$$\frac{10}{3}s + 4\left(\frac{10}{3}\right)g = 48$$

If we multiply this equation by 3, and then apply the same reasoning to the second and third fields, we get the three equations

$$\begin{aligned} 10s + 40g &= 144 \\ 10s + 90g &= 189 \\ 24s + 432g &= 18n \end{aligned}$$

Since the first two equations involve only the two unknowns s and g, we might solve them simultaneously as the first stage of our solution.

Divide and conquer. Stage 1

$$\begin{aligned} 10s + 40g &= 144 \\ 10s + 90g &= 189. \end{aligned}$$

Subtract the first from the second to obtain

$$\begin{aligned} 50g &= 45 \\ g &= \frac{9}{10} \end{aligned}$$

Substitute this value of g in either of the two equations and solve to find that $s = \frac{54}{5}$.

Divide and conquer. Stage 2

Knowing the values of s and g, we can now return to the third equation.

$$24\left(\frac{54}{5}\right) + 24(18)\left(\frac{9}{10}\right) = 18n.$$

This is easily solved; multiply by 10.

$$48(54) + 24(18)(9) = 180n$$
$$n = 36.$$

The third pasture can feed 36 oxen for 18 weeks.

Summary

The techniques for using more than one letter to represent unknown quantities in a problem are entirely similar to those used with just one unknown. If you choose to use two unknowns, then you will need to determine two equations relating them to one another.

Cathleen Morawetz

Sometimes the mastery of a particular method of solution is what dictates the kind of problem one works on. Cathleen Morawetz reflected in an interview for *More Mathematical People* [Albers, Alexanderson, and Reid, Harcourt, Brace and Jovanovich, 1990] on how she got started in the area of mathematics where she ultimately made her reputation.

> In mathematics there is frequently the problem of kicking up one's enthusiasm. Another time when I really didn't know what to work on and was struggling with a paper someone had given me about singular ordinary differential equations with some pathological behavior, Jürgen Moser came by and said, "Ach, it's ridiculous to work on that problem." So I threw the paper away. What got me working on the wave equation, which later dominated a lot of my work, was a lecture by Joe Kelleo on unsolved problems. As I was sitting there, I saw that the technique I had used on mixed equations ought to have application to Keller's problems—they were elliptic this way and hyperbolic that way—and that really worked out. I don't believe that I ever again had that feeling of floundering.

One does indeed get the impression that Professor Horawetz has not floundered. Though her career in mathematics started at a time when people consciously placed barriers in the way of women entering the field, she persisted and ultimately became the first woman invited by the American Mathematical Society to give the prestigious Gibbs lecture at its annual meeting (Her lecture was titled, "The mathematical approach to the sonic barrier."), she has been the director of New York University's Courant Institute (a center for applied mathematical research and graduate education), and she serves at the time of this writing in 1994 as President of the American Mathematical Society.

Problem Set 2.6

1. The sum of two numbers is 17. If you take two-thirds of the first and subtract the second, you will get 8.
   (a) If x and y are the two numbers, what is $x + y$?
   (b) What algebraic expression describes two-thirds of the first, subtract the second?
   (c) Find the two numbers.
2. Two numbers are so related that their sum is 35, and three times the one number is twice the other number. Find the numbers.
3. A boy has some quarters and nickels with a total value of $6. There are three times as many nickels as quarters.
   (a) Let n be the number of nickels, q the number of quarters. Measured in cents, what is the value of his nickels? Of his quarters? The sum of these values must be 600.
   (b) Express algebraically that there are three times as many nickels as quarters.
   (c) How many coins of each kind does he have?
4. The price of admission tickets to a certain theater was $2.00 for adults and $1.50 for children. If 410 tickets were sold and the total receipts were $765, how many adults and how many children attended the theater?
5. A grocer has some Aromatic coffee worth $3.84 per pound and some Caffineo coffee worth only $2.64 per pound. She wishes to mix them so as to get 10 pounds worth $3.18 per pound.
   (a) Let x be the number of pounds of Caffineo, y the number of pounds of Aromatic to be used in the mix. What is $x + y$?
   (b) Measured in cents, what is the value of the Caffineo put into the mix? The value of the Aromatic in the mix? The value of the mix?
   (c) How much of each coffee should she use?
6. A solution that is 40 percent alcohol is to be mixed with one that is 90 percent alcohol to obtain 100 liters of 60 percent alcohol solution. How much of each should be used?
7. If the length of a rectangle is decreased by 10 meters and the width is increased by 8 meters, the resulting figure will be a square whose area is the same as that of the rectangle.
   (a) If the length and the width of the given rectangle are l and w, respectively, what expressions describe the length and width of the new rectangle?
   (b) What algebraic equation states that the new rectangle is a square?
   (c) What algebraic equation states that the rectangles have equal area?
   (d) What are the dimensions of the given rectangle?
8. Professor Witquick, who had visited on his way to the office with a man digging a ditch, also stopped on the way home. Noting that the man's head was now below ground level, Witquick quipped, "Well, I see you've gotten a head in your work." Appreciating the pun, the worker replied that the top of his head was now as far below ground level as it had been above ground that morning and that his feet were now twice as far underground. "Ah," said Witquick, "but you'll have to tell me your height if you expect me to figure the depth of the ditch." "I'm 5 feet 9 inches," replied the worker. How deep is the ditch?

9. A rowing crew that has been practicing on a river would like to know its rate in still water. The crew was able to row 16 miles downstream in 1 hour but took 2 hours to row back.
   (a) If x is the rate at which the crew rows in still water and y is the rate at which the stream carries the boat, what is the rate going downstream? Upstream?
   (b) Use $D = RT$ to get an equation describing the trip downstream. Do the same for the trip upstream.
   (c) Find the rate at which the crew rows in still water.

10. A fisherman motored upstream at a steady pace for 4 hours and then turned around and went back in 2 hours. The next day he needed 6 hours to travel 48 miles upstream. Assuming that the boat was going at full throttle both days and that the rate of the current remained constant, find out how fast the boat can go in still water.

11. Susan Sharp paid $2400 for some dresses and coats—$20 for each dress and $50 for each coat. She sold the dresses at a 20 percent markup and the coats at a 50 percent markup and made in all a profit of $900. How many dresses and how many coats did she buy?

12.* Workers in a certain factory are classified in two groups, depending upon the skill required for their jobs. Group A workers are paid $12.00 per hour; group B workers are paid $7.00. In negotiations for a new contract, the union demands that the hourly wages for group B workers be brought up to two-thirds of those for group A workers. The company has 55 employees in group A and 40 in group B, all of whom work a 40-hour week. If the company's offer raises its weekly payroll by $4540, what hourly rates were proposed for each class of workers?

13.* A woman leaves her fortune to her children. The first child receives $1000 plus one-tenth of what is left; the second receives $2000 plus one-tenth of what is left; the third receives $3000 plus one-tenth of what is left; and so on. It turns out that each of the children receives the same amount. How many children are there, and how much does each receive?

14.* Three mutually tangent circles have centers A, B, and C and radii a, b, and c, respectively. The lengths of segments AB, BC, and CA are 17, 23, and 12, respectively. Find the lengths of the radii.

15.* Solve the problem **PAT vs. BUN** that opened this section.

Hints to Selected Problems

8. Draw two pictures, one for the morning, one for the afternoon. Let x be the height of the worker's head above ground in the morning and y the depth of his feet below ground at that time. Then $2y$ is the depth of the ditch now.

10. Let s be the speed of the boat in still water and w be the rate at which the water carries the boat along. The distance upstream of $(s + w)(2)$ equals the distance downstream of (what?)

14. AB passes through the point where the circles centered at A and B are tangent, so $a + b = 17$.

chapter 3

Reflecting on Solutions

The reader who offers solutions in the strict sense only (this is what was asked, and here is how it goes) will miss a lot of the point, and he will miss a lot of the fun. Do not just answer the question, but try to think of related questions, of generalizations . . . and of special cases . . . . What makes an assertion true? What would make it false?

PAUL HALMOS

Depending on Clarity, Not Charity

> "The study of mathematics is wrongly divorced from the practice of speaking and writing good English. For the essence of good mathematics, like good language, is first to know what to say, and then to be able to frame words which express exactly that and nothing else."
>
> E. Cunningham
> Eureca No. 5 (1941) p. 9

It is clear that for a student in an undergraduate mathematics class who goes on to become a mathematician, great importance attaches to the stuff of the course: the definitions and the theorems; the tricks and the techniques; the accumulation of facts. But, we sometimes ask ourselves as teachers, what lasting contribution can we make to those students who do not become mathematicians? One answer is this: we can have as a principal goal the development of skill in lucid writing.

As opposed to terms used in politics, economics, religion, and other disciplines, the terms of mathematics are free from ambiguity and the content is not emotionally charged. We are, therefore, better able to focus our entire attention on the argument itself, and when something is not said correctly, counterexamples are usually at hand to succinctly illustrate the difficulty. Our writing should depend on the clarity, not the charity of the mind of our reader. But Galileo has said it another way:

> . . . for just as in nature itself there is no middle ground between truth and falsehood, so in rigorous proofs one must either establish his point beyond doubt, or else beg the question inexcusably. There is no chance of keeping one's feet by involving limitations, distinctions, verbal distortions, or other mental acrobatics. One must with a few words and at the first assault become Caesar or nothing at all.

Galileo's statement clearly puts a premium on being concise as well as precise. Stripped of superfluous comments, the proof exposes the heart of the argument in succinct, grammatically correct paragraphs that the reader finds attractive as well as compelling. For many mathematicians, the writing of such an argument is itself an aesthetically satisfying experience.

> "Gauss always strove to give his investigations the form of finished works of art. He did not rest well until he had succeeded, and hence he never published a work until it had achieved the form he wanted. He used to say that when a fine building was finished, the scaffolding should no longer be visible."
>
> Sartorius Von Katterhauser

> "You know that I write slowly. This is chiefly because I am never satisfied until I have said as much as possible in a few words, and writing briefly takes far more time than writing at length."
>
> Carl Gauss

In an essay that praises brevity, perhaps we have written too much. We want to emphasize, however, that this matter of writing is not a secondary goal in our teaching. This is one way in which mathematical training can make a significant contribution to anyone's education.

3.1 Is It Correct?

First in, Last out

In her office, Ms. Rusher will at various times of the day drop a letter to be typed in her secretary's tray. She numbers these letters as she drops them in. When there is time, the secretary takes the top letter from her basket and types it. On a day when six letters got dropped in and typed, which of the following could not be the order in which the secretary typed them.

(a) 546321 (b) 425631 (c) 354261 (d) 246531

All problem solvers make mistakes. We overlook completely some possibility that needs consideration. We make errors of logic. We make errors in calculation.

It follows that we all need to check our work. The traditional way to do this in mathematics courses is to compare our answers with those in the back of the book. It is a time-honored method that has no other honor. The problems we encounter in life do not have answers in any book to which we are privy, and if our goal is to prepare for life, we need to develop habits and methods for checking on ourselves, just as surely as we develop methods to solve problems in the first place.

There is nothing wrong, of course, with just checking over one's work a second time; we might go so far as to say it's a good practice. Most of us know from experience, however, that once a mistake has been made, there is a tendency for us to make it over again on a second pass. We need to develop alternative methods.

Compare with Estimates, Original Problems

In the very first section of Chapter 1, we met Homer growing impatient with Professor Branebom's answers to his problem because he had a reasonable expectation of an answer, and she wasn't coming close. We usually have or can quite easily make some estimate of what answer is to be expected. It is not a bad idea to begin by writing down one's initial estimate. This presumes, of course, that you will take the time to check it against whatever answer you get later, and that if your guess is quite off the mark, you will take further time to ascertain why this is so.

This is not to say that a reasoned answer that differs significantly from an estimate is necessarily wrong. We often jump to an "obvious" conclusion that turns out to be wrong, even in quite simple problems. Try this one on a friend not taking this wonderful course.

A sales representative drives 150 miles to make a call, averaging 60 mph. On her way back, she encounters several delays so that she averages only 40 mph. What was her average speed on the trip?

When careful consideration and calculation shows the answer to be 48 mph, the person who made the likely guess of 50 mph should think about it until he or she sees why that first estimate was wrong.

Sometimes, of course, it is possible to check an answer to see if it meets the conditions of the problem. If it is proposed that $t = 48$ is a solution to

$$\frac{1}{3}(t + 4) = 23,$$

you only need to check whether or not

$$\frac{1}{3}(48 + 4) = 23.$$

Since the left side is not an integer, it certainly is not 23. That lays aside a lot of uncertainty; $t = 48$ is not a solution to the given equation.

Suppose that in trying to solve a problem, you decide to set up and then solve an equation involving an unknown. The solution obtained should then be checked, not in the equation you set up, but in the problem as first stated. In Section 2.5, we solved as an illustration the following problem.

How many liters of pure sulfuric acid should be added to 5 liters of 15 percent sulfuric acid to obtain a solution that is 50 percent sulfuric acid?

The answer obtained was 3.5 liters. Try it. If we add 3.5 liters of acid to the 5 liters of 15 percent solution, then the 8.5 liters will contain $.15(5) + 3.5 = .75 + 3.5 = 4.25$ liters of sulfuric acid, and that is 50 percent of 8.5. We know in both senses of the word that we have the required solution.

In Section 1.3 we began with the relationship

$$lw = 2l + 2w$$

and took several algebraic steps to arrive at

$$l = \frac{2w}{w - 2}.$$

If after performing such calculations, you question whether your work is correct, remember that the unknowns represent numbers, so any number or set of numbers that make one equality true should make any derived equality true, and conversely. In the example at hand, it is clear from the second equation that if we set $w = 1$, $l = -2$. Substituting these numbers in the first equation gives

$$(-2)(1) = 2(-2) + 2(1).$$

This kind of check, while not foolproof (an example that works does not constitute a proof), buoys up our confidence in the work done.

Verify That Symmetries Are Maintained

There is another point to be made from our last example. Notice that the original expression

$$lw = 2l + 2w$$

is symmetric in l and w; that is, if l is everywhere replaced by w and w is everywhere replaced by l, the expression says the same thing. The same should be true, then, of any derived expression. If

$$l = \frac{2w}{w - 2}$$

then it should also be true that

$$w = \frac{2l}{l - 2}$$

Verify that it is so by multiplying this last expression by $(l - 2)$. This same symmetry can be seen in the table of values first computed in Section 1.3.

w	l
3	6
4	4
6	3

Sometimes the presence of symmetry in the original data combined with the knowledge that we should find this same symmetry in the conclusion gives a strong hint as to how we might find the solution.

The three boxes pictured below each contain six colored chips: one contains six black chips, one contains six red chips, and one contains three of each color. Unfortunately, all boxes are labeled incorrectly. You may draw a chip from any box, then a second chip, etc. The object is to draw as few chips as possible before you can with certainty put correct labels on all the boxes. What is the maximum number of chips you may have to draw?

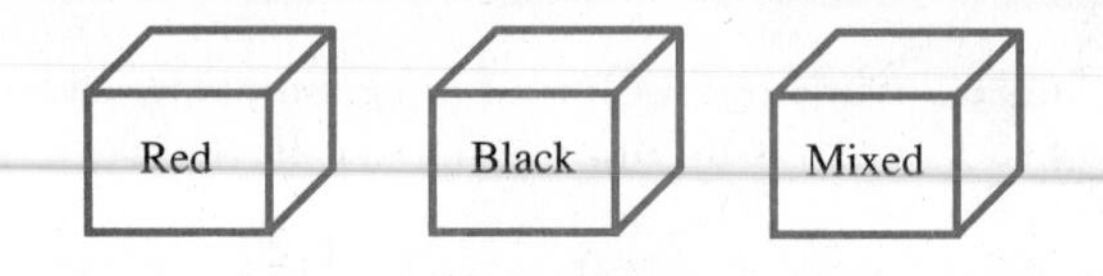

There is no "tilt" in the problem toward either red or black; i.e. we have complete symmetry. To maintain this state of affairs, at least consider as a first step drawing one chip from the box marked "Mixed." The drawn chip will of course be either red or black. With this clue, you should be well on your way to a solution.

Have All the Relevant Data Been Used?

In our solution in Section 2.5 of the gas mileage problem, we made a false start, obtaining at one point an equation in which the letter representing the quantity to be found dropped out, leaving us with a statement that was true but not helpful. Review of the attempted solution will also show that up to that point, we had omitted a key piece of information, the fact that the new car went 15 miles further on a gallon of gas than did the old one.

One has to make a judgment as to whether the omitted fact might actually be irrelevant. We might have asked ourselves if the 15 miles per gallon was likely to be irrelevant. Suppose the improvement had been 30 miles per gallon. Would that have affected the outcome? Surely it would have. Then the number of miles per gallon by which the new car improves must be important, and it could have been predicted that its omission would get us into trouble.

Sometimes, however, as was pointed out in Section 1.1, there really is more information than is needed, and not using it is actually part of stripping away irrelevant information. Consider the following.

By leaving early to get ahead of the traffic, Ms. Didwell can average 54 mph on her 37-mile trip to work, but when returning over the same route in afternoon traffic, she averages 42 mph. What is her average for the entire round trip?

From the formula $d = rt$ or $t = \frac{d}{r}$, we see that

$$\text{time going} = \frac{d}{54}$$

$$\text{time returning} = \frac{d}{42}$$

A student who was good enough at fractions to correctly add

$$\frac{d}{54} + \frac{d}{42} = \frac{d}{6 \cdot 9} + \frac{d}{6 \cdot 7} = \frac{7d + 9d}{6 \cdot 7 \cdot 9} = \frac{16d}{6 \cdot 7 \cdot 9} = \frac{8d}{189}$$

realized that the total distance was $2d$ and proceeded this way.

$$\text{average rate} = \frac{\text{total distance}}{\text{total time}}$$

$$= \frac{2d}{\left(\frac{8d}{189}\right)} = \frac{189(2d)}{8d} = \frac{189}{4} = 47.25$$

The solution nowhere used the fact that $d = 37$. Though it was not apparent when we started, this information turned out to be irrelevant.

If you are not entirely convinced, or if in the spirit of what we have said previously in this chapter, you wish to check what was just done, try this check. Work the problem using your own data. Since the distance is supposed to be irrelevant, use $d = 6(7)(9) = 378$ so that both 54 and 42 will divide in evenly and you can use integers throughout. See if you get an overall average of 47.25.

One useful check in deciding whether or not a piece of information is irrelevant is this. Ask yourself whether halving it or doubling it would make an obvious difference in the answer. If it would, then you had better look for the reason that you somehow seemed not to need it in your solution.

Does It Work for Known Special Cases?

While working on a problem involving parallelograms, Amy remembered that there was some theorem that related the squares of the

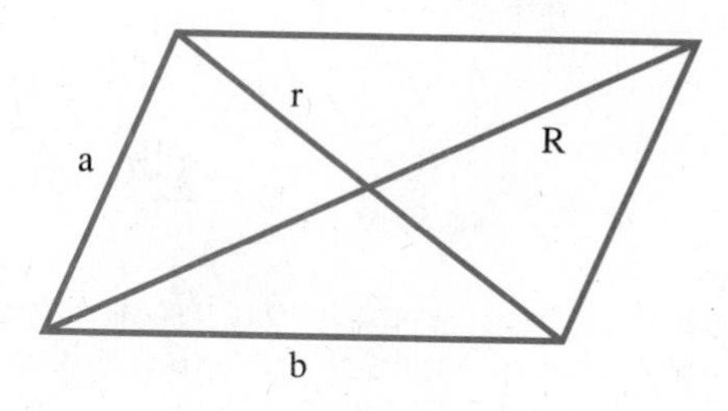

lengths of the sides a and b and the short and long diagonals r and R respectively. She thought it might be

$$a^2 + b^2 = r^2 + R^2$$

How can she test her guess?

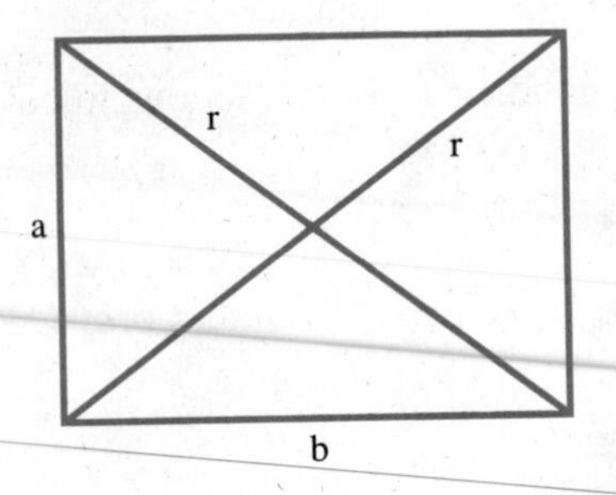

One easy idea would be to examine it in the case where the parallelogram is a rectangle. In this case, the proposed relation would say

$$a^2 + b^2 = r^2 + r^2 = 2r^2$$

This violates the Theorem of Pythogoras, however, which says for the right triangle formed by two adjacent sides and the connecting diagonal that $a^2 + b^2 = r^2$. Amy then decides that perhaps the theorem she was trying to remember had a 2 on the left side; that is, perhaps

$$2a^2 + 2b^2 = r^2 + R^2$$

What do you think?

We have already encountered and we will prove in Section 4.2 that there is a formula for the sum of the first n integers:

$$1 + 2 + 3 + \cdots + n = \frac{n(n+1)}{2}$$

Is it right? Try some special cases.

$$n = 1 \qquad 1 = \frac{1 \cdot 2}{2}$$

$$n = 2 \qquad 1 + 2 = \frac{2 \cdot 3}{2}$$

$$n = 3 \qquad 1 + 2 + 3 = \frac{3 \cdot 4}{2}$$

It seems to be working. Again we insert our usual warning that examples do not prove theorems or establish formulas, but they do build up one's hopes that something just might be right.

Summary

No solution to a problem should go unchecked. A good check does not simply repeat work done the first time. It compares the result with estimates made prior to actually working it out. It takes into consideration the original information. It verifies symmetries and whether all relevant data has been used. A good check will see if the same method will give what is known to be a right answer for special cases. Good checking exercises the imagination, and calls for careful attention to details.

Problem Set 3.1

1. Verify the assertion made in this section that a person who averages 60 mph one way, 40 mph the other way on a 150-mile trip will have averaged 48 mph on the entire trip.
2. Does the trip average of 48 mph in Problem 1 depend on the length of the trip? That is, if the average of 60 and 40 were for a 200-mile trip, would the trip average still be 48? Guess first; then calculate.
3. The following problem was given as Problem 9 in Problem Set 1.3. Three drivers happen to leave Plainview for the monotonous freeway drive of 96 miles to Westflats. All set their cruise control and left it at a constant speed for the entire trip. When the first driver arrived, she was 6 miles ahead of the next car and 10 miles ahead of the third. How far behind was the third car when the second car arrived? Without looking back at your solution, answer this question. Would the distance between the second car and the third car have been more or less than 4 miles when the second car arrived? Why? Now check the actual solution.
4. "Super Bread" rises in such a way that it doubles in volume every minute. A loaf of this dough completely fills a certain oven in 20 minutes. In how long will two such loaves fill the same oven?
5. Grandma Uphill was making 25 flowerlike patterns, each flower having 20 petals. She began by sewing 4 petals together for all 25 flowers, and when finished announced that she was $\frac{1}{5}$ through with sewing petals together. She wasn't, of course. What fraction of the sewing of one petal to another had she really finished?

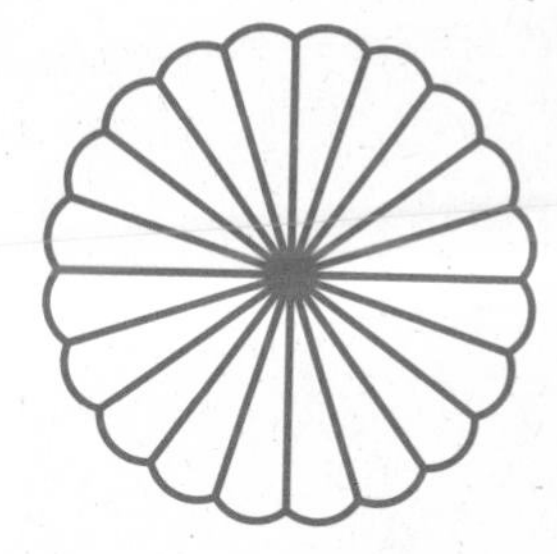

6. If you checked the table in Section 1.3 for w and $l = \dfrac{2w}{w-2}$, you found that calculations were made not only for w equal to 3, 4, and 6, but also for $w = 5$. Verify the symmetry for this value also.
7. Complete the puzzle about the incorrectly labeled boxes. What number of chips must be drawn for certain identification of all boxes?
8. Prove that for a parallelogram with sides of lengths a and b and diagonals of r and R,

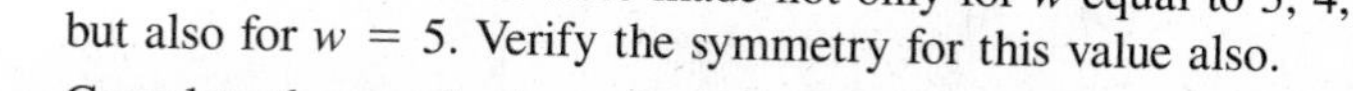

$$2a^2 + 2b^2 = r^2 + R^2$$

9. Refer to Section 2.4, Problem 10. Does the number of wheels matter? Could you work the problem using just one wheel?
10. Answer the question asked in **First in, Last out.**

Hints to Selected Problems

1. The average for the entire trip is found by dividing the total distance of the round trip by the time going plus the time returning.

4. Consider the situation after 18 minutes, first with one loaf, then with two.

5. When 5 sets of four petals have been made, will the work be done?

6. When $w = 5$, $l = \frac{10}{3}$. What is w when $l = \frac{10}{3}$?

7. The first chip drawn determines the color of all the chips in the (mis)labeled box "Mixed". Now, the other two boxes are also known to be mislabeled.

8.

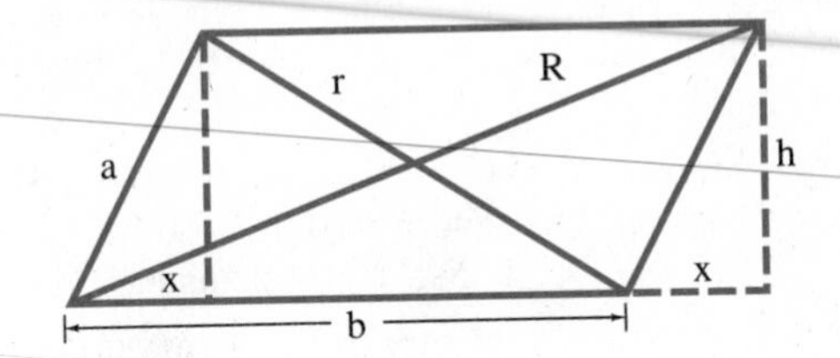

10. Under the numbers 1 2 3 4 5 6, place checks in the order that letters get typed. What rules guide proper placement of the checks?

3.2 Is the Method the Best Possible?

Sprinklers

Sixteen circular sprinklers, each with a stream reaching 10 feet, could be used to water a square plot that is 80 feet on a side, leaving certain regions uncovered. The same plot could be watered by 4 circular sprinklers, each with a stream reaching 20 feet, or 1 sprinkler that waters a circular area of radius 40 feet, again leaving some regions uncovered. Which system covers the most total area?

What would make one method of solution better than another? Being a matter of taste, it's hard to say, but some guidelines exist. As a general rule, a short solution is preferred over a long one unless the long one has redeeming features. These might be providing greater clarity and insight into why things are as they are, making connections with other parts of mathematics or its applications, or suggesting generalizations that lead to interesting new problems.

Can It Be Shortened?

In Section 1.1, we saw evidence that led us to guess that the sum of the first n odd integers might be n^2. If we notice that nth odd number is $2n - 1$ (try it for $n = 1, 2, 3, \ldots$), the guess can be written in the form

$$1 + 3 + 5 + \cdots + (2n - 1) = n^2.$$

There are several ways to prove this, once it has been guessed.

Proof 1 This formula is frequently used to illustrate a method of proof called mathematical induction. Your instructor may choose to show you the basic idea. We shall say no more about it, as it would take us too far afield, and few students find that they understand the formula any better for having seen an induction proof.

Proof 2 This proof makes use of the formula

$$1 + 2 + 3 + \cdots + m = \frac{m(m + 1)}{2}$$

mentioned at the end of the last section, and a rather clever (though very routinely used) idea. Notice that if we begin with the sum of the first n odd integers, and add in and then immediately subtract out all the missing even integers, we get

$$\begin{aligned} &1 + 3 + 5 + \cdots + (2n - 1) \\ &\quad = 1 + 2 + 3 + 4 + 5 + \cdots + (2n - 1) + 2n \\ &\qquad\; - 2 \qquad - 4 \qquad\qquad\qquad\qquad - 2n \\ &\quad = (1 + 2 + 3 + \cdots + 2n) - 2(1 + 2 + 3 + \cdots n). \end{aligned}$$

Using our formula twice, the first time with $m = 2n$ and then with $m = n$, gives us

$$\begin{aligned} 1 + 3 + 5 + \cdots + (2n - 1) &= \frac{2n(2n + 1)}{2} - 2\frac{n(n + 1)}{2} \\ &= n(2n + 1) - n(n + 1) \\ &= 2n^2 + n - n^2 - n \\ &= n^2. \end{aligned}$$

This proof, besides depending on what must seem to the beginning mathematician as an inspired idea of adding and subtracting the same integers in and out, uses another formula that we have pretty well accepted, but not yet proved.

This discussion sets the stage for a proof that illustrates how a proof can be shorter and easier to follow, while also revealing just why the thing would have to be true.

Proof 3

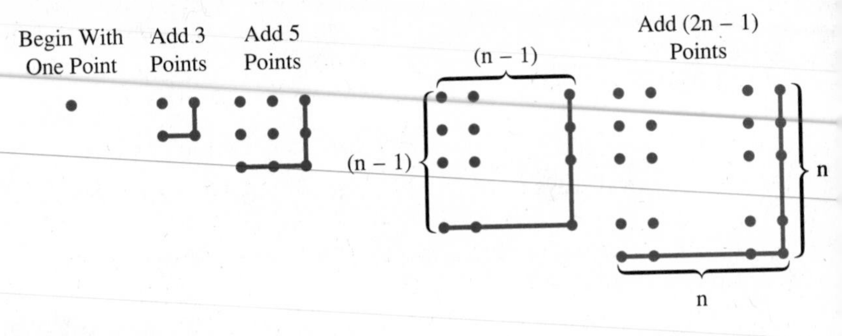

The last two pictures really tell the whole story. To a square array of dots, $(n - 1)$ on a side, we can add $(2n - 1)$ dots to make a new square that has n dots on a side.

Can It Be Clarified?

The example just concluded demonstrates that a proof can sometimes be shortened and clarified at the same time. There are other times, however, when clarification comes at the price of added length.

A lone goose was flying in the opposite direction from a flock of geese. He cried: "Hello, 100 geese!" The leader of flock answered: "We aren't 100! If you take twice our number and half our number, and add a quarter of our number, and finally add you, the result is 100, but . . . well, you figure it out."

This is the kind of problem that begs for easy solution by setting up and solving an algebraic equation. Let g be the number of geese in the flock. Then the statement of the leader of the flock translates directly to

$$2g + \frac{g}{2} + \frac{g}{4} + 1 = 100$$

Then . . . well, you figure it out.

The problem comes, however, from a Russian story said to have been told in homes and schools for over 50 years. We finish the story here because it nicely illustrates how we might clarify the meaning of our algebraic symbols. [Kordemsky, Boris A., *The Moscow Puzzles,* p. 95]

The one goose flew on, but could not find the answer. Then he saw a stork on the bank of a pond looking for frogs. Now among the birds the stork is the best mathematician. He often stands on one leg for hours, solving problems. The goose descended and told his story.

The stork drew a line with his beak to represent the flock. Then he drew a second line of the same length, a third line half as long, another line a fourth as long, and a very small line, rather like a dot, to represent the goose.

"Do you understand?" the stork asked.

"Not yet."

The stork explained the meaning of the lines: The first and the second represented the flock, the third half the flock, the next a quarter of the flock, and the dot stood for the goose. He rubbed out the dot, leaving lines that now represented 99 geese. "Since a flock contains four quarters, how many quarters do the four lines represent?"

Slowly the goose added 4 + 4 + 2 + 1. "Eleven," he replied.

"And if 11 quarters make 99 geese, how many geese are in a quarter?"

"Nine."

"And how many in the entire flock?"

The goose multiplied 9 by 4 and said:

"Thirty-six."

"Correct! But you couldn't get the answer yourself, could you? You . . . goose!"

Can It Be Confirmed by Another Method?

Here is a question used in the Minnesota State High School Mathematics League competition.

> I think you get good at mathematics by trying to see how many different ways you can solve the same problem.
>
> *Mary Hu*
> *(Minnesota's 1993 Math League Top Scorer)*

Find the single integer that is equal to

$$1^2 - 2^2 + 3^2 - 4^2 + \cdots - (124)^2 + (125)^2$$

Discussion of its solution gives an opportunity to review a number of our problem-solving guidelines. To start with, since there is an obvious pattern, we can examine the same pattern, but not carried so far. We note that the pattern ends with an odd square.

Start with a small example.

(1) $1^2 = 1 = 1 \cdot 1$

(2) $1^2 - 2^2 + 3^2 = 6 = 3 \cdot 2$

(3) $1^2 - 2^2 + 3^2 - 4^2 + 5^2 = 15 = 5 \cdot 3$

(4) $1^2 - 2^2 + 3^2 - 4^2 + 5^2 - 6^2 + 7^2 = 28 = 7 \cdot 4$

Organize the information.

Several facts beckon for attention.

(a) The last odd integer in the series has (so far) always been a factor of the sum. The other factor has been the "count" of the number of lines written down.
(b) In going from one line to the next, we use the previous total and add on the difference between the square of the next odd integer and its preceeding even integer.

Make a guess.

Based on (a) we might guess that the fifth line would be $9 \cdot 5 = 45$ and the sixth line would sum to $11 \cdot 6 = 66$. Then demonstrate that these guesses are correct.

(5) $1^2 - 2^2 + 3^2 - 4^2 + 5^2 - 6^2 + 7^2 + (-8^2 + 9^2) = 45 = 9 \cdot 5$

(6) $1^2 - 2^2 + 3^2 - 4^2 + 5^2 - 6^2 + 7^2 + (-8^2 + 9^2) + (-10^2 + 11^2) = 66 = 11 \cdot 6$

We are ready to make a guess, are we not, on the original problem. Imagine our list continued until the last term is 125^2. Since 125 is the sixty-third odd integer, this would be the sixty-third line, and so our guess is that the sum might be $125(63) = 7875$.

That's a solution, arrived at by observing patterns and guessing. Can we obtain it in a way that proves the result beyond the shadow of a doubt? Sometimes the easiest way to do that is to prove a generalized result for which your problem is a particular example. That is the route we will follow.

Can It Be Generalized?

The nth odd integer is $2n - 1$; the even number just before it is $2n - 2$. The problem above suggests looking at the sum of

$$1^2 - 2^2 + 3^2 - 4^2 + 5^2 - 6^2 + 7^2 - \cdots - (2n-2)^2 + (2n-1)^2$$

n	$2n - 2$	$2n - 1$
1	0	1
2	2	3
3	4	5
4	6	7
⋮	⋮	⋮

Represent the information in new ways.

Observation (b) above suggests writing the sum in a way that emphasizes that in each step, we add on the difference between the squares of the next odd integer and its preceeding even integer. That is, we write it as

$$1^2 + [3^2 - 2^2] + [5^2 - 4^2] + [7^2 - 6^2] + \cdots + [(2n-1)^2 - (2n-2)^2]$$

Simplifying the terms within each bracket gives

$$1 + [5] + [9] + [13] + \cdots + [(2n-1)^2 - (2n-2)^2]$$

We now have two questions to answer.

(i) Is the desired sum really the sum of every other odd integer?
(ii) Can we find a formula for the sum of every other odd integer?

Divide and conquer.

$$\begin{array}{r} 2n - 1 \\ 2n - 1 \\ \hline -2n + 1 \\ 4n^2 - 2n \\ \hline 4n^2 - 4n + 1 \end{array} \qquad \begin{array}{r} 2n - 2 \\ 2n - 2 \\ \hline -4n + 4 \\ 4n^2 - 4n \\ \hline 4n^2 - 8n + 4 \end{array}$$

Part (i) The nth term in the sum is

$$\begin{aligned}(2n-1)^2 - (2n-2)^2 &= (4n^2 - 4n + 1) - (4n^2 - 8n + 4) \\ &= 4n - 3\end{aligned}$$

This conclusion can be confirmed by comparing a table of values of $4n - 3$ with the differences observed for $n = 1, 2, 3, 4$. The sum we want is indeed

$$1 + 5 + 9 + 13 + \cdots + (4n - 3)$$

n	$4n - 3$
1	1
2	5
3	9
4	13

Part (ii) The average term in the sum is

$$\frac{1 + (4n-3)}{2} = \frac{2(2n-1)}{2} = 2n - 1$$

There are n terms. The sum is $n(2n - 1)$. The values obtained for the choices from $n = 1$ to $n = 6$ are exactly the values we obtained when we carried out the complete calculations for these cases. The value for $n = 63$ gives the solution to our original problem.

The idea of generalizing one's solution to a particular problem is instilled in a mathematician as a part of the regular thought pattern. It is a key to finding new problems to work on. It is a key to survival in mathematical research.

n	$n(2n - 1)$
1	1
2	6
3	15
4	28
5	45
6	66
63	7875

Summary

Getting an answer to a nice problem is not the end of the matter; it is the beginning of an effort to clarify, to give new insights, to generalize. These components of a solution, taken together and presented in polished rhetorical form, perhaps define that elusive thing called mathematical elegance. One should not leave a problem until satisfied that it is correct, understood, and presented as elegantly as possible.

Problem Set 3.2

1. The **Sprinklers** problem suggests at least the following questions.
   (a) What would need to be the radius of a single sprinkler if all the plot is to be covered, and what would be the area of the "extra regions" watered?

First I solve a problem. Then I try to understand why my method worked. Then I look around to see what other problems I might have solved. Five years or more after I've solved a problem, I begin to feel that I understand it.

L. C. Young

(b) Ask the question raised in (a), and add to it a question about the area of regions being multiply covered if 4 sprinklers are used; 16 sprinklers.

(c) This problem can be thought of as one about inscribing first one circle, then four, then sixteen, in a square. Can you get comparable results about the area left uncovered using other geometrical shapes (equilateral triangles, hexagons, . . .)?

(d) What similarities and differences do you see between this problem and **Planting Trees, Envisioning a Forest** in Section 1.3?

2. How many variations can you think of on the game of picking up pennies described in Section 1.3? How do these games compare with NIM?

3. In Section 2.1 we encountered a surprising result that related the difference of areas of circles circumscribed about and inscribed within a fixed regular polygon of side length 1. What about the area of regular polygons circumscribed about and inscribed within a fixed circle?

4. Would **The Handshake Problem** discussed in Section 1.2 work if one posed it for more couples at the party?

5. **Pictures Perfect** (Section 2.1), to be solved for n pictures, requires finding the total number of subsets of a set of n elements. Has this been done somewhere? Can you solve this problem for n pictures?

6. The first illustrative problem in Section 1.1 (Homer's problem about selling theater tickets) required that we find integer solutions to $3s + 6g = 310$. It had no such solutions. There are equations of the form $ax + by = c$ that do have solutions in integers. Can you learn any tricks for easily identifying those that do have solutions? and if so, can you also learn how to find the integer solutions?

7. A conference of 4 schools A, B, C, D can easily set up a schedule using just three Saturdays that provides for each school's team to play every other team

Sat 1	AB	CD
Sat 2	AC	BD
Sat 3	AD	BC

How many Saturdays are needed for every team to play every other team if one more school joins the conference? What generalizations of this problem suggest themselves?

8. Consider the **First In, Last Out** problem that opened Section 3.1.

(a) Can you write down a general rule that tells which sequences are possible, which are not?

(b) Would your rule work just as well for n letters?

Hints to Selected Problems

1. (a) The area of extra regions watered would be

$$\pi[(40\sqrt{2})^2] - 80^2.$$

1. (b) One sprinkler covers too much by $\pi[(20\sqrt{2})^2] - 40^2$; the four together cover too much by 4 times this. How does this compare with the answer to 1(a)? Show that half of this extra watering is being used for double coverage.

4. Try it for Dick, Jane, and four other couples; then again for five other couples. Do you see a pattern?

6. Problems of this type are discussed in Section 12.4.

part II
Finding Order

People seem to appreciate patterns, whether they are found in the mosaics of a great artist, the web of a common spider, or the molecules of a salt crystal. This appreciation extends to number patterns; most folks are at least momentarily intrigued when it is pointed out to them that

$$1^3 + 2^3 = (1 + 2)^2$$
$$1^3 + 2^3 + 3^3 = (1 + 2 = 3)^2$$
$$\vdots \qquad \vdots$$

There are people, of course, who resent too much organization or structure in their lives, who seemingly thrive on chaos and disharmony. And there are philosophers who believe that our world is ultimately chaotic, that it has no rhyme or reason. Consciously or unconsciously, most scientists reject this view. They can hardly do otherwise. The aim of science is to explain the universe—to explain it in terms of principles, rules, and laws. Asked how he arrived at the theory of relativity, Einstein replied that he had discovered it because he was so "firmly convinced of the harmony of the universe." In writing (with Infield) a popular book on science, he said that he had tried "to give some idea of the eternal struggle of the inventive human mind for a fuller understanding of the laws governing physical phenomena."

Mathematicians are firmly committed to the same struggle. The prince of mathematicians, Carl Gauss, took as his motto, "Thou, nature, art my goddess; to thy laws my services are bound." Yet there is a difference between mathematics and science. The mathematicians's laboratory is not so much the physical world of birds, bees, and molecules as it is the world of numbers, geometric shapes, and algebraic formulas. And here we find structure, laws, and patterns which may surpass in beauty those of the physical world itself.

But patterns, whether in the score of a symphony or in the arrangement of numbers in a sequence, are more easily recognized when we are trained to look for them. And so in Part II, we take as our theme the task of discovering patterns, of looking for order where at first none seems to exist. In our search for ordered phenomena, we move across many branches of mathematics—including sequences, counting, probability, statistics, and networks.

Problem solving continues to be important. Every section begins with a problem; it inevitably ends with a host of others. There will be abundant opportunity to practice the strategies we learned in Part I. But now we have a loftier goal. We do not simply solve a single problem; we look at all its relatives. We try to discern a principle that underlies a whole class of problems. In short, we search for order in all that we do.

The chief aim of all investigations of the external world should be to discover the rational order and harmony which has been imposed on it by God and which he revealed to us in the language of mathematics.

JOHANNES KEPLER

Born in Switzerland to a minister who wanted him to study theology, Leonhard Euler instead took up mathematics, completing his formal study at age 15. He was soon publishing papers and at age 19 won a prize from the French Academy of Sciences for his work on the masting of ships. Thus was launched the career of the most prolific writer on mathematics who ever lived. His collected works, when completed, will fill 74 volumes.

Euler made contributions to all fields of pure and applied mathematics. His interests spanned algebra, geometry, number theory, calculus, physics, astronomy, music, and such practical subjects as ship design, cartography, insurance, and canal building. His prodigious memory allowed him to write over 400 articles after he was totally blind. By the time of his death he had been accorded universal respect, and it was said that all the mathematicians of Europe were his pupils. In his breadth of interests and in his uncanny ability to find pattern and order in every subject, Leonhard Euler best represents the spirit of Part II.

Leonhard Euler
(1707–1783)

chapter 4

Numerical Patterns

"The northern ocean is beautiful," said the Orc, "and beautiful the delicate intricacy of the snowflake before it melts and perishes, but such beauties are as nothing to him who delights in numbers, spurning alike the wild irrationality of life and the baffling complexity of nature's laws."

J. L. SYNGE

The Fascination of Numbers

There is a story, well known to mathematicians, about a visit that the world-famous mathematician G. H. Hardy paid to his brilliant collaborator Ramanujan in the hospital. Hardy observed that the number on the taxicab from which he had just emerged was 1729, a rather dull number. Ramanujan replied, "No Hardy! No Hardy! It is a very interesting number. It is the smallest number expressible as the sum of two cubes in two different ways."

It is true that professional mathematicians are, as a group, fascinated with numbers and notice a variety of properties that escape more casual observers. The mathematician Lagrange (1736–1813) proved that every positive integer is the sum of four or fewer perfect squares. He also proved what had been talked about for over *a* 100 years before he proved it, namely that the number $(1)(2)(3)\ldots(p-1)(p)+1$ would be divisible by p if and only if p is a prime. (Try it for the non-prime 6 and the prime 7). In our own time, mathematicians talk about the conjecture of Goldbach (1690–1764) who speculated on the basis of having looked at a hundred or so examples that every even integer greater than 2 can be written as the sum of two primes, for example, $50 = 19 + 31$ and $180 = 71 + 109$. No one has yet proved that.

It is also true that many people who are not involved in mathematics in a professional way are still fascinated by the properties of numbers. Amateurs have had no better luck than the professionals in their efforts to prove Goldbach's conjecture, but their efforts are more entertaining to read, and have merited an entire chapter in U. Dudley's book, Mathematical Cranks, MAA (1992).

Ancient philosophers associated odd numbers with male characteristics, even with female, and found in $2 \cdot 3 = 6$ the symbol of perfection "for it is made of the union of the two sexes." Euclid studied 6 as a number that is perfect because it is the sum of all of its divisors, save 6 itself; that is $6 = 1 + 2 + 3$. St. Augustine wrote "Six is a number perfect in itself, and not because God created all things in six days; rather the inverse is true. God created all things in six days because this number is perfect."

Once 6 has been singled out as perfect number because it is the sum of its divisors, it is natural that people will wonder if there are other perfect numbers, what the next one might be, or how many there might be. There are, as of this writing, 31 perfect numbers that are known. You might find the next one after 6; you won't find all 31 without some serious study of number theory.

When one tires of the integers, there are plenty of other numbers to ponder, the number π suggesting itself for first consideration. It is approximated by 3 in the Bible (I Kings 7:23), Archimedes had it between $3\frac{1}{7}$ and $3\frac{10}{71}$, and its approximation by the methods of calculus occupied both Newton and Leibniz. The aforementioned Ramanujan developed powerful methods for computing π to a high degree of accuracy in very efficient ways. Modern computers have in recent years enabled people not yet tired of π to compute it to over 400 million decimal places.

The number π is of interest, not only because of its role in circles and spheres, but because it turns up in wholly unexpected ways. Throw a one inch stick into a table ruled with parallel lines two inches apart. The probability of the stick falling so as to intersect one of the ruled lines is $\frac{1}{\pi}$. Mathematicians in Newton's generation got interested in the sum

$$1 + \frac{1}{2^2} + \frac{1}{3^2} + \frac{1}{4^2} + \cdots + \frac{1}{n^2}$$

They ultimately decided that the larger n became (that is, the more terms they added into their sum), the closer they came to 1.6649. It took the insight of the great Leonard Euler to recognize that sum was really getting closer to $\frac{\pi^2}{6}$. Petr Beckman had no trouble finding material for a book [*The History of Pi,* St. Martin's Press, 1971].

People are fascinated by numbers.

4.1 Number Sequences

Getting to the Root of the Problem

Instruction	*Example 1*	*Example 2*	*Example 3*
Let A be any non-negative real number	$A = 3$	$A = 4$	$A = 5$
Let $a_1 = \frac{1}{2}A$	$a_1 = 1.5$	$a_1 = 2$	$a_1 = 2.5$
Let $a_2 = \frac{1}{2}\left(a_1 + \frac{A}{a_1}\right)$	$a_2 = \frac{1}{2}\left(1.5 + \frac{3}{1.5}\right)$ $a_2 = 1.75$	$a_2 = \frac{1}{2}\left(2 + \frac{4}{2}\right)$ $a_2 = 2$	$a_2 = \frac{1}{2}\left(2.5 + \frac{5}{2.5}\right)$ $a_2 = 2.25$
Let $a_3 = \frac{1}{2}\left(a_2 + \frac{A}{a_2}\right)$	$a_3 = \frac{1}{2}\left(1.75 + \frac{3}{1.75}\right)$ $a_3 = 1.7321$	$a_3 = \frac{1}{2}\left(2 + \frac{4}{2}\right)$ $a_3 = 2.0000$	$a_3 = \frac{1}{2}\left(2.25 + \frac{5}{2.25}\right)$ $a_3 = 2.2361$

Questions

1. How does a_3 relate to A in each example?
2. Will the process work for any choice of A?
3. Why does this work (or work as well as it does)?

The title of the problem posed above is meant to be, in its own way, suggestive. The number a_3 obtained in each example is, accurate to four decimal places, the square root of the initial choice A, an observation that answers question 1.

Will the process work for any choice of A? Our problem-solving guidelines are clear here. We begin the investigation of such a question by trying a few more choices of A. The next obvious case is $A = 6$. The three numbers generated are $a_1 = 3$, $a_2 = 2.5$, $a_3 = 2.4500$. Surprise; it's not quite $\sqrt{6} = 2.4495$; but it's pretty close, and since another of our guidelines is to look for patterns, a modest amount of curiosity will lead us to find a_4 by writing

$$a_4 = \frac{1}{2}\left(2.45 + \frac{6}{2.45}\right) = 2.4495.$$

This is the square root of 6 accurate to four decimal places. Indeed, the last computation, if rounded to six places, gives 2.449490 which is the square root of 6 accurate to six decimal places.

The process we have been using is called **recursion.** Begin with a number, or with several numbers. Use these numbers and some fixed rule to generate a new number. Then, using this new number, use the same fixed rule to generate another number, etc. In the process above, we began with A and $a_1 = \frac{1}{2}A$. The rule we used to generate the next number could be written as

$$a_n = \frac{1}{2}\left(a_{n-1} + \frac{A}{a_{n-1}}\right)$$

This rule for generating the next number is called a **recursion formula,** and the numbers generated

$$a_1, a_2, a_3, \ldots$$

form what is called a **sequence** of numbers, or a number sequence.

Most students first meet number sequences on intelligence tests. Given a few numbers arranged according to some pattern, young scholars are encouraged to exhibit their mental powers by filling in a few more numbers—say two more, as in the problems we pose here.

(a) 2, 4, 6, 8, □, □, . . .
(b) 2, 4, 8, 16, □, □, . . .
(c) 1, 8, 27, 64, □, □, . . .
(d) 2, 5, 10, 17, □, □, . . .
(e) 1, $\frac{1}{2}$, $\frac{1}{3}$, $\frac{1}{4}$, □, □, . . .

Keep It Simple

The principle of looking for a simple pattern that explains a given phenomenon is a guiding principle in science.

"Our experience justifies us in believing that nature is the realization of the simplest conceivable mathematical ideas."

A. Einstein

Strictly speaking there is no definite answer to these problems. Different people, looking at the same arrangement, see diffeent patterns; in fact, the same person looking at the same arrangement may see different patterns at different times. The Rorschach Test in psychology is given with the idea that people reveal something about their mental outlook in telling what they see when looking at an inkblot. Number sequences are used in intelligence tests with the idea that people will reveal something about their mental ability in telling what they see when looking at number patterns. It is assumed in such tests that people will look for as simple a pattern as possible and (as opposed to their grandmother's phone number) a pattern that others are likely to recognize.

Consider, for example, sequence d:

$$2, 5, 10, 17, \square, \square, \ldots$$

The chances are that most people, noting the pattern of adding successively larger odd integers (i.e., 3, 5, and 7) will conclude that the numbers in the boxes should be 26 and 37. But a more imaginative soul may observe that 3, 5, 7 are the first three odd primes and conclude that the answers are

$$17 + 11 = 28$$
$$28 + 13 = 41$$

The principle of preferring the simple solution suggests that we use the first answers; but if the person giving the second answers is forceful as well as imaginative, it will be hard to say the second answer is wrong—because it isn't.

Formulas for Sequences

A general sequence can be written:

$$a_1, a_2, a_3, a_4, \ldots$$

with a_1 designating the first term, a_2 the second, etc. Note that the subscript indicates the position of the term within the sequence. In example (a) above,

$$a_1 = 2 \qquad a_2 = 4 \qquad a_3 = 6 \qquad a_4 = 8 \qquad \ldots$$

Now we ask a question. What is a_{21}? The keen observer, noting that in this sequence you can get the value of a term by doubling its subscript, answers 42.

Mathematicians express such observations in succinct symbolic sentences called formulas. Let n be a counting number. What is the value of a_n? That is, can we give an algebraic formula for the nth term of our sequence? For the example discussed above, we saw that

$$a_n = 2n$$

There, in that short line, is all the information you need to construct the sequence. No longer need we speak of the sequence 2, 4, 6, 8, ... We can speak of the sequence $a_n = 2n$, and we will have no more trouble writing down the 99th or 121st term than in writing the 21st.

$$a_{99} = 2 \cdot 99 = 198 \qquad a_{121} = 2 \cdot 121 = 242$$

We call a formula like this an **explicit formula.** It tells us how a_n is related to n, that is, how the value of the term a_n is related to its subscript.

There is another way of describing the pattern we observe in a sequence. We can observe how a term relates to previous terms. In our example, we always add 2 to a term to find the next one. A concise way of saying this is

$$a_n = a_{n-1} + 2$$

For example, the fourth term equals the fourth minus one term (or third term) plus 2. Now, knowing that $a_1=2$, we can find all the other terms provided we list them in order. Thus

$$a_2 = a_1 + 2 = 2 + 2 = 4$$
$$a_3 = a_2 + 2 = 4 + 2 = 6$$
$$a_4 = a_3 + 2 = 6 + 2 = 8$$

A formula that tells how the general term is related to the previous term (or terms) is called a **recursion formula.**

Recursion is a fundamental idea in the operation of a computer. A calculator produces a number in response to numbers and instructions entered by an operator, whereas a computer performs recurring operations, taking its output from one set of operations as input to the next set of operations. The introductory problem in this section indicates how (and Problem 25 indicates why) recursion can be used by a computer to find square roots.

In Table 4-1 below, we give both recursion and explicit formulas for the sequences (a) through (e) introduced above. The reader should verify his or her understanding by checking carefully.

Table 4-1

Sequence	*Recursion Formula*	*Explicit Formula*
a. 2, 4, 6, 8, . . .	$a_n = a_{n-1} + 2$	$a_n = 2n$
b. 2, 4, 8, 16, . . .	$b_n = 2b_{n-1}$	$b_n = 2^n$
c. 1, 8, 27, 64, . . .	$c_n = ?$	$c_n = n^3$
d. 2, 5, 10, 17, . . .	$d_n = d_{n-1} + (2n - 1)$	$d_n = n^2 + 1$
e. $1, \frac{1}{2}, \frac{1}{3}, \frac{1}{4}, \ldots$	$e_n = ?$	$e_n = 1/n$

We have left two question marks where the formulas are not obvious. Sometimes it seems impossible to find either a recursion formula or an explicit formula (e.g., no one knows a formula for the *n*th prime). But the challenge is always there; find a formula. Sometimes, as in many computer-related problems, recursion formulas are important. In other cases, explicit formulas are the most useful; they allow us to answer almost any question concerning terms of the sequences right away. What is the fourteenth term in sequence d? $d_{14} = 14^2 + 1 = 197$. What is the sum of the fifth and sixth terms in sequence c?

$$c_5 + c_6 = 5^3 + 6^3 = 125 + 216 = 341$$

Triangular and Square Numbers

The Greeks were fascinated by numbers that arise in a geometric way. Consider the following diagrams.

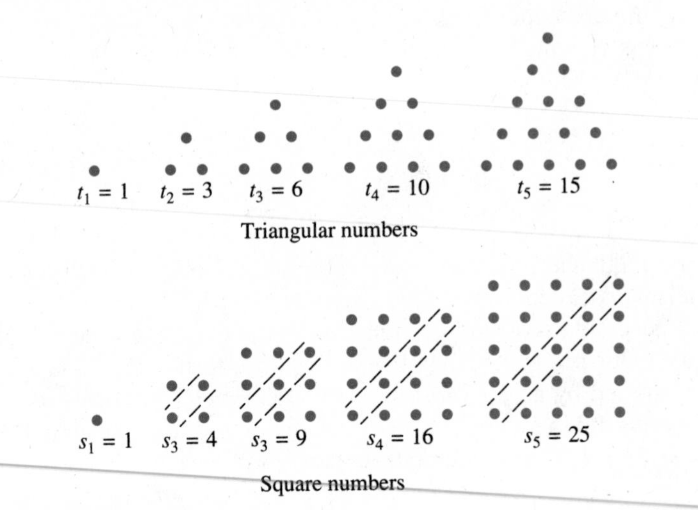

Triangular numbers

Square numbers

We seek explicit formulas for both t_n and s_n. The latter is easy; $s_n = n^2$. The former requires some ingenuity. A particularly nice way to obtain a formula for t_n is to observe a relationship between the pictures of the triangular and square numbers. Look, for example, at the t_3 and s_4 diagrams. Above and below the indicated diagonal in the s_4 diagram, we have the t_3 diagram. There are four dots on the diagonal itself. Thus, $s_4 = 2 \cdot 6 + 4 = 2t_3 + 4$. A similar look at s_2, s_3, s_5, etc., leads to

$$s_2 = 2t_1 + 2$$
$$s_3 = 2t_2 + 3$$
$$s_4 = 2t_3 + 4$$
$$s_5 = 2t_4 + 5$$

and, in general,

$$s_{n+1} = 2t_n + (n + 1)$$

But

$$s_{n+1} = (n + 1)^2 = n^2 + 2n + 1$$

and so

$$n^2 + 2n + 1 = 2t_n + n + 1$$
$$n^2 + n = 2t_n$$
$$\frac{n(n + 1)}{2} = t_n$$

Now we can find the 100th triangular number. It is just

$$t_{100} = 100(101)/2 = 5050$$

Here is a problem that challenged the great mathematician Leonhard Euler. Are there any square triangular numbers, that is, numbers that are both square and triangular? A tedious check (yes, even mathematicians find calculating tedious) reveals the four numbers 1, 36, 1225, and 41,616. Are there more? The answer is an emphatic yes; Euler found that there are infinitely many and, as early as 1730, he gave formulas for them. Here are his results, one an explicit formula and the other a recursion formula:

$$u_n = \frac{(17 + 12\sqrt{2})^n + (17 - 12\sqrt{2})^n - 2}{32}$$

$$u_n = 34u_{n-1} - u_{n-2} + 2$$

We expect that most readers will, if they pause to think about it, wonder how Euler discovered these formulas. But once they have been given, it is easy to check that they work, at least for small values of n.

Summary

A number sequence $a_1, a_2, a_3, \ldots$ is an ordered arrangement of numbers. We find it convenient if the sequence can be described by a formula which can take one of two forms. An explicit formula (e.g., $a_n = 3n$) relates the value of the nth term a_n directly to n. A recursion formula (e.g., $a_n = a_{n-1} + 3$) relates the nth term to the previous term or terms. Explicit formulas are especially useful, an important one being that for the nth triangular number, namely, $t_n = n(n + 1)/2$.

Problem Set 4.1

1. Try to find a pattern and use it to fill in the boxes.
   (a) 3, 6, 9, 12, □, □, . . .
   (b) 21, 18, 15, 12, □, □, . . .
   (c) 1, $\frac{1}{2}$, $\frac{1}{4}$, $\frac{1}{8}$, □, □, . . .

2. Fill in the boxes.
   (a) 1, 3, 5, 7, □, □, . . .
   (b) 2, 8, 32, 128, □, □, . . .
   (c) 2, $\frac{2}{3}$, $\frac{2}{9}$, $\frac{2}{27}$, □, □, . . .
3. In each case, an explicit formula is given. Find the indicated terms.
   (a) $a_n = 3n + 2$; $a_6 =$ □, $a_{10} =$ □
   (b) $a_n = 1/2n$; $a_9 =$ □, $a_{20} =$ □
   (c) $a_n = (\frac{1}{2})^n$; $a_3 =$ □, $a_6 =$ □
   (d) $a_n = 15 - 2n$; $a_6 =$ □, $a_{12} =$ □
4. Find the indicated terms.
   (a) $a_n = 2n - 9$; $a_3 =$ □, $a_{12} =$ □
   (b) $a_n = (-1)^n(2n)$; $a_2 =$ □, $a_5 =$ □
   (c) $a_n = (\frac{2}{3})^n$; $a_4 =$ □, $a_7 =$ □
5. Give an explicit formula for each sequence in Problem 1. Hint: Try listing the terms of the sequence in a row, each term directly under a general term, as illustrated in the margin for the sequence 1, 4, 9, 16, . . . Then try to relate the value of the term to the subscript above it.

$a_1\ a_2\ a_3\ a_4 \cdots$
$1^2\ 2^2\ 3^2\ 4^2 \cdots$
$a_n = n^2$

6. Give an explicit formula for each sequence in Problem 2.
7. In each case, an initial term (or terms) and a recursion formula are given. Find a_5. Hint: First find a_2, a_3, a_4.
   (a) $a_1 = 1$; $a_n = 3a_{n-1}$
   (b) $a_1 = 2$; $a_n = a_{n-1} + 3$
   (c) $a_1 = 4$; $a_n = \frac{1}{2}a_{n-1}$
   (d) $a_1 = 1$; $a_2 = 2$; $a_n = 2a_{n-1} - a_{n-2}$
8. Find a_6 for each of the following.
   (a) $a_1 = 1$; $a_n = 2a_{n-1} + 1$
   (b) $a_1 = 4$; $a_n = 3a_{n-1}$
   (c) $a_1 = 6$; $a_n = \frac{1}{3}a_{n-1}$
   (d) $a_1 = 2$; $a_2 = 3$; $a_n = a_{n-1} + a_{n-2}$
9. Give a recursion formula for each of the sequences in Problem 1.
10. Give a recursion formula for each of the sequences in Problem 2.
11. Find a pattern in each of the following sequences and fill in the boxes.
    (a) 2, 6, 18, 54, 162, □, □, . . .
    (b) 11, 9, □, 5, 3, □, . . .
    (c) $\frac{1}{2}$, $\frac{2}{3}$, $\frac{3}{4}$, $\frac{4}{5}$, $\frac{5}{6}$, □, □, . . .
    (d) 1, $\frac{1}{8}$, $\frac{1}{27}$, $\frac{1}{64}$, $\frac{1}{125}$, □, □, . . .
    (e) 6, 3, □, $\frac{3}{4}$, $\frac{3}{8}$, □, . . .
    (f) 2, 2, 4, 6, 10, □, □, . . .
    (g) 2, 2, 4, 8, 14, 26, □, □, . . .
    (h) 2, 3, 5, 7, 11, 13, 17, □, □, . . .
    (i) 1, 3, 6, 10, 15, 21, □, □, . . .
12. In each case in Problem 11, try to find a recursion formula or an explicit formula (or both, if you can) for the sequence.
13. Let $A_n = a_1 + a_2 + a_3 + \cdots + a_n$. Find A_6 for each of the following. Hint: Begin by finding a_1, a_2, a_3, a_4, a_5, a_6.
    (a) $a_n = 2^n$
    (b) $a_n = n + 2$

(c) $a_n = (-1)^n$
(d) $a_n = n^2 - n$
(e) $a_1 = 8,\ a_n = \frac{1}{2}a_{n-1}$
(f) $a_1 = 1,\ a_2 = 2,\ a_n = a_{n-1} - a_{n-2}$

14. Let $a_n = (\frac{1}{2})^n$ and $A_n = a_1 + a_2 + \cdots + a_n$.
(a) Find a_1, a_2, a_3, a_4, a_5.
(b) Find A_1, A_2, A_3, A_4, A_5. (A_n is defined in Problem 13.)
(c) Guess at an explicit formula for A_n.

15. Let a_n be the nth digit in the decimal expansion of $\frac{1}{7}$. Using long division, obtain $a_1, a_2, \ldots, a_{14}$. Do you observe a pattern? Find a_{53}. Do the same for the decimal expansion of $\frac{5}{13}$.

16. Find $1 + 2 + 3 + \cdots + 1000$. Hint: It's a triangular number.

17.* Consider the sequence $1, 1 + 3, 1 + 3 + 5, \ldots$. If we call the nth term a_n, guess at an explicit formula for a_n. Now look at the diagrams for the square numbers. Do you see anything?

18.* The number 1, 5, 12, . . . are called pentagonal numbers (see diagrams at the right). Call the resulting sequence p_n. Find p_4, p_5, and p_6. Can you guess at an explicit formula for p_n. If so, what is p_{100}?

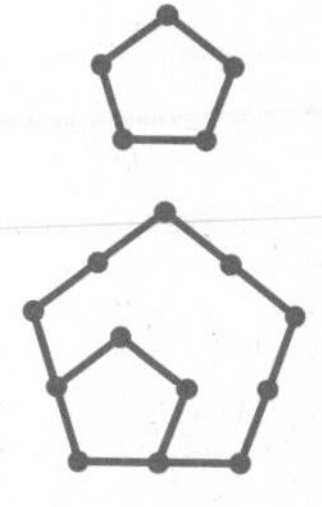

19.* If $a_1 = 3$, $a_2 = 5$, and $a_n = 3a_{n-1} - 2a_{n-2}$, find an explicit formula for a_n.

20.*

$$\begin{aligned} 1 &= 1 \\ 2 + 3 + 4 &= 1 + 8 \\ 5 + 6 + 7 + 8 + 9 &= 8 + 27 \\ 10 + 11 + 12 + 13 + 14 + 15 + 16 &= 27 + 64 \end{aligned}$$

Guess the general law suggested by these examples and try to write it using good mathematical notation.

21.* The recursion formula for the square triangular numbers is $u_n = 34u_{n-1} - u_{n-2} + 2$. Use this formula and the fact that $u_1 = 1$ and $u_2 = 36$ to calculate u_3, u_4, u_5, and u_6.

22. The explicit formula for the square triangular numbers is given near the end of this section. Use it to calculate u_3, u_4, u_5, and u_6. Compare your answers to those obtained in Problem 21. Can you account for any discrepancies?

23. Consider the recursion formula introduced in this section for finding $\sqrt{A}$. We saw in Example 2 that for $A = 4$, $a_1 = a_2 = a_3 = \cdots = 2$. Does it happen whenever A is a perfect square that $a_1 = a_2 = a_3 = \cdots = \sqrt{A}$?

24. Try the recursion formula for finding $\sqrt{A}$ on $A = 20$; on $A = 200$. What do you guess happens as A increases?

25. To get some insight into why the recursion formula for $\sqrt{A}$ works, assume that the formula $a_n = \dfrac{1}{2}\left(a_{n-1} + \dfrac{A}{a_{n-1}}\right)$ does indeed produce a sequence $a_1, a_2, \ldots, a_n, \ldots$ having the property that the larger n gets, the closer a_n gets to some fixed number r. It would follow then that $r = \dfrac{1}{2}\left(r + \dfrac{A}{r}\right)$. Solve this algebraic equation for r.

For Research and Discussion

Discuss the use on intelligence tests of sequences in which the next term is to be guessed. What do you make of our warning that an imaginative person might answer differently from what might be commonly expected? Might working on material such as is presented in this section improve one's score on such tests? Should one, by studying a certain topic, be able to improve his or her IQ? Why are sequences commonly used? Should they be used?

4.2 Arithmetic Sequences

Round and Round She Goes

Elizabeth's resolve to begin daily running after over-eating at Thanksgiving paid off, so that she was easily running 2 miles (32 laps around the track in Macalester's old gym) each day by term's end. Encouraged, she decided to begin the new year by running 2 miles each day in the first week, and then to increase the distance by a fixed number of laps each week for the rest of the 16-week term. How many laps would she have to add each week if, running only weekdays, she wants to become a member of Macalester's *Run to Madison* club, for those whose total miles over the term reach at least the 265 miles between St. Paul and Madison?

Suppose that each week, Elizabeth adds L laps (that's $\frac{L}{16}$ miles) to her daily run. Thus, in week 1, she runs 2 miles each day; in week 2 she runs $2 + \frac{L}{16}$ each day; etc. This seems to be a problem where it might help to organize the information in a table.

Organize the information.

Week	*Miles per Day*	*Miles per Week*
1	2	$5(2) = 10$
2	$2 + \dfrac{L}{16}$	$5\left(2 + \dfrac{L}{16}\right) = 10 + \dfrac{5L}{16}$
3	$2 + 2\dfrac{L}{16}$	$5\left(2 + 2\dfrac{L}{16}\right) = 10 + (2)\dfrac{5L}{16}$
⋮	⋮	⋮
16	$2 + 15\dfrac{L}{16}$	$5\left(2 + 15\dfrac{L}{16}\right) = 10 + (15)\dfrac{5L}{16}$

If w_n is the total number of miles run in week n of the term, then

$$w_1 = 10,\ w_2 = 10 + \frac{5L}{16},\ w_3 = 10 + (2)\frac{5L}{16}, \ldots,$$

$$w_{16} = 10 + (15)\frac{5L}{16}$$

This is a sequence in which each term is obtained by adding added $\frac{5L}{16}$ to the preceding one. It is easy to write both a recursion and an explicit formula for w_n

$$w_n = w_{n-1} + \frac{5L}{16}, \qquad w_n = 10 + (n - 1)\frac{5L}{16}$$

A sequence $a_1, a_2, a_3, \ldots$ in which each term is obtained by adding some constant d to the preceding term is called an **arithmetic sequence.** The constant d is called the **common difference.** It has the recursion formula

$$a_n = a_{n-1} + d,$$

The diagram at the right should help you discover the explicit formula. Note that the number of d's to be added is 1 less than the subscript of the term; that is

$$a_n = a_1 + (n - 1)d$$

The problem with which we began this section gave rise in a natural way to an arithmetic sequence. We shall return to it after we have learned more about arithmetic sequences in general.

"Did you ever notice that remarkable coincidence?" F. Scott Fitzgerald wrote in 1928 to British writer Shane Leslie. "Bernard Shaw is 61 years old, H. G. Wells is 51, G. K. Chesterton is 41, you're 31 and I'm 21—all the great authors of the world in arithmetic progression."

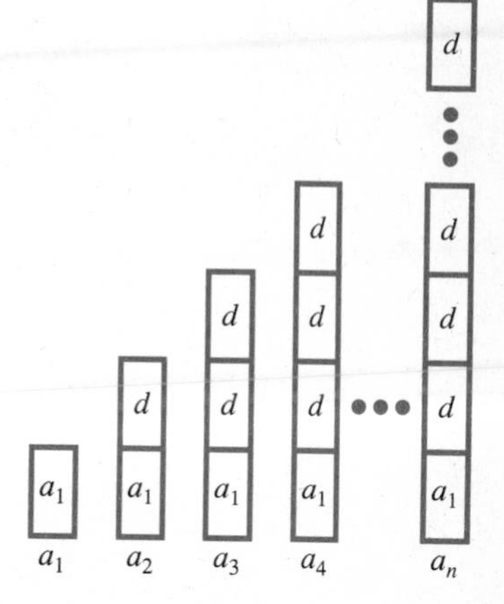

Sums of Arithmetic Sequences

Here is an old problem. Calculate the sum

$$1 + 2 + 3 + \cdots + 98 + 99 + 100$$

Without testifying to its accuracy, we pass along an anecdote about it taken from E. T. Bell's book, *Men of Mathematics* (New York: Simon and Schuster, 1937). Carl Gauss was 10 years old when he was admitted to his first arithmetic class. His teacher, one Buttner by name, was in the habit of obtaining an hour or so of quiet for himself by requiring the class to do long addition problems. One day he asked his pupils to add all the numbers from 1 to 100. Hardly had he made the assignment when young Carl flung his slate on the teacher's desk with the correct answer on it. This is how he thought about the problem:

$$\begin{aligned}
\text{Sum} &= 1 + 2 + 3 + \cdots + 98 + 99 + 100 \\
\text{Sum} &= 100 + 99 + 98 + \cdots + 3 + 2 + 1 \\
2\ \text{Sum} &= 101 + 101 + 101 + \cdots + 101 + 101 + 101 \\
\text{Sum} &= \frac{100\ (101)}{2}
\end{aligned}$$

Carl F. Gauss (1777–1855)
Carl Gauss was born to humble parents in Brunswick, Germany. An infant prodigy who discovered a mistake in his father's accounts at age 3, he became the greatest mathematician of the nineteenth century. He gave the first proof of the Fundamental Theorem of Algebra and contributed to all branches of mathematics as well as physics. He and Weber invented the telegraph.

Bell observes with admirable restraint, "[This trick] is very ordinary once it is known, but for a boy of ten to find it instantaneously by himself is not so ordinary."

The same trick that Gauss used works, with slight modification, to find the sum A_n of the first n terms of any arithmetic sequence.

$$\begin{aligned} A_n &= a_1 + a_2 + \cdots + a_{n-1} + a_n \\ A_n &= a_n + a_{n-1} + \cdots + a_2 + a_1 \\ \hline 2A_n &= (a_1 + a_n) + (a_2 + a_{n-1}) + \cdots + (a_{n-1} + a_2) + (a_n + a_1) \end{aligned}$$

Note that

$$\begin{aligned} a_2 + a_{n-1} &= (a_1 + d) + [a_1 + (n-2)d] \\ &= a_1 + [a_1 + (n-1)d] = a_1 + a_n \end{aligned}$$

Indeed, each of the n columns on the right sums to $a_1 + a_n$.

$$2A_n = n(a_1 + a_n)$$

$$A_n = \frac{n}{2}(a_1 + a_n)$$

We are now ready to decide how many laps Elizabeth must add to her run each week in order to achieve her goal of running 265 miles in 16 weeks. Using the notation introduced at the beginning of this section and the formula just derived, the total distance run will be

$$\begin{aligned} \frac{16}{2}[w_1 + w_{16}] &= 8\left[10 + \left(10 + 15\frac{5L}{16}\right)\right] \\ &= 8\left[20 + \frac{75L}{2 \cdot 8}\right] = 160 + \frac{75L}{2} \end{aligned}$$

Adding $L = 2$ laps each week will only give her $160 + 75 = 235$ miles; $L = 3$ will give her $160 + 75\ (1.5) = 272$, putting her over the top.

Some Applications

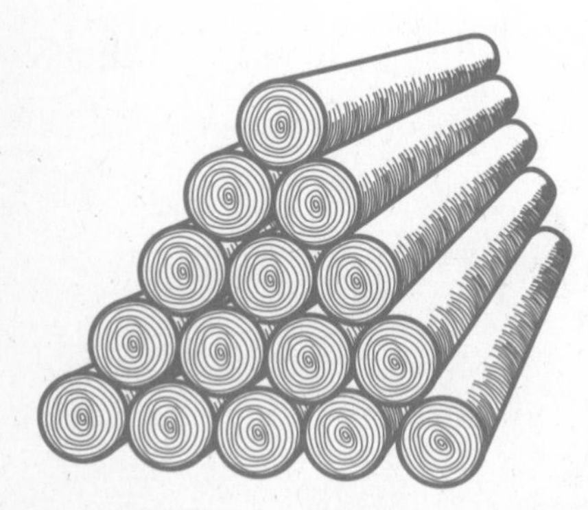

Imagine a mammoth pile of logs stacked in the manner shown at the left. Let us suppose that there are 113 logs in the bottom row. How many logs are there in the whole pile? Note that this amounts to adding

$$1 + 2 + 3 + \cdots + 113$$

The formula above gives us

$$A_{113} = \frac{113}{2}(1 + 113) = \frac{113}{2}(114) = 6441$$

Consider next a salary problem. Suppose that you have been offered a starting annual salary of \$18,000 with annual increments of \$1500 for each of the succeeding 39 years of your career. How much will you make during the last year of work and how much during your whole 40-year career?

The sequence to be considered is

$$18{,}000,\ 19{,}500,\ 21{,}000,\ \ldots$$

For it we must calculate a_{40} and A_{40}

$$a_{40} = a_1 + (40 - 1)d = 18{,}000 + 39(1500) = \$76{,}500$$

$$A_{40} = \frac{40}{2}(a_1 + a_{40}) = 20(18{,}000 + 76{,}500) = \$1{,}890{,}000$$

Here is a trickier problem. Willie Hitit must make a difficult decision. Both the Podunk Possums and the Tooterville Toads have offered him a 5-year baseball contract.

Podunk contract: Starting salary of \$40,000 per year with annual increases of \$8000 each.

Tooterville contract: Starting salary of \$40,000 per year with semiannual increases of \$2000 each.

Willie likes the Tooterville manager best but considers Podunk's offer to be superior financially. Should he opt for the manager or the money? The answer is—he should do some mathematics. Actually, Tooterville is offering more money and the best manager.

To see why, draw two 5-year time lines indicating the salary obtained during each period (salary indicated in thousands).

Podunk: 40, 48, ..., 72

Tooterville: 20, 22, 24, 26, ..., 38

Both sequences are arithmetic. Thus last-period salaries and 5-year totals can be obtained from our formulas for the nth term a_n and the sum of the first n terms A_n. Note that we must use $n = 5$ in the case of annual raises and $n = 10$ in the case of semiannual raises.

$$\text{Podunk total: } \frac{5}{2}(40{,}000 + 72{,}000) = \$280{,}000$$

$$\text{Tooterville total: } \frac{10}{2}(20{,}000 + 38{,}000) = \$290{,}000$$

Summary

A sequence that increases (or decreases) by always adding the same number d is called an arithmetic sequence. For such a sequence, there are two important formulas, one giving the nth term a_n and the other giving the sum A_n of the first n terms:

$$a_n = a_1 + (n - 1)d$$

$$A_n = a_1 + a_2 + \cdots + a_n = \frac{n}{2}(a_1 + a_n)$$

Problem Set 4.2

1. Fill in the boxes.
(a) 1, 3, 5, 7, □, □, . . .
(b) 5, 8, 11, 14, □, □, . . .
(c) 100, 96, 92, 88, □, □, . . .

2. Fill in the boxes.
(a) 17, 15, 13, □, □, . . .
(b) 2.0, 2.5, 3.0, □, □, . . .
(c) 100, 150, 200, □, □, . . .

3. Determine the common difference d and the fortieth term for each sequence in Problem 1. Hint: To find the fortieth term use $a_n = a_1 + (n - 1)d$ with $n = 40$.

4. Determine d and the thirtieth term for each sequences in Problem 2.

5. Find A_{40}, B_{40}, and C_{40} for the sequences in Problem 1. Hint: To find A_{40}, recall that you found a_{40} in Problem 3, and then use $A_n = (n/2)(a_1 + a_n)$ with $n = 40$.

6. Find A_{30}, B_{30}, and C_{30} for the sequences in Problem 2.

7. Calculate:
(a) $2 + 4 + 6 + \cdots + 100$
(b) $1 + 3 + 5 + \cdots + 99$
(c) $3 + 6 + 9 + \cdots + 99$

8. Calculate:
(a) $1 + 3 + 5 + \cdots + 199$
(b) $5 + 10 + 15 + \cdots + 600$
(c) $8 + 12 + 16 + \cdots + 64$

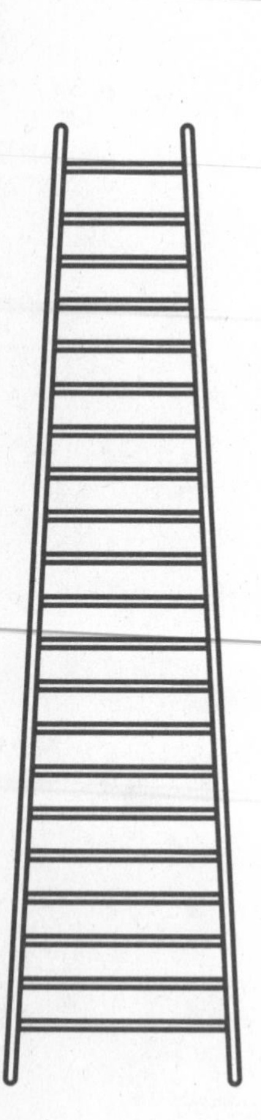

9. A ball falls 16 feet during the first second, 48 feet during the second, 80 feet during the third, etc. How far does it fall during the tenth second? What is the total distance it falls in 10 seconds?

10. How many integers are there between 100 and 300 that are multiples of 7? What is the sum of these integers?

11. The bottom rung of a tapered ladder is 60 centimeters long; the top one is 40 centimeters long. If there are 21 rungs, how much rung material was needed to make the ladder?

12. The terms 2, a, b, c, 3.2, . . . form an arithmetic sequence. Determine a, b, and c.

13. If Sarah saves 7 cents the first week of the year, 14 cents the second, 21 cents the third, etc., how much will she have after one year (52 weeks)?

14. A contractor who fails to complete a building in a specified amount of time must pay a penalty of $100 for each of the first 8 days of extra time required. After that the penalty is increased by $10 per day. If he goes 30 days over, what is his total penalty?

15. At a club meeting with 200 people present, everyone shook hands with every other person exactly once. How many handshakes were there? Suggestion: Number the people from 1 to 200. Let person 1 shake hands with each of the 199 other people. Now let person 2 shake hands—with how many people? etc.

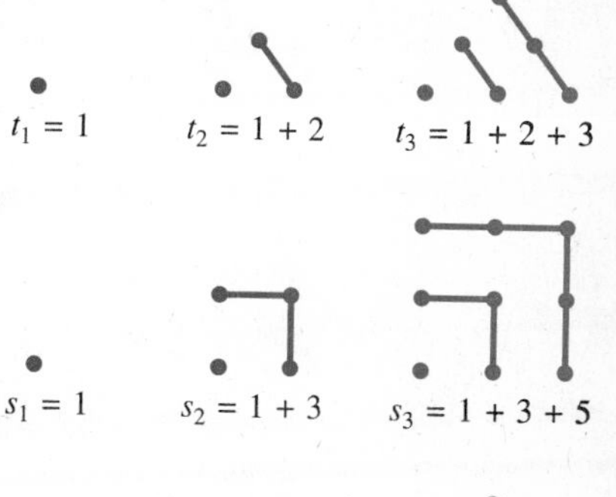

16. A clock strikes once at one o'clock, twice at two o'clock, etc. How many times does it strike between 8:15 A.M. on Monday morning and the same time on the next day?

17. A pile of logs has 120 logs in the first layer, 119 in the second layer, 118 in the third, etc. How many logs are there in the pile if there are 57 layers?

18. Consider the diagrams of the first three triangular, square, and pentagonal numbers shown at the right. Note the appearence of arithmetic sequences and use this fact to find explicit formulas for t_n, s_n, and p_n.

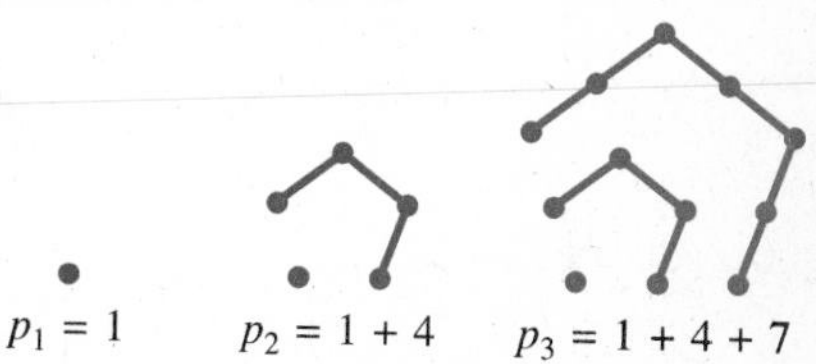

19. Consider the list

$$\begin{aligned} 1 &= 1 \\ 2 + 3 + 4 &= 1 + 8 \\ 5 + 6 + 7 + 8 + 9 &= 8 + 27 \\ 10 + 11 + 12 + 13 + 14 + 15 + 16 &= 27 + 64 \end{aligned}$$

What does the nth row look like? Consider adding all these equalities down through the nth row. The left sum is $1 + 2 + 3 + \cdots + n^2$. What is the right sum? Can you use this to obtain an explicit formula for $1^3 + 2^3 + 3^3 + \cdots + n^3$?

20. Consider the list

$$\begin{aligned} 1 &= 1^3 \\ 3 + 5 &= 2^3 \\ 7 + 9 + 11 &= 3^3 \\ 13 + 15 + 17 + 19 &= 4^3 \end{aligned}$$

What does the nth row look like? Now add all these equalities through the nth row. Simplify. Compare with Problem 19.

21. The first three hexagonal numbers, pictured in the margin, are 1, 7, and 19. Find a formula for the nth hexagonal number.

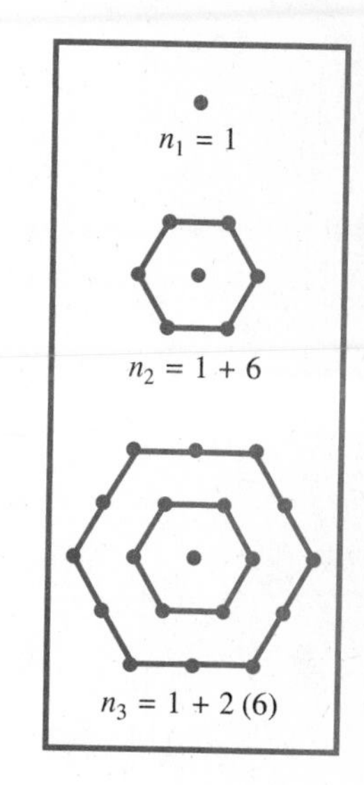

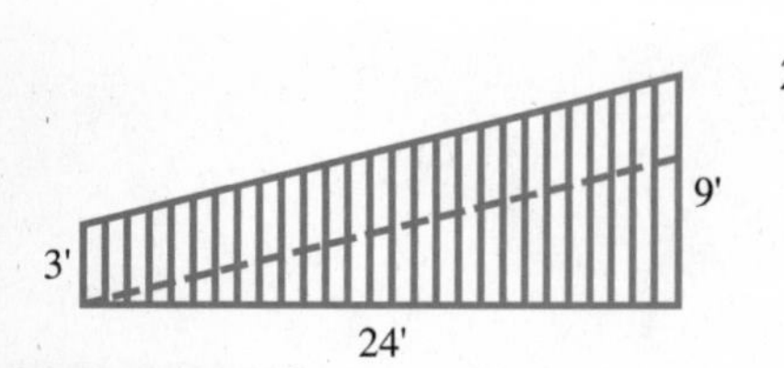

22. Suppose that $x_0 = 0$ and that for every $n > 1$, $x_n = n - x_{n-1}$. Find $x_1 + x_2 + x_3 + \cdots + x_{100}$.

23. The figure in the margin shows the wooden handrail on a ramp in the center of a shopping mall. For decorative purposes, it is supported by iron rods at 3-inch intervals. The two end pieces are 9 feet and 3 feet, respectively, and they are 24 feet from each other. How many feet of iron rod was used altogether?

For Research and Discussion

Increase the laps you run by a fixed number each week, increase your savings by a fixed amount each pay period, estimate the number of logs in a pile (or the number of seats in a tiered auditorium), figure the amount of material needed for a project like that described in Problem 23: these are examples you can find every day where arithmetic sequences occur. See how many examples you can find around you. Describe them. Use what has been said in this section to answer natural questions that might be asked about them.

4.3 Geometric Sequences

Double Your Money

With the advent of paper money, Silas, who used to like to run his fingers through his gold coins, now likes to contemplate his fortune piled up in crisp, new $1 bills. Poor Silas started with just $1, but he has found a way to double his money each year. How high a pile does he have after 25 years? (The bank tells us that a package of 100 dollar bills from the mint is one-half inch high.)

Double your money–the dream of the aspiring entrepreneur, the promise of the perspiring promoter. How likely is it over an extended period of time? How likely is it that you can, with a \$1 stake, build in 25 years a fortune that, in \$1 bills, reaches over 2 miles into the sky (9.6 times the height of the Sears Tower in Chicago)? Equally likely, it turns out. Let's see why.

Beginning with \$1, Silas would have 2 bills at the end of the first year, $2 \cdot 2 = 2^2$ at the end of two years, $2 \cdot 2^2 = 2^3$ at the end of three years, and 2^{25} at the end 25 years. That's

$$\frac{2^{25}}{100}\left(\frac{1}{2}\right)\frac{1}{12} = 13{,}981 \text{ feet}$$

One mile is 5280 feet; the Sears Tower is 1454 feet.

We have just considered a *doubling sequence.* Try to discover a similar feature in the following sequences and then fill in the boxes.

a. 2, 6, 18, 54, □, □, . . .
b. 16, 8, 4, 2, □, □, . . .
c. 1, −2, 4, −8, □, □, . . .

Each of these is a **geometric sequence.** Any term in such a sequence is obtained from its predecessor by multiplication by a fixed number (3, $\frac{1}{2}$, −2, respectively, in the three examples). It has the general form a_1, a_2, a_3, . . . with the recursion formula

$$a_n = ra_{n-1}$$

The fixed number r is called the **common ratio.**

We seek an explicit formula for a_n. Note that

$$\begin{aligned} a_1 &= a_1 \\ a_2 &= ra_1 \\ a_3 &= ra_2 = r(ra_1) = r^2a_1 \\ a_4 &= ra_3 = r(r^2a_1) = r^3a_1 \\ &\vdots \end{aligned}$$

$$\boxed{a_n = r^{n-1}a_1}$$

It is easy to put this formula to use. For the three sequences above, we have

$$\begin{aligned} a_7 &= 3^6(2) = 729(2) = 1458 \\ b_{11} &= \left(\frac{1}{2}\right)^{10}(16) = \frac{1}{2^{10}} \cdot 2^4 = \frac{1}{2^6} = \frac{1}{64} \\ c_8 &= (-2)^7(1) = -128 \end{aligned}$$

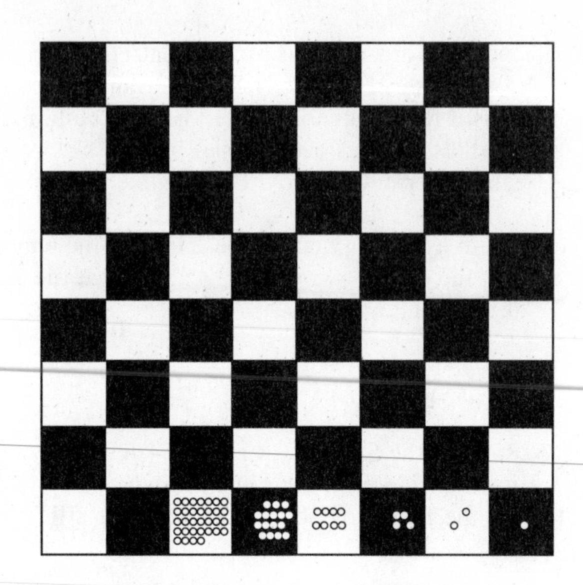

Sums of Geometric Sequences

There is a legend associated with geometric sequences and chessboards. When Sessa invented the game of chess, the king of Persia was so pleased that he promised to fulfill any request—to the half of his kingdom. Sessa, with an air of modesty, asked for one grain of wheat for the first square of the chessboard, two for the second square, four for the third, etc. The king was amused at this odd request but immediately commanded one of his servants to get a sack of wheat and begin counting it out. It was soon apparent that the king could not fulfill his promise. It would have taken the world's total production of wheat of several centuries. And to count the required number of grains at the rate of five per second would have taken 100 billion years.

What is the mathematics behind Sessa's request? He asked for

$$1 + 2 + 2^2 + 2^3 + \cdots + 2^{63}$$

grains of wheat, the sum of a geometric sequence with a common ratio of 2.

Let's look at the general problem. Consider $A_n = a_1 + a_2 + \cdots + a_n$, where $a_1, a_2, \ldots$ is a geometric sequence with a common ratio of r. Multiply A_n by r and subtract the product from A_n as indicated:

$$\begin{aligned} A_n &= a_1 + ra_1 + r^2a_1 + \cdots + r^{n-1}a_1 \\ rA_n &= ra_1 + r^2a_1 + \cdots + r^{n-1}a_1 + r^na_1 \\ \hline A_n - rA_n &= a_1 + 0 + 0 + \cdots + 0 - r^na_1 \end{aligned}$$

$$A_n(1 - r) = a_1(1 - r^n)$$

$$A_n = \frac{a_1(1 - r^n)}{1 - r} = \frac{a_1(r^n - 1)}{r - 1}$$

Thus Sessa's request was for

$$A_{64} = \frac{1(2^{64} - 1)}{2 - 1} = 2^{64} - 1 \approx 2^{64} \approx 2^4(1000)^6$$
$$= 18{,}500{,}000{,}000{,}000{,}000{,}000$$

grains of wheat.

For the Musically Inclined

For the Greeks, music and mathematics were inseparably linked. Plato and Aristotle taught that one must study the *quadrivium,* consisting of arithmetic, geometry, astronomy, and music in order to understand the laws of the universe. Mathematics and music still appear close to each other in most college catalogs, but this is due more to administrative respect for the alphabet than educational philosophy. Discussing geometric sequences affords us the opportunity to demonstrate a genuine interconnection.

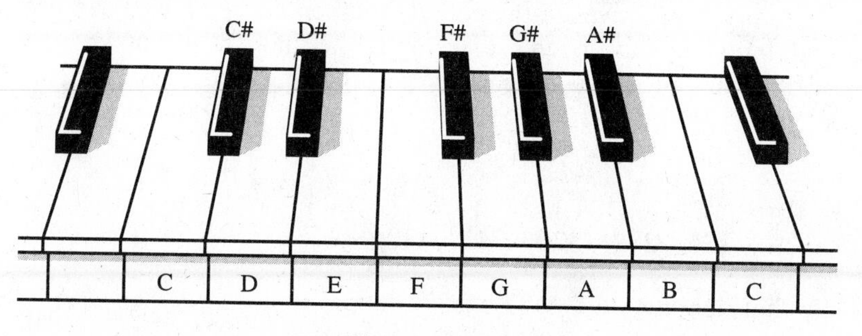

Piano keyboard

In the equally tempered scale, to which all keyed instruments have been tuned since the days of Bach (1685–1750), the intervals between the 12 halftones are said to be equal. What this really means is that the frequencies of vibration for these 12 tones form a geometric sequence. If middle C has frequency f (f is usually chosen to be about 262), the other frequencies are as follows.

C	C♯	D	D♯	E	F	F♯	G	G♯	A	A♯	B	$\overline{\text{C}}$
f	rf	r^2f	r^3f	r^4f	r^5f	r^6f	r^7f	r^8f	r^9f	$r^{10}f$	$r^{11}f$	$r^{12}f$

Moreover, the frequency of high C, labeled $\overline{\text{C}}$ is just twice that of middle C, so that $r^{12} = 2$, or

$$r = \sqrt[12]{2} = 1.05946\ldots$$

You don't have to be a music major to know that when two piano keys are struck together you may or may not hear a sound that you like. What is it that makes some concords (intervals) pleasing while others are positively jarring? It turns out that the concords most pleasant to the human ear are those with frequencies that form small-integer ratios (e.g., $\frac{2}{1}, \frac{3}{2}, \frac{4}{3}, \frac{5}{4}$). Here are some pleasant concords familiar to all music students.

Interval Name	*Interval*	*Ratio of Frequencies*
Octave	C to $\bar{C}$	$(1.05946)^{12} = 2 = \frac{2}{1}$
Perfect fifth	C to G	$(1.05946)^{7} = 1.49831 \approx \frac{3}{2}$
Perfect fourth	C to F	$(1.05946)^{5} = 1.33484 \approx \frac{4}{3}$
Major third	C to E	$(1.05946)^{4} = 1.25992 \approx \frac{5}{4}$

Actually none of the ratios, other than that of the octave, is exactly a small-integer ratio. For example, the perfect fifth has a ratio of 1.49831 which is a bit off from $\frac{3}{2} = 1.50000$. It is these slight discrepancies that make it difficult for a violinist to play with a piano, since the violinist is naturally inclined to play a perfect fifth so that the ratio involved is exactly $\frac{3}{2}$.

Many alternatives to the equally tempered scale have been proposed, but none considered satisfactory. One that is well-known is the so-called diatonic scale which has the following frequencies (for the white keys):

C	D	E	F	G	A	B	$\bar{C}$
256	288	320	$341\frac{1}{3}$	384	$426\frac{2}{3}$	480	512

Note that the interval C to G has the ratio

$$\frac{384}{256} = \frac{3(128)}{2(128)} = \frac{3}{2}$$

while the interval C to F has the ratio

$$\frac{341\frac{1}{3}}{256} = \frac{\frac{1024}{3}}{256} = \frac{1024}{3(256)} = \frac{4(256)}{3(256)} = \frac{4}{3}$$

This looks promising. However, equal tempering has been sacrificed (i.e., the interval between successive notes is not constant), and this means that transpositions between different keys will be troublesome. The reader who is interested in other possible scales can consult H. Steinhaus, *Mathematical Snapshots* (New York: Oxford University Press, 1960), pp. 40–43.

Summary

A geometric sequence is built by multiplying any given term by a fixed number r to find the next term. Two important formulas for geometric sequences are

$$a_n = r^{n-1}a_1$$

$$A_n = a_1 + a_2 + \cdots a_n = \frac{a_1(1 - r^n)}{1 - r} = \frac{a_1(r^n - 1)}{r - 1}$$

Problem Set 4.3

1. Fill in the boxes.
   (a) 2, 8, 32, 128, □, □, . . .
   (b) 27, 9, 3, 1, □, □, . . .
   (c) 3, −3, 3, −3, □, □, . . .
2. Fill in the boxes.
   (a) 3, 6, 12, □, □, . . .
   (b) 4, 2, 1, □, □, . . .
   (c) 7, 0, 0, □, □, . . .
3. Write an explicit formula for the fortieth term in each sequence in Problem 1; that is, write formulas for a_{40}, b_{40}, and c_{40}.
4. Write an explicit formula for each of the sequences in Problem 2.
5. Use a formula we have derived to find the sum of the first five terms in each sequence in Problem 1.
6. Find the sum of the first seven terms in each sequence in Problem 2.
7. A certain culture of bacteria doubles every 24 hours. If there are 1000 bacteria initially, how many will there be after 10 full days?
8. "As I was going to St. Ives,
   I met a man with seven wives.
   Every wife had seven sacks,
   Every sack had seven cats,
   Every cat had seven kits.
   Kits, cats, sacks and wives,
   How many were going to St. Ives?"
9. In Problem 8, how many were coming from St. Ives?
10. Take a large sheet of paper and consider folding it in half 40 times (actually this is physically impossible, but we can still consider it mentally). If the paper is 0.01 inch thick, approximately how high will the resulting stack of paper be?
11. Suppose that the sheet of paper in Problem 10 winds up having an area of 1 square inch after 40 folds. What was its area originally? Approximate.
12. A tropical water lily grows so rapidly that each day it covers a surface double in size the area it covered the day before. At the end of 30 days it completely covers a lake. How long would it take two of these lilies to cover the same lake?
13. Let $a_n = (1.023)^n$. Find a_{40} and A_{40}.
14. Let $b_n = (.921)^n$. Find b_{30} and B_{30}.
15. A home is increasing in value by 4% each year (that is, its value is 4% more at the end of each year than it was at the beginning of the year). How much will a home worth $75,000 today be worth at the end of 10 years?

16. Suppose that inflation is running at 4% per year so that an item costing \$1.00 today will cost \$1.04 one year from now. Thus in constant value dollars, \$1.00 will be worth 1.00/1.04 = \$.9615 one year from now. How much will today's dollar be worth 20 years from now?

17. One-third of the air in a container is removed with each stroke of a vacuum pump. What fraction of the original amount of air remains in the container after five strokes? How many strokes are necessary to remove 99% of the air?

18. A ball is dropped from a height of 10 feet. At each bounce it rises to two-thirds of its previous height. Suppose that (due to imperfect elasticity) the ball comes to rest when it drops less than one-eight inch. How far will it have traveled altogether (up and down) when it comes to rest.

19.* Recall the formula for the sum of a geometric sequence?

$$A_n = \frac{a_1(1 - r^n)}{1 - r} = \frac{a_1}{1 - r} - \frac{a_1 r^n}{1 - r}$$

If r is between 0 and 1, then as n gets larger and larger, A_n gets closer and closer to $a_1/(1 - r)$. Why? To what number does A_n get closer and closer in each of the following?

(a) $A_n = \frac{3}{10} + \frac{3}{100} + \cdots + \frac{3}{10^n}$

(b) $A_n = \frac{7}{10} + \frac{7}{100} + \cdots + \frac{7}{10^n}$

(c) $A_n = \frac{9}{10} + \frac{9}{100} + \cdots + \frac{9}{10^n}$

Do your answers seem reasonable? Reconsider Problem 18 in light of this problem.

20.* The ideas in Problem 19 lead to a great deal of advanced mathematics. If $a_1, a_2, \ldots$ is a geometric sequence with a ratio r between 0 and 1, we say that its total sum is $a_1/(1 - r)$. Use this fact to calculate the values of the following infinite repeating decimals.

(a) 0.3333 . . . Note: $0.3333\ldots = \frac{3}{10} + \frac{3}{100} + \frac{3}{1000} + \cdots$.

(b) 0.1111 . . .

(c) 0.5555 . . .

(d) 0.3232323232 . . .

(e) 0.134134134 . . .

Can you demonstrate that any infinite repeating decimal is a rational number (a ratio of two integers)?

21.* Suppose the government spends an extra billion dollars without raising taxes. What is the total increase in spending due to this action? Assume that each business and individual saves 20% of its income and spends the rest, so that of the initial \$1 billion spent by the government 80% is respent by individuals and businesses, and then 80% of that, etc. In economics, this is called the multiplier effect.

For Research and Discussion

Discuss the risk taken by the conservative person who, not trusting banks, keeps dollars hidden in a safe place or by the one who does put dollars in a savings account "where the principal will always be safe." How have savings interest rates and the rate of growth of the value of stocks compared over various periods in U.S. history?

4.4 Population Growth

Maxing Out

The following assertions can all be found in literature discussing world population.

(1) World population in 1986 was 5 billion.
(2) World population is increasing at about 2% per year.
(3) It takes about 1 acre of land to provide food for one person.
(4) The world has 8 billion acres of arable land.

Based on this information, when will we reach the maximum population that can be fed?

Statements (1) and (2) taken together mean that with a population of $P_1 = 5$ billion in 1986,

the population in 1987 should have been P_2
$$= P_1 + (.02)P_1 = (1 + .02)P_1$$
the population in 1988 should have been $P_3 = P_2 + (.02)P_2$
$$= (1 + .02)P_2 = (1 + .02)^2P_1$$
the population in 1989 should have been $P_4 = P_3 + (.02)P_3$
$$= (1 + .02)P_3 = (1 + .02)^3P_1$$

It follows that the world population n years after 1986 should be $(1 + .02)^nP_1$

Since the world can support a population of only 8 billion people, our problem is to determine n so that

$$5(1.02)^n = 8$$

$$(1.02)^n = \frac{8}{5} = 1.6$$

Table 4-2	
n	$(1.02)^n$
5	1.104
10	1.219
15	1.346
20	1.486
25	1.641
30	1.811
35	2.000
40	2.208
45	2.438
50	2.692
55	2.972
60	3.281
65	3.623
70	4.000
75	4.416
80	4.875
85	5.383
90	5.943
95	6.562
100	7.245
105	7.998
110	8.831
115	9.750

In Table 4-2, we find that $(1.02)^{25} = 1.6$, so n is approximately 25. In $1986 + 25 = 2011$, the world will reach maximum population. Of course, long before that mass starvation is likely to occur as a result of inequities in food distribution. In Problem Set 4.4, we explore the effect of changes in assumptions. The conclusions are alarming under any assumptions that are realistic. Unchecked population growth is not a good idea; but then neither are most of the checks (war, famine, and plagues) used so far.

Linear Growth vs Exponential Growth

Most people tend to think of growth as a linear process. A quantity is growing **linearly** if in each time period (of specified duration) it increases by a fixed amount. If town A starts with a population of 100 and adds 25 people each year, its population is growing linearly. However, a quantity is growing **exponentially** if during each time period it increases by a fixed multiple of the amount present at the beginning of that period. If town B starts with a population of 100 and has at the end of 1 year $100(1.25) = 125$ people, at the end of 2 years $125(1.25) \approx 156$ people, etc., its population is growing exponentially.

A graphical comparison will illustrate the difference between these two types of growth in a dramatic way. While both towns grow the same amount during the first year, the exponentially growing town actually grows more than three times as much over a 12-year period. As a matter of fact, in 12 years a town growing exponentially at only 10% per year will overtake a town growing linearly at the rate of 25%. An exponentially growing curve will always eventually conquer a linearly growing curve.

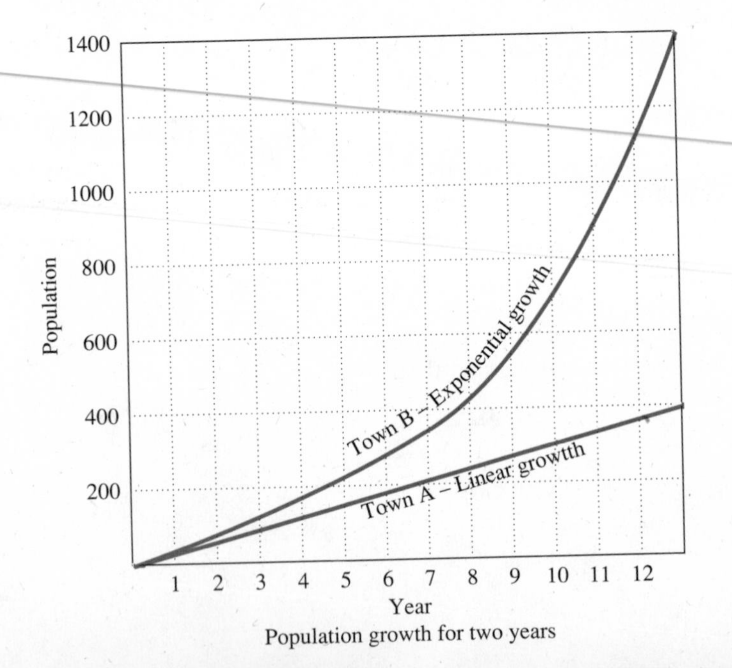

Population growth for two years

We should note an assumption we have made in connection with population growth. Both in our discussion and in drawing nice, smooth curves, we have implicitly assumed that population grows continuously with time. Of course, it really doesn't; it grows at distinct times and by unit amounts. But here, as throughout all of science, we are attracted by simplicity and elegance. We are willing to sacrifice exactness in order to obtain a model we can work with easily. A model that allowed for discontinuous jumps at random times would be intolerably complicated and certainly not appropriate for our present purposes.

Relation to Earlier Work

We have already met the ideas discussed above. If we look at the population of town A at yearly intervals, we obtain the sequence

$$100, 125, 150, 175, \ldots$$

which is an arithmetic sequence. The corresponding sequence for town B is

$$100, 100(1.25), 100(1.25)^2, 100(1.25)^3, \ldots$$

which is a geometric sequence. Linear corresponds to arithmetic, exponential to geometric. The words "linear" and "exponential" are generally used to describe continuous processes; the adjectives "arithmetic" and "geometric" are applied to sequences.

Doubling Times and Half-Lives

Perhaps you have already noticed a characteristic feature of exponential growth. If an exponentially growing body doubles in size in n years, it will double again in the next n years. We call n its **doubling time.** If the growth rate for world population is 2% per year, the doubling time according to Table 4-2 is 35 years. Thus a population of 5 billion in 1986 will grow to 10 billion by 2021, 20 billion by 2056, 40 billion by 2091, etc. This again emphasizes the spectacular nature of exponential growth.

But not everything grows; some things decline or decay exponentially, in which case we talk not about doubling time, but rather halving time—or half-life. The **half-life** of a given substance is the time it will take for half of what exists to disappear. Thus one-half of a substance will disappear during its first half-life, one-half of the remainder during its second half-life (leaving one-fourth of the original), etc. Notable examples are radioactive elements, all of which are believed to decay exponentially. A given amount of radium, for example, decays so that after 1690 years one-half of it remains, at the end of $2(1690) = 3380$ years only one-fourth of it remains, etc. We say that radium has a half-life of 1690 years. This law of decay is used as the basis for dating old objects. If an object contains radium and lead (radium changes to lead when it decays) in the proportion $1:3$, it is believed that an amount of pure radium has decayed to one-fourth of its original amount. The object is two half-lives (3380 years) old.

We should emphasize that radioactive dating methods are based on assumptions (e.g., that decay is exactly exponential over long periods of time). Recent findings suggest that the problem of dating old objects is not as simple as the above description may imply.

Growth of Human Population

Most efforts to model the growth of a population, whether of humans or of a bacteria, whether in a particular country or in a petri dish, begin with the reasonable assumption that the rate of growth will be proportional to the population present. This leads to exponential growth as the appropriate model, a highly useful model used with success in many areas of science.

How well does the exponential model work in a purely human population? Specific projections have not been notably successful over spans of more than 30 years or so. For example, Pearl and Reed [On the rate of growth of the population of the United States since 1870 and its mathematical representation, Proceedings of the National Academy of Sciences, Vol. 6, No. 6, June 15, 1920, pp. 275–288], accepted the assumption that leads to the exponential model but modified it with the assumption of a natural bound so as to avoid the otherwise unavoidable idea that the population will grow without bound forever. After analyzing the US census figures at their disposal, they projected that the maximum population for the US would be 197,274,000. The 1990 population was 248,709,873.

I said that population, when unchecked, increased in a geometrical ratio, and subsistence for man in an arithmetical ratio.

Thomas Malthus, 1798

The failure of efforts to make specific projections is not much emphasized in the writing of doomsayers on the subject. Dating at least to Thomas Malthus (1798), economists and demographers have with varying degrees of urgency made dire predictions. As always, sweeping generalizations are easier to make than specific projections, and few readers of mathematically based projections realize that all models rest on assumptions, a subject we shall say more about in Chapter 10.

Population growth must inevitably level off. The only question is; Will it be because of high death rates or low birth rates?

All projections agree on the trend; up. And it is sobering to list the reasons that hold population in check: disease, famine, natural disaster, war, birth control. So far, neither mathematical predictions nor graphs depicting actual data seem to have affected very much people's zeal for multiplying.

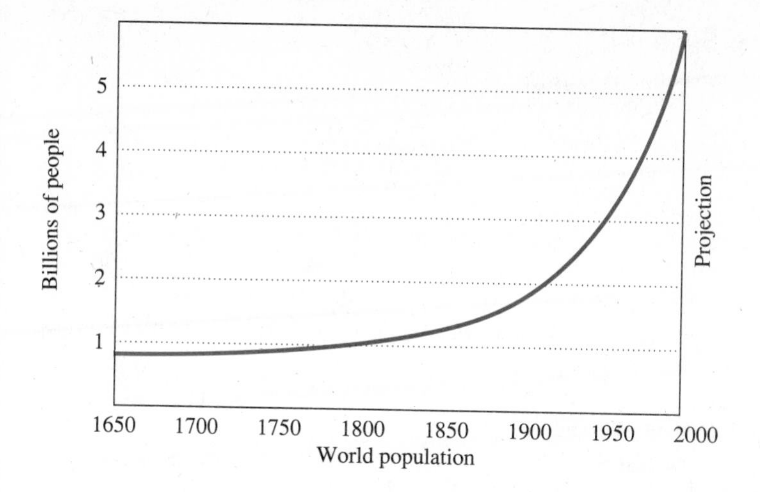

Summary

Many natural processes exhibit exponential growth (or decay). This means that the amounts of material present at the ends of consecutive time periods form a geometric sequence. There is evidence that the world's population (at least in the 1900–1975 period) is growing exponentially at about 2% per year. It is a matter of simple mathematical calculation to determine the dire consequences if this pattern continues.

> One hears young people asking for a cause. The cause is here. It is the biggest single cause in history: simply because history has never presented us with such a danger . . . . No more people than the earth can take. That is the cause.
>
> *C. P. Snow (1969)*

Problem Set 4.4

1. A population is growing linearly at 10% per year. If it has 1000 members today, how many will it have after 5 years?
2. A population is growing exponentially at 10% per year. If it has 1000 members today, how many will it have after 5 years? Approximate.
3. A bacterial culture is growing exponentially. If it now contains 1000 bacteria and doubles its number every 3 hours, how many bacteria will there be after 6 hours? Nine hours? Thirty hours?
4. A manufacturing company adds 100 new workers each year. Is it growing exponentially? Linearly? If it has 10,000 workers today, how many will it have after 10 years? Twenty years? How long will it take to double in size? Quadruple in size?
5. Does the human body grow exponentially or linearly or neither over its lifetime? Explain.
6. A certain city grew so that its population was 1000, 1100, 1300, 1600, and 2000 in 1930, 1940, 1950, 1960, and 1970, respectively. Is it growing linearly or exponentially or neither? Do you see any pattern in its growth?
7. The radioactive element polonium has a half-life of 140 days. If the Chemtom Company has 200 grams today, approximately how much will it have after 3 years and 25 days?
8. Carbon 14 decays exponentially with a half-life of 5730 years. The concentration of carbon 14 in a piece of wood of unknown age is one-fourth of that in a modern piece of wood. How old do you estimate the old piece of wood to be?
9. The number of bacteria in a culture grew (exponentially) from 100 to 800 in 24 hours. How many bacteria were there after 8 hours?
10. Consider the problem on world population in this section. Suppose that through the use of fertilizers, insecticides, etc., food production can be increased so that $\frac{1}{4}$ acre will support one person. When will the world reach maximum population?
11. As in Problem 10 suppose $\frac{1}{4}$ acre can support a person and suppose further that 9 billion acres can be cultivated. When will the world reach maximum population?
12. Suppose we reduce the growth rate of the world's population to 1% per year, that $\frac{1}{4}$ acre can support a person, that there are 12 billion acres of arable land. When will the world reach maximum population?
13. The population of a certain country is 14,600,000 today and it is growing exponentially at 1.5% per year. What will its population be at the end of 20 years?
14. How long will it take the country of Problem 13 to double in population? Hint: You must solve the equation $(1.015)^n = 2$. Experiment.

15. Town A is growing linearly at 5% per year while town B is growing exponentially at 1.5% per year. Both have a population of 1000 today.
(a) Calculate the populations of both towns after 40, 60, 80, and 100 years.
(b) About when will the population of town B overtake that of town A?

For Research and Discussion

Write a report on several serious efforts to predict the growth of a human population (in the world, a country, a town, or a school district). Do the writers make clear their assumptions? Do you agree with them? In the case of older articles, how are their predictions holding up?

4.5 Compound Interest

A Story of Interest

The Teacher's Insurance and Annuity Association, in its November, 1992 Participant Newsletter, posed the following problem for young professors considering when to begin saving for retirement.

Alice and her husband David both started teaching careers at age 25. Alice began immediately to put $3000 each year into a Supplemental Retirement Annuity (SRA) that paid an annual return of 7%. At age 35, having contributed a total of $30,000, she took a job in private industry and was unable to make further contributions.

It was then, at age 35, that David began making annual $3000 contributions to his SRA, also paying 7%. When he reached 65, he had contributed $90,000.

Both Alice and David retired at 65. How much had accumulated in each of their SRA accounts?

Actually, the principle of compound interest is very simple. We already have all the necessary mathematical machinery. What we need to learn is vocabulary, starting with the distinction between the principle we are learning and the **principal** (the money) we are investing. Next, we need to realize that when interest is compounded annually, it means that one is paid interest earned not only on the principal originally invested, but also on the interest earned by that principal in previous years.

Suppose that you invest \$1.00 in a bank that pays interest at 4% compounded annually. At the end of 1 year, the bank will add \$0.04 to the account. During the second year, the resulting \$1.04 will draw interest. Thus, the bank will add (0.04)(1.04) to the account, bringing the total to

$$1.04 + (0.04)(1.04) = (1 + 0.04)(1.04) = (1.04)^2$$

at the end of two years. The process will continue, yielding the geometric sequence

$$1.00,\ 1.04,\ (1.04)^2,\ (1.04)^3,\ (1.04)^4,\ \ldots$$

After 25 years, the \$1.00 will be worth $(1.04)^{25} = \$2.67$. To calculate numbers like this, one of course needs a calculator.

Let's consider one more possibility. Suppose the \$1.00 is invested at 4% compound interest, with interest compounded quarterly. Interest rates are usually stated as annual rates; 4% interest compounded quarterly really means 1% per quarter. At the end of 3 months, the bank will add (0.01)(1.00) or \$0.01 to the account, and \$1.01 will draw interest during the second quarter. The sequence generated by this process is

$$1.00,\ 1.01,\ (1.01)^2,\ (1.01)^3,\ \ldots$$

In this case, after 25 years (100 quarters), the \$1.00 will be worth $(1.01)^{100} = \$2.70$

These two examples illustrate the general result. If savings P (the principal) are invested in an account that pays r% annually, and if interest is paid at the end of n equal time intervals during the year, the amount A that will have accumulated after t years is

$$A = P\left(1 + \frac{r}{n}\right)^{nt}$$

Regular Savings

Suppose you decide to be a consistent saver. At the end of each year for the next 30 years you plan to put \$1000 in a plan that pays 7% interest compounded annually. How much will your account be worth after you have made your final deposit?

Let's do the problem for \$1 payments and then multiply the final result by 1000. In the accompanying diagram, we keep track of each \$1 payment by means of time lines.

Year Saving of $1 at 7% Interest

Year	1	2	3	29	30
Payment Number					
1	1	1.07	$(1.07)^2$	$(1.07)^{28}$	$(1.07)^{29}$
2		1	(1.07)	$(1.07)^{27}$	$(1.07)^{28}$
3			1	$(1.07)^{26}$	$(1.07)^{27}$
29				1	(1.07)
30					1
Savings at end of year		2.07	3.21		

The value of the account after the thirtieth payment is made is obtained by adding the entries in the last column.

$$1 + 1.07 + (1.07)^2 + \cdots + (1.07)^{29}$$

This is just a problem of summing a geometric sequence, and for this we developed a formula in Section 4.3

$$A_{30} = \frac{(1 - r^{30})}{1 - r} = \frac{[1 - (1.07)^{30}]}{1 - 1.07} = \frac{(1 - 7.61225504)}{-0.07}$$
$$= \$94.46078632$$

The savings program outlined in the problem will accumulate to $94,406.86.

The computations showing the value of saving $1 at the end of each year for 30 years, if multiplied by 3000, give the value at retirement of David's annuity in the situation posed in the problem that opened this section. A similar computation for ten years will give the value of Alice's annuity when she leaves teaching. This amount, compounded at 7% for the next 30 years, gives the value of her annuity when she retires. Compare the results of her $30,000 in contributions with his $90,000 in contributions. Draw a conclusion. Resolve to act upon it when you begin your own career. Never call this course worthless.

House Payments

So far we have talked about saving. But most of us seem to be more concerned with borrowing.

Suppose that you decide to purchase a house for $80,000 on which you put $20,000 down. The balance will be paid in equal installments at the end of each month for the next 25 years (300 months). If interest is at 9% compounded monthly, what will your payments be?

For which of you intending to build a tower, sitteth not down first, and counteth the cost, whether he have sufficient to finish it?

Luke 14:28

The first thing to realize is that $1.00 paid today and $1.00 paid 1 month from today have quite different values in the world of business. Why? Simply because $1.00 paid today could be put in the bank, draw interest for a month, and therefore be worth more than $1.00 a month from now. Put slightly differently, we ask what is today's value (the so-called present value) of $1.00 that is going to be paid a month from now? It is less than $1.00, but by how much? Recall that, if a principal of P is put in the bank today, it is worth $P(1 + i)$ dollars a month from now. Thus, if $1/(1 + i)$ dollars is put in the bank today, it will be worth

$$\frac{1}{1 + i}(1 + i) = \$1.00$$

a month from now. Note that 9% annual interest corresponds to $i = \frac{3}{4}\% = 0.0075$ per month. Therefore the present value of a payment of $1.00 a month from now is only $1/(1 + i) = \$0.9926$. Similarly, a payment of $1.00 two months from today is worth only $1/(1 + i)^2$ dollars today.

Payments of m dollars at the end of each of 300 months have the present values shown on the time line below.

$$\begin{array}{ccccc} \dfrac{m}{1+i} & \dfrac{m}{(1+i)^2} & \dfrac{m}{(1+i)^3} & \cdots & \dfrac{m}{(1+i)^{299}} & \dfrac{m}{(1+i)^{300}} \\ 1 & 2 & 3 & & 299 & 300 \end{array}$$

We therefore want

$$60{,}000 = \frac{m}{1 + i} + \frac{m}{(1 + i)^2} + \frac{m}{(1 + i)^3} + \cdots + \frac{m}{(1 + i)^{299}} + \frac{m}{(1 + i)^{300}}$$

The right side is again geometric; it can be summed by the formula in Section 4.3. We obtain

$$60{,}000 = \frac{\dfrac{m}{1+i}\left[1 - \left(\dfrac{1}{1+i}\right)^{300}\right]}{1 - \dfrac{1}{1+i}} = \frac{m\left[1 - \left(\dfrac{1}{1+i}\right)^{300}\right]}{i}$$

More generally, if a principal P is to be paid off in N periods with equal payments of size m assuming interest at i per period, then

$$P = \frac{m[1 - (1+i)^{-N}]}{i}$$

Solved for the payment m, it says that

$$m = \frac{iP}{1 - (1+i)^{-N}}$$

When we substitute $P = 60{,}000$, $i = 0.0075$, $N = 300$ and use a pocket calculator, we obtain $m = \$503.52$

To get the answer to a complicated problem is an achievement; to understand its implications is more significant. Spreading the payments over 25 years rather than paying \$60,000 cash today has a whopping effect on what you actually pay. For in 25 years, you will make 300 payments amounting to 300(503.56) = \$151,056. Behold how the trivial difference between \$0.9926 today and \$1.00 a month from now grows when compounded over 25 years.

Finally, we remark on the tremendous effect a reduction in the interest rate can make. At 5% interest rather than 9% ($i = 0.004167$ rather than $i = 0.0075$), the monthly payment would only be \$350.82. And the total amount over 25 years would be 300(350.82) = \$105,246.

Summary

To make wise financial decisions, an investor or a borrower should understand compound interest. This is ultimately a matter of understanding geometric sequences. For if a dollar is invested at compound interest at the rate of i per period (for example, 8% interest compounded quarterly corresponds to $i = 0.02$), the value of that dollar now and at the end of succeeding periods is

$$1,\ 1 + i,\ (1+i)^2,\ (1+i)^3,\ (1+i)^4,\ \ldots$$

Problem Set 4.5

1. Find:
   (a) $(1.02)^{50}$ (b) $(1.12)^{15}$
   (c) $(1.08)^{9}$ (d) $100\ (1.08)^{20}$

2. Find:
   (a) $(1.01)^{100}$ (b) $(1.02)^{25}$
   (c) $(1.04)^{25}$ (b) $1000\ (1.08)^{9}$

3. If you put \$100 in the bank today, how much will it be worth after 25 years.
   (a) At 4% compound interest, compounded annually?
   (b) At 4% compound interest, compounded quarterly?

4. If you put \$1000 in the bank today, how much will it be worth after 10 years
   (a) At 3% compound interest, compounded annually?
   (b) At 3% compound interest, compounded every 2 months?

5. If you put \$100 in the bank today at 3% compound interest, compounded monthly, how much will it be worth after 100 months?

6. If you put \$50 in the bank today at 4% compound interest, compounded quarterly, how much will it be worth after $12\frac{1}{2}$ years?

7. How long (approximately) does it take money to double
   (a) At 4% compound interest, compounded annually?
   (b) At 4% compound interest, compounded every 2 months?
   (c) At 4% compound interest, compounded monthly?

8. How long (approximately) does it take money to double
   (a) At 4% compound interest, compounded annually?
   (b) At 4% compound interest, compounded quarterly?

9. Suppose a consumer buys a television set for \$200, using her charge account. If the store adds 1% of the unpaid balance to the bill each month it is not paid, and if she pays nothing until just after 12 charges have been made (that is, after 1 year),
   (a) How much does she then owe the firm?
   (b) How much interest (considered as simple interest) is she being charged?

10. Redo Problem 9, this time assuming there is no interest for the first month and only 11 charges are made.

11. Calculate:
    (a) $350\left(1 + \frac{0.13}{12}\right)^{24}$ (b) $475\left(1 + \frac{0.11}{365}\right)^{365}$

12. What interest problems do the expressions in Problem 11 represent?

13. If Horace puts \$560 in the bank today at 4% interest, how much will it be worth at the end of 2 years if
    (a) Interest is compounded annually?
    (b) Interest is compounded monthly?
    (c) Interest is compounded daily?

14. Hugo invested $100 in a $2\frac{1}{2}$ year money market certificate that pays 7% interest compounded daily. How much will it be worth when it matures?

15. Mary invested $750 in a $2\frac{1}{2}$ year money market certificate that pays 6% interest compounded daily. How much will it be worth when it matures?

16.* If Susan Sharp saves $200 at the end of each month for the next 25 months and receives 4% compound interest, compounded monthly, how much will she be worth at the end of the 25 months? First write an expression for her worth and then evaluate it.

17.* In order to buy a new car, Rachel Rinker decides to put $400 in the bank every 3 months for the next 4 years. If she receives 3% compound interest, compounded quarterly, what will she have at the end of this 4-year period? Assume the first deposit is made at the end of 3 months.

18.* If Professor Witquick buys a house for $90,000 with $10,000 down and the rest to be paid in equal monthly payments over 30 years (360 payments), and if the interest is 9%, compounded monthly, what are his payments?

19.* John and Mary bought a house for $80,000, paying $10,000 down. The balance of $70,000 will be paid in equal installments at the end of each month for the next 30 years. What will these payments be if interest is 8% compounded monthly?

20.* David and Evelyn are financing a mortgage of $60,000 by making equal payments at the end of each month for 25 years. If interest is 8.5% compounded monthly, what will their payments be?

21.* Homer bought a watch costing $20.00 which he will pay for on the installment plan at $1.00 per month for 24 months. The salesperson pointed out that he will pay only $4.00 in interest, or $2.00 per year. On $20.00 this is 10% interest.

(a) What is wrong with the salesperson's reasoning? Hint: Will Homer really use the full $20.00 for 2 years?

(b) Table 4-4 shows the present value of an item at $1.00 per month. For example, an item paid for in 12 monthly installments at 15% interest should have a price tag of $11.08. What interest rate does Homer really pay?

22.* If an item is paid for in monthly installments of $1.00 for 20 months and has a price tag of $18.05, what is the actual interest rate (see Problem 21)?

23.* Rodney Roller bought a jalopy worth $160.30 but agreed to pay for it at $10.00 per month over 18 months. What interest rate did he pay? Hint: This is equivalent to buying an item worth $16.03 at $1.00 per month for 18 months (cf. Problem 21).

24.* Complete the problem posed at the beginning of this section.

Table 4-4
Installment Buying—Present Value of an Item at $1.00 a Month

Number of Months	*Interest*		
	12%	**15%**	**18%**
1	0.99	0.99	0.99
2	1.97	1.96	1.96
3	2.94	2.93	2.91
4	3.90	3.88	3.85
5	4.85	4.82	4.78
6	5.80	5.75	5.70
7	6.73	6.66	6.60
8	7.65	7.57	7.49
9	8.57	8.46	8.36
10	9.47	9.35	9.22
11	10.37	10.22	10.07
12	11.26	11.08	10.91
13	12.13	11.93	11.73
14	13.00	12.77	12.54
15	13.87	13.60	13.34
16	14.72	14.42	14.13
17	15.56	15.23	14.91
18	16.40	16.03	15.67
19	17.23	16.82	16.43
20	18.05	17.60	17.17
21	18.86	18.37	17.90
22	19.66	19.13	18.62
23	20.46	19.88	19.33
24	21.24	20.62	20.03

For Research and Discussion

Compare interest rates as advertised by savings institutions in your area. How do they determine "annual effective rate?" Also examine installment purchase plans available to you. Convert interest rates announced on a monthly basis to annual rates.

4.6 Fibonacci Sequences

Strike Up the Band

To the organizer of a parade, a unit is a float, a motorcycle brigade, a marching band, a celebrity in an open car, etc. One rule for the organizer is that two marching bands cannot be adjacent to each other. How many ways can bands be arranged in a parade that involves n units?

Clarify the question

It doesn't say how many bands are to be included in the n units. How do we get started on this one? We are well served by the advice that tells us to begin with small cases, draw some diagrams, and try in this way to really understand the problem.

What would be the situation with a parade of one unit? Well, if there was a band, that would be it; and if there was some unit other than a band, that too would have to be it. Using B to represent a band, and N to represent a non-band, the two short parades (the principle does say to start small) could be represented by

B, N (2 ways when $n = 1$)

That was easy. Let's try a parade with two units. They can't both be bands. With one band, it can go first or second, and with no bands, there is no problem. These parades can be represented by

BN NB NN (3 ways when $n = 2$)

We are making sense of this, apparently, without knowing ahead of time how many bands there are in the n units. With growing confidence, we march into the case of three units. The only way to fit two bands into this parade is to put them first and last, with a non-band between them. Otherwise there will have to just one, or perhaps no bands.

BNB BNN NBN NNB NNN (5 ways when $n = 3$)

Maybe we can drum up a pattern by looking at one more parade, this time with $n = 4$. Speaking of patterns, it behooves us to get some organization to the way we go at listing these parades. Let's consider first a parade that begins with a band. It must have a non-band in second place, leaving two more spots to be filled. Since the non-band in second place imposes no restrictions on the next two spots, we can after BN list all the cases we noted for $n = 2$: BNBN, BNNB, BNNN. That takes care in an organized way of all the four unit parades that begin with a band. Now what about those that begin with a non-band. Since the lead unit in that case imposes no restrictions on what follows, we may list after the initial N all the permissible arrangements of a parade of three units. In this way, listing first the parades that begin with a band, and those that begin with a non-band, we get

BNBN, BNNB, BNNN, NBNB, NBNN, NNBN, NNNB,NNNN
(8 ways when $n = 4$)

Any mathematician who notices a sequence starting with the numbers 2, 3, 5, 8, . . . will think immediately of the Fibonacci sequence, a sequence that is famous precisely because it turns up in so many places. Study of this sequence is associated with Leonardo Fibonacci, a mathematician of the middle ages who purportedly considered the following problem about rabbits

A pair of rabbits is mature enough to reproduce another pair after 2 months and will do so every month thereafter. If each new pair of rabbits has the same reproductive habits as its parents and none of them dies, how many pairs will there be at the end of 4 months? Thirty months? n months?

The Rabbit Habit

Number of months	Solid figures represent mature pairs	Number of pairs
1		1
2		1
3		2
4		3
5		5
6		8

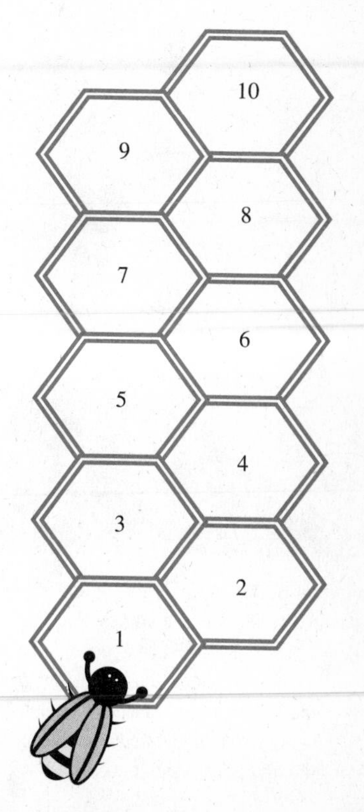

The results are shown in the table on page 130. Descending down its right side is the Fibonacci Sequence.

Suppose we are asked how many ways a bee can crawl over the hexagonal cells shown at the right. Naturally he can crawl only to an adjacent cell, and it is assumed that he insists that each move be a step up in the world. With these constraints, there is only one way he can reach cell 2, but there are two ways to reach cell 3 (directly from cell 1, or by way of cell 2). How many paths are there to cell 4? To cell 5?

Table 4-5	
n	f_n
1	1
2	1
3	2
4	3
5	5
6	8
7	13
8	21
9	34
10	55
11	89
12	144
13	233
14	377
15	610
16	987
17	1597
⋮	⋮
30	832040

Make a chart showing the number of paths from cell 1 to cell n for $n = 2, 3, 4, 5, 6$. If you don't get stuck on this honey of a problem, you will see that once again you are being entangled in the Fibonacci web.

The Fibonacci Sequence was first studied by Leonardo Fibonacci in 1202. The reader who has decided that the next two numbers are 13 and 21 may already be caught in the web surrounding this intriguing sequence. Some enthusiasts have made a life-long hobby of studying this sequence; they have formed the Fibonacci Society, and they regularly publish the *Fibonacci Quarterly*.

Some Formulas

The crucial observation to make about the Fibonacci Sequence is that any term (after the first two) is the sum of the preceding two. In fact, we take this as our formal definition. The **Fibonacci Sequence** is the sequence $f_1, f_2, f_3, f_4, \ldots$ which has its first two terms f_1 and f_2 both equal to 1 and satisfies thereafter the recursion formula

$$f_n = f_{n-1} + f_{n-2}$$

With this formula we can gradually build up Table 4-5. If Fibonacci were really correct in his assumptions about rabbit reproduction, one pair would result in 832,040 pairs after 30 months.

Note again the difficulty with a recursion formula. To use it, we must calculate all the terms preceding the one we want. It works, but it is time-consuming. Is there an explicit formula for f_n? Yes. It was apparently discovered by the Swiss mathematician Daniel Bernoulli in 1724.

$$f_n = \frac{1}{\sqrt{5}}\left[\left(\frac{1+\sqrt{5}}{2}\right)^n - \left(\frac{1-\sqrt{5}}{2}\right)^n\right]$$

It is too complicated to be of much use except for theoretical purposes, and we have no intention of trying to prove it. Perhaps you would like to check it for $n = 1, 2, 3$.

We turn now to the problem of finding the sum of the first n Fibonacci numbers. Let

$$F_n = f_1 + f_2 + f_3 + \cdots + f_n$$

and let's see if we can find another formula. Observe that

$$\begin{aligned} F_1 &= 1 &&= 2 - 1 = f_3 - 1 \\ F_2 &= 1 + 1 = 2 &&= 3 - 1 = f_4 - 1 \\ F_3 &= 1 + 1 + 2 = 4 &&= 5 - 1 = f_5 - 1 \\ F_4 &= 1 + 1 + 2 + 3 = 7 &&= 8 - 1 = f_6 - 1 \end{aligned}$$

A pattern is emerging. It suggests the formula

$$F_n = f_{n+2} - 1$$

Leonardo Fibonacci (ca. 1170–1250) Leonardo of Pisa was perhaps the greatest mathematician of the Middle Ages. After traveling in North Africa and the Orient, he wrote *Liber Abaci* which collected a host of arithmetic and algebraic facts and introduced the Hindu-Arabic numeration system to western Europe.

which is further confirmed by nothing that

$$F_9 = 88 = 89 - 1 = f_{11} - 1$$

But of course we don't claim to have proved it, only to have accumulated some evidence for Research and Discussion at the end of this section.

The Golden Ratio

Surely one of the most remarkable properties of the Fibonacci Sequence is that the ratio of two consecutive terms is alternately larger or smaller than an important number, the so-called **golden ratio,** a number of great interest to the Greeks and symbolized by one of their letters, namely, ϕ (phi). Its value is

$$\phi = \frac{1 + \sqrt{5}}{2} = 1.618033989 \ldots$$

As n gets larger and larger, the ratio of f_{n+1} to f_n gets closer and closer to this value. This fact can be demonstrated using the explicit formula for f_n that we gave earlier.

The golden ratio takes its name from its progenitor, the golden rectangle, which is constructed as follows. Beginning with a square *GBCD* of side length 2, locate the midpoint *P* of *GB* as shown in the margin. Use center *P* and radius *PC* (which by the Pythagorean Theorem has length $\sqrt{5}$) to draw an arc, thus determining a point *O* on the extension of *GB*. Finally, locate *L* so that *GOLD* forms a rectangle, a so-called **golden rectangle** in which the ratio of the length to the width is the aforementioned golden ratio ϕ. The golden rectangle enchanted the Greeks and appears over and over in ancient architecture and art, one prominent example relating to the dimensions of the Parthenon. That the golden rectangle has genuine geometric significance can be seen by studying a modern geometry book, for example, H. S. M. Coxeter, *Introduction to Geometry* (New York: Wiley, 1962).

Table 4-6

n	f_n	f_{n+1}/f_n
1	1	1
2	1	2
3	2	1.3
4	3	1.66
5	5	1.60
6	8	1.625
7	13	1.615
8	21	1.619
9	34	1.618
10	55	1.618

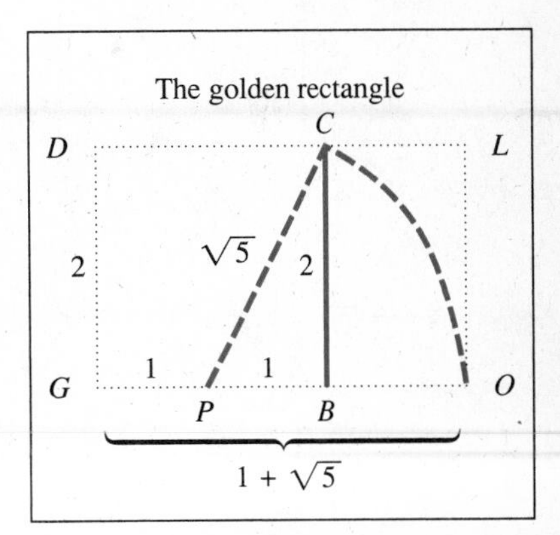

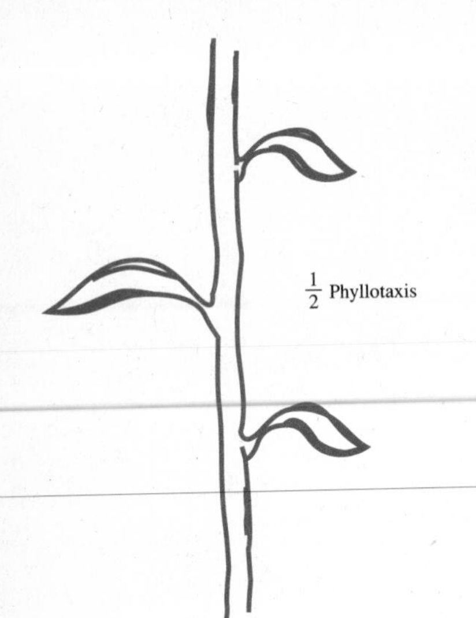

Botanical Applications

The Fibonacci numbers appear mysteriously in nature in a phenomenon called **phyllotaxis** (literally "leaf arrangement"). The leaves of many plants spiral around and up the branch or stem. On the branches of certain trees (e.g., elm and basswood), the leaves occur alternately on one side and then the other, that is, leaves occur at one-half turns around the branch. We call this $\frac{1}{2}$ phyllotaxis. In other trees (e.g., beech and hazel), the leaves occur at one-third turns around a branch ($\frac{1}{3}$ phyllotaxis). Oak and apricot exhibit $\frac{2}{5}$ phyllotaxis, poplar and pear $\frac{3}{8}$, willow, and almond $\frac{5}{13}$, etc. The remarkable fact is that nature displays a fascinating tendency to pick ratios that involve Fibonacci numbers.

Another example can be found in the heads of certain flowers, especially the sunflower. The seeds are distributed over the head in spirals, one set unwinding clockwise and the other counterclockwise. If one counts the numbers of these two kinds of spirals, they are almost always Fibonacci numbers. Small heads may have 13 : 21 or 21 : 34 combinations, large heads 34 : 55, 55 : 89, or even 89 : 144 combinations.

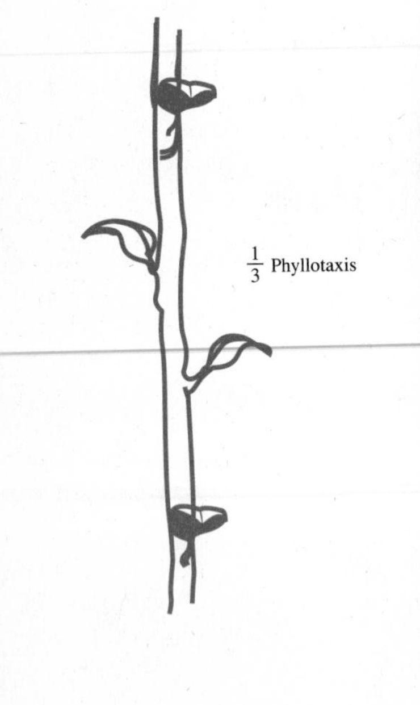

Summary

The Fibonacci Sequence has fascinated people in all walks of life since its discovery about 1200 A.D. It is determined by the recursion formula $f_n = f_{n-1} + f_{n-2}$, together with the condition that $f_1 = f_2 = 1$. Among its many occurrences are some, perhaps fanciful, relating to bees and rabbits; others, less artificial, are found in geometry and botany.

Problem Set 4.6

1. Figure f_{18}, f_{19}, f_{20} (Use Table 4-5 to save yourself some work).
2. Let $e_n = e_{n-1} + e_{n-2}$, but with $e_1 = 0$, $e_2 = 1$. Find e_3, e_4, e_5, e_6, e_7, e_8, e_9. How does this sequence compare with the Fibonacci Sequence?
3. Recall that $F_n = f_1 + f_2 + \cdots + f_n$. Thus $F_2 = f_1 + f_2 = 1 + 1 = 2$ and $F_3 = f_1 + f_2 + f_3 = 1 + 1 + 2 = 4$. Find F_4, F_5, F_6, F_7, F_8, F_9. Do your results agree with the formulas $F_n = f_{n+2} - 1$?
4. Let $E_n = e_1 + e_2 + \cdots + e_n$ (see Problem 2). Find E_4, E_5, E_6, E_7, E_8, E_9. Can you guess a formula for E_n?
5. Fill in the boxes.
   (a) $f_1 = \square$
   (b) $f_1 + f_3 = \square$
   (c) $f_1 + f_3 + f_5 = \square$
   (d) $f_1 + f_3 + f_5 + f_7 = \square$
   (e) $f_1 + f_3 + \cdots + f_{2k-1} = \square$
6. Fill in the boxes.
   (a) $f_2 = \square$
   (b) $f_2 + f_4 = \square$
   (c) $f_2 + f_4 + f_6 = \square$
   (d) $f_2 + f_4 + f_6 + f_8 = \square$
   (e) $f_2 + f_4 + \cdots + f_{2k} = \square$
7. **The Bee Line**
   A male bee has only a mother; a female has both a mother and a father. Thus Buzzy has one parent, two grandparents, three third-order parents, etc. How many tenth-order parents does he have?

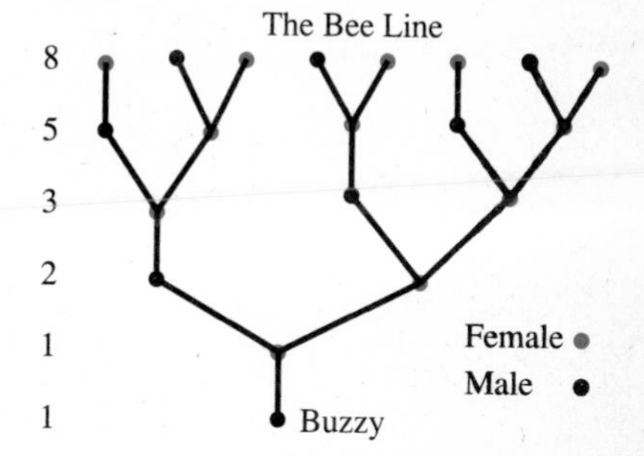

8. Look at the bee line again (Problem 7)
   (a) Write the first few terms in the sequence that gives the number of male ancestors at each level above Buzzy.
   (b) Do the same for female ancestors. What do you notice?
9. The bee line problem is based on some big assumptions that were not explicitly stated. Can you identify some of them?
10. Fill in the boxes.

$$f_n: 1, 1, 2, 3, 5, 8, 13, 21, \ldots$$
$$(f_n)^2: 1, 1, 4, \square, \square, \square, \square, \square, \ldots$$
$$(f_n)^2 + (f_{n+1})^2: 2, 5, 13, \square, \square, \square, \square, \square, \ldots$$
$$f_{2n+1}: 2, 5, 13, \square, \square, \square, \square, \square, \ldots$$

Guess at a relationship.

11. Fill in the boxes.
$1^2 + 1^2 = 1 \cdot 2$
$1^2 + 1^2 + 2^2 = 2 \cdot 3$
$1^2 + 1^2 + 2^2 + 3^2 = \square \cdot \square$
$1^2 + 1^2 + 2^2 + 3^2 + 5^2 = \square \cdot \square$
Conjecture a formula for $f_1^2 + f_2^2 + \cdots + f_n^2$.

12. Fill in the boxes.
$2 \cdot 1 - 1^2 = 1$
$3 \cdot 1 - 2^2 = -1$
$5 \cdot 2 - 3^2 = \square$
$8 \cdot 3 - 5^2 = \square$
$13 \cdot 5 - 8^2 = \square$
Conjecture a formula for $f_{n+1}f_{n-1} - (f_n)^2$.

13. The Lucas Sequence is determined by $g_1 = 1$, $g_2 = 3$, and $g_n = g_{n-1} + g_{n-2}$.
(a) Write out the first 10 terms for g_n.
(b) Fill in the boxes.
f_n: 1, 1, 2, 3, 5, 8, . . .
g_n: 1, 3, 4, 7, 11, 18, . . .
$f_{n-1} + f_{n+1}$: ?, 3, 4, $\square$, $\square$, $\square$, . . .
f_{2n}: 1, 3, 8, $\square$, $\square$, $\square$, . . .
$f_n g_n$: 1, 3, 8, $\square$, $\square$, $\square$, . . .
Conjecture two relationships.

14. Let $h_1 = a$, $h_2 = b$, and $h_n = h_{n-1} + h_{n-2}$. We call h_n a generalized Fibonacci Sequence.
(a) Write out the first 10 terms of h_n.
(b) Show that $(h_1 + h_2 + \cdots + h_{10}) = 11h_7$.

15. Each of the following is a generalized Fibonacci Sequence. Fill in the boxes.
(a) 9, 10, $\square$, $\square$, . . .
(b) 6, $\square$, 14, $\square$, . . .
(c) 11, $\square$, $\square$, 43, . . .

16. A sequence whose terms are the sum of the preceding three is somewhat facetiously called a "Tribonacci" sequence.
(a) If the first three terms are all 1, find the next five terms.
(b) Write the recursion formula in mathematical symbols.

17.* Let $k_1 = 1$ and $k_n = 1 + 1/k_{n-1}$. Write out the first eight terms of k_n. Do you observe anything?

18.* Let $\phi = (1 + \sqrt{5})/2$ be the golden ratio. Show that $\phi = 1 + 1/\phi$ and that $\phi^2 - \phi - 1 = 0$.

For Research and Discussion

It is one thing to guess at a result and quite another to demonstrate rigorously that it is correct. Try to prove each of the facts below. One of your main tools will be Mathematical Induction. This is really a form of deduction and is one of the basic tools of mathematicians. Let P_n be a statement about the counting number n. If P_1 is true and if the truth of $P_1, P_2, \ldots, P_k$ implies the truth of P_{k+1}, then P_n is true for every counting number n.

(a) $F_n = f_1 + f_2 + \cdots + f_n = f_{n+2} - 1$

(b) $f_1 + f_3 + \cdots + f_{2n-1} = f_{2n}$ (Problem 5)

(c) $f_2 + f_4 + \cdots + f_{2n} = f_{2n+1} - 1$ (Problem 6)

(d) $f_{2n+1} = (f_n)^2 + (f_{n+1})^2$ (Problem 9)

(e) $(f_1)^2 + \cdots + (f_n)^2 = f_n f_{n+1}$ (Problem 10)

(f) $f_{n+1}f_{n-1} - (f_n)^2 = (-1)^n$ (Problem 11)

(g) $(f_n)^3 + (f_{n+1})^3 - (f_{n-1})^3 = f_{3n}$

(h) $f_n = \dfrac{1}{\sqrt{5}}\left[\left(\dfrac{1 + \sqrt{5}}{2}\right)^n - \left(\dfrac{1 + \sqrt{5}}{2}\right)^n\right]$

(i) $\dfrac{f_{n+1}}{f_n} \to \dfrac{1 + \sqrt{5}}{2}$ as $n \to \infty$

(j) $(f_{n+2})^2 - (f_{n+1})^2 = f_n f_{n+3}$

Suggested Reading

For more information on the Fibonacci numbers, we recommend the *Fibonacci Quarterly* published by the Fibonacci Association, especially the February 1963 and April 1963 issues. To learn more about the golden ratio, see Martin Gardner's column, "Mathematical Games," *Scientific American,* August 1959, and David Bergamini and the editors of *Life, Mathematics,* Time-Life Science Library, New York, 1963.

chapter 5

Systematic Counting

The deriving of shortcuts from basic principles covers some of the finest achievements of the greatest mathematicians.

M. H. A. NEWMAN

Imagination

Did anyone get that last problem? The class falls silent; the mathematics teacher waits hopefully. Slowly, with a mixture of uncertainty and pride based on hope, a student not usually at the head of the class when tests are returned admits to having what seems to be a solution. The class is clearly skeptical as the student moves toward the board to describe the proposed solution.

Starting off in an unlikely direction not even considered by those better steeped in the methods in vogue in the course, the work seems almost unrelated to the question. But then suddenly, like a flickering coal that suddenly bursts into a bright flame, the approach puts the matter into a new light where everything is seen clearly. "How," ask the admiring few who really tried hard to solve the problem, "did you ever think of doing that"

How indeed? The question has been put to mathematicians through the ages. Answers are seldom more satisfying than the one offered by Newton who explained that he was able to solve problems by continually thinking about them until finally the solution was apparent.

Imagination is, apparently, a necessary ingredient in the makeup of anyone who is a successful problem solver.

> "To raise new questions, new possibilities, to regard old problems from a new angle, requires creative imagination and marks real advance in science."
>
> Leopold Infeld and Albert Einstein
> *Evolution of Physics,* page 95

In his *Philosophical Dictionary,* Voltaire wrote that ". . . there was far more imagination in the head of Archimedes than in that of Homer." Havelock Ellis, not a mathematician himself, says in *The Dance of Life* that "it is here [that is, in mathematics] that the artist has the fullest scope for his imagination."

This being true, it is surprising that so many people, in other ways well-educated, think of mathematics as a discipline that calls for such things as concentration, rigor, ability to proceed methodically, a knack for manipulating abstract symbols, . . . but not imagination. This is no doubt one indication of the way in which mathematicians fail to communicate to outsiders those aspects of their work that can make it interesting.

Mathematics is a wonderful context in which to see and appreciate the imagination of others. Questions can be asked which are clear enough so that we can see what needs to be done, and we can proceed to try doing it. Then, when someone comes along, uses an approach we never thought of, and obtains an answer that we can see is correct, our admiration knows no bound.

Several years ago, at the annual meet of the American Regions Mathematics League which traditionally brings together about 1500 of the most mathematically talented students in the secondary schools of the United States, the contestants had been as usual challenged with eight questions; and as usual, the participants were complaining to each other between rounds about the difficulty of the questions, about the fact that the ordinary techniques they all had mastered seemed wholly inadequate to solve such problems.

Then something unusual for these competitions happened. When the papers had been collected and the answers given, the person in charge asked if anyone had turned in all eight correct answers. This drew laughs from most of the people in the large auditorium, but then people began to point, and we saw that there was, incredibly, someone indicating that he had been able to solve all eight. There was stunned silence as those present, all of whom had tried and tried hard, began to grasp the fact that someone had been able to do such a thing. There followed the kind of standing ovation that is reserved for those few occasions when we believe we have witnessed a truly marvelous performance.

Just as a course in music appreciation is offered with the intent of helping people to understand and appreciate the work of gifted composers, so a course in mathematics can be designed to help you understand and appreciate the imaginative approach that some gifted people have been able to exercise in approaching an interesting problem. You should not be discouraged or feel intimidated by the realization that you would never have thought of some solution, any more than you would feel that way after listening to a wonderful symphony or seeing a great sculpture.

It is a mark of the educated person, to be able to recognize and appreciate the work of an imaginative artist, whatever his or her venue may be.

5.1 Principles of Counting

A Course for Amy

As she selects courses for her sophomore year, Amy is considering a major in either English or History. She decides that she will take one course in each of these departments. The college timetable lists five courses in the English Department that would be open to her, and four such courses in History. Of these, two of the English courses conflict with two of the history courses. In how many ways might Amy choose two courses?

Organize the information.

A	A	A	
b	c	d	
B	B	B	
a	c	d	
C	C	C	C
a	b	c	d
D	D	D	D
a	b	c	d
E	E	E	E
a	b	c	d

The numbers in the case of Amy's problem seem small enough so that we need not begin with a smaller case. What is needed is a methodical way to list the possibilities, and for that we need identifying labels for the courses. Let the English courses be A, B, C, D, E and the History courses be a, b, c, d, with designations chosen so that it is A that conflicts with a, B that conflicts with b.

Now if Amy selects A from English, then her choices in History are b, c, or d. A similar list of three possibilities occurs if Amy chooses B.

Things change if Amy selects C; there are no conflicts, making four choices available in history. Choices D or E in English similarly open four possibilities in History.

We have solved our problem; Amy can choose her courses in $6 + 12 = 18$ ways. It remains, in the words of our problem-solving heuristics, to reflect on what we have done.

The problem naturally fell into two cases: (1) choose an English course that conflicts with a History course; (2) choose an English course that does not conflict with a History course. In either case, once the English course is chosen, it's all History. In case (1), there were two ways to choose the English course, after which there were three ways to choose the History course, giving six different schedules. In case (2), there were three ways to choose the English course, after which there were four ways to choose the History course, giving us twelve more schedules. Both cases illustrate our first principle of counting.

Can we observe a general method?

MULTIPLICATION PRINCIPLE Let M be a set of m ways in which a first task can be accomplished, and let N be a set of n ways in which a second task can be accomplished. Then the two tasks together, the first followed by the second, can be accomplished in mn ways.

We digress to illustrate this fundamental principle. A person in St. Paul wishing to visit Purdue University will in all likelihood take one of five nonstop flights to Indianapolis, from there: the choices are: be picked up by your host, catch a connecting flight into West Lafayette, take the shuttle bus, or rent a car. With 5 options for getting to Indianapolis followed by 4 options for finishing the trip, the traveler has $5 \times 4 = 20$ ways to make the trip.

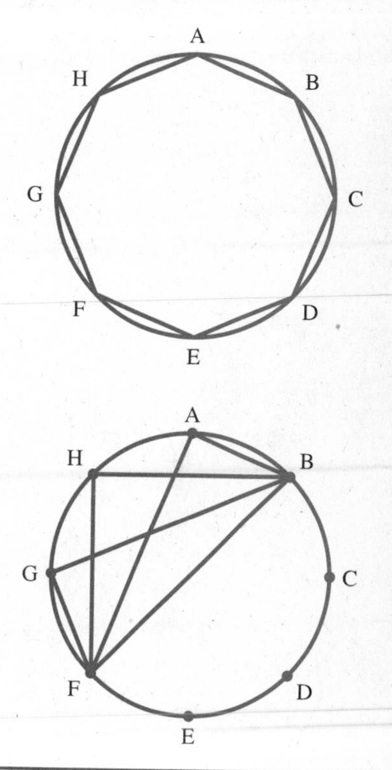

How many distinct right triangles can be inscribed in the circle in the margin using the vertices of the regular octagon as vertices? Begin by recalling that a right triangle inscribed in a circle will always have a diameter as the hypotenuse. We can choose a diameter in 4 ways. Then choose a third vertex; this can be done in 6 ways. We can, by the multiplication principle obtain $4 \times 6 = 24$ distinct right triangles.

Minnesota license plates consist of three letters, followed by three digits. How many cars can be licensed using this system? The multiplication principle, extended in the obvious way to a six step process, gives $26 \times 26 \times 26 \times 10 \times 10 \times 10 = 17{,}576{,}000$, enough to keep Minnesota's wheels rolling for years to come.

Back to our reflections on the solution to Amy's scheduling problems. Cases (1) and (2) are mutually exclusive; that is, they cannot both occur. Case (1) led to 6 schedules, case (2) to 12. Since Amy could do either, her total choices were counted by adding, and this is our second principle.

ADDITION PRINCIPLE Let M be a set of m ways in which a first task can be accomplished, and let N be a set of n ways in which a second task can be accomplished. Then if there is no way to accomplish them simultaneously, that is if $M \cap N = \varnothing$, then one may do either the first task or the second in $m + n$ ways.

We illustrate this principle by returning to the traveler from St. Paul to Purdue, assuming now that time is of the essence, forcing the decision to fly all the way. Of the five flights to Indianapolis, just two have reasonable connecting flights to West Lafayette. Another option is to take one of the four flights to Chicago that has connecting flights. In

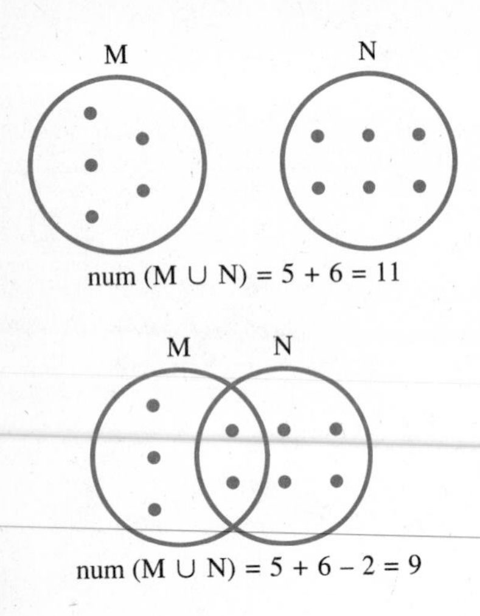

this case, there are two ways to proceed through Indianapolis, four ways through Chicago. The ways are mutually exclusive; the traveler has a total of $2 + 4 = 6$ options.

Attention must be paid in using the addition principle to the requirement that $M \cap N = \varnothing$. If there are three dinner flights and five evening flights from St. Paul to Chicago, can we conclude that we are talking about $3 + 5 = 8$ flights? No, because a dinner flight may also be counted among the evening flights. Without more information, we can't tell the total number of flights. The addition principle does not apply.

Suppose, however, that we know that two of the dinner flights are also evening flights. The number of flights is then $3 + 5 - 2 = 6$. Let num(S) be the number of members in set S. We are using here the fact that if M and N are sets that have elements in common, that is sets for which $M \cap N \neq \varnothing$, then num $(M \cup N) = \text{num}(M) + \text{num}(N) - \text{num}(M \cap N)$. This is our final principle of counting.

INCLUSION-EXCLUSION PRINCIPLE Let M be a set of m ways in which a first task can be accomplished, and let N be a set of n ways in which a second task can be accomplished. If there are some ways to accomplish both tasks simultaneously, that is if $M \cap N \neq \varnothing$, then one may do either the first task or the second in $m + n - \text{num}(M \cap N)$ ways.

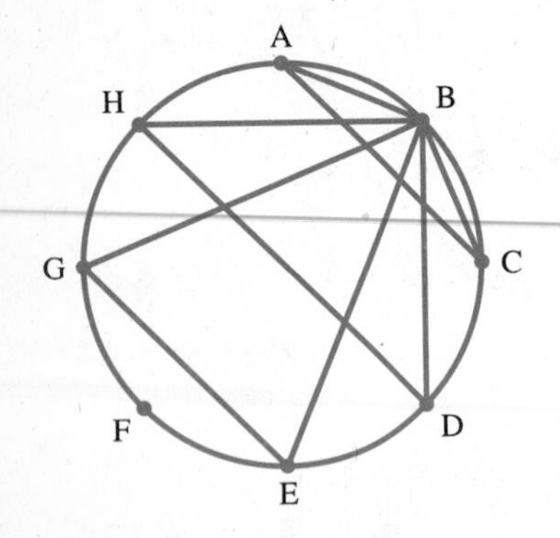

The case in which we choose B, then draw three isosceles triangles for which it is the apex.

We illustrate this principle first by returning to the vertices of a regular octagon located on the circumference of a circle. This time we ask how many triangles having these points as vertices are either right triangles or isosceles triangles (triangles with two equal sides.) Let M be the set of ways to form right triangles; we saw above that $m = 24$. Let N be the set of ways to form isosceles triangles. Find n by choosing any point (8 ways) and noting that this point is the apex (the point where the equal sides meet) of three distinct isosceles triangles. Thus, $n = 8 \times 3 = 24$.

We're not through, however, At each point, one of the isosceles triangles (the middle one in the drawing in the margin) is also a right triangle. Thus, $\text{num}(M \cup N) = 8$, and by the inclusion-exclusion principle, the number of triangles that are either right or isosceles is $24 + 24 - 8 = 40$.

To Add or to Multiply

Let us recapitulate. Suppose that M is a set of m ways that a first task can be accomplished and that N is a set of n ways that a second task can be accomplished. If the occurrence of the first event does not change the number of ways that the second event can occur, then the accomplishment of *both* the first *and* second task can occur in mn ways. If M and N are disjoint sets, that is if there is no way that both tasks can be accomplished simultaneously, then *either* one *or* the other can be accomplished in $m + n$ ways. The rule of thumb is that we multiply when interested in accomplishing *both . . . and;* we add when interested in accomplishing *either . . . or.*

Suppose we are to select a president, vice-president, and secretary from a class consisting of 5 boys and 4 girls. If there are no restrictions as to sex, this can be done in $9 \cdot 8 \cdot 7 = 504$ ways. But suppose the president is to be of one sex and the other two officers of the opposite sex. That is, suppose we are to select a female president and male vice-president and male secretary or a male president and female vice-president and female secretary. This can be done in

$$4 \cdot 5 \cdot 4 + 5 \cdot 4 \cdot 3 = 140$$

ways (see diagram in the margin).

Pres.	VP	Sec.
4 choices	5 choices	4 choices
F	M	M

or

5 choices	4 choices	3 choices
M	F	F

The Factorial Symbol

In how many ways can n names be placed in a list? The multiplication principle gives us an immediate answer. We multiply the number n of choices for the first position, times the number $(n - 1)$ of choices for the second position, continuing on down to the factor of 1 corresponding to the number of choices for the last position on the list.

The product of the integers from 1 to n comes up repeatedly in counting problems and is given a special name, ***n* factorial,** and a special symbol, $\boldsymbol{n!}$. Thus,

$$n! = n(n - 1)(n - 2) \cdots 2 \cdot 1$$
$$7! = 7 \cdot 6 \cdot 5 \cdot 4 \cdot 3 \cdot 2 \cdot 1$$

Factorials are tiresome to calculate; so we have provided the table at right. Note that we have defined 0! to be 1, a convenience for later work. Also observe how rapidly the sequence $a_n = n!$ grows. Just as geometric growth ultimately surpasses arithmetic growth, so factorial growth will ultimately conquer geometric growth. For example, $n!$ will for large enough n exceed 100^n. Can you see why?

It is important for us to be able to manipulate factorials with ease. Be sure you understand the following calculations.

$$\frac{6!}{4!} = \frac{6 \cdot 5 \cdot \not{4} \cdot \not{3} \cdot \not{2} \cdot \not{1}}{\not{4} \cdot \not{3} \cdot \not{2} \cdot \not{1}} = 6 \cdot 5 = 30$$

$$\frac{8!}{5!} = \frac{8 \cdot 7 \cdot 6 \cdot \not{5} \cdot \not{4} \cdot \not{3} \cdot \not{2} \cdot \not{1}}{\not{5} \cdot \not{4} \cdot \not{3} \cdot \not{2} \cdot \not{1}} = 8 \cdot 7 \cdot 6 = 336$$

$$\frac{n!}{(n-2)!} = \frac{n(n-1)(n-2)(n-3) \cdots 3 \cdot 2 \cdot 1}{(n-2)(n-3) \cdots 3 \cdot 2 \cdot 1} = n(n-1)$$

$$\frac{9!}{4!\,5!} = \frac{9 \cdot 8 \cdot 7 \cdot 6 \cdot \not{5} \cdot \not{4} \cdot \not{3} \cdot \not{2} \cdot \not{1}}{4 \cdot 3 \cdot 2 \cdot 1 \cdot \not{5} \cdot \not{4} \cdot \not{3} \cdot \not{2} \cdot \not{1}}$$

$$= \frac{9 \cdot \overset{2}{\not{8}} \cdot 7 \cdot \not{6}}{\not{4} \cdot \not{3} \cdot \not{2} \cdot \not{1}} = 9 \cdot 2 \cdot 7 = 126$$

Table 5-1

n	$n!$
0	1
1	1
2	2
3	6
4	24
5	120
6	720
7	5,040
8	40,320
9	362,880
10	3,628,800

On the other hand, the factorial symbol can be used to condense certain expressions:

$$7 \cdot 6 \cdot 5 \cdot 4 = \frac{7 \cdot 6 \cdot 5 \cdot 4 \cdot 3 \cdot 2 \cdot 1}{3 \cdot 2 \cdot 1} = \frac{7!}{3!}$$

$$2 \cdot 4 \cdot 6 \cdot 8 \cdot 10 = (2 \cdot 1)(2 \cdot 2)(2 \cdot 3)(2 \cdot 4)(2 \cdot 5) = 2^5 \cdot 5!$$

Summary

Three principles guide us in systematic counting. Suppose that M is a set of m ways to accomplish task I, and that N is a set of n ways to accomplish task II. Then

(1) the two tasks done in sequence, task I followed by task II, may be done in mn ways.
(2) if there is no way to accomplish the tasks simultaneously, that is, if $M \cap N = \varnothing$, then one may do either task I or task II in $m + n$ ways.
(3) if there are some ways to accomplish both tasks simultaneously, that is if $M \cap N \neq \varnothing$, then one may do either task I or task II in $m + n - \text{num}(M \cap N)$

Order Form

Salad

Pizza

Dessert

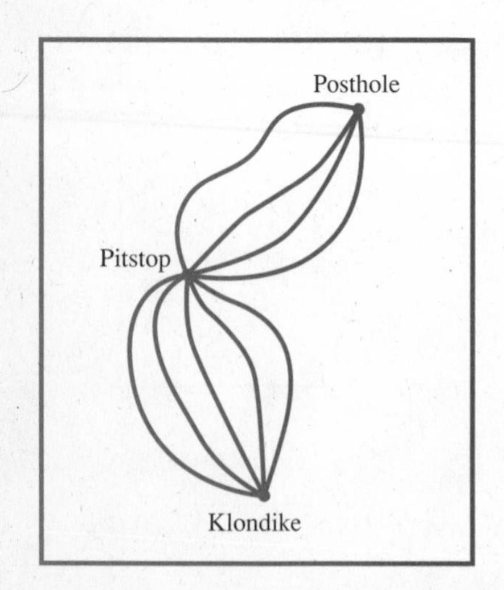

Problem Set 5.1

1. A pizzeria offers 3 choices of salad, 10 kinds of pizza, and 4 different desserts. How many different three-course meals can be ordered?

2. The Fair Game Dating Service has cards for 60 men and 80 women. How many different dates can it arrange?

3. Amy plans to drive from Posthole to Klondike but will make an intermediate stop at Pitstop to pick up her friend Beth. There are four roads from Posthole to Pitstop, and five from Pitstop to Klondike. How many different routes can she choose?

4. A club with 20 members wants to choose a president and a vice-president. In how many ways can this be done?

5. Homer has six bow ties, eight regular ties, and four shirts. How many different tie-shirt combinations can he wear?

6. A man can go from Springfield to Chicago by one of three busses or one of two trains. From Chicago to New York he will pick one of four air flights. How many different choices does he have for his trip from Springfield to New York?

7. There are five different roads from Podunk to Tubville. In how many ways can Homer make a round trip between the two cities? A round trip returning by a different road?

8. A combination lock with 10 digits on its face is opened by turning clockwise to a digit, counterclockwise to a second digit, and then clockwise to a third digit (but digits may be repeated, for example, 3-3-1). How many different combination locks of this type are possible?

9. The letters RATS are written one on each of four cards. How many code words can be made with these cards? Hint: Code words can have on letter, two letters, three letters, or four letters.

10. The letters STRONG are written one on each of six cards. How many code words can be made with these cards?

11. The Chicago Cubs have 25 players (3 catchers, 9 pitchers, 7 infielders and 6 outfielders). In how many ways can the manager fill the nine positions (assuming, for example, that an infielder can play any of the four infield positions)?

Cubs Lineup	
Catcher	
Pitcher	
1st base	
2nd base	
SS	
3rd base	
LF	
CF	
RF	

12. After the manager of the Cubs has chosen the nine players to start a game, he must turn in the batting order. How many different batting lineups are possible? How many if the pitcher bats last?

13. Telephone numbers consist of seven digits (e.g., 641-2256). Assuming any choice of digits can be used, how many different telephone numbers are possible? How many if the first digit cannot be 0?

14. A man wants to have a picture taken of himself with his four boys and two girls all lined up in a row. In how many ways can the photographer line them up if the man insists on being in the middle with a girl adjacent to him on either side?

15. How many three-digit numbers are there? (Be careful; the first digit cannot be 0). How many are there without repeated digits?

16. How many even three-digit numbers are there? How many of these have no repeated digits?

17. How many three-digit numbers are there ending in 3, 4, or 5 that are greater than 400? Greater than 432?

18. How many numbers greater than 5000, with no repeated digits, can be formed using only, 0, 2, 3, 4, 5, and 6?

19. Calculate the easy way (i.e., first cancel between the numerator and denominator).

(a) $\frac{7!}{2!}$ (b) $\frac{8!}{4!}$

(c) $\frac{7!}{4!\,3!}$ (d) $\frac{9!}{5!\,2!\,2!}$

20. Calculate:

(a) $\frac{8!}{5!}$ (b) $\frac{9!}{5!\,4!}$

(c) $\frac{100!}{98!}$ (d) $\frac{10!}{5!\,3!\,2!}$

21. Simplify:

(a) $\frac{(n-1)!}{n!}$ (b) $\frac{(n!)^2}{(n-1)!\,(n-2)!}$

22.* How many seating arrangements are possible for a class of 15 students if there are 30 chairs in a room, arranged in 6 rows of 5 each and bolted to the floor?

23.* In how many ways can nine different books be divided among three people
(a) If each person is to get three books?
(b) If each person is to get at least one book?
(c) If there are no restrictions on the number of books each person gets?

24. Minnesota license plates used to consist of two letters followed by four digits before being changed to the system described in this section of three letters followed by three digits. How many more plates can be produced using the new system.

Problems 25–28 give you a chance to experiment with your calculator. In each case, try $n = 5, 10, 15, \ldots$ and continue until you are convinced of the answer.

25. Which grows faster as n increases, $n!$ or 10^n?

26. Which grows faster as n increases, $n!$ or n^{10}?

27. Which grows faster, $n!$ or $(n/2)^n$?

28. Which grows faster, $n!$ or $(n/3)^n$?

29. In a survey of 317 people regarding preference in soft drinks, 155 liked Coke, 203 liked Pepsi, and 124 liked 7-Up. 87 liked both Coke and Pepsi, 60 liked Pepsi and 7-Up, and 72 liked Coke and 7-Up. There were 35 that liked all three. How many didn't like any?

30. In a survey of 42 students to determine interest in professional sports, 24 followed baseball, 22 followed football, and 19 followed basketball. 13 followed both baseball and football, 11 followed football and basketball, and 15 followed baseball and basketball. There were 9 that followed all three. How many didn't followed any? (We resisted asking how many were spoilsports.)

31.* How many integers between 100 and 800 are odd numbers having distinct digits?

32.* How many integers between 99 and 801 are even integers having distinct digits?

33.* A group of 5 young women decide to attend a conference on women in science. When they call the hotel, they find that only 3 rooms are left, one desirable, one acceptable, and one undesirable.
(a) In how many ways can they be assigned to the rooms if they take (and by inference) use all of them?
(b) Given that all 5 could stay in any one of the rooms, that they could just take two of the rooms, or that they could use all three of the rooms, in how many ways might they arrange themselves in the rooms?

For Research and Discussion

By examining a number of books in your library with titles like *Discrete Mathematics, Finite Mathematics, or Introductory Combinatorics,* you can identify certain classes of problems that are used to illustrate the principle of inclusion/exclusion. Find and solve several of these prototype problems.

5.2 Permutations

Piles for Pennies

Hoping to keep his young daughter Toni occupied for a while so he can read the paper, David tells her that he will give her a penny for every distinct way she can pile up her 8 blocks. If the blocks are identical, except that 2 are red, 3 are green, and 3 are white, how much is the big spender risking?

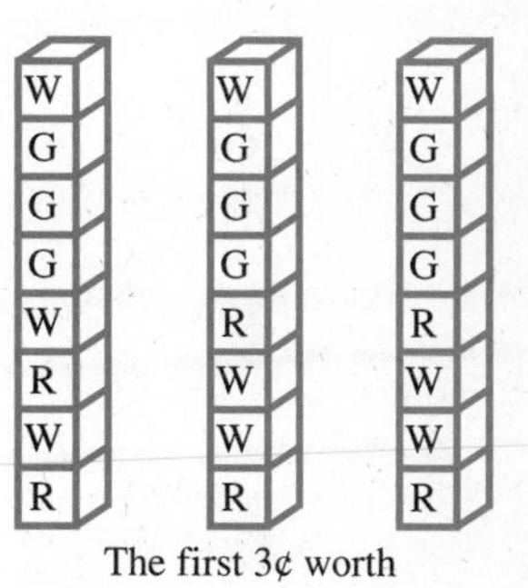

The first 3¢ worth

To permute a set of objects is to rearrange them. A permutation of a set of objects is just an arrangement of those objects in a particular order. Thus, H O M E R is one permutation of the letters {E, H, M, O, R}; R H M E O and M O R E H are two more. In the parlance of mathematicians who are not interested in what a particular ordering of the letters might spell, but are interested in counting all possible orderings (permutations) of the letters, each permutation, whether it makes sense or not, is called a word.

So—how many words can we make from the five letters of HOMER? We can consider this a problem of using the five letters at our disposal to fill five positions.

□ □ □ □ □

We can fill the first position in five ways. Having done this we can fill the second in four ways, etc. From the multiplication principle, the answer is $5! = 5 \cdot 4 \cdot 3 \cdot 2 \cdot 1 = 120$.

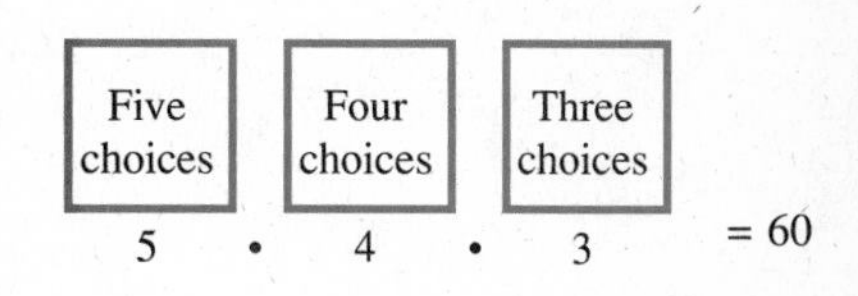

But now suppose we ask how many three-letter words we can make from the letters HOMER. Again we are faced with a problem of

filling positions, three of them. We can fill the first position in five ways, the second in four ways, and the third in three ways. The answer (by the multiplication principle) is $5 \cdot 4 \cdot 3 = 60$.

This is such a common problem that we introduce special language for it. When we have n distinguishable objects and we select r of them to arrange in a row, we call the resulting arrangement a **permutation of n things taken r at a time.** The question of interest is: How many of these permutations are there? The answer is designated by the symbol ${}_nP_r$. Thus ${}_5P_3$ denotes the number of permutations of five things taken three at a time; in particular, it represents the number of three-letter words that can be made from the five-letter word HOMER. It has the value

$${}_5P_3 = 5 \cdot 4 \cdot 3 = 60$$

Similarly,

$${}_6P_4 = 6 \cdot 5 \cdot 4 \cdot 3 = 360$$
$${}_{10}P_2 = 10 \cdot 9 = 90$$

In general, ${}_nP_r$ is the product of the numbers starting with n and working down until there are r factors. We can even give a formula:

$${}_nP_r = n(n-1)(n-2)\cdots(n-r+1)$$

You should check that this formula gives the right answer in each of the examples immediately above. Note that, when $r = n$,

$${}_nP_n = n(n-1)(n-2)\cdots 3 \cdot 2 \cdot 1 = n!$$

There is another formula for ${}_nP_r$ which we will need later. Let's see if we can work up to it by means of examples. Note that

$$_5P_3 = 5 \cdot 4 \cdot 3 = \frac{5 \cdot 4 \cdot 3 \cdot 2 \cdot 1}{2 \cdot 1} = \frac{5!}{2!}$$

$$_7P_3 = 7 \cdot 6 \cdot 5 = \frac{7 \cdot 6 \cdot 5 \cdot 4 \cdot 3 \cdot 2 \cdot 1}{4 \cdot 3 \cdot 2 \cdot 1} = \frac{7!}{4!}$$

The general formula that these special cases suggest is

$$_nP_r = \frac{n!}{(n-r)!}$$

It is correct, as you can verify by writing out both numerator and denominator and canceling.

Here is a good problem dealing with permutations. How many words of any length can be made from HOMER? We translate this into: How many five-letter words or four-letter words or three-letter words or two-letter words or one-letter words can be made? The answer (using the addition principle) is

$$\begin{aligned} &_5P_5 + {_5P_4} + {_5P_3} + {_5P_2} + {_5P_1} \\ &= 5 \cdot 4 \cdot 3 \cdot 2 \cdot 1 + 5 \cdot 4 \cdot 3 \cdot 2 + 5 \cdot 4 \cdot 3 + 5 \cdot 4 + 5 \\ &= 120 + 120 + 60 + 20 + 5 = 325 \end{aligned}$$

In how many ways can we scramble SCRAMBLE? The answer is $_8P_8 = 8 \cdot 7 \cdot 6 \cdot 5 \cdot 3 \cdot 2 \cdot 1$. How many four-letter words can be made from SCRAMBLE? We can make $_8P_4 = 8 \cdot 7 \cdot 6 \cdot 5$ or 1680 of them.

It is important to notice that the letters of HOMER and SCRAMBLE are all different. In fact, in our discussion thus far, we have assumed that we were talking about permutations of distinguishable objects. But what happens if some of the objects are alike? This is our next subject.

Permutations with Some Objects Alike

Having imagined a character named Homer, can you go further and imagine a girlfriend named HESTER AMANDA GOODROAD? It is a name more suited to illustrate concepts we have in mind than to please a modern young woman, but there are things to do with even the worst of names. Take HESTER, for example. With just the slightest bit of juggling, it becomes ESTHER which isn't really bad at all. And that leads us to the question we wanted to ask. How many permutations (words) can be made from HESTER using all six letters exactly once? If it weren't for the fact that there are two E's, it would be easy. The answer would be $6! = 720$. But there are two E's, and unfortunately HESTER and HESTER (the second has the two E's interchanged) spell exactly the same word.

HESTER

$\frac{6!}{2!}$ words

Let's pretend we can distinguish between the two E's. We could write one in black ink and the other in red but, better yet, let's tag the two E's with subscripts, so that we have E_1 and E_2. Now HE_1STE_2R and HE_2STE_1R look different. In fact, if we maintained this distinction, we could make $6! = 720$ different arrangements. But of course we have really counted every word twice corresponding to the interchange of the two E's. There are only 720/2 or 360 genuinely different words that can be made.

AMANDA

$\frac{6!}{3!}$ words

Take AMANDA next. Suppose we tag the three A's with subscripts so that we can distinguish between them. Then $A_1MA_2NDA_3$ can be arranged in $6! = 720$ different ways. But every word is counted $3! = 6$ times, because the letters $A_1A_2A_3$ can be arranged in $3 \cdot 2 \cdot 1$ ways. Thus the number of different six-letter words that can be made from AMANDA is

$$\frac{6!}{3!} = 6 \cdot 5 \cdot 4 = 120$$

GOODROAD

$\frac{8!}{2!3!}$ words

GOODROAD is even more interesting, for it has two D's and three O's. If we pretend that we can distinguish among the two D's and three O's, we can make 8! words. But of course we really can't, so we must divide 8! by 2! and by 3! to compensate for the repeated letters.

In each of the above cases, we supposed that each letter was available for use only the number of times it appeared in the given word. The words "permute" and "scramble" imply this. But what happens if each letter is available in unlimited supply? This is our final subject.

Arrangements with Unlimited Repetitions Allowed

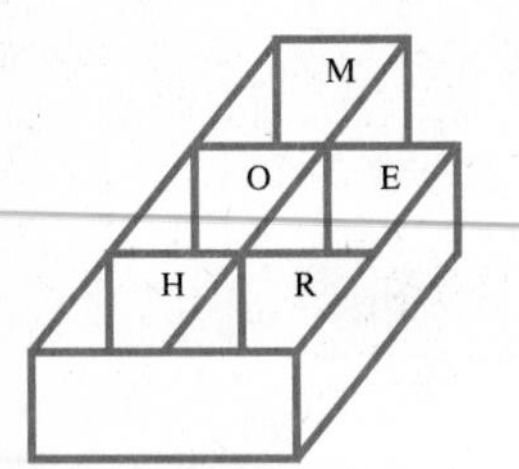

We suppose now that we have five bins filed with the letters H, O, M, E, and R, respectively. Thus we can make words like ROMER, or even RRRRR. How many five-letter words can we make altogether? Again we are faced with filling five positions. But after filling the first position with one of the five letters, we still have five choices for the second position, etc. By the multiplication principle, we can create $5 \cdot 5 \cdot 5 \cdot 5 \cdot 5 = 3125$ words. How many four-letter words can we make? Answer: $5 \cdot 5 \cdot 5 \cdot 5 = 625$.

The general situation is this. We have n items, each available in unlimited supply. How many different ordered sets of length m can we make? The answer is $n \cdot n \cdot n \cdots n$($m$ times); that is, it is n^m.

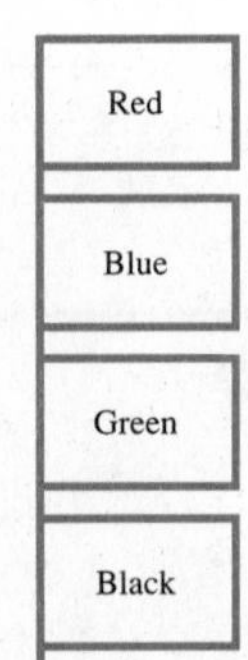

Flags run up on a ship have been used since ancient times to send signals. If plain flags are available in six colors, how many signals can be made up using just four of them at a time? We are faced with an immediate interpretation problem. Is there just one flag of each color? If so, it is a permutation problem and the answer is ${}_6P_4 = 6 \cdot 5 \cdot 4 \cdot 3 = 360$. Or are there several (at least four) flags of each color? In this case the answer is $6 \cdot 6 \cdot 6 \cdot 6 = 1296$.

This example nicely illustrates one of the chief difficulties in counting problems, namely, the difficulty in interpreting the problem correctly. And it reinforces the maxim of Section 1.1 Before you try to solve a problem, *be sure you know precisely what the problem is.* Until this is settled, you are just spinning your wheels.

Yet you should not only think of your problem in some vague way, you should face it, you should see it clearly, you should ask yourself; *What do you want?*

George Polya

Summary

We have considered three quite different arrangement problems in this section: (1) permutations of distinct objects, (2) permutations with some objects alike, and (3) arrangements with unlimited repetitions allowed. Let's put them together using the idea of bins. Suppose there is just one letter in each bin on page 148. How many three-letter words can we make? Equivalently, how many permutations of BAN are there? Answer: $3! = 6$. Suppose there are three A's, two N's, and one B in the bins. How many six-letter words can we make; equivalently, how many permutations are there of BANANA? Answer $6!/3!2! = 60$. Finally, suppose the bins are full of letters. How many four-letter words can we make? Answer: $3 \cdot 3 \cdot 3 \cdot 3 = 81$.

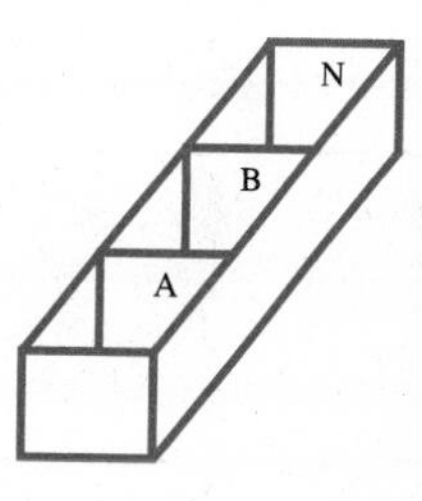

Problem Set 5.2

1. How many different words can be made using all the letters of
   (a) MEAT (b) CREAM (c) ORANGES
2. In how many ways can we permute the letters of
   (a) FORD (b) HORNET (c) MUSTANG
3. Suppose the letters MAVERICK are written on eight cards. How many four-letter words can be made? Five-letter words?
4. The letters PONTIAC are written on seven cards. How many six-letter words can be made? Five-letter words?
5. How many different words can be made using all the letters of
   (a) NIXON (b) HOOVER
   (c) KENNEDY (d) EISENHOWER
6. How many different words can be made using all the letters of
   (a) MAINE (b) KANSAS
   (c) WISCONSIN (d) MISSISSIPPI
7. How many words of all lengths can be made from PONIES if letters cannot be repeated?
8. How many three- or four- or five-letter words can be made from MONKEYS if letters cannot be repeated?
9. How many six-letter words can be made from CHEESE if:
   (a) Letters can be used only as often as they appear in CHEESE?
   (b) Letters can be used an unlimited number of times?
10. Homer has 12 poker chips; 5 are white, 4 are red, and 3 are blue. In how many different-looking stacks of 12 chips can he pile them?

11. If 10 horses are entered in a race, in how many ways can the first three places (win, place, and show) be taken?

12. Thirty people enter a contest offering scholarships to the first- and second-place winners. In how many ways can the two winners be chosen?

13. A coin is to be tossed 10 times. How many possible results are there? Hint: Think of this as making 10-letter words using H's and T's.

14. A die is to be tossed five times. How many possible results are there?

15. Homer has nine flags: four red, three green, and two white. How many nine-flag signals can he run up a pole?

16.* With the flags in Problem 15, how many seven-flag signals can Homer run up a pole?

17.* The World Series is won by the team that first gets four wins in a maximum of seven games. In a Dodgers-Twins series, count all possible outcomes (e.g., DDDD and DDTTDD are two possible ways for the Dodgers to win).

18. How many code words of length 10 or less can be made from the 26 letters of the alphabet if letters cannot be repeated in a word?

19. Answer the question of Problem 18 if letters can be repeated.

20. Solve the problem that opened this section.

For Research and Discussion

Investigate the number of telephone numbers that can be assigned under various realistic constraints. For example,

(a) at Macalester College, all extensions are identified by a four-digit number that begins with 6.

(b) No local seven-digit number ever begins with 0. Are there other constraints? Within the constraints, how many different telephone numbers can be assigned within an area code?

(c) all telephones in the greater metropolitan area of Chicago had the area code 312 until 1992 when suburban locations were given the code 708. What rule is used in deciding when to introduce a new area code?

(d) Futurists talk of the day when every citizen will have a small telephone carried at all times (like a watch) and a unique number, enabling us to dial up people rather than locations. Would the present system of ten-digit phone numbers (counting three area code digits) support such a system?

5.3 Combinations

Making a Short List

A search committee for a new City Manager has narrowed the field to eight applicants. The committee's charge is to submit to the City Council a list, in alphabetical order, of three candidates that they believe to be the best qualified. In how many ways might they up this final list?

Strip away irrelevant details.

One need not think too long about the question posed above before realizing that the requirement that the candidates be listed in alphabetical order is irrelevant. The question really is, in how many ways can three finalists be selected from eight candidates. After the names are chosen, they can be alphabetized by a machine.

Can we solve a related problem?

So the order in which we choose the three finalists does not matter. That's too bad, really, because if it mattered, we would know how to solve the problem. We can easily find the number of permutations of the three objects chosen from eight; it would be $8 \times 7 \times 6 = 336$.

A	A	G	G	T	T
G	T	T	A	A	G
T	G	A	T	G	A

The problem is that this counts not only the alphabetical listing of **A**dparks, **G**rownomore, and **T**axit; it counts all six ways of listing them. That's a key observation, however, for the same thing could be said for the selection of any three candidates. The selection gets counted not once, but six times. Evidently the answer to the question posed is $\frac{336}{6} = 56$.

Can we generalize, or develop a general method?

Now, does the method we have used suggest a general method for selecting r objects from n possibilities when order doesn't matter? It does! First count them as if order did matter (as if we were still in Section 5.2). That will give us ${}_nP_r$. Then observe that we will have counted $r!$ selections where we really wanted to count only one. Clearly the answer we wanted is $\dfrac{{}_nP_r}{r!}$.

An unordered subset of a given set of objects is called a **combination** of the objects. If we select r objects from a set of n distinguishable objects, the resulting subset is called a **combination of n things taken**

***r* at a time.** We use the symbol ${}_nC_r$ to denote the number of such combinations. Thus

$$ {}_{20}C_3 = \frac{{}_{20}P_3}{3!} \qquad {}_{20}C_4 = \frac{{}_{20}P_4}{4!} \qquad {}_{10}C_6 = \frac{{}_{10}P_6}{6!} $$

and, in general,

$$ {}_nC_r = \frac{{}_nP_r}{r!} = \frac{n(n-1)(n-2)\cdots(n-r+2)(n-r+1)}{r(r-1)(r-2)\cdots 2\cdot 1} $$

The formula above is both correct and useful. We remember how to use it by the following device. First write out the denominator. Then fill in the same number of factors in the numerator, starting with n and working down. Finally, cancel common factors, Thus

$$ {}_{15}C_4 = \frac{15 \cdot \overset{7}{\cancel{14}} \cdot 13 \cdot \cancel{12}}{\cancel{4} \cdot \cancel{3} \cdot \cancel{2} \cdot 1} = 15 \cdot 7 \cdot 13 = 1365 $$

and

$$ {}_{10}C_6 = \frac{10 \cdot \overset{3}{\cancel{9}} \cdot \cancel{8} \cdot 7 \cdot \cancel{6} \cdot \cancel{5}}{\cancel{6} \cdot \cancel{5} \cdot \cancel{4} \cdot \cancel{3} \cdot \cancel{2} \cdot \cancel{1}} = 210 $$

For theoretical purposes, another formula is very useful. Recall from Section 5.2 that

$$ {}_nP_r = \frac{n!}{(n-r)!} $$

Thus

$$ {}_nC_r = \frac{{}_nC_r}{r!} = \frac{n!}{r!\,(n-r)!} $$

We turn to several applications. Keep in mind that combinations are just subsets of a given collection of objects, subsets in which the order of the objects does not matter. A class of 15 students wants to choose a social committee consisting of 4 of its members. In how many ways can this be done? Is order important in this problem? No. Therefore the answer is

$$ {}_{15}C_4 = 1365 $$

(see calculation above). The same class wishes to choose a president, vice-president, secretary, and treasurer. In how many ways can this be done? Is order important? Yes. Now there are four specific slots to fill. Therefore the answer is

$$ {}_{15}P_4 = 32{,}760 $$

A 4-card hand is to be dealt from a suit of 13 cards. In how many ways can this be done? Is order important in a hand of cards? No. It is the cards that are held, not the order in which they are held, that is significant. Thus the answer is

$$_{13}C_4 = \frac{13 \cdot \not{12} \cdot 11 \cdot \overset{5}{\not{10}}}{\not{4} \cdot \not{3} \cdot \not{2} \cdot 1} = 715$$

Permutations and Combinations

If asked to count the number of ways that r things can be selected from $n > r$ possibilities, one decides whether it is a matter of counting permutations or combinations by asking whether the order of selection matters or not. This sometimes takes a little care, as in the case of our lead problem in this section where the requirement of alphabetical listing appears as a possible distracter, but there is usually no great difficulty in answering the question. There are problems that involve both permutations and combinations, however, and these can cause confusion. We conclude this section with two examples where alternate procedures may occur to different people.

In how many ways might you appoint a committee of 5, designating a chair and a secretary, from a group of 12 people? One person might reason that there would be $_{12}C_5$ ways to appoint the committee, after which there would be $_5P_2$ ways to designate a chair and secretary. A second person might reason that you could begin by choosing a chair and secretary in $_{12}P_2$ ways, after which you could choose the other 3 committee members in $_{10}C_3$ ways. The happy fact that you can check is that $_{12}C_5 \cdot {_5P_2} = {_{12}P_2} \cdot {_{10}C_3}$.

Let's try one more. A certain district in Minnesota was entitled to send 4 delegates to the Democratic Party Convention. The 15 people who showed up at the caucus were divided into three factions of 5 people each on a key issue, so it was decided that each faction would get to choose one delegate, and that the fourth position would be decided by lot. How many distinct delegations of 4 might be chosen in this way.

One solution might argue that there are 5 ways for each faction to choose a delegate, meaning there are 5^3 ways to choose the first three delegates. The fourth delegate can then be chosen from among the remaining 12, giving a total of $(125)(12) = 1500$ possible ways to choose a delegation.

A second solution might reason that in the end, one faction will have two delegates, and that they can be chosen in $_5C_2 = 10$. Since this faction can be chosen in 3 ways, and since the other two factions can choose their delegate in 5 ways, the delegation can be chosen in $3 \cdot 10 \cdot 5^2 = 750$ ways.

Clearly these answers are not the same. What is wrong? It may take you a while to discover that the first method, the one that seems so straightforward, is counting delegations more than once. See it this way.

Let the factions be {A, B, C, D, E}, {K, L, M, N, P}, and {V, W, X, Y, Z}. Think now about listing all 125 delegations of three that might result from the step of having each faction select a delegate. In this list you would find AKV, AKW, and 123 others. Now when, starting with AKV, you list the 12 delegations that might be formed by the casting of lots, AKVW will be listed. Unfortunately, when you list the 12 possible delegations that begin with AKW, the same delegation will be listed again, this time as AKWV.

Reflect on what you have done.

Counting is an area of mathematics where it is particularly important to heed the advice of Chapter 3. Do look for several ways to solve a problem so that you can compare the answers. Do try your method of solution on an example small enough to allow listing all the answers as a check.

$$_nP_r = \frac{n!}{(n-r)!}$$

$$_nC_r = \frac{n!}{(n-r)!\,r!}$$

Summary

The number of ways to select r items from n choices is called a

permutation $_nP_r$ if the order of selection matters;

combination $_nC_r$ if the order of selection does not matter.

It is always a good idea to reflect on your answers. It is a particularly good idea in counting problems.

Problem Set 5.3

1. Recall the easy way to calculate $_nC_r$. First write out $r!$ in the denominator. Then put the same number of factors in the numerator, starting with n and working down. Cancel and calculate. Thus

$$_8C_3 = \frac{8 \cdot 7 \cdot \not{6}}{\not{3} \cdot \not{2} \cdot 1} = 56$$

Evaluate each of the following.

(a) $_{10}C_3$ (b) $_9C_4$ (c) $_{15}C_3$
(d) $_{100}C_2$ (e) $_8C_5$ (f) $_8P_5$

2. Evaluate:

(a) $_7C_3$ (b) $_5C_4$ (c) $_{50}C_3$
(d) $_{11}C_2$ (b) $_9C_3$ (f) $_9P_3$

3. The local union wants to select a committee of 3 to represent it in contract negotiations. If there are 50 people in the union, in how many ways can this be done?

4. The faculty of 30 at Podunk University plans to select a 4-member educational policies committee. In how many ways can it make its selection?

5. From a penny, a nickel, a dime, a quarter, and a half-dollar, how many different sums can be formed of:
(a) Three coins each?
(b) Four coins each?
(c) At least three coins each? Hint: Use the addition principle.

6. Given eight points on a sheet of paper, no three on the same line, how many triangles can be drawn using three of these points as vertices?

7. From a suit of 13 playing cards, how many different 5-card hands can be dealt? How many of these hands include the queen?

8. In how many ways can Clement select 4 of 12 friends to invite to lunch? In how many ways can she do it if she doesn't want to invite Euodias and Syntyche together?

9. From a group of six girls and nine boys, how many five-member committees can be formed involving
(a) Three boys and two girls? Hint: First select boys, then girls, and use the multiplication principle.
(b) Three girls and two boys?
(c) Five boys or five girls? Hint: Use the addition principle.
(d) At least three boys?

10. From a bag containing six white balls and ten black balls, in how many ways can we draw a group of six balls consisting of
(a) Three white balls and three black balls?
(b) Six white balls or six black balls?
(c) At least three white balls?
Note: Assume the balls are distinguishable; for example, they may have numbers on them.

11. In how many ways can a group of 12 people be divided up so that 4 play tennis, 4 play golf, and 4 go hiking?

12. In how many ways can the 52 cards in a standard deck be dealt to four players? Write the answer in combination symbols. It's too big to evaluate.

13. A woman helping to set up displays at the county fair is given 4 model airplanes, 3 model boats, and 4 model cars to set along a shelf. If she keeps the airplanes, boats and cars grouped together, in how many ways can she arrange the exhibit?

14. A test consists of 5 questions in Part I, and 3 questions in Part II. Students are to answer 3 questions from Part I and 1 question from Part II. In how many ways might a student select the questions he or she will answer.

15. A man trying to decide which, if any, books he wishes to borrow from a library, sets aside 5 that he is considering. If you know that he made his final selection from among these 5, how many different sets might he have finally taken with him, understanding that none or all are to be counted as possibilities?

16. From a group of 6 men and 4 women, how many ways are there to form a committee of 5 members if two of the men refuse to serve together?

17. The executive committee of the student government, ten members in all, must select five of their committee to attend a conference. One married couple in the group will only go if they can go together, and two other members refuse to both go. In how many ways can they choose the 5-person delegation?

18.* A TV news team covering an economic summit called by the governor of their state wants to engage participants from the state's largest three companies in a round table discussion. The room available for the filming will comfortably seat six people. They decide that they must pick at least one representative from each company, but that with that done, they can fill the other three spots with whomever seems most likely to make an interesting and balanced group. How many ways are there to choose the six participants?

19.* Using only pennies, nickels, and dimes, in how many ways can we make 16 cents? Twenty cents? Thirty-five cents? Sixty cents? Begin by trying to fill in the table below (p = pennies; n = nickels; d = dimes). Look for a pattern.

Table 5-2

Amount	*1*	*2*	*3*	*4*	*5*	*6*	*7*	*8*	*9*	*10*	*11*	*12*	*13*	*14*	*15*	*16*	*20*	*35*
Number of ways with p and n	1	1	1	1	2	2	2	2	2	3						4	5	
Number of ways with p, n, and d										4								

If a_n is the number of ways to make n cents using pennies and nickels, b_n the number of ways to make n cents using pennies, nickels, and dimes, find a formula connecting a_n and b_n.

20. Consider a standard deck of 52 cards consisting of 4 suits (spades, clubs, hearts, diamonds) each with 13 cards (2, 3, 4, . . . , 10, jack,

queen, king, ace). A bridge hand consists of 13 cards.
(a) How many different bridge hands are there?
(b) How many of them have exactly 3 aces?
(c) How many of them have no aces?
(d) How many of them have cards from just 3 suits?
(e) How many of them have only honor cards (aces, kings, queens, jacks)?
(f) How many of them have one card of each kind (one ace, one king, one queen, etc.)?
(g) How many of them have all cards from just one suit?

For Research and Discussion

Two nonnegative integers are said to have the same **parity** if both are even or both are odd. Otherwise they have opposite parity. This simple notion is surprisingly useful in problem solving, a fact illustrated (but not exhausted) in a chapter titled "It's Combinatorics That Counts" in Ross Honsberger's book, *Mathematical Gems I* (The Mathematical Association of America, 1973). The problems he solves using parity arguments include:

(a) The integer 3 has two divisors: 1 and 3. The integer 4 has three divisors: 1, 2, and 4. The number 12 has six divisors: 1, 2, 3, 4, 6, and 12. Thus, 3 and 12 have an even number of divisors, while 4 has an odd number of divisors. What property is shared by all numbers having an odd number of divisors.
(b) We saw in Section 2.4, Problem 15 that a checkerboard with diagonally opposite corner squares removed could not be covered by thirty-one 2×1 dominoes. The argument hung on the fact that the two squares removed would be the same color. Well, can we remove one square of each color in such a way that the remaining 62 squares still cannot be covered by the 31 dominoes?
(c) A jailer walks down a long row of n cells, unlocking each cell as he goes. Returning to his starting point, he again passes down the row, this time relocking every other cell, starting with cell 2. He does it a third time, this time turning the lock of every third cell, starting with cell 3. He does this until he has made n passes. Which cells, if any, remain unlocked?

See how many problems you can find that can be solved using parity arguments.

5.4 The Binomial Theorem

The Course to Amy

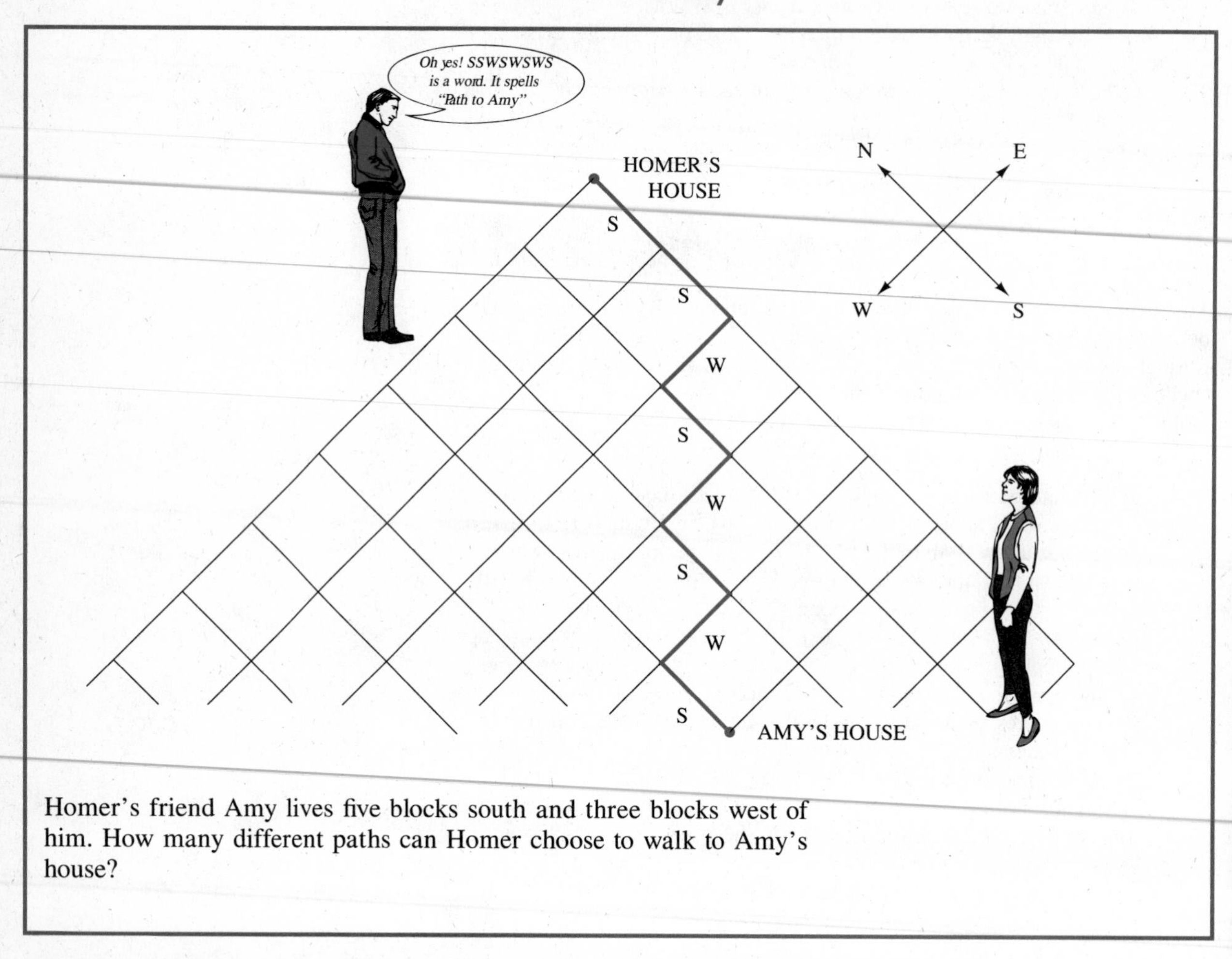

Homer's friend Amy lives five blocks south and three blocks west of him. How many different paths can Homer choose to walk to Amy's house?

Clarify the question.

First we need to sharpen the question. What is a path? Can Homer cut across backyards? No; the neighbors have had their fill of Homer's shortcuts. Does he wander around the neighborhood on his way to Amy's? No. Homer has had his fill of visiting with the neighbors. He is looking for the shortest route to Amy's house, but there are many such paths. In fact, Homer must walk exactly five blocks south and three blocks west, but he can do this in any order. If S denotes south and W west, SSWSWSWS does spell "path to Amy"; so does WWWSSSSS.

How many paths are there from Homer's house to Amy's house? Exactly the number of eight-letter words consisting of five S's and three W's. We have just dressed a new subject in a set of old clothes. And in the old clothes, we recognize it. It is the problem of permutations with some objects alike (Section 5.2). The answer is

$$\frac{8!}{5!\,3!} = \frac{8 \cdot 7 \cdot 6}{3 \cdot 2 \cdot 1} = 56$$

Homer can travel a different path on every Saturday night date for over a year. Another observation is more relevant. The answer can be given as ${}_8C_5$, since its value is $8!/(5!3)$.

We can obtain the answer ${}_8C_5$ by another kind of reasoning. In his eight-block walk, Homer must select five blocks on which he will walk south (which can be done in ${}_8C_5$ ways); on the other three, he will have to walk west.

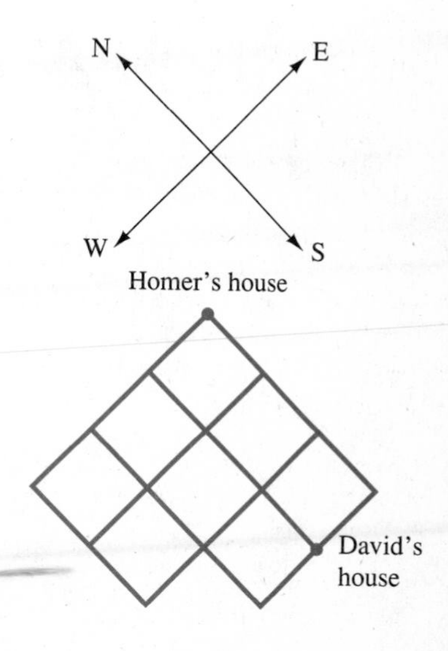

How many paths are there from Homer's house to David's house (three blocks south, one block west)? As many as there are four-letter words consisting of three S's and one W. And this is $4!/(3!1!)$ which is of course also equal to ${}_4C_3$; both have the value 4. This time the answer is easy to check. The four paths correspond to the words SSSW, SSWS, SWSS, and WSSS.

Our general claim is this. The number of paths from Homer's house to any intersection is ${}_{m+n}C_m$, where m is the number of blocks south and n the number of blocks west. Here it must be understood that ${}_nC_0 = 1$. We can nicely summarize our discussion in a triangular arrangement. The number at an intersection is the number of paths from the vertex to that point.

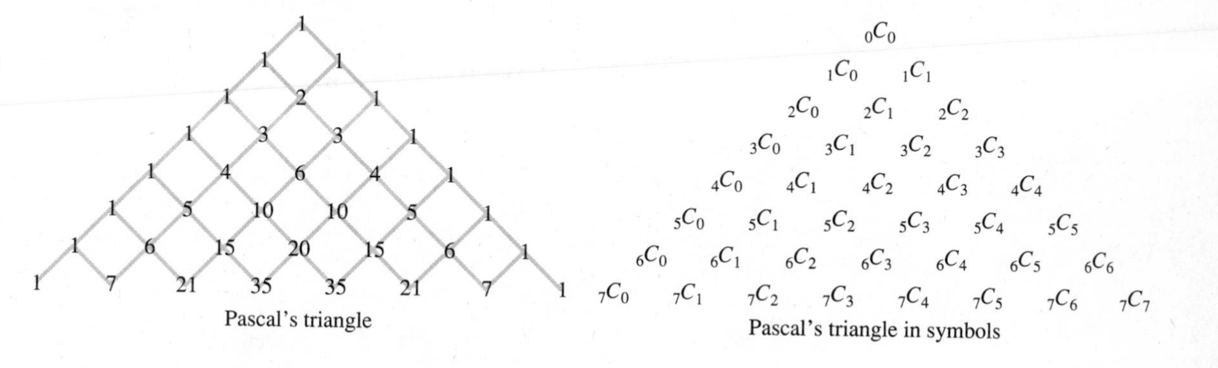

Pascal's triangle

Pascal's triangle in symbols

Blaise Pascal (1623–1662)
In 39 short years, this French mathematician discovered an important theorem in geometry, helped found the theory of probability, planted the seeds of the calculus, wrote a treatise on the "arithmetical triangle," invented a computing machine and authored *Penses,* a profound religious work.

Pascal's Triangle

The triangular array which appears on p. 159 has been called **Pascal's Triangle** after the great mathematician-philosopher Blaise Pascal. No other array of numbers has so intrigued mathematicians. It is loaded with nuggets awaiting discovery. Let's uncover a few of them.

First, note the symmetry of the array. If we fold the triangle across its altitude, the numbers match. For example,

$$_6C_2 = {_6C_4}$$
$$_7C_3 = {_7C_4}$$
$$_7C_1 = {_7C_6}$$

and, in general,

$$_nC_r = {_nC_{n-r}}$$

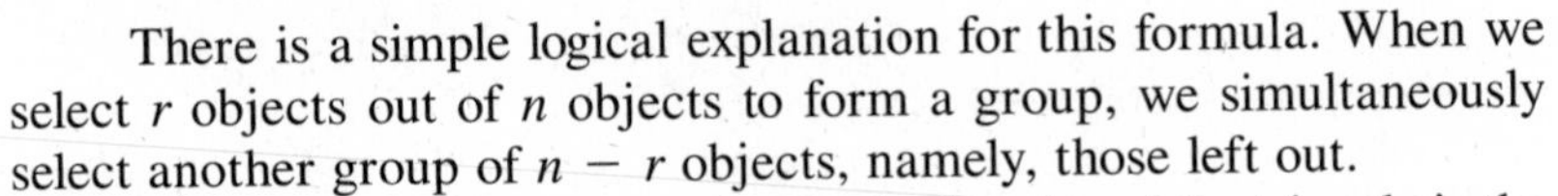

There is a simple logical explanation for this formula. When we select r objects out of n objects to form a group, we simultaneously select another group of $n - r$ objects, namely, those left out.

Second, note that any number in the interior of the triangle is the sum of the two neighbors immediately above it. For example,

or, equivalently, reading from the right diagram,

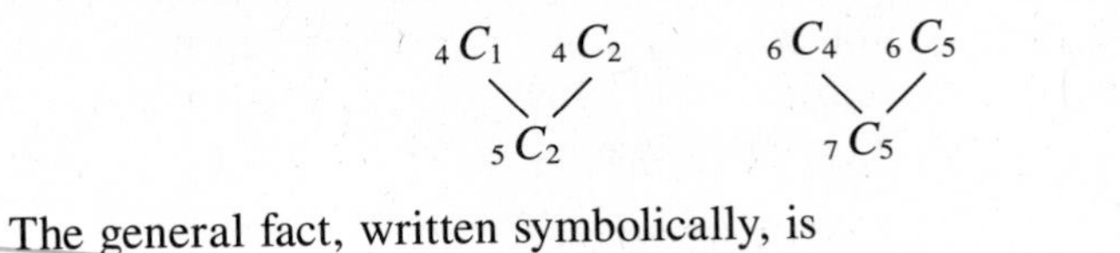

The general fact, written symbolically, is

$$_nC_r = {_{n-1}C_{r-1}} + {_{n-1}C_r}$$

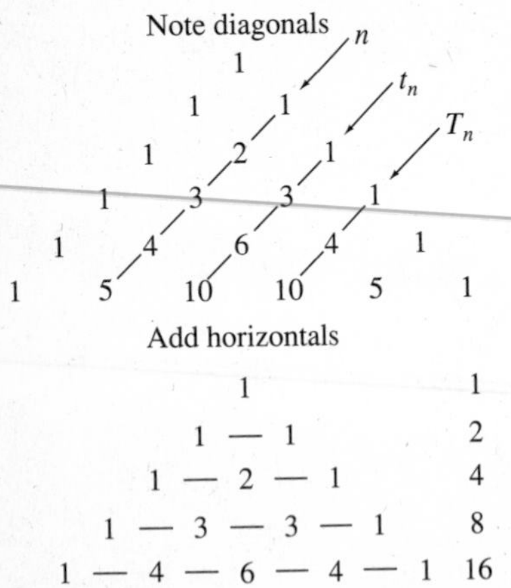

Features of Pascal's triangle

It is a recursion formula. We explain how to establish it algebraically in Problem 16.

Look at the second diagonal now (see figure titled Features of Pascal's Triangle). It is the sequence of counting numbers n. The third diagonal gives the triangular numbers t_n (see Section 4.1), and the fourth the summed triangular numbers $T_n = t_1 + t_2 + \cdots + t_n$ (often called tetrahedral numbers. Why?)

Next, add the horizontal rows. We get powers of 2. Algebraically, this means that (cf. Problems 9 and 10)

$$_nC_0 + {_nC_1} + {_nC_2} + \cdots + {_nC_n} = 2^n$$

Look around when you have got your first mushroom or made your first discovery: they grow in clusters.

George Polya

This has an important interpretation. Recall that $_nC_r$ was originally introduced as the number of combinations of n things taken r at a time.

Thus the formula says that the total number of subsets of all sizes of a set with n elements is 2^n. For example, the set {A, B, C, D} ought to have $2^4 = 16$ subsets. Let's check by listing them (see the table below).

Table 5-3
Subsets of {A, B, C, D}: 1 + 4 + 6 + 4 + 1 = 2^4

Zero-Member Set	*One-Member Subsets*	*Two-Member Subsets*	*Three-Member Subsets*	*Four-Member Subsets*
The empty set { }	{A} {B} {C} {D}	{A, B} {A, C} {A, D} {B, C} {B, D} {C, D}	{A, B, C} {A, B, D} {A, C, D} {B, C, D}	{A, B, C, D}
${}_4C_0 = 1$	${}_4C_1 = 4$	${}_4C_2 = 6$	${}_4C_3 = 4$	${}_4C_4 = 1$

Can you find the Fibonacci numbers hidden in Pascal's Triangle? They are there, though we must admit that they are some-what disguised (Problem 11).

The Binomial Formula

By far the most important fact related to Pascal's triangle is an algebraic theorem dealing with raising a binomial, say $x + y$, to an integral power.

$$\begin{aligned}
(x+y)^0 &= 1\\
(x+y)^1 &= x + y\\
(x+y)^2 &= x^2 + 2xy + y^2\\
(x+y)^3 &= x^3 + 3x^2y + 3xy^2 + y^3\\
(x+y)^4 &= x^4 + 4x^3y + 6x^2y^2 + 4xy^3 + y^4\\
&\vdots\\
(x+y)^n &= {}_nC_0x^ny^0 + {}_nC_1x^{n-1}y^1 + \cdots + {}_nC_{n-1}x^1y^{n-1} + {}_nC_nx^0y^n
\end{aligned}$$

The last displayed formula is the **Binomial Formula.** We have obtained it by observing the pattern of coefficients we get as we raise $x + y$ to higher and higher powers. An observation is hardly a proof. Nevertheless it is correct (we suggest a way to prove it in Problem 15); and we shall feel free to use it from now on. Because of this formula the numbers ${}_nC_r$ are often called the binomial coefficients.

In the Binomial Formula, we may substitute any numbers for x and y. For example, the substitution $x = 1$, $y = 1$ leads to the formula displayed near the bottom of page 160.

The Binomial Formula provides a good way of calculating numbers near 1, such as the numbers that occur in compound interest tables (see Section 4.5). Since

$$\begin{aligned}(x + y)^{12} &= x^{12} + 12x^{11}y + 66x^{10}y^2 + 220x^9y^3 + \cdots \\ (1.01)^{12} &= 1^{12} + 12(1)^{11}(0.01) + 66(1)^{10}(0.01)^2 + 220(1)^9(0.01)^3 \\ &\quad + \cdots \\ &= 1 + 0.12 + 0.0066 + 0.000220 + \cdots \\ &\approx 1.12682\end{aligned}$$

Note that, if we are interested in only five-decimal-place accuracy, there is no need to calculate all the terms in the expansion. Only the first few contribute to the first five decimal places.

Summary

To specify a path from A to B (B being southwest of A) on a rectangular grid, one need only say how many blocks south and west, in some order, a person should walk. The number of such paths is then the number of words that can be formed with S's and W's used the appropriate number of times. If B is m blocks south and n blocks west of A, there are ${}_{n+n}C_m$ paths from A to B.

The numbers ${}_nC_r$ can be displayed in an array called Pascal's triangle which exhibits many beautiful and surprising patterns. These numbers also appear in the Binomial Formula which tells us how to expand $(x + y)^n$. Once again, we offer an example to remind the reader of a simple way to calculate them.

$${}_{15}C_3 = \frac{\overset{5}{\cancel{15}} \cdot \overset{7}{\cancel{14}} \cdot 13}{\cancel{3} \cdot \cancel{2} \cdot \cancel{1}} = 35 \cdot 13 = 455$$

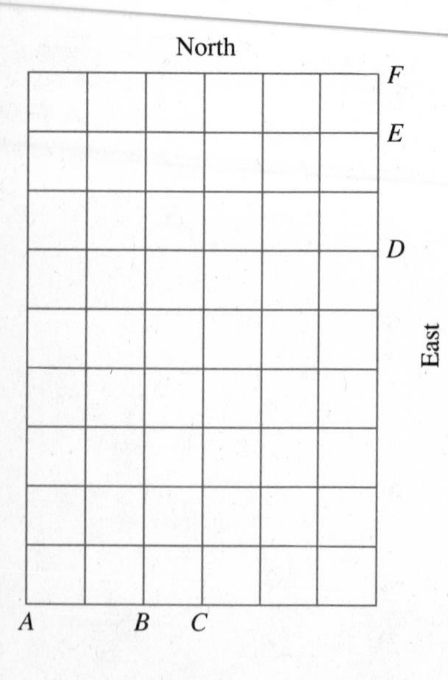

Problem Set 5.4

1. Evaluate:
   (a) ${}_5C_3$ (b) ${}_9C_3$
   (c) ${}_{10}C_4$ (d) ${}_{12}C_4$
2. Evaluate:
   (a) ${}_6C_3$ (b) ${}_{12}C_3$
   (c) ${}_9C_4$ (d) ${}_{40}C_3$
3. List all the five-letter words that can be made with three E's and two N's. Are there ${}_5C_3$ of them?
4. List all the six-letter words that can be made with three E's and three N's. Are there ${}_6C_3$ of them?
5. How many paths (shortest routes) are there, in the diagram at the left, from
   (a) C to D? Hint: Three E's and six N's
   (b) C to F?
   (c) A to D?

6. How many paths are there in the diagram from
(a) B to E?
(b) B to D?
(c) B to F?

7. Calculate:
(a) ${}_{11}C_2$ and ${}_{11}C_9$.
(b) ${}_{8}C_3$ and ${}_{8}C_5$.
What formula do these examples illustrate?

8. Recall the formula

$${}_{n}C_{r} = \frac{n!}{r!\,(n-r)!}$$

Use it to show that

$${}_{n}C_{r} = {}_{n}C_{n-r}.$$

9. Consider the set $\{a, b, c, d, e\}$. Finish filling in the table below.

Table 5-4		
Zero-member subset	{ }	${}_5C_0 = 1$
One-member subsets	$\{a\}$ $\{b\}$ $\{c\}$ $\{d\}$ $\{e\}$	${}_5C_1 = 5$
Two-member subsets		${}_5C_2 = 10$
Three-member subsets		${}_5C_3 = 10$
Four-member subsets		${}_5C_4 = 5$
Five-member subsets		${}_5C_5 = 1$

Notice that the total is

$${}_5C_0 + {}_5C_1 + {}_5C_2 + {}_5C_3 + {}_5C_4 + {}_5C_5 = 32 = 2^5$$

10. Again consider the set $\{a, b, c, d, e\}$. Suppose we wish to pick a subset out of it. We have five decisions to make, each with two choices. Will a go in the subset or not? Will b go in the subset or not? And so on. By the multiplication principle, we can make these decisions in $2 \cdot 2 \cdot 2 \cdot 2 \cdot 2$ ways. Looked at from this perspective, how many subsets does $\{a, b, c, d, e, f, g\}$ have?

11. Fill in the boxes. What do you observe?

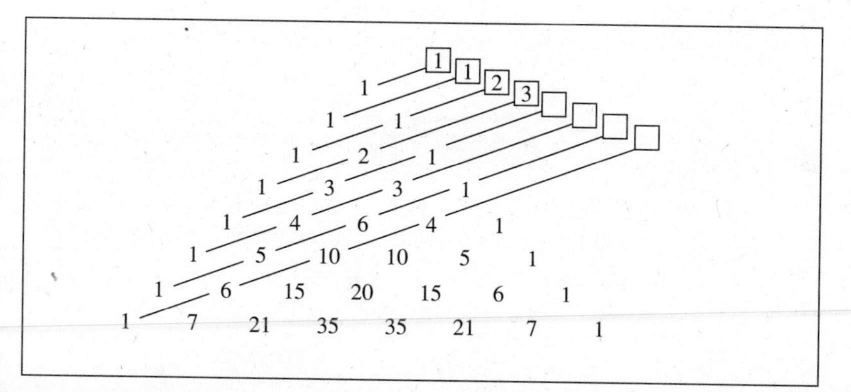

12. Find 11^2, 11^3, 11^4 and 11^5. What do you observe?

13. Expand each of the following:
(a) $(a + b)^4$
(b) $(c - d)^5$ Hint: $c - d = c + (-d)$.
(c) $(u + 2v)^6$

14. Expand:
(a) $(1 + x)^7$
(b) $(2 + y)^5$
(c) $(2x + y)^5$

15.* Here is an outline of a proof of the Binomial Theorem. Try to understand it.
(a) $(x + y)^n = (x + y)(x + y)(x + y) \cdots (x + y)$, ($n$ factors).
(b) When expanded, each term arises from picking x's from r of the factors and y's from the remaining factors and then multiplying these x's and y's together.
(c) Many of the terms are alike and can be collected together. When this is done, each resulting term has the form $Ax^r y^{n-r}$.
(d) The coefficient A is the number of ways of picking r x's out of the n factors, i.e., $A = {}_nC_r$.

16.* Remember that

$$ {}_nC_r = \frac{n!}{r!\,(n-r)!} \qquad {}_{n-1}C_r = \frac{(n-1)!}{r!\,(n-1-r)!} \cdots $$

Use these facts and some algebra to show that

$$ {}_nC_r = {}_{n-1}C_{r-1} + {}_{n-1}C_r $$

17.* Convince yourself that

$$ {}_nC_r = {}_{n-1}C_{r-1} + {}_{n-1}C_r $$

is correct by filling in the details in the following argument. Consider a set of n objects, one of which we label A. To form a subset consisting of r objects we may choose to include A or not include it. In how many ways can we form the subset in each of these two cases?

18. Use your calculator to check the accuracy of the estimate of $(1.01)^{12}$ obtained at the end of this section.

19. Use the first four terms of the Binomial Formula to estimate $(1.02)^{15}$. Then use your calculator to evaluate this number directly.

For Research and Discussion

Try to discover other patterns in Pascal's Triangle. Here is one you can try. Take any number a in the interior of the triangle and consider its six neighbors.

$$\begin{matrix} & s & t & \\ x & a & u \\ & w & v & \end{matrix}$$

Calculate suw and tvx. Do this for several choices of a and then guess a result. Can you prove it? An excellent source to get you started on mathematical discovery as it can be experienced studying Pascal's triangle is the book by T. M. Green and C. L. Hamberg, *Pascal's Triangle*, Dale Seymour Publications, Palo Alto, CA.

5.5 Partitions

Picking a Dozen

Anticipating that the editors would have a long night getting the student paper ready for printing, Beth decided to lay in a supply of a dozen cans of soft drinks. When she went to the machine she found it well stocked with six choices. She could get 2 of each, or 12 of one kind, or 7 Cokes and 1 of each of the others, or . . . . Well, we can't list all the possibilities because there are too many. How many?

A little experimenting confirms the implication of the statement of the problem. There are too many possibilities to list.

Examine a smaller case.

Perhaps we should cut the problem in half. Suppose Beth wanted to buy 6 cans and was confronted with 3 choices that we will for convenience call A, B, and C. One way to organize our counting is to list all the possibilities if no A is chosen, then all the possibilities when one A is chosen, etc,

Keep systematic records.

Choose 0 A's		*Choose 1 A*		....	*Choose 5 A's*		*Choose 6 A's*	
B	C	B	C		B	C	B	C
0	6	0	5		0	1	0	0
1	5	1	4		1	0		
2	4	2	3					
3	3	3	2					
4	2	4	1					
5	1	5	0					
6	0							
(7 ways)		(6 ways)			(2 ways)		(1 way)	

The solution to our smaller problem is

$$7 + 6 + \cdots + 2 + 1 = \frac{8 \cdot 7}{2} = 28.$$

Does this solution suggest a general method?

Has the solution suggested a general method? You might now try listing the ways to buy 6 cans if there are 4 choices, or 7 cans if there are 3 choices. In any of these cases, systematic recordkeeping enables one to get an answer. In none of these cases does a general method jump out. At least no inspiration jumps out to this author.

Inspiration in mathematics, like inspiration in music or any creative human activity, is hard to come by. It is hard to explain how someone else came by it. Sometimes it is even harder to convince someone that something is the result of inspiration. One needs to know a lot about music, even a lot about the history of music of a certain era, in order to appreciate why a certain piece, written at a certain time, is regarded as an inspired work.

An inspired solution to a problem in mathematics is not likely to be appreciated by someone who has not worked on the problem. A good solution, after all, will be clear (obvious is the word frequently used) and direct (no tour of plausible blind alleys explored by the first solver). The one who has not tried may be excused for passing over a truly inspired solution with a shrug. "I see; yes, I see not only that it works, but that it will work on all problems of this kind. But then I expected it to. You are a mathematician, this is a mathematics course, and I am here to learn the tricks of the trade. Why should I be surprised when shown that something works?"

The problem posed at the beginning of this section has an inspired solution. It is almost a shame to show it to you so soon. Be a sport. Go back and spend some real time trying to solve this problem in a way that will give you insight into solving a more general problem.

Partitioning an Integer n

How many nonnegative integer solutions are there to $x_1 + x_2 + \cdots + x_r = n$?

$$x_1 + x_2 + x_3 + x_4 = 7$$

has the solution

$$(x_1, x_2, x_3, x_4) = (2, 0, 1, 4).$$

Another solution is (2, 2, 2, 1).

First note that this is equivalent to Beth's problem. Number the six choices 1, 2, 3, 4, 5, 6, and let x_i be the number of cans of choice i. Since the problem says that she intends to come back with 12 cans, she must choose x_i so that $x_1 + x_2 + \cdots + x_6 = 12$. To go back with two cans of each would be to choose $x_1 = x_2 = \cdots = x_6 = 2$. If x_1 is the number of Cokes, then choosing 7 Cokes and 1 of everything else would correspond to the solution $x_1 = 7, x_2 = \cdots = x_6 = 1$.

We say that an object is partitioned if it is broken up into nonoverlapping pieces that, in their totality, constitute the whole. The reason for interest in counting the ways of partitioning an integer n into the sum of some fixed number r of nonnegative integers is precisely because the solution to this problem gives us a solution to so many other problems.

It is equivalent, for example, to the problem that asks how many ways there are to distribute n identical objects into r distinct containers. In this case, think of the containers as being numbered 1, 2, . . . r and let x_i represent the number of objects that are to be put in container i. This is, in fact, the point of view that sets up the promised elegant solution. Let o's represent the n identical objects, and indicate their arrangement in containers by vertical lines |. Thus, in the given problem involving 12 cans and 6 choices, the choice of 2 cans of each would be indicated by

Transform the problem. Ask a question that, if answered, will answer the original problem.

oo | oo | oo | oo | oo | oo

Taking 7 of the first choice and 1 of each of the others would be represented by

ooooooo | o | o | o | o | o

Taking 3 each of choices 2, 3, 5 6 would be represented by

| ooo | ooo | | ooo | ooo

The key observation is that any possible solution can be represented by an arrangement of 12 o's and 5 |'s. And in how many ways can we arrange in order a total of 17 things (o's and |'s) when 12 are identical and another 5 are identical? That's easy. It's a permutation with some objects identical: The problem that opened this section is completely solved in the margin.

$$\frac{17!}{5! \cdot 12!} = 6188$$

Moreover, our method immediately gives us a solution to the problem of partitioning an interger n into the sum of r nonnegative integers. We have seen that such a partition can always be represented by n o's and $r - 1$ |'s, and that there will be $\frac{(n + r - 1)!}{n! \cdot (r - 1)!}$ ways to do this.

The equation $x_1 + x_2 + \cdots + x_r = n$ will for $n > r$ have $\frac{(n + r - 1)!}{n!\,(r - 1)!}$ nonnegative integer solutions

---

You may wish to verify that the formula just obtained does in fact give the correct answer in the case we worked out for 6 cans to be selected when there are 3 choices.

$$\frac{(6 + 3 - 1)!}{6!\,(3 - 1)!} = \frac{8 \cdot 7}{2} = 28$$

Summary

We have examined three equivalent counting problems.

1. How many nonnegative integer solutions are there to

$$x_1 + x_2 + \cdots + x_r = n?$$

2. How many ways may we select n objects from r choices if the same choice may be repeated, and the order of selection is immaterial?
3. In how many ways can n identical objects be distributed among r distinct containers?

Problem Set 5.5

1. Write out all the nonnegative integer solutions of $x_1 + x_2 + x_3 + x_4 = 6$. Do you get the number of solutions given by the formula?

2. Write out all the nonnegative integer solutions of $x_1 + x_2 + x_3 = 10$. Do you get the number of solutions given by the formula?

3. Each day the proprietor of the *Cookie Jar* bakes large quantities of 6 kinds of cookies. In how many ways may I choose a dozen cookies?

4. A neophyte at preparing for our 3-day sailing trip, I was told to distribute the 24 cans of soft drinks in several different "holds" (bins I found under seats, above the bunks, etc.) so they would be handy from several locations on the boat. I found 5 holds. In how many ways could I have distributed the cans?

5. The chef had prepared 5 desserts, each of which was available in ample supply. A waiter, wanting to make his dessert tray as appealing as possible, decided he could carry a total of 12 servings and that he wanted to have at least one serving of each available choice. In how many ways might he choose 12 desserts?

6. In the game of Yahtzee, you begin by rolling five dice at one time. How many distinct outcomes are possible if
(a) the five dice are all different in color?
(b) the five dice are identical?

7. How many ways are there to put 18 soft drink cans into five distinguishable boxes if
(a) the cans are all different?
(b) the cans are identical?
(c) there are 10 cans of one kind, 8 of another kind?

8.* Determine the number of nonnegative integer solutions to the pair of equations

$$x_1 + x_2 + x_3 + x_4 = 7$$
$$x_1 + x_2 + x_3 + x_4 + x_5 + x_6 + x_7 = 13.$$

For Research and Discussion

We digressed a bit in this section to discuss the nature of "inspired work." It is no doubt defined differently in different fields. Talk to faculty members in several disciplines in which you have interest. Ask them if they can give you examples of "inspired work" in their field, and tell you why they would so classify it. Put the same question to several faculty members in your mathematics department. What similarities do you notice? what differences?

chapter 6

The Laws of Chance

There are laws of chance. We must avoid the philosophically intriguing question as to why chance, which seems to be the antithesis of all order and regularity, can be described at all in terms of laws.

WARREN WEAVER

Predicting the Unpredictable

I was posing to some friends at a small dinner party a problem that had been sent to me that day. It told of a baseball umpire who really was blind, and was calling balls and strikes in a completely random manner. What, the problem asked, were the chances of a batter who swung at no pitches being walked on four balls before being called out on three strikes? I thought it was a very cute problem, and had figured out to the customary three decimal places what should be the batting average of a player before he should be allowed to swing at all in such a situation.

One of the guests, a nonmathematician, was incredulous. "Let me get this straight. The umpire is totally blind, and as whimsy dictates, calls each pitch a ball or strike in a totally unpredictable way." I agreed that this was the situation. "And from this information, you compute to three decimal places the likelihood of the batter being walked on four balls before there are three strikes." I admitted to having done it.

My friend persisted in thinking that I had misstated the problem or was unconsciously making some assumption that had not yet come to light in our conversation. If each call is unpredictable, unrelated to any previous call (that is, we assume that the umpire not only has no sight, but also no memory), then how in the world can the outcome be predictable—to three decimals yet?

It does seem incredible that we can find patterns in events that are happening in a random, totally unpredictable way. Yet this is precisely what the science of probability and statistics not only attempts, but does with such success as to have become an indispensible tool in areas that range from agriculture to medicine, from quality control in a modern factory to the overbooking of seats in an airplane.

The yields of different fields planted with different seeds are compared. Do the variations prove one seed better than another, or are they well within what might be expected from random events like birds eating seeds, seeds falling on stones or areas where they cannot take root, etc.? To be able to answer such questions, we must first have some idea of the kind of patterns that will emerge from a series of random events.

There has been a long history of this kind of work in mathematics. In recent years, the even more unlikely sounding topics of chaos and of fuzzy sets (sets in which the boundary between points in and points not in the set is not clear) have found their way into the literature of serious mathematics.

Mathematicians do indeed find useful patterns in places where common sense might indicate that no pattern is to be expected. It is one more facet of the discipline that is hard to explain to a public that already views not only the discipline, but also the practitioners themselves, to be unpredictable.

6.1 Probability Defined

The Carnival Game

In a game often encountered at carnivals, a player standing behind a restraining rail tries to toss a penny onto a table that is ruled by lines parallel to the edges of the table into one inch squares. If the penny falls entirely inside a square, the player receives a token to be accumulated toward a prize. Wherever it falls (including the possibility of the floor, which by observation seems to happen to about one penny in seven), the operator keeps all the pennies thrown. What should be the value of the token if this is to be a fair game?

Our mechanic tells us that the work we are paying for will *probably* take care of the problem; our physician says the medicine being prescribed will *probably* clear up our ailment. We hope for certainty, but life is uncertain, so outcomes that are expected (or desperately hoped for) by all parties concerned are described as probable.

Probability is the very guide of life.

Cicero

In other situations, we say something is probable if there seems to be anything better than a 50% chance. A student says she will *probably* neturn that book tomorrow; her professor says the papers will *probably* be returned at the next class meeting.

"Two inches of partly cloudy."

Formal probability theory tries, like the weather forecaster tries, to quantify the popular idea that something is probable. "There is a 60% chance of rain tomorrow," conveys a different message than if it is said to be a 90% chance. Like the mechanics and physicians, the weather forcasters leave a little hedge, almost never saying there is a 100% chance of rain, and from this observation we make our first point.

The probability of an event (your car starting tomorrow, your rash clearing up, rain falling down) is expressed as either a percent or its decimal equivalent between 0 and 1, with 0 meaning "no way" and 1 meaning "absolutely certain."

Probability does in fact have its roots in the study of games of chance that are characterized by having a finite number of equally likely outcomes (rolling a red and a white die can result in the 36 outcomes shown below), some of which are regarded by prior agreement (perhaps having the sum of the dots equaling 7) to be successes. In such a case,

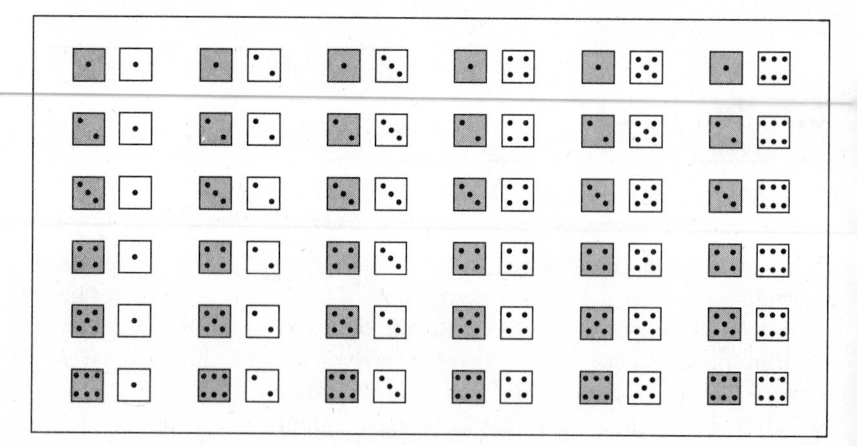

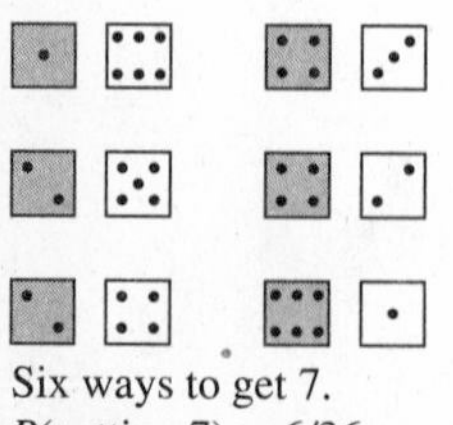

Six ways to get 7.
P(getting 7) = 6/36

the definition of probability is just what you would expect:

$$\text{probability of success} = \frac{\text{number of successful outcomes}}{\text{number of possible outcomes}}$$

We emphasize that this definition is predicated on the assumption that one outcome is as likely to occur as another. On a given day, you either will be involved in an automobile accident or you won't. That doesn't mean that your probability of having an accident on any given day is $\frac{1}{2}$. The outcomes are not, for most sober drivers, equally likely.

There are many situations in which the assumption of equal likehood seems plausible. Toss a standard die. Is there any reason to think that one side has an edge in the battle to show its face? We think not. Draw a card from a well-shuffled deck. Nature is perfectly democratic. It gives each card the same chance of being drawn.

We are ready to state a general definition of probability.

> Logically cautious readers may have noticed a disturbing aspect of this definition of probability. Since it speaks of "equally likely," i.e., equally probable, events, the definition sits on its own tail, so to speak, defining probability in terms of probability.
>
> *Warren Weaver*

If an experiment can result in any one of n equally likely outcomes and if exactly m of them are included in the event E, we say that the **probability of E** is m/n. We write this as

$$P(E) = \frac{m}{n}$$

A Pair of Dice

We return to the 36 possible outcomes pictured on page 172 that come from rolling a red and a white die. The fact that we can picture all possible outcomes enables us to get some practice at applying the definition directly.

$P(\text{getting a total of } 11) = \frac{2}{36} = \frac{1}{18}$
$P(\text{getting a total of } 12) = \frac{1}{36}$
$P(\text{getting a total greater than } 7) = \frac{15}{36} = \frac{5}{12}$
$P(\text{getting a total less than } 13) = \frac{36}{36} = 1$
$P(\text{getting a total of } 13) = \frac{0}{36} = 0$

Two ways to get 11.
$P(\text{getting } 11) = 2/36$

The last two are worthy of comment. The event "getting a total less than 13" is certain to occur; we call it a **sure event.** The probability of a sure event is always 1. However, the event "getting a total of 13" cannot occur; it's called an **impossible event.** The probability of an impossible event is 0.

Consider the event "getting 7 or 11" which is important in the dice game called craps. From the picture, we note that 8 of the 36 outcomes give a total of 7 or 11, so the probability of this event is $\frac{8}{36}$. But note that we could have calculated this probability by adding together the probability of getting 7 and the probability of getting 11:

$$P(7 \text{ or } 11) = P(7) + P(11)$$
$$\frac{8}{36} = \frac{6}{36} + \frac{2}{36}$$

It appears that we have found a very useful property of probability. However, a different example dampens our enthusiasm. Note that

$$P(\text{odd } or \text{ over } 7) \neq P(\text{odd}) + P(\text{over } 7)$$

since

$$\frac{27}{36} \neq \frac{18}{36} + \frac{15}{36}$$

Why is it that we can add probabilities in one case and not in the other? The reason is a simple one. The events "getting 7" and "getting 11" are disjoint (they can't both happen). But the events "getting an odd total" and "getting over 7" overlap (a number such as 9 satisfies both conditions).

Considerations like these lead us to the main properties of probability:

1. P(impossible event) $= 0$; P(sure event) $= 1$.
2. $0 \leq P(A) \leq 1$ for any event A.
3. $P(A \text{ or } B) = P(A) + P(B)$, provided A and B are disjoint, i.e., can't both happen at the same time.

From these properties, another follows. The events "A" and "not A" are certainly disjoint. Thus, from property 3,

$$P(A \text{ or not } A) = P(A) + P(\text{not } A)$$

But the event "A *or* not A" must occur; it is a sure event. Hence, from property 1,

$$1 = P(A) + P(\text{not } A)$$

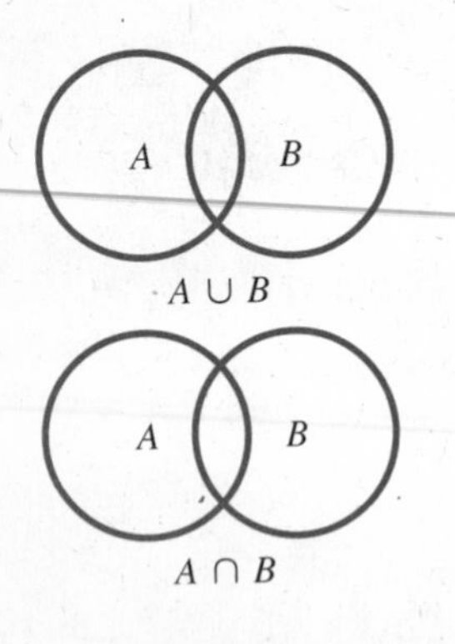

or

4. $P(A) = 1 - P(\text{not } A)$.

Let us illustrate this fourth property. If A is the event "getting a total less than 11," not A is the event "getting a total of at least 11." Thus

$$P(\text{total} < 11) = 1 - P(\text{total} \geq 11) = 1 - \frac{3}{36} = \frac{33}{36} = \frac{11}{12}$$

Relation to Set Language

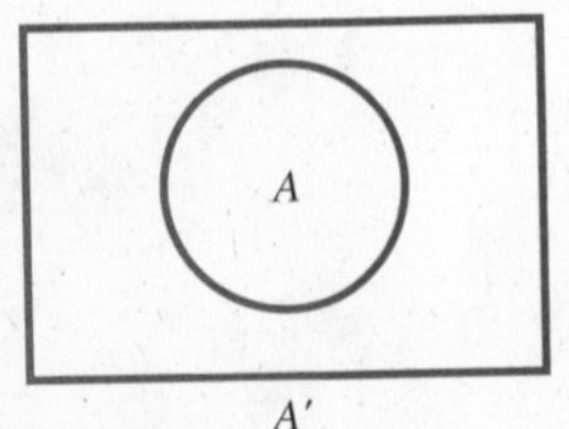

Most American students are introduced to the language of sets in the early grades. Even so, a brief review may be helpful. In everyday language, we talk of a bunch of grapes, a class of students, a herd of cattle, a flock of birds, or perhaps even a team of toads, a passel of possums, or a gaggle of geese. Why are there so many words to express the same idea? Mathematicians use one word—**set.** The objects that make up a set are called **elements** or **members.**

Sets can be put together in various ways. We have $A \cup B$ (read "A **union** B"), which consists of the elements in A *or* B. There is $A \cap B$ (read "A **intersection** B"), which is made up of the elements in both A *and* B. We have the notion of an **empty set** $\emptyset$ and of a **universe** S (the set of all elements under discussion). Finally, we have A' (read "A **complement**"), which is composed of all elements in the universe that are *not* in A.

The language of sets and the language used in probability are very closely related, as the list below suggests.

Probability Language	*Set Language*
Outcome	Element
Event	Set
Or	Union
And	Intersection
Not	Complement
Impossible event	Empty set
Sure event	Universe

If we borrow the notation of set theory, we can state the laws of probability in a very succinct form:

1. $P(\emptyset) = 0$; $P(S) = 1$.
2. $0 \leq P(A) \leq 1$.
3. $P(A \cup B) = P(A) + P(B)$, provided $A \cap B = \emptyset$.
4. $P(A) = 1 - P(A')$.

Difficulties with probability are usually caused, not by the simple rules that are used, but by the problem of trying to count the total number of outcomes and also those outcomes that lead to the event in question. This is one reason for the inclusion of a chapter on counting in this book (Chapter 5).

Poker Hands

We assume the reader is familiar with a standard deck of 52 cards. There are 4 suits (spades, clubs, hearts, and diamonds), each with 13 cards (2, 3, . . . , 10, jack, queen, king, and ace). A poker hand consists of 5 cards. Now recalling that ${}_{52}C_5$ is the number of combinations of 52 things taken 5 at a time, we see that there are

$${}_{52}C_5 = \frac{52 \cdot 51 \cdot 50 \cdot 49 \cdot 48}{5 \cdot 4 \cdot 3 \cdot 2 \cdot 1} = 2{,}598{,}960$$

different possible poker hands. Suppose we plan to deal a 5-card poker hand from a well-shuffled deck.

What is the probability of a diamond flush (i.e., 5 diamonds)? Of the ${}_{52}C_5$ possible outcomes, ${}_{13}C_5$ give a diamond flush. Thus

$$P(\text{diamond flush}) = \frac{{}_{13}C_5}{{}_{52}C_5} \approx .0005$$

What is the probability of a flush of any kind? By property 3,

$$\begin{aligned} P(\text{flush}) &= P(\text{diamond flush}) + P(\text{heart flush}) + P(\text{club flush}) \\ &\quad + P(\text{spade flush}) \\ &\approx .0005 + .0005 + .0005 + .0005 \\ &= .002 \end{aligned}$$

What is the probability of getting at least 1 ace? Here it is much easier to look at the complementary event "get no aces" and apply property 4. Note that there are 48 nonaces in a deck.

$$P(\text{at least 1 ace}) = 1 - P(\text{no aces})$$
$$= 1 - \frac{{}_{48}C_5}{{}_{52}C_5} \approx 1 - .66 = .34$$

The Carnival Game

We pointed out above that probability theory grew out of the study of games of chance having a finite number of equally likely outcomes, and our definition of probability was stated in terms of these assumptions. Now notice that the problem that opens this section does not fit this situation; if we focus on the center of the penny, we see that it may occupy any position on the table. That's not a finite number of outcomes. What is to be done?

Draw a picture. Make a physical model.

Let intuition be your guide. Place a penny on a one-inch square. Measure the penny. Since its diameter is $\frac{3}{4}$ inches the penny will lie entirely inside the square so long as the center of the penny is no closer than $\frac{3}{8}$ of an inch from any border. Put another way, of the pennies having their center in this square, the "successful pennies" will be the ones having their center in a square having the same center, drawn so its sides are parallel to but $\frac{3}{8}$ of an inch inside the original square. The smaller square evidently has sides of length $\frac{1}{4}$ and an area of $\frac{1}{16}$. Intuition suggests that if the penny lands with its center in the large square, the probability of its center being in the small square (so that the penny lies entirely within the large square) is $\frac{1}{16}$.

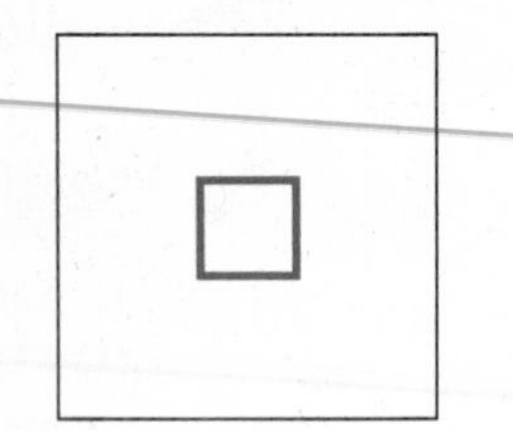

We cannot count the points in either square, but the concept of area seems to be a reasonable alternative. Suppose we look not at our definition of probability, but at the four properties of probability as they are stated for sets. If we think of S as the large square (the set of all centers of circles we have been considering), then Area(S) = 1; and of course the empty set $\varnothing$ has an area of 0; that is Area ($\varnothing$) = 0. If A is any subset of S, then $0 \leq \text{Area(A)} \leq 1$. Similarly observe that the last two properties of probability are satisfied if we interpret the probability of an event A to be the area of that subset A in S that corresponds to a success.

We are skipping over some subtleties. If we toss a coin onto a table, the center of the coin is more likely to fall in some places on the table than in others, and the general theory of probability demands more care than we have given to establishing that our definition in terms of a finite number of outcomes is equivalent to a definition in terms of the four properties.

Nevertheless, what we have hinted at is correct, and it nicely illustrates a feature of mathematics about which we shall have more to say in Part IV of this book. It is quite common to take consequences observed in an elementary setting, show that these consequences can themselves be used as a beginning point, and that when this is done, the resulting theory applies to a much broader class of problems. The process is called abstraction.

Let us finish our discussion of the carnival game. Having been told from observation that one in seven bounces or rolls off the table, we may regard the outcome as a two-step process in which the penny comes to rest on the table with probability $\frac{6}{7}$ and falls entirely within one of the ruled squares with probability $\frac{1}{16}$. As we shall see in the next section, this means that the probability of the events in sequence is $\frac{6}{7} \cdot \frac{1}{16} = \frac{6}{112} = .0536$. A player has about 1 chance in 20 (since $\frac{1}{20} = .05$) of winning, less if the ruling lines are drawn thicker so as to cut down the area of the large square. To make the game fair, the token for winning should be worth at least 20 cents; otherwise you really are throwing money way.

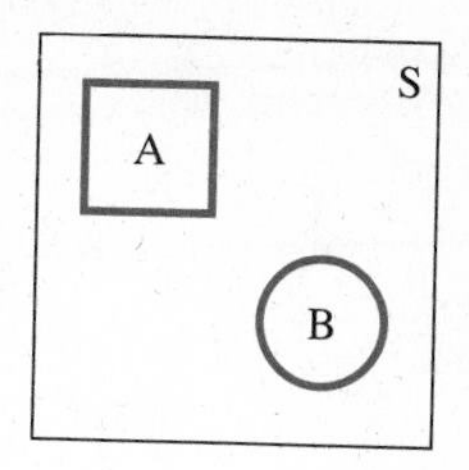

1. Area (∅) = 0;
   Area (S) = 1
2. 0 ≤ Area (A) ≤ 1
3. Area (A ∪ B)
   = Area (A) + Area (B)
   provided A ∩ B = ∅
4. Area (A) = 1 − Area (A′)

Summary

If an experiment can result in n equally likely outcomes of which m are in the event A, the probability of A is given by

$$P(A) = \frac{m}{n}$$

From this definition follow the four main properties or laws of probability which were stated twice in this section. Some problems can be worked by appealing directly to the definition. To calculate a probability is then a matter of counting all the outcomes and counting those that lead to the event in question. Often this is difficult, and it is better to break the problem into disjoint parts. Then, one of the properties of probability allows us to add together the probabilities of these pieces. Occasionally, it is easier to calculate the probability of the complementary event and then subtract it from 1. This is justified by another of the properties. In any case, it is important to have a clear notion of the experiment to be performed. This means, in particular, that one must be able to enumerate the (equally likely) outcomes in a systematic way.

Problem Set 6.1

1. An ordinary die is tossed. What is the probability that the number of spots on the upper face will be
(a) Three? (b) Greater than 3? (c) Less than 3?
(d) An even number? (e) An odd number?

2. Nine balls, numbered 1, 2, . . . , 9, are in a bag. If one is drawn at random, what is the probability that its number is
(a) Nine? (b) Greater than 5? (c) Less than 6?
(d) Even? (e) Odd?

3. A penny, a nickel, and a dime are tossed. List the eight possible outcomes of this experiment. What is the probability of
(a) Three heads? (b) Exactly two heads?
(c) More than one head?

4. A coin and a die are tossed. Suppose that one side of the coin has a 1 on it, and the other a 2. List the 12 possible outcomes of this experiment, e.g., (1, 1), (1, 2), (1, 3). What is the probability of
(a) A total of 4? (b) An even total? (c) An odd total?

5. Two ordinary dice are tossed. What is the probability of
(a) A double (both showing the same number)?
(b) The number on one of the dice being twice that on the other?
(c) The numbers on the two dice differing by at least 2?

6. A letter is chosen at random from the word PROBABILITY. What is the probability that it will be (treating Y as a vowel)
(a) P? (b) B? (c) M? (d) A vowel?

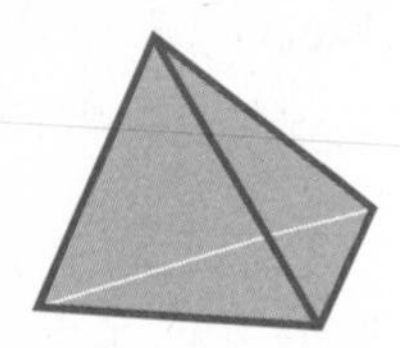

7. Two regular tetrahedra (tetrahedra have four identical equilateral triangles for faces) have their faces numbered 1, 2, 3, and 4. Suppose they are tossed and we keep track of the outcome by listing the numerals on the bottom faces, e.g., (1, 1), (1, 2).
(a) How many outcomes are there?
(b) What is the probability of a sum of 7?
(c) What is the probability of a sum less than 7?

8. Two regular octahedra (polyhedra having eight identical faces) have their faces numbered 1, 2, . . . , 8. Suppose they are tossed and we record the outcomes by listing the numerals on the bottom faces.
(a) How many outcomes are there?
(b) What is the probability of a sum of 7?
(c) What is the probability of a sum less than 7?

9. What is wrong with each of the following statements?
(a) Since there are 50 states, the probability of being born in Wyoming is $\frac{1}{50}$.
(b) The probability that a person smokes is .45, and that the he drinks, .54; therefore the probability that he smokes or drinks is .54 + .45 = .99.
(c) The probability that a certain candidate for president of the United States will win is $\frac{3}{5}$, and that she will lose, $\frac{1}{4}$.
(d) Two football teams A and B are evenly matched; therefore the probability that A will win is $\frac{1}{2}$.

10. During the past 30 years, Professor Witquick has given only 100 A's and 200 B's in Math 13 to the 1200 students who registered for the class. Based on these data, what is the probability that a student who registers next year
(a) Will get an A or a B?
(b) Will not get either an A or a B?

11. A poll was taken at Podunk University on the question of requiring students to demonstrate certain competencies in mathematics in order to graduate, producing the following results.

Table 6-1

	Administrators	*Faculty*	*Students*	*Total*
For	4	16	100	120
Against	3	32	100	135
No opinion	3	2	40	45
TOTAL	10	50	240	300

On the basis of this poll, what is the probability that
(a) A randomly chosen faculty member will favor a requirement
(b) A randomly chosen student will be against a requirement
(c) A person selected at random at Podunk University will favor a requirement
(d) A person selected at random at Pondunk University will be a faculty member who is against a requirement

12. The well-balanced spinner shown at the left is spun. What is the probability that the pointer will stop at
(a) Red? (b) Green?
(c) Red or green? (d) Not green?

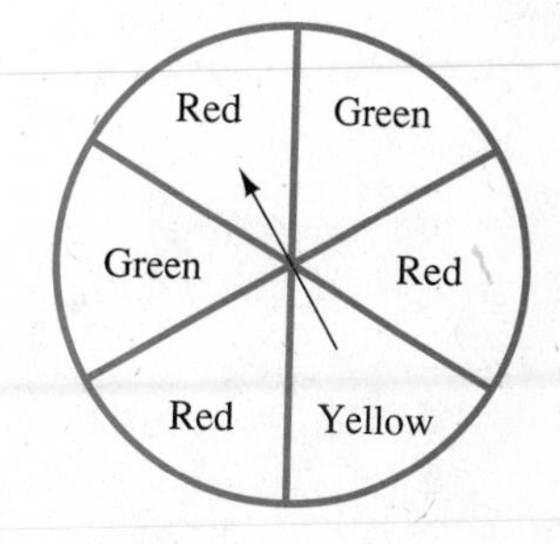

13. Four balls numbered 1, 2, 3, and 4 are placed in a bag, mixed, and drawn out, one at a time. What is the probability that they will be drawn in the order 1, 2, 3, 4?

14. A five-volume set of books is placed on a shelf at random. What is the probability they will be in the right order?

15. From an ordinary deck of 52 cards, 1 card is drawn. What is the probability that it will be
(a) Red? (b) A spade? (c) An ace?
Note: Two of the suits are red, and two are black.

16. From an ordinary deck of cards, two cards are drawn. This can be done in ${}_{52}C_2$ ways. What is the probability that both will be
(a) Red? (b) Of the same color? (c) Aces?

17. From an ordinary deck of 52 cards, 3 cards are drawn. What is the probability that
(a) All will be red? (b) All will be diamonds?
(c) Exactly 1 will be a queen? (d) All will be queens?

18. If three men and two women are seated at random in a row, what is the probability that
(a) Men and women will alternate?
(b) The men will be together?
(c) The women will be together?

19. A die has been doctored so that the probabilities of getting 1, 2, 3, 4, 5, and 6 are, respectively, $\frac{1}{3}$, $\frac{1}{4}$, $\frac{1}{6}$, $\frac{1}{12}$, $\frac{1}{12}$, and $\frac{1}{12}$. Assuming the rules for probabilities are still valid, find the probability of throwing
(a) An even number.
(b) A number less than 5.
(c) An even number or a number less than 5.

20. On a history exam, eight events are to be matched with eight dates, with each item used just once. Homer is sure of four dates and matches the others at random. What is the probability that he will be right on
(a) All eight dates? (b) At least six dates?
(c) Just four dates?

21. The third law of probability is $P(A \cup B) = P(A) + P(B)$, provided $A \cap B = \varnothing$. Show that the following extension always holds.
$P(A \cup B) = P(A) + P(B) - P(A \cap B)$

22.* Three dice are tossed. What is the probability that 1, 2, and 3 all will appear?

23.* In poker, what is the probability that a player will be dealt a straight flush (five consecutive cards in the same suit; an ace counts both high and low)?

24.* A careless secretary typed four letters and four envelopes, and then inserted the letters randomly into the envelopes. Find the probability that
(a) None went into the right envelope.
(b) At least one went into the right envelope.
(c) All went into the right envelope.
(d) Exactly two went into the right envelope.

25. If a poker hand (5 cards) is dealt from a standard deck (52 cards), the probability of a diamond flush (that is, all diamonds) is

$$\frac{{}_{13}C_5}{{}_{52}C_5} = \frac{\dfrac{13 \cdot 12 \cdot 11 \cdot 10 \cdot 9}{5 \cdot 4 \cdot 3 \cdot 2 \cdot 1}}{\dfrac{52 \cdot 51 \cdot 50 \cdot 49 \cdot 48}{5 \cdot 4 \cdot 3 \cdot 2 \cdot 1}}$$

Calculate this value but be sure to do all the cancelling you can before using your calculator.

26. Calculate the probability of getting a poker hand consisting of all honor cards (aces, kings, queens, jacks).

27. Calculate the probability of getting a poker hand consisting of all kings and queens.

28. Calculate the probability of getting a bridge hand (13 cards) consisting completely of honor cards.

29. Calculate the probability of getting a poker hand with no honor cards.

For Research and Discussion

The penny tossing problem is an example of what is called geometric probability. Here is another such problem for you to try.

> The plan is that the northbound bus and the westbound bus will meet at a certain corner at 4:45, but of course neither one ever shows up exactly on time; in fact, they arrive randomly between 4:40 and 4:50. The rule they use is that if a driver arrives and does not see the other bus, he or she is to wait 2 minutes, then proceed. What percentage of the time may a rider on one bus expect to catch the other bus?

Perhaps the most famous problem in geometric probability is the Buffon Needle problem. It's solution requires an understanding of calculus, but anyone can understand its statement, and can see that the connection of this problem with the number π is quite surprising. Look up the problem; one source is [F. Mosteller, *Fifty Challenging Problems with Probability Solutions* (Dover, 1987)]. Use the conclusion as the basis of your own experiment to determine the value of π.

6.2 Independent Events

The Car and the Goats

The following problem appeared in the "Ask Marilyn" column of *Parade* (a Sunday newspaper supplement) on September 9, 1990.

A TV host shows you three numbered doors, one hiding a car (all three equally likely) and the other two hiding goats. You get to pick a door, winning whatever is behind it. You choose door #1, say. The host, who knows where the car is, opens one of the other two doors to reveal a goat, and invites you to switch your choice if you so wish. Assume he opens door #3. Should you switch to #2?

A perfectly balanced coin has shown nine tails in a row. What is the probability that it will show a tail on the tenth flip? Some people argue that a mystical law of averages makes the appearance of a head practically certain. It is as if the coin had a memory and a conscience; it must atone for falling on its face nine times in a row. Such thinking is pure, unadulterated nonsense. The probability of showing a tail on the tenth flip is $\frac{1}{2}$, just as it was on each of the previous nine flips. The outcome of any flip is independent of what happened on previous flips.

Here is a different question, not to be confused with the one just answered. If one plans to flip a coin 10 times, what is the probability of getting all tails? To answer, we reason that there are two possibilities on the first flip, two on the second, etc. By the multiplication principle for counting (Chapter 5), there are $2 \cdot 2 \cdot 2 \cdot 2 \cdot 2 \cdot 2 \cdot 2 \cdot 2 \cdot 2 \cdot 2$ or 1024 possible outcomes of this experiment, only one of which consists of all tails. The probability of 10 tails is $1/1024$, a very unlikely event indeed.

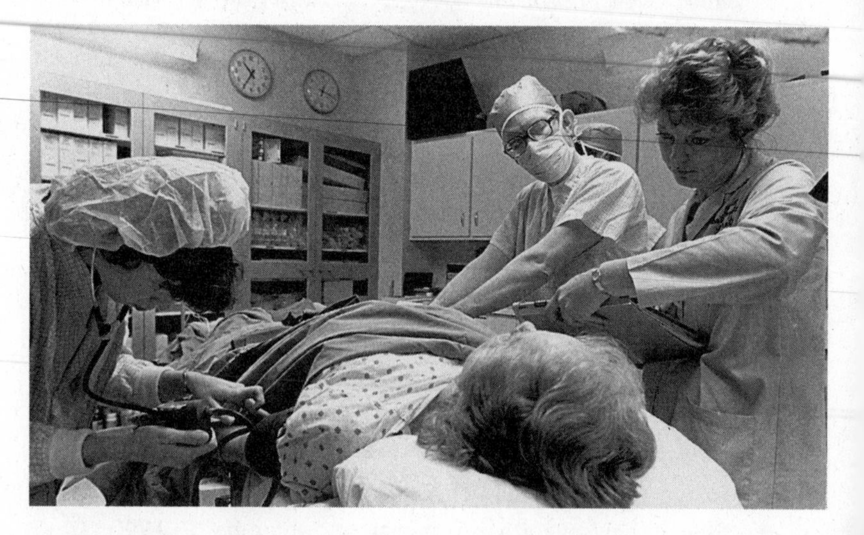

There is no other branch of mathematics in which intelligent people make such foolish mistakes.

Consider now the worried patient in the photograph. Most of us would not (and should not) find the doctor's logic very comforting. In fact, if we make the assumption that the outcome of the tenth operation is completely independent of the first nine, the probability of a tenth failure is still $\frac{9}{10}$. The assumption of independence may be questioned. Perhaps doctors improve with experience; but this particular patient had better hope for a miracle.

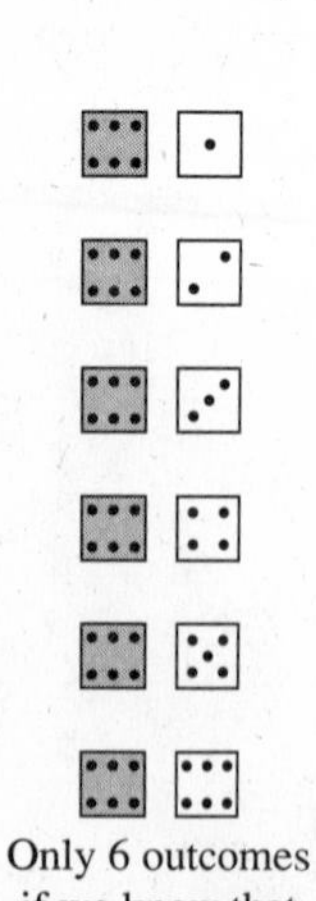

Only 6 outcomes if we know that brown die shows 6

Dependence Versus Independence

Perhaps we can make the distinction between dependence and independence clear by describing an experiment in which we again toss two dice—one brown and the other white. Let A, B, and C designate the following events:

A: brown die shows 6
B: white die shows 5
C: total on the two dice is greater than 7

Consider first the relationship between B and A. It seems quite clear that the chance of B occurring is not affected by our knowledge that A has occurred. In fact, if we let $P(B \mid A)$ denote the **probability of B, given that A has occurred,** then (see illustration in margin)

$$P(B \mid A) = \frac{1}{6}$$

But this is equivalent to the answer we get if we calculate $P(B)$ without any knowledge of A. For then we look at all 36 outcomes for two dice and note that 6 of them have the white die showing 5, that is,

$$P(B) = \frac{6}{36} = \frac{1}{6}$$

We conclude that $P(B \mid A) = P(B)$, just as we expected.

The relation between C and A is very different; C's chances are greatly improved if we know that A has occurred. From the marginal illustration, we see that

$$P(C \mid A) = \frac{5}{6}$$

However, if we have no knowledge of A, and calculate $P(C)$ by looking at the 36 outcomes (page 172) for two dice, we find

$$P(C) = \frac{15}{36} = \frac{5}{12}$$

Clearly $P(C \mid A) \neq P(C)$.

This discussion has prepared the way for a formal definition. If $P(B \mid A) = P(B)$, we say that A and B are **independent events.** If $P(B \mid A) \neq P(B)$, A and B are **dependent events.**

> **WARNING**
> *Do not confuse the words "independent" and "disjoint." Roughly speaking, independent events are events that do not influence each other. Disjoint events are events that cannot happen simultaneously.*

And now, recalling that $A \cap B$ means A and B, we can state the multiplication rule for probabilities:

$$P(A \cap B) = P(A)P(B \mid A) = P(B)P(A \mid B)$$

In words, the probability of both A and B occurring is equal to the probability that A will occur multiplied by the probability that B will occur, given that A has already occurred. In the case of independence, this takes a particularly elegant form:

$$P(A \cap B) = P(A)P(B)$$

Recall our solution to the problem, **The Carnival Game.** We had been given that for event A, having the penny land on the table, $P(A) = \frac{6}{7}$, and we had decided that for event B, the penny coming to rest entirely inside a square, $P(B) = \frac{1}{6}$. We then used the rule above to conclude that

$$P(A \cap B) = \frac{6}{7} \cdot \frac{1}{16} = \frac{6}{112}$$

To illustrate these rules in less complicated circumstances, consider first the problem of drawing two cards one after another from a well-shuffled deck. What is the probability that both will be spades? Based on previous knowledge (i.e., Section 6.1), we respond:

$$P(\text{two spades}) = \frac{{}_{13}C_2}{{}_{52}C_2} = \frac{\dfrac{13 \cdot 12}{2 \cdot 1}}{\dfrac{52 \cdot 51}{2 \cdot 1}} = \frac{13 \cdot 12}{52 \cdot 51}$$

But here is another approach. Consider the events:

A: getting a spade on the first draw
B: getting a spade on the second draw

Our interest is in $P(A \cap B)$. According to the rule above, it is given by

$$P(A \cap B) = P(A)P(B \mid A) = \frac{13}{52} \cdot \frac{12}{51}$$

which naturally agrees with our earlier answer.

Here is a different but related problem: Suppose we note the character of the first card, replace it, shuffle the deck, and draw a second card. Now A and B are independent.

$$P(A \cap B) = P(A)P(B) = \frac{13}{52} \cdot \frac{13}{52}$$

6 RED AND
4 GREEN BALLS.

Urns and Balls

For reasons not entirely clear, teachers have always illustrated the central ideas of probability by talking about urns (vases) containing colored balls. Most of us have never seen an urn containing colored balls, but it won't hurt to use a little imagination.

9 RED AND
1 GREEN BALLS.

Consider two urns labeled A and B, A containing six red balls and four green balls, and B containing nine red balls and one green ball. If a ball is drawn from each urn, what is the probability that both will be red?

$$P(\text{red from A } \textit{and} \text{ red from B}) = P(\text{red from A}) \cdot P(\text{red from B})$$

$$= \frac{6}{10} \cdot \frac{9}{10} = \frac{54}{100} = \frac{27}{50}$$

As a second problem, suppose we choose an urn at random and then draw one ball. What is the probability that it will be red? In other words, what is the probability of the event

"choose A *and* draw red" *or* "choose B *and* draw red"?

The events in quotation marks are disjoint, so their probabilities can be added. We obtain the answer

$$\frac{1}{2}\cdot\frac{6}{10}+\frac{1}{2}\cdot\frac{9}{10}=\frac{15}{20}=\frac{3}{4}$$

The reader should note the procedure we use. We describe the event using the words "and" and "or." When we determine probabilities, "and" translates into "times," and "or" into "plus."

A Historical Example

Two French mathematicians, Pierre de Fermat (1601–1665) and Blaise Pascal (1623–1662), are usually given credit for originating the theory of probability. This is how it happened: The famous gambler, Chevalier de Méré, was fond of a dice game in which he would bet that a 6 would appear at least once in four throws of a die. He won more often than he lost for, though he probably didn't know it,

$$P(\text{at least one } 6) = 1 - P(\text{no } 6\text{'s})$$
$$1-\frac{5}{6}\cdot\frac{5}{6}\cdot\frac{5}{6}\cdot\frac{5}{6}=1-.48=.52$$

Growing tired of this game, he introduced a new one played with two dice. Méré then bet that at least one double 6 would appear in 24 throws of two dice. Somehow (perhaps he noted that $\frac{4}{6}=\frac{24}{36}$) he thought he should do just as well as before. But he lost more often than he won. Mystified, he proposed it as a problem to Pascal who in turn wrote to Fermat. Together they produced an explanation. Here it is:

He [Méré] is very intelligent but he is not a mathematician: this as you know is a great defect.

Pascal, in a letter to P. de Fermat

$$\begin{aligned} P(\text{at least one double } 6) &= 1 - P(\text{no double } 6\text{'s}) \\ &= 1-\left(\frac{35}{36}\right)^{24} = 1 - .51 = .49 \end{aligned}$$

Out of this humble and slightly disreputable origin grew the science of probability.

Summary

We must carefully distinguish between independent and disjoint events. Two events are independent if the occurrence of one does not influence the occurrence of the other. They are disjoint if they cannot occur simultaneously. The corresponding laws of probability are

$$P(A \cap B) = P(A)\cdot P(B) \quad \text{(if } A \text{ and } B \text{ are independent)}$$
$$P(A \cup B) = P(A) + P(B) \quad \text{(if } A \text{ and } B \text{ are disjoint)}$$

Both laws have extensions which are always valid:

$$P(A \cap B) = P(A)P(A \mid B)$$
$$P(A \cup B) = P(A) + P(B) - P(A \cap B)$$

The last law was introduced in Problem 21 in Section 6.1

Problem Set 6.2

1. Toss a balanced die three times in succession. What is the probability of all 1's?
2. Toss a fair coin four times in succession. What is the probability of all heads?
3. Spin the two spinners pictured below. What is the probability that
(a) Both will show red?
(b) Neither will show red?
(c) Spinner A will show red and spinner B not red?
(d) Spinner A will show red and spinner B red or green?
(e) Just one of the spinners will show green?

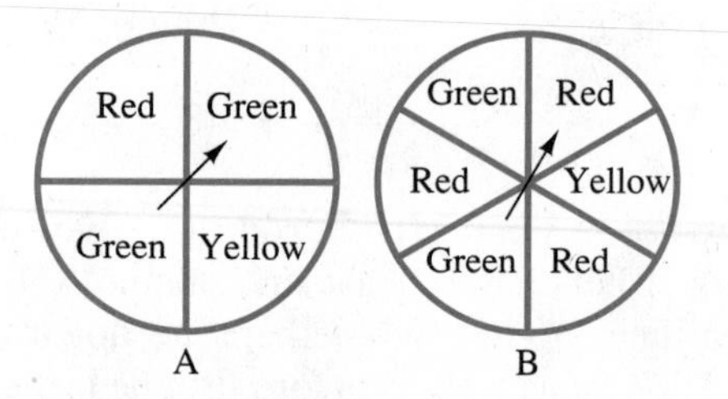

4. The two boxes shown below are thoroughly shaken and a ball drawn from each. What is the probability that
(a) Both will be 1's?
(b) Exactly one of them will be a 2?
(c) Both will be even?
(d) Exactly one of them will be even?
(e) At least one of them will be even?

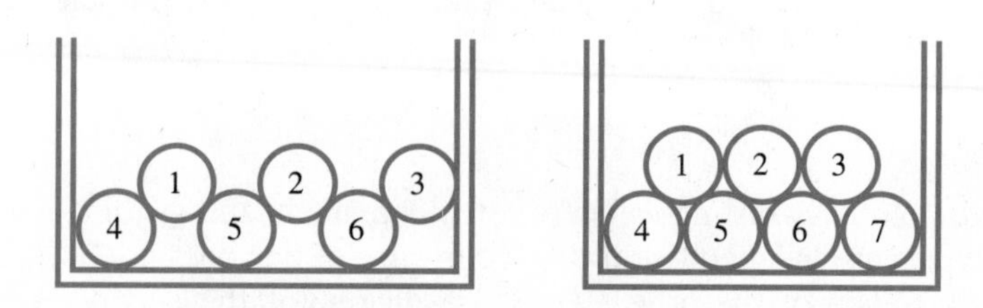

5. An urn contains five red balls and seven black balls. Two balls are drawn in succession. What is the probability of drawing two red balls if
(a) The first ball is replaced before the second is drawn?
(b) The first ball is not replaced before the second is drawn?

6. Answer the questions in Problem 5 if the urn contains five red balls, six black balls, and seven green balls.

7. Given that $P(A) = .8$, $P(B) = .5$, and $P(A \cap B) = .4$, find
(a) $P(A \cup B)$ (b) $P(B \mid A)$ (c) $P(A \mid B)$

8. Given that $P(A) = .8$, $P(B) = .4$, and $P(B \mid A) = .3$, find
(a) $P(A \cap B)$ (b) $P(A \cup B)$ (c) $P(A \mid B)$

9. In each case indicate whether or not the two events seem independent to you. Explain.
(a) Getting an A in physics and getting an A in math.
(b) Getting an A in physics and winning the tennis match.
(c) Getting a new shirt for your birthday and stubbing your toe the next day.
(d) Going to Harvard and being an American Indian.
(e) Being a woman and being a doctor.
(f) Walking under a ladder and having an accident the next day.

10. In each case indicate whether the two events are disjoint.
(a) Getting an A in Physics 101 and getting a B in Physics 101.
(b) The sun shining on Tuesday and raining on Tuesday.
(c) In tossing two dice, getting an odd sum and getting the same number on each die.
(d) In tossing two dice, getting an odd sum and getting a six on one of the dice.
(e) Getting an A in physics and getting an A in math.
(f) Not losing the football game and not winning the football game.

11. A machine produces bolts which are put in boxes. It is known that 1 box in 10 will have at least one defective bolt in it. Assuming that the boxes are independent of each other, what is the probability that a customer who ordered 3 boxes will get all good bolts?

12. Suppose the probability of being hospitalized during a year is .20. Assuming independence of family members, what is the probability that no one in a family of five will be hospitalized this year? Do you think the assumption of independence is reasonable?

13. Suppose that 4% of males are color blind, that 1% of females are color blind, and that males and females each make up 50% of the population. If a person is chosen at random, what is the probability that this person will be color blind?

14. Consider two urns, one with three red balls and seven white balls, and the other with six red balls and four white balls. If an urn is chosen at random and then a ball drawn, what is the probability that it will be red?

15. A committee of three is chosen at random from among a group of six boys and four girls. What is the probability that it will consist of all boys?

16. A coin is tossed eight times. What is the probability of getting at least one head?

17.* Let us suppose that, in a World Series, the probability that team A will win any given game is $\frac{2}{3}$. Then $\frac{1}{3}$ is the probability that team B will win a given game. What is the probability that
(a) A will win the series in four games?
(b) B will win in four games?
(c) The series will end in four games?
(d) A will win in five games?
(e) The series will end in five games?

18.* Fifteen girls went to the beach. Five got sunburned, eight got bitten by mosquitoes, and seven returned without a mishap. Find the probability that
(a) A girl was both burned and bitten.
(b) A burned girl was bitten.
(c) A bitten girl was burned.

19.* Among families with four children known to have at least one boy, what percentage actually have two boys? What assumptions do you have to make to do this problem?

For Research and Discussion

Probability can generate a lot of argument, even among professional mathematicians. What was your answer to the opening problem, "The car and the goats?" Marilyn asserted that you should switch, provoking thousands of letters, not just a few from college faculty, saying she was wrong. The controversy went on for several months (see her columns of December 2, 1990, February 17, 1991). Mathematicians were not distinguished for their agreement nor their agreeableness. They were in fact chastised for the way some of them jumped into the public dispute by the President of the Mathematics Association of America [Leonard Gillman, "The car and goat fiasco," *Focus* (the Newsletter of the Mathematical Association of America), 11 (June 8, 1991)].

Read and summarize the published letters; also read and explain to your class the solution that finally appeared in the mathematical literature [L. Gillman, "The Car and the Goats," *Math. Monthly* (January, 1992) pp. 3–7].

6.3 The Binomial Distribution

Yahtzee

In the game Yahtzee, which involves rolling five dice, it is very desirable to get several 6's. What is the probability of getting exactly three 6's on the first roll? Four 6's?

The manufacturer of Yahtzee chooses to make all five dice of the same color, usually white. This is a matter of convenience; surely the reader will agree that a little point on each of the dice does not affect the way they roll. Let's imagine that the five dice have been painted black, green, purple, red, and white, respectively. This will allow us to keep track of each die separately. Now call getting a 6 a success (S), and getting anything else a failure (F). Thus, for each die,

$$P(S) = \frac{1}{6} \qquad P(F) = \frac{5}{6}$$

If we roll five dice, there are 10 ways of getting exactly three 6's (see chart in the margin). Because of the independence of the five dice, we can calculate the probability of each of these events by multiplication. For example,

$$P(\text{SSSFF}) = \frac{1}{6} \cdot \frac{1}{6} \cdot \frac{1}{6} \cdot \frac{5}{6} \cdot \frac{5}{6} = \left(\frac{1}{6}\right)^3 \left(\frac{5}{6}\right)^2$$

Color of Die				
B	*G*	*P*	*R*	*W*
S	S	S	F	F
S	S	F	S	F
S	S	F	F	S
S	F	S	S	F
S	F	F	S	S
S	F	S	F	S
F	S	S	F	S
F	S	F	S	S
F	S	S	S	F
F	F	S	S	S

and

$$P(\text{SSFSF}) = \frac{1}{6} \cdot \frac{1}{6} \cdot \frac{5}{6} \cdot \frac{1}{6} \cdot \frac{5}{6} = \left(\frac{1}{6}\right)^3 \left(\frac{5}{6}\right)^2$$

In fact, each of the 10 disjoint events has this same probability. Consequently,

$$P(\text{three 6's in rolling five dice}) = 10\left(\frac{1}{6}\right)^3 \left(\frac{5}{6}\right)^2 \approx .03$$

We need to know why the number 10 appears in this problem. Look at the chart again. A row is determined as soon as we decide where to put the three S's, that is, as soon as we select from the set {B, G, P, R, W} three dice to classify as S's. Now, we can choose three objects from five in ${}_5C_3$ ways; and recall that

$${}_5C_3 = \frac{5 \cdot 4 \cdot 3}{3 \cdot 2 \cdot 1} = 10$$

There is another way to view this problem. We are really asking for the number of five-letter words that can be made using three S's and two F's. From Section 5.2, we know that there are 5!/3!2! such words. But

$$\frac{5!}{3!2!} = {}_5C_3$$

No matter how we look at it, there are ${}_5C_3$ ways of getting three 6's in a roll of five dice. Thus

$$P(\text{three 6's in rolling five dice}) = {}_5C_3\left(\frac{1}{6}\right)^3 \left(\frac{5}{6}\right)^2$$

and, by similar reasoning.

$$P(\text{four 6's in rolling five dice}) = {}_5C_4\left(\frac{1}{6}\right)^4 \left(\frac{5}{6}\right)^1$$

We make one final remark about rolling five dice. Whether we roll five dice at once (as in Yathzee) or roll one die five times is of no significance in probability questions. From now on we adopt the second point of view.

The Binomial Distribution

The general situation we have in mind is this. Suppose that the outcomes of an experiment fall into two categories. One we call a success (S) and the other a failure (F). The probabilities S and F are presumed to be known, say

$$P(\text{S}) = p \qquad P(\text{F}) = q$$

It should be clear to the reader that $q = 1 - p$.

For example, if p were .3, then q would be .7. The experiment is repeated n times. What is the probability of getting exactly k successes?

The answer, based on the same reasoning as in the five-dice problem, is

$$P(k \text{ successes in } n \text{ trials}) = {}_nC_k p^k q^{n-k}$$

This result is closely related to the Binomial Formula we met in Section 5.4. Recall, for example, that

$$\begin{aligned}(p+q)^5 &= (q+p)^5\\ &= {}_5C_0p^0q^5 + {}_5C_1p^1q^4 + {}_5C_2p^2q^3 + {}_5C_3p^3q^2 + {}_5C_4p^4q^1\\ &\quad + {}_5C_5p^5q^0\\ &= q^5 + 5pq^4 + 10p^2q^3 + 10p^3q^2 + 5p^4q + p^5\end{aligned}$$

The terms in the last row are, respectively, the probabilities of zero, one, two, three, four, and five successes in an experiment repeated five times (e.g., the Yahtzee problem). For this reason, the number of successes in n trials of an experiment is said to exhibit a **Binomial Distribution.**

The case $p = q = \frac{1}{2}$ is particularly interesting. Suppose that a perfectly balanced coin is tossed nine times. What is the probability of getting exactly one head? Two heads? Three heads?

$$P(\text{one head in nine tosses}) = {}_9C_1\left(\frac{1}{2}\right)^1\left(\frac{1}{2}\right)^8 = 9 \cdot \frac{1}{512} \approx .02$$

$$P(\text{two heads in nine tosses}) = {}_9C_2\left(\frac{1}{2}\right)^2\left(\frac{1}{2}\right)^7 = 36 \cdot \frac{1}{512} \approx .07$$

$$P(\text{three heads in nine tosses}) = {}_9C_3\left(\frac{1}{2}\right)^3\left(\frac{1}{2}\right)^6 = 84 \cdot \frac{1}{512} \approx .16$$

The bar graph below shows the probability for any number of heads from zero to nine. We have superimposed on it the famous normal curve (see Section 7.3), which is often used to approximate the Binomial Distribution.

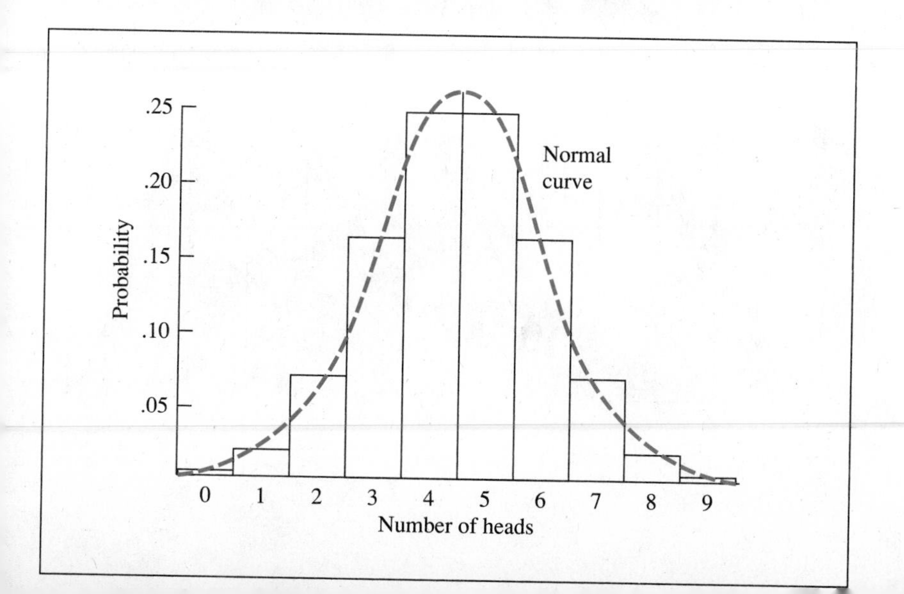

If you toss a coin 9 times, you may of course get anything from zero to nine heads. But if you repeat this experiment 100 or 1000 times and record the proportion of times you get zero, one, two, . . . , nine heads, you should expect a distribution something like that in the chart.

The Way the Ball Bounces

A very simple device called a Hexstat allows one to do the equivalent of tossing a coin nine times over and over again with very little effort. It consists of a device in which balls are dropped through a triangular-shaped maze, which should remind you of the path problem and Pascal's Triangle (Section 5.4).

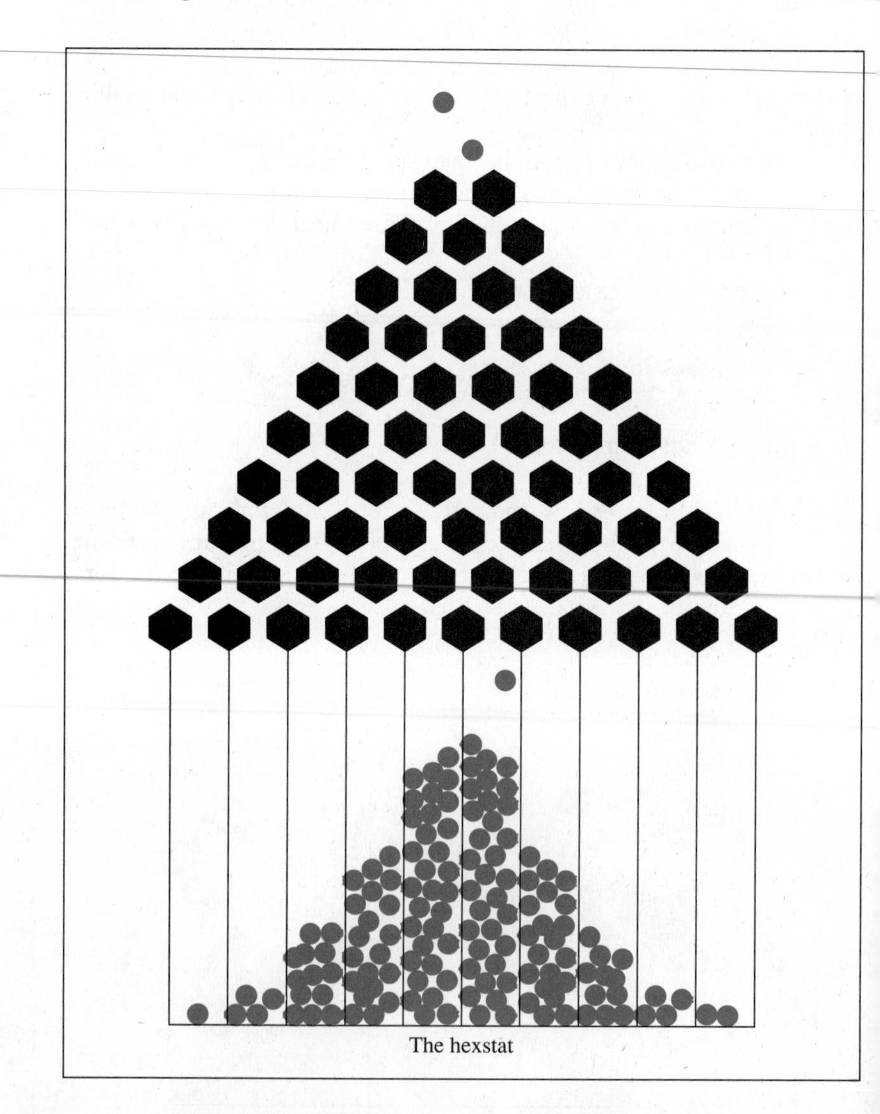

The hexstat

Notice that a ball dropped through the top slot has two "choices." It can go right or left, each with probability $\frac{1}{2}$. It continues to choose right or left at each of the nine stages of the maze, which is exactly like tossing a coin nine times. Thus the probability is $(\frac{1}{2})^9$ that a ball will wind up in the left column, ${}_9C_1\,(\frac{1}{2})^9 = 9\,(\frac{1}{2})^9$ that it will wind up in the next column, etc. Dropping 100 balls through this maze is equivalent to tossing a coin nine times on 100 occasions.

Two Practical Applications

A certain rare disease has been studied for over 50 years, and it is known that 30% of the people afflicted with it will eventually recover without treatment; the rest will die. A drug company has discovered what it claims is a miracle cure, citing as evidence the fact that 8 of the 10 people on whom it was tested recovered. Of course, this might have happened by chance even if the drug is absolutely worthless. We would like to know the probability that 8 or more of the 10 patients would have recovered without treatment.

> It is remarkable that a science which began with the consideration of games of chance should have become the most important object of human knowledge. . . . The most important questions of life are, for the most part, really only problems of probability.
>
> *Pierre Simon de Laplace*

$$P(8 \text{ would have recovered}) = {}_{10}C_8(.3)^8(.7)^2 \approx .0014$$
$$P(9 \text{ would have recovered}) = {}_{10}C_9(.3)^9(.7)^1 \approx .0001$$
$$P(10 \text{ would have recovered}) = {}_{10}C_{10}(.3)^{10}(.7)^0 \approx .0000$$

Therefore the probability of 8 or more recovering naturally is $.0014 + .0001 = .0015$. Either we have observed a very rare event or the drug is useful. The latter conclusion seems likely; thus the drug merits further experimentation.

A company that manufacturers $\frac{1}{2}$-inch steel bolts advertises that at most 1% of its bolts will break under a force of 10,000 pounds. To maintain quality control, it tests a random sample of 100 bolts from each day's output and keeps track of the number of defective ones. A typical record for 10 days is: 0, 2, 0, 1, 0, 3, 0, 0, 0, 1. The question the company faces is to decide when its manufacturing process is out of control. For example, does getting 3 defective bolts on the sixth day establish that something is wrong? Not necessarily. For by chance, one may get 3 or 4 or even 10 defective bolts in a random sample of 100 even if the day's total production has only 1% defective.

Based on many years' experience, the company has established the following rule. When samples from two consecutive days yield 3 or more defective bolts each, the manufacturing process is shut down and the equipment carefully checked. For if the process were under control (i.e., 1% or fewer defective bolts),

$$P(0 \text{ defective}) = {}_{100}C_0(.01)^0(.99)^{100} \approx .37$$
$$P(1 \text{ defective}) = {}_{100}C_1(.01)^1(.99)^{99} \approx .37$$
$$P(2 \text{ defective}) = {}_{100}C_2(.01)^2(.99)^{99} \approx .18$$

Thus

$$P(3 \text{ or more defective}) = 1 - P(\text{less than 3 defective})$$
$$\approx 1 - (.37 + .37 + .18) = .08$$

The probability of 3 or more defective bolts being produced on two consecutive days is $(.08)(0.8) = .0064$. An event with this small a probability is so rare that it is unlikely to have occurred by chance; it seems likely that something is wrong with the process.

Summary

Consider an experiment in which the outcomes can be put into two categories, one of which we call success and the other failure. Suppose that the probability of success is p, while that of failure is $q = 1 - p$. When this experiment is repeated n times, the probability of exactly k successes occurring is given by

$$P(k \text{ successes in } n \text{ trials}) = {}_nC_k p^k q^{n-k}$$

For example, if we toss a fair die 12 times and consider getting a 6 to be a success, the probability of getting exactly four 6's is

$$_{12}C_4\left(\frac{1}{6}\right)^4\left(\frac{5}{6}\right)^8$$

Problem Set 6.3

1. Calculate:
   (a) $_6C_1$ (b) $_6C_2$ (c) $_6C_3$
2. Calculate:
   (a) $_8C_1$ (b) $_8C_2$ (c) $_8C_3$
3. A balanced coin is tossed six times. Calculate the probability of getting
   (a) No heads.
   (b) Exactly one head.
   (c) Exactly two heads.
   (d) Exactly three heads.
   (e) More than three heads.
4. A fair coin is tossed eight times. Calculate the probability of getting
   (a) No tails.
   (b) Exactly one tail.
   (c) Exactly two tails.
   (d) Exactly three tails.
   (e) At most three tails.

5. Experiments indicate that, for an ordinary thumbtack, the probability of falling head down is $\frac{1}{3}$ and head up, $\frac{2}{3}$. Write an expression (but do not evaluate) for the probability in 12 tosses of its falling
(a) Head up exactly four times.
(b) Head up exactly six times.

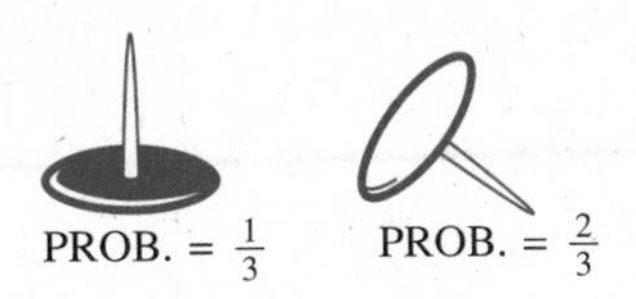

6. On a true-false test of 20 items, Homer estimates the probability of his getting any one item right at $\frac{3}{4}$. Write an expression for the probability of his getting
(a) Exactly 19 right.
(b) At least 19 right.

7. Extensive tables have been developed to calculate binomial probabilities. Use the table at the left to calculate
(a) ${}_{10}C_6(.25)^6(.75)^4$
(b) ${}_{10}C_3(.35)^3(.65)^7$

8. Veterinarians know that German shepherd pups will die from a certain disease with a probability of $\frac{1}{4}$. In a litter of 10 pups, what is the probability that
(a) Exactly two will die?
(b) At most two will die?

9. Major Electronics sells transistors to the United States government in lots of 1000. The government takes a random sample of 10 from each lot, puts them through a rigorous test, and accepts the lot if no more than 3 of the 10 break down. Major Electronics feels certain that at least three-fourths of its transistors will pass government tests. If they are correct, what is the probability that any given lot will be accepted by the government?

10. Assuming that men and women are equally likely to turn up in Math 412, what is the probability that Professor Witquick will have a class of nine women and one man?

11. Slugger Brown has a batting average of .350. In a three-game series, he expects to have 10 official at-bats. What is the probability that he will get three or more hits? What assumptions did you make in getting your answer?

12. A multiple-choice test has 10 questions, each with 4 alternative answers. If Homer uses the method of pure guessing, what is the probability he will get at least 5 right?

13.* In the game of Yahtzee, five dice are thrown at once. What is the probability on one throw of getting
(a) Five of a kind?
(b) Four of a kind?
(c) Three of a kind?
(d) A full house (two of one kind and three of another)?
(e) A large straight (1, 2, 3, 4, and 5 or 2, 3, 4, 5, and 6)?

Table 6-2
Table of ${}_{10}C_k p^k q^{10-k}$

k	$p = .25$	$p = .35$	$p = .50$
0	.0563	.0135	.0010
1	.1877	.0725	.0098
2	.2816	.1757	.0439
3	.2503	.2522	.1172
4	.1460	.2377	.2051
5	.0584	.1536	.2461
6	.0162	.0689	.2051
7	.0031	.0212	.1172
8	.0004	.0043	.0439
9	.0000	.0005	.0098
10	.0000	.0000	.0010

14.* A player in Yahtzee has reached his last turn to play and needs at least three 6's to achieve a top score. He has three chances to get it. On the first throw he tosses all five dice, but on the second and third he tosses only those that do not already show 6's. What is the probability that he will succeed?

15. A fair die is tossed 20 times. Calculate the probability of getting 3 or less sixes.

16. A coin is loaded so that the probability of getting a head is .469. Calculate the probability of getting exactly 6 heads when this coin is tossed 10 times.

17. A company's records indicate that the probability it will make a seriously defective TV set is .0124. Calculate the probability that in a production run of 100 sets, more than 2 will be seriously defective.

18. In a certain country, 104 boys are born for every 100 girls. In that country, what percentage of families with 8 children should have 4 boys and 4 girls?

For Research and Discussion

The game of Yahtzee has come up in this section in the introductory problem, in the illustrative discussion, and in the problem set. One who plays the game is confronted continually with decisions that can be informed by analysis of the type introduced in this section. A player with 2, 3, 4, 6, 6, showing on five dice after the first roll has a need on her scorecard for three 6's, but also needs a large straight (five numbers in succession, meaning either 1, 2, 3, 4, 5, or 2, 3, 4, 5, 6). For which should she try, given that she has two more turns to pick up any group of dice on the table and roll them again? Develop a list of numerous situations that arise, together with the probability of going from that position to a desired conclusion.

6.4 Some Surprising Examples

Your Hat, Sir?

When no tips were put on the counter, the angry hatcheck boy distributed the hats at random. He hoped that nobody would get the right hat. What is the probability that he succeeded?

The problem posed by the angry hatcheck boy is surprisingly difficult; that's why we are going to follow the advice given in Section 2.1 to begin with a simple case. If there were only three men, their hats could be distributed in $3! = 6$ ways (ABC, ACB, BAC, *BCA*, *CAB*, and CBA).

Two of these ways (those in italics) result in wrong hats on every head. The probability of this event is $\frac{2}{6} = .333$. With four men, there are $4! = 24$ ways of distributing the four hats and, as the reader can check by making a list, nine of them assign a wrong hat to everyone. The probability has increased to $\frac{9}{24} = .375$. By the time we get to six men, the problem of the ill-fitting hats has become a large headache. But with determined effort, we can list all $6! = 720$ ways of distributing the six hats and discover that 265 of them fail to put even one hat on the right head. The probability has dropped slightly to .368.

What if there were 100 men and 100 hats? Surely there is a better method than listing all 100! arrangements. There is. It was discovered by M. de Montmart in 1708, and the answer is easy to understand though the method of deriving it is not. Montmart showed that the probability P_n of no matches in a distribution of n hats to n men is given by

$$P_n = 1 - \frac{1}{1!} + \frac{1}{2!} - \frac{1}{3!} + \frac{1}{4!} - \frac{1}{5!} + \frac{1}{6!} + \cdots \pm \frac{1}{n!}$$

Table 6-3

n	P_n
1	.000
2	.500
3	.333
4	.375
5	.367
6	.368
7	.368
8	.368
9	.368
10	.368
100	.368
1000	.368

Most people think that, with 100 people, it would be almost certain that at least one person would get the right hat; that is, P_{100} would be very small. But contrary to intuition $P_{100} = .368$ (correct to three decimal places); it is almost exactly the same as P_6. In fact, P_n gets closer and closer to $1/e$, e being a mysterious number like π which occurs frequently in advanced mathematics. (e is an irrational number with a value of 2.71828. . . .)

Here is a different version of the same problem. A mischievous mailperson has 10 letters, 1 for each of 10 houses. She distributes them at random. What is the probability that at least 1 letter will be delivered to the right house? The event "at least one letter will go to the right house" is complementary to "no letter will go to the right house." The probability of the latter is P_{10} which we find in our table. Thus the probability we desire is

$$1 - P_{10} = 1 - .368 = .632$$

The Birthday Problem

Here is an old problem that has confused many a student and an occasional professor. Suppose there are 40 students in Math 13; let's call them $A_1, A_2, \ldots, A_{40}$. What is the probability that at least two of them have the same birthday? With 365 days to choose from, one might think it quite unlikely that there would be a match. Yet we shall show that the probability is .89; it is very likely that at least one match will occur.

To see this, we first make three simplifying assumptions which, if not quite true, are at least approximately correct.

1. There are 365 days in a year.
2. One day is as likely as another for a birthday.
3. Our 40 students were born ("chose" their birthdays) independently of each other.

Now let B be the event that at least two have the same birthday. Consider the complementary event B' that no two have the same birthday. We can think of B' as occurring this way. First A_1 chooses a birthday, then A_2 chooses a birthday different from that of A_1, next A_3 chooses a birthday different from both that of A_1 and A_2, etc. Using our three assumptions, we obtain

$$P(B') = 1 \cdot \frac{364}{365} \cdot \frac{363}{365} \cdot \cdots \cdot \frac{326}{365} \approx .11$$

Consequently,

$$P(B) = 1 - P(B') = 1 - .11 = .89$$

Here is a related problem. How large a class is needed to give a 50% chance (that is, probability $\frac{1}{2}$) of at least one pair of matching birthdays? To answer this and other birthday problems, we let

$$Q_n = P(\text{at least two in a class of } n \text{ people having the same birthday})$$

By looking at the complementary event as in our first problem, we can calculate any Q_n we wish. Some values are indicated in the table in the margin. Note that a class of 25 has well over a 50% chance of a match of birthdays (a class of 23 has almost exactly a 50% chance).

Table 6-4

n	Q_n
5	.03
10	.12
15	.25
20	.41
25	.57
30	.71
35	.81
40	.89

Probability Misused

The following story quoted from the April 26, 1968, issue of *Time* magazine illustrates in a vivid way how an uncritical use of probability theory can lead one astray.

> After an elderly woman was mugged in an alley in San Pedro, Calif., a witness saw a blonde girl with a ponytail run from the alley and jump into a yellow car driven by a bearded Negro. Eventually tried for the crime, Janet and Malcolm Collins were faced with the circumstantial evidence that she was white, blonde, and wore a ponytail while her Negro husband owned a yellow car and wore a beard. The prosecution, impressed by the unusual nature and number of matching details, sought to persuade the jury by invoking a law rarely used in a courtroom—the mathematical law of statistical probability.

The jury was indeed persuaded, and ultimately convicted the Collinses (*Time,* Jan. 8, 1965). Small wonder. With the help of an expert witness from the mathematics department of a nearby college, the presecutor explained that the probability of a set of events actually occurring is determined by multiplying together the probabilities of each of the events. Using what he considered "conservative" estimates (for example, that the chances of a car's being yellow were 1 in 10, the chances of a couple in a car being interracial 1 in 1,000), the prosecutor multiplied all the factors together and concluded that the odds were 1 in 12 million that any other couple shared the characteristics of the defendants.

Only one couple. The logic of it all seemed overwhelming, and few disciplines pay as much homage to logic as do the law and math. But neither works right with the wrong premises. Hearing an appeal of Malcolm Collins' conviction, the California Supreme Court recently turned up some serious defects, including the fact that not even the odds were all they seemed.

To begin with, the prosecution failed to supply evidence that "any of the individual probability factors listed were even roughly accurate." Moreover, the factors were not shown to be fully independent of one another as they must be to satisfy the mathematical law; the factor of a Negro with a beard, for instance, overlaps the possibility that the bearded Negro may be part of an interracial couple. The 12 million to 1 figure, therefore, was just "wild conjecture." In addition, there was not complete agreement among the witnesses about the characteristics in question. "No mathematical equation," added the court, "can prove beyond a reasonable doubt (1) that the guilty couple in fact possessed the characteristics described by the witnesses, or even (2) that only one couple possessing those distinctive characteristics could be found in the entire Los Angeles area."

Improbable Probability. To explain why, Judge Raymond Sullivan attached a four-page appendix to his opinion that carried the necessary math far beyond the relatively simple formula of probability. Judge Sullivan was willing to assume it was unlikely that such a couple as the one described existed. But since such a couple did exist—and the Collinses demonstrably did exist—there was a perfectly acceptable mathematical formula for determining the probability that another such couple existed. Using the formula and the prosecution's figure of 12 million, the judge demonstrated to his own satisfaction and that of five concurring justices that there was a 41% chance that at least one other couple in the area might satisfy the requirements.

"Undoubtedly," said Sullivan, "the jurors were unduly impressed by the mystique of the mathematical demonstration but were unable to assess its relevancy or value." Neither could the defense attorney have been expected to know of the sophisticated rebuttal available to them. Janet Collins is already out of jail, has broken parole and lit out for parts unknown. But Judge Sullivan concluded that Malcolm Collins who is still in prison at the California Conservation Center, had been subjected to "trial by mathematics" and was entitled to a reversal of his conviction. He could be tried agian but the odds are against it.

Time, April 26, 1968, p. 41. Reprinted by permission from *Time,* The Weekly News Magazine; © 1968 Times Inc.

Summary

All of us are guided to some extent by intuition; and a well-developed intuition can be invaluable even in mathematics. However, in problems involving probability, most people are easily deceived by intuition. It is therefore especially important to think through such problems in a very careful systematic way. We have tried to do this for two problems that many people find both intriguing and mystifying, the hat check problem and the birthday problem. The first few questions in the problem set are related to these two problems. But we also consider several new problems which many readers may find challenging and surprising.

Problem Set 6.4

1. Consider four bottles labeled A, B, C, and D.
   (a) List all 24 ways of arranging these bottles in a row.
   (b) How many of these arrangements have every letter out of its normal position?
   (c) If the four bottles are arranged in a row at random, what is the probability none of them will be in its normal position?
   (d) Calculate:

$$1 - \frac{1}{1!} + \frac{1}{2!} - \frac{1}{3!} + \frac{1}{4!}$$

   It should agree with your answer to part (c).

2. A playful secretary typed eight different business letters, addressed the eight corresponding envelopes, and then distributed the letters among the envelopes at random. Use one of the tables in this section to calculate the probability that
   (a) No one will get the right letter.
   (b) At least one person will get the right letter.

3. For a room of 20 people, calculate the probability that no 2 have the same birthday. Use a table in this section.

4. For a room of 30 people, calculate the probability that at least 2 of them have the same birthday.

5. A mischievous stock clerk took 24 cans of pineapple and marked them at random using a labeling device which can print any number of cents from 0 to 99. Write an expression for the probability that no 2 cans were marked with the same price.

6. The stock clerk in Problem 5 marked 120 cans of orange juice at random using the same labeling device. What is the probability that at least 2 were marked with the same price?

7. Who is right in the following problem? Peter and Paul are equally skilled at playing marbles, but Peter has two marbles and Paul only one. They roll to see who comes closer to a stake in the ground.

Peter reasons that there are four possibilities: Both his marbles may be better than Paul's, or his first may be better and his second worse, or his second may be better and his first worse, or both may be worse. Since three of the four cases lead to Peter's winning, Peter wins with a probability of $\frac{3}{4}$.

Paul looks at it this way. Each of the three marbles has an equal chance of winning. Two of the marbles are Peter's. Therefore Peter wins with a probability of $\frac{2}{3}$.

8. Here is an old folklore problem that admittedly has sexist overtones. A sultan wishes to increase the proportion of women (making larger harems possible). He decrees, "As soon as a mother gives birth to a male child, she shall be sterilized. If she bears a female, she may continue to have children." Will this law achieve the desired end?

9. Criticize the following imaginary conversation.

Susan: What is the probability of there being no horses on Mars?
Sally: Not knowing anything about it, I'd have to say $\frac{1}{2}$.
Susan: And what is the probability of no cows on Mars?
Sally: Again, I'd have to say $\frac{1}{2}$.
Susan: And of no dogs?
Sally: $\frac{1}{2}$.

(This conversation continued until Susan had named seven more forms of life.)

Susan: Very well. By the laws of probability, this means that the probability there are no horses, nor cows, nor dogs, nor any of the seven other forms of life is $\frac{1}{2} \cdot \frac{1}{2} \cdot \frac{1}{2} \ldots = 1/1024$. Thus the probability of life on Mars is at least $1 - (1/1024) = 1023/1024$. It's almost a certainty.

10. Professor Witquick and his wife each play a fair game of chess, but an honest observer would have to admit that the professor usually loses to his wife. One day, their son asked his father for $10 for a Saturday night date. After puffing his pipe for a moment, the professor replied:

"Today is Wednesday. You will play a game of chess tonight, another tomorrow, a third on Friday. Your mother and I will alternate as opponents. If you can win two games in a row you will get the $10."

"Whom do I play first, you or Mom?"

"Take your choice," said the professor.

If you were son, whom would you choose?

11. Suppose that, in flight, airplane engines fail independently of each other with a probability of .001 and that an airplane must have at least half of its engines working to fly successfully. Calculate the probability that a plane will make a successful flight if it has
(a) Two engines. (b) Four engines.

12.* Suppose in Problem 11 that the probability of failure for an engine is q. Show that a four-engine plane is safer if $q < \frac{1}{3}$, but a two-engine plane is safer if $q > \frac{1}{3}$.

13.* (St. Petersburg paradox.) A coin is tossed until a head appears. When a head appears on the first toss, the bank pays the player $1. When a head appears for the first time on the second toss, the bank pays $2. When a head first appears on the third toss, the bank pays $4, on the fourth toss $8, etc. How much should a player be willing to pay for the privilege of playing this game.

For Research and Discussion

Think of between five and ten situations where you will be in group of twenty-five or more people (but small enough to get their cooperation in a little experiment). Explain the birthday problem to them and ask them to please put on a sheet of paper that you will circulate their birth dates. Do your results conform to what theory predicts? Note that the theory itself requires some analysis on your part. For example, if you perform this experiment in each of five classes of sizes (let us say) 20, 30, 30, 35, and 70, how many times might you expect to find common birthdays? What if you treat all your data as one experiment with a total of (to continue our example) 185?

chapter 7

Organizing Data

When you can measure what you are speaking about and express it in numbers you know something about it; but when you cannot express it in numbers, your knowledge is of a meagre and unsatisfactory kind.

LORD KELVIN

Statistical thinbking will one day be as necessary for efficient citizenship as the ability to read and write.

H. G. WELLS

Not Really Good at That

"You're a mathematician; you keep score."

"You're a mathematician? Oh, wow! I can't even balance my checkbook."

A mathematician must be careful in social settings. One's friends invariably feel that you should be the one to sort out who owes what when the bill is brought to your table in the restaurant. You should quickly figure for them the effect on their taxes if the state's new tax bill gets approved, or whether an investment they are considering promises a decent return.

It's not that you don't want to help them. It's just that you're not as good at these things as is generally assumed. You are not a human computer. Your neighbor's well intentioned question about whether all these computers are likely to put you out of work is not on the top of your list of worries. The effect of the new tax bill on the university's budget is a bigger concern. Your spouse may manage the household checkbook because you forget to keep adequate records.

A mathematician may be interested in the number of ways that three toppings for the pizza can be chosen from the available options, or in the implications of Arrow's Theorem on rank orderings for the group process of choosing the toppings. The possibility of cutting the pizza into n pieces of equal area without cutting through any of the slices of pepperoni could be of interest. Straightening out the bill will almost certainly not be of interest.

The mathematician Euler was interested when a friend, noting that it is better to receive interest compounded quarterly rather than annually, and better yet if compounded daily, posed the puzzle of what might happen if it was compounded hourly, or minute by minute, or even continuously. Few mathematicians, however, are interested in computing the interest on their own bank account. Like everybody else, they take the banker's word for it.

Mathematicians think of their work as that of finding an effective, preferably elegant, way to solve a problem that at first looks untractable, of finding patterns in data, of guessing and perhaps proving that a collection of accepted statements have certain implications that are not at all evident, and of generalizing work they have done in ways that might make it applicable to new situations. They do not see themselves as wizards who can do arithmetic computations in their heads faster or more accurately than other people. They are reminded on a daily basis that their self-perception is not the popular perception.

7.1 Getting the Picture

What's Wrong Here?

This picture appeared with a story reporting that the federal government was setting fuel economy standards to be met by automobile manufacturers, beginning with 18 miles per gallon in 1978 and moving in steps to 27.5 miles per gallon in 1985. What do you see that is wrong with this graphic description of information?

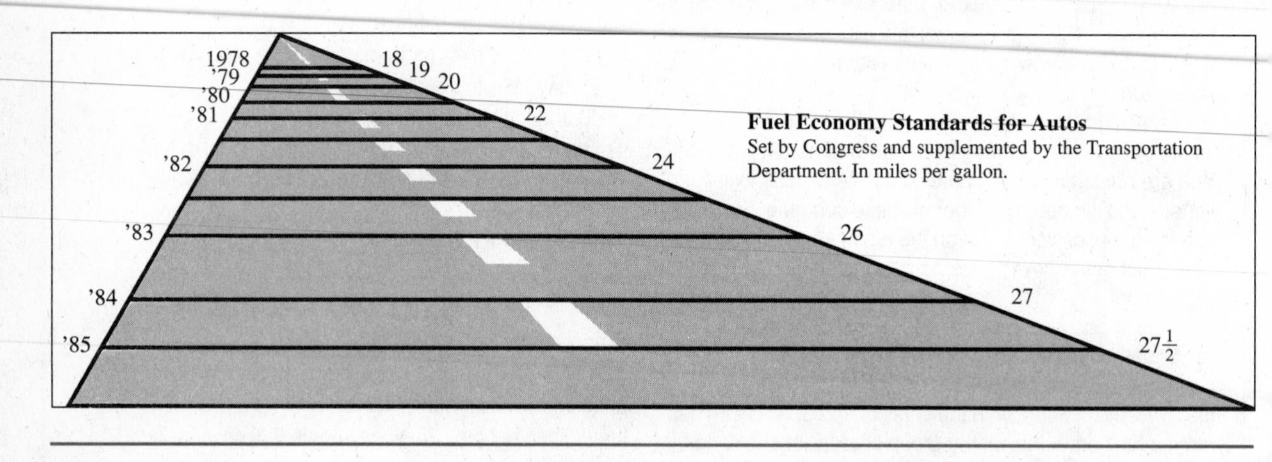

A dry cubic foot of white oak weighs about 47 pounds. A package (1.69 ounces) of M&M plain chocolate candies contains 230 calories and 10 grams of fat. The Farmer's Home Administration made or guaranteed $20.5 billion in loans to buy farmlands, tractors, livestock, seeds, or chemicals from 1980 through 1992.

Collected data, not put into some sort of context, are usually dull, frequently meaningless. It is only when data have been condensed, analyzed, and interpreted that they begin to have significance. Statistics is the science of organizing, interpreting, and drawing inferences from numerical data.

An old proverb says that one picture is worth a thousands words. Statisticians believe this maxim. Faced with a mass of data, a statistician tries to think of a clear way to display it. The picture used depends on what is deemed significant or at least what the statistician wants us to believe is significant about the data. Some commonly used devices are shown on the facing page. We call them pie graphs and bar graphs.

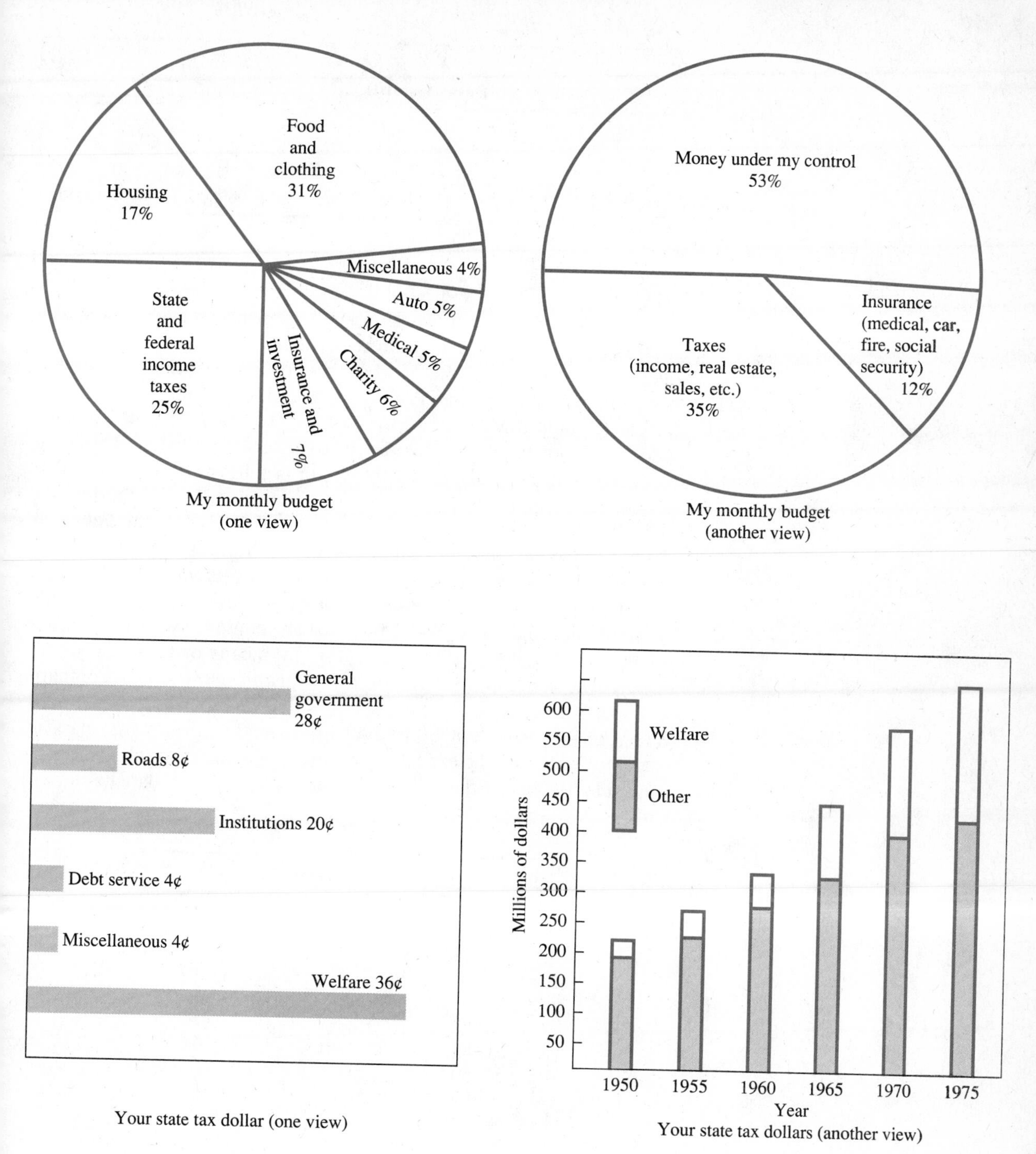

My monthly budget (one view)

My monthly budget (another view)

Your state tax dollar (one view)

Your state tax dollars (another view)

Table 7-1
Data: What Do They Mean?

32	34	39	40	42	40	50	55	60	62	60	50
42	44	46	42	48	50	50	58	64	62	62	56
48	51	52	52	60	61	55	57	80	60	60	58
51	53	52	55	60	54	60	60	80	70	70	66

Frequency Charts and Bar Graphs

Consider the set of data above. How shall we display it in a pictorial way that will bring out its main features? Clearly one cannot answer this question without some idea of what the data represent.

Let us suppose that Professor Witquick has just received the SMOB (Standard Measure of Brainpower) scores for the 48 students who have registered for his freshmen calculus course. They are recorded in the table above. The professor is delighted to note two 80's, perfect scores on the SMOB, but he also sees several scores in the 30's and 40's. Under his breath, he damns the high schools for failing to teach mathematics as they ought to. But before making further judgments, he decides to analyze the data more carefully.

One good idea is a **frequency chart.** Professor Witquick groups the data into 10 classes (8 to 15 classes are commonly used), keeping track of how many fall into each class by means of tallies. Then he displays the same information on a **bar graph.** Now he can compare this year's class with last year's, for which he has kept a bar graph. When he does this, maybe he will discover that this year's class is actually better than he expected. In any case, the bar graph gives him a much better idea of the ability of his class than the raw data he started with.

Frequency Chart (SMOB Scores)

Class Range	*Class Tally*	*Frequency*
31–35	II	2
36–40	III	3
41–45	IIII	4
46–50	~~IIII~~ II	7
51–55	~~IIII~~ ~~IIII~~	10
56–60	~~IIII~~ ~~IIII~~ II	12
61–65	~~IIII~~	5
66–70	III	3
71–75		0
76–80	II	2

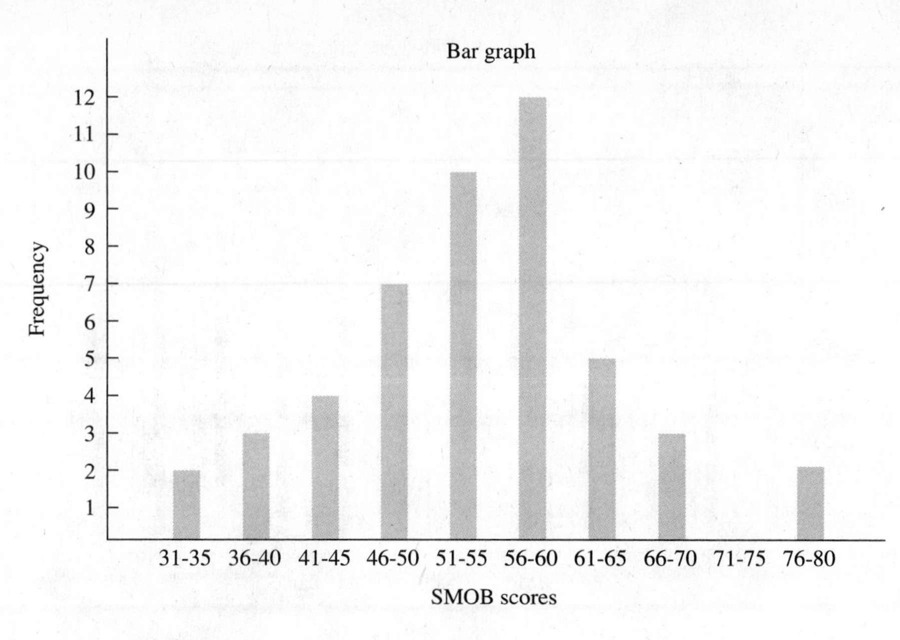

Trend Lines and Comparative Bar Graphs

Consider the same data once more (page 208). Suppose now that the first row represents the monthly car sales of Chrysler City during 1990 the second row shows sales during 1991, etc. It makes no sense to treat the sales data as we did the SMOB scores. What interests the sales manager is a chart showing trends over time, or perhaps some kind of

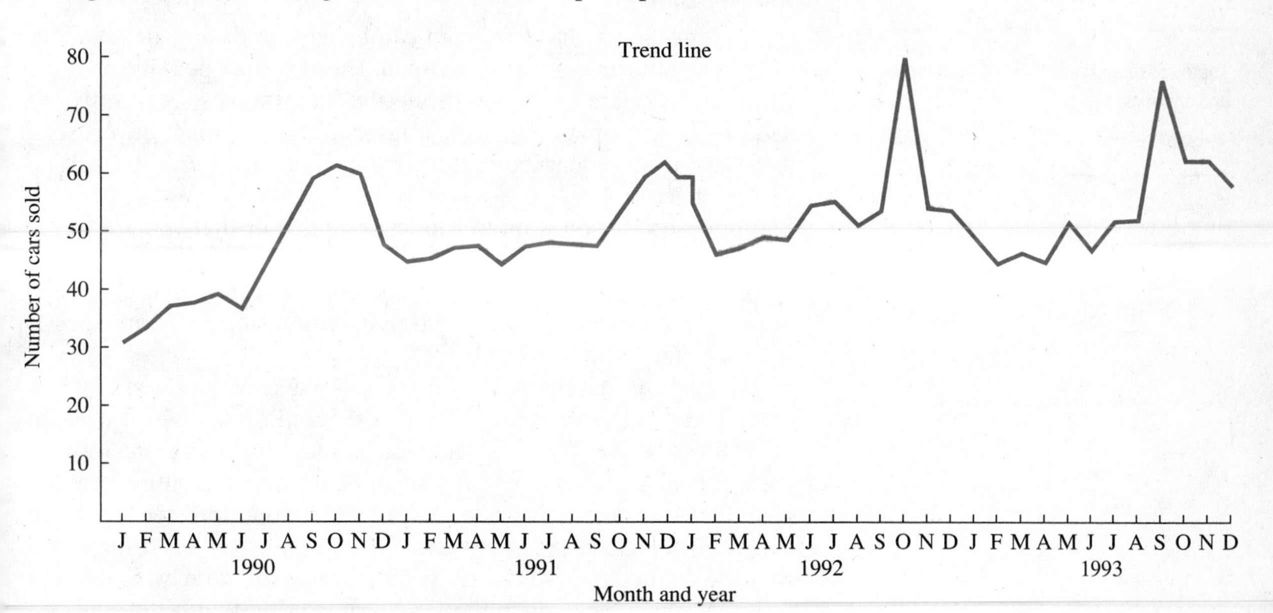

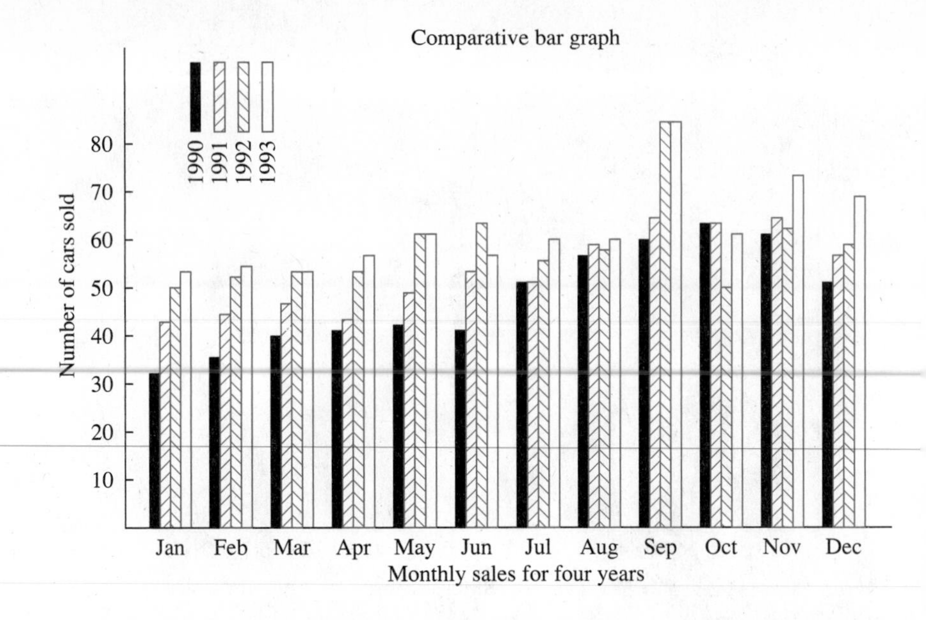

chart that compares one year with another. We can accomplish the first job with a so-called **trend line,** and the second with a **comparative bar graph** (see accompanying diagrams). Note that, while great fluctuations are present, the trend is upward; and this makes the sales manager happy. Can you account for the spurt in sales each year during early fall?

Distorting Data

> There are three kinds of lies: lies, damned lies and statistics.
>
> *Disraeli*

We have seen that a chart or graph can be used to display data so that their true significance is easily grasped. But it is also possible to use a graph to distort data or to exaggerate the importance of carefully selected bits of information. This is a favorite trick of some advertisers, and most of us have learned to discount their claims even if backed up by fancy charts.

Suppose that you want to win an argument or that you are trying to sell something, say stocks in the ABC Company. Naturally you want to display the stock price history of the company in a favorable way; perhaps you are tempted to use a bit of careful distortion. Some possibilities are shown on page 211.

Or suppose that the ABC Company is engaged in oil refining and has doubled production from 1975 to 1985 and then doubled it again from 1985 to 1990. This performance is good, but there are ways to make it appear even better. Draw a small oil drum representing production in 1975, another twice as high for 1985, and a third twice as high again for 1990. So that the drums will look like real oil drums, you double the diameter as well as the height. You can't quite be accused of cheating; yet you have given the visual impression that oil production has grown eightfold at each stage (see Problems 11 and 12). A blazing title encourage the impression.

Of course, you want your clients to extrapolate into the future. If oil production doubled in 10 years, and then again in 5 years, it can be expected to double again in $2\frac{1}{2}$ years, then again in $1\frac{1}{4}$ years, etc. Of course you hope that this vague impression of spectacular growth will stay just that way; vague. The client who habitually explores the implications of this thinking when carried to long periods of time will see that it leads to absurdity.

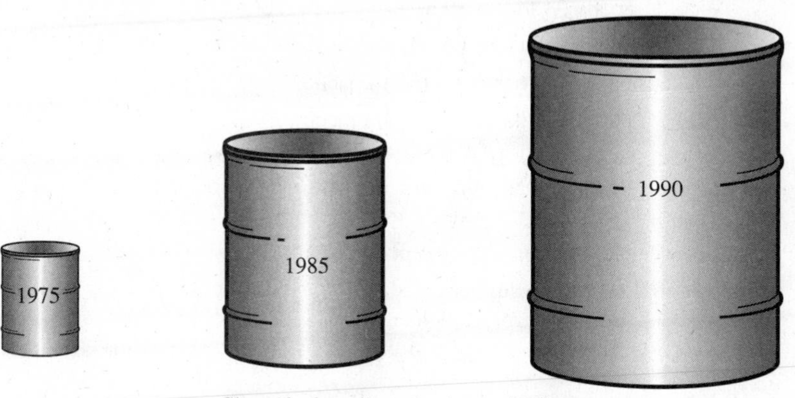

Skyrocketing oil production at ABC

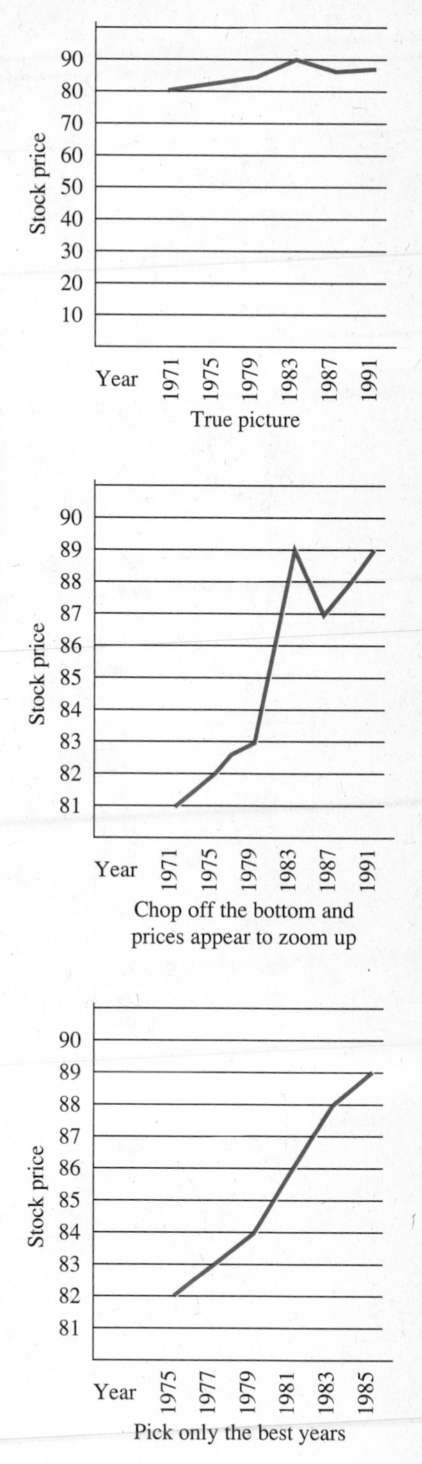

True picture

Chop off the bottom and prices appear to zoom up

Pick only the best years

Mark Twain recognized this fallacy and wrote about it in *Life on the Mississippi,* (New York: Harper & Row, 1901), p. 136:

> In the space of one hundred and seventy-six years the Lower Mississippi has shortened itself two hundred and forty-two miles. That is an average of a trifle over one mile and a third per year. Therefore, any calm person, who is not blind or idiotic, can see that in the Old Oolitic Silurian Period, just a million years ago last November, the Lower Mississippi River was upward of one million three hundred thousand miles long, and stuck out over the Gulf of Mexico like a fishing rod. And by the same token any person can see that seven hundred and forty-two years from now the Lower Mississippi will be only a mile and three-quarters long, and Cairo and New Orleans will have joined their streets together, and be plodding comfortably along under a single mayor and a mutual board of aldermen. There is something fascinating about science. One gets such wholesale returns of conjecture out of such a trifling investment of fact.

Summary

We have sampled a few of the many possibilities for pictorial display of data. A statistician needs both common sense and imagination—common sense to grasp what is important in the data, and imagination to think of a way to display the data clearly and without distortion. A dose of integrity would be a nice adornment. Data, clearly displayed and fairly interpreted, can lead to wise decisions. But they can also be manipulated by a deceptive person who wants to justify a course of action already decided on.

Problem Set 7.1

Data Set A		
25	24	30
40	50	48
50	52	54
81	81	83
84	88	94
60	62	72
59	64	72
57	59	68
49	54	66
43	50	54
42	45	50
30	35	42

Data Set B		
80	83	92
62	70	81
63	71	80
50	55	54
36	34	40
32	36	40
24	30	34
25	24	30
30	36	40
55	59	64
81	90	93
92	92	94

1. At Podunk University the student congress allocated its $40,000 budget as follows: student newspaper $15,000; literary magazine $4000; social committee $8000; student government $3000; student organizations $5000; special projects $2000; reserve fund $1500; miscellaneous $1500. Make a pie graph to display these data.

2. Make a bar graph to display the data in Problem 1.

3. Make a frequency chart for data set A shown in the margin, using the class intervals 24 to 29, 30 to 35, 36 to 41, etc.

4. Follow the directions in Problem 3 for data set B.

5. Make a bar graph based on the frequency chart in Problem 3.

6. Make a bar graph based on the frequency chart in Problem 4.

7. Which do you think is more likely? Data set A represents the scores on a physics exam, or it represents the ages of welfare recipients in Jackson Township. Justify your answer.

8. Answer the question in Problem 7 for data set B.

9. Suppose that the columns in data set A represent the monthly tractor sales of Smith's Farm Implements for 1990, 1991, and 1992, respectively. Display this data in two good ways. What conclusions can you draw?

10. Suppose that the columns in data set B represent the monthly snowmobile sales of the Olson Manufacturing Company for 1990, 1991, and 1992, respectively. Make a comparative bar graph displaying these data. What conclusions can you draw?

11. If you double the dimensions of a square, what happens to its area? If you double the dimensions of a cube, what happens to its volume?

12. If you triple the radius of a circle, what happens to its area? If you triple the radius of a sphere, what happens to its volume?

13. Which of the following statistical arguments seem valid to you? Explain your answer.

(a) In 1992, 492 men and 301 women drivers had auto accidents in Mudville. Therefore women are safer drivers than men.

(b) Ohio had more traffic fatalities than Rhode Island in 1990. Therefore it is safer to drive in Rhode Island.

(c) Ninety percent of all the cars sold in the United States by a certain foreign country are still on the road. No American manufacturer can make this claim. Therefore these foreign cars last longer than American cars.

(d) All the students in a class failed a certain physics test. It must be a dumb class.

14. The total sales of a certain company were $10,000, $20,000, $40,000, and $80,000 in 1988, 1989, 1990, and 1991, respectively. A stockbroker claims that it will certainly have total sales of $1 million by 1995. How would you analyze this claim?

15.* Roll a pair of dice 100 times, keeping track of the total number of spots showing after each roll. Make a frequency diagram and a bar graph for these data.

16.* Data set C represents the total car sales of Watson Pontiac during each of the years 1981 to 1990. Make a bar graph that displays this information in an objective manner. Now make a bar graph that distorts this information in such a way as to make the sales appear to be going up dramatically. Suggestions: Combine years, chop off the bottom, etc.

Data Set C	
1981	90
82	81
83	90
84	83
85	92
86	82
87	90
88	85
89	89
1990	88

For Research and Discussion

The graphic display of gas mileage standards, taken originally from the *New York Times,* is reproduced in E. R. Tufte, *The Visual Display of Quantitative Information* (Chesire, Conn.: Graphics Press, 1983), p. 57, together with the following observations.

1. The line representing 18 miles per gallon in 1978 is .6 inches long. The line representing 27.5 miles per gallon (a 53% increase over 18) is represented by a line 5.3 inches long (a 783% increase over .6).

2. When a road is used to depict a time line, the present is usually at our feet and the future recedes into the horizon. This display reverses the convention, greatly exaggerating the severity of the new standards.

3. While the dates on the left remain constant in size as they move toward us, the numbers on the right increase, apparently to make a point.

Consult the book by Tufte, or other books [D. Huff, *How to Lie with Statistics* (Norton, 1954)] that help recognize common tricks. Then look in current periodicals for examples that seem designed to carry editorial comment as well as information.

7.2 On the Average

The Average Salary

The owner of XYZ Enterprises likes to point out that the average salary at his company is $26,000. He does not mention, however, that the average for him and his son is $147,000. What is the average of the other 21 employees?

Few people can look at a mass of data and make sense out of it. This is why statisticians search for ways to condense and summarize data. The goal of a summary is to pick out what is most significant. Summaries do not tell everything; but they ought to give the main idea.

Let's carry the notion of summarizing data to its absolute extreme. For a given set of numerical data, what *one* number best represents these data? That is the theme of this section.

Compensation at XYZ Enterprises

The average salary of, shall we say, the average worker at XYZ is easy to calculate using the methods of algebra (let A be the average of the 21 employees, so that their total earnings are 21A), but it might be more consistent with a labor negotiator's viewpoint to begin with the observation that the company's total payroll is 23($26,000) = $598,000, of which 2($147,000) = $294,000 goes to the owner and his son. That leaves $304,000 for the rest of them. Then

$$\frac{\$304,000}{21} = \$14,476$$

is their average salary.

Strip away the irrelevant details.

Here are the actual salary data at XYZ, data that the owner would much prefer to keep secret. President Hornblower makes $216,000; the Vice-President, Hornblower's son gets $78,000. There are three supervisors at $20,000 each, eight skilled workers at $16,000 each, nine laborers at $12,000 each, and a student assistant at $8,000. One way of displaying these data is simply to list them from highest to lowest with repetitions as they occur:

216,000
78,000
20,000; 20,000; 20,000
16,000; 16,000; 16,000; 16,000; 16,000; 16,000; 16,000; 16,000
12,000; 12,000; 12,000; 12,000; 12,000; 12,000; 12,000; 12,000; 12,000
8000

We call this a **distribution,** and the actual numbers that occur are called values. Thus the distribution above consists of 23 values.

The Mean, Median, and Mode

The **mode** of a distribution is the most frequently occurring value. In the salary example, the mode is 12,000. It is possible for several different values to occur with the maximum frequency. Each of them is then called a mode.

The **median** of a distribution is its middle value. If there is an odd number of values, this makes good sense. For example, in the distribution of 23 salaries, the middle value is the twelfth from the top (or bottom), which happens to be 16,000. When there is an even number of values, two of them will vie for the middle. In this case, we take the median to be the number halfway between these two; that is, we add them and divide by 2. In the distribution of 14 quiz scores shown at the right, the median is $(7 + 6)/2 = 6.5$. Incidentally, this distribution of scores has two modes, namely, 5 and 7. We say it is bimodal.

Quiz Scores
10
9, 9
8
7, 7, 7
6, 6
5, 5, 5
4, 4

Median: 6.5
Modes: 5 and 7
Mean: 6.57

The **arithmetic mean** (often called simply the mean or average) is the number most commonly used to represent a distribution. If x_1, $x_2, \ldots, x_n$ are the values of the distribution (including repetitions), the mean $\bar{x}$ (read "x bar" is defined by

$$\bar{x} = \frac{x_1 + x_2 + \cdots + x_n}{n}$$

Thus, for the quiz scores,

$$\bar{x} = \frac{4 + 4 + 5 + 5 + 5 + 6 + 6 + 7 + 7 + 7 + 8 + 9 + 9 + 10}{14}$$

$$= \frac{92}{14} \approx 6.57$$

With regard to salaries at XYZ Enterprises, President Hornblower was right. The mean salary (average salary) is $26,000. But who can blame the workers for thinking that this is a mean way to figure the average pay?

There is another way to calculate the mean, which is especially useful when the data are grouped into categories. Let us suppose that x_1, $x_2, \ldots, x_m$ are *distinct* (i.e., different) values of the distribution and that they occur with frequency $f_1, f_2, \ldots, f_m$, respectively. Then

$$\bar{x} = \frac{f_1x_1 + f_2x_2 + \cdots + f_mx_m}{f_1 + f_2 + \cdots + f_m}$$

Looked at this way, the mean of the quiz scores is

$$\bar{x} = \frac{2(4) + 3(5) + 2(6) + 3(7) + 1(8) + 2(9) + 1(10)}{2 + 3 + 2 + 3 + 1 + 2 + 1}$$

$$= \frac{92}{14} \approx 6.57$$

Of course, the answer is the same as before.

If you don't understand the average, you're about average.

Which of these three numbers—mean, median or mode—should we use to represent a distribution? There is no easy answer to this question, except to say that one should understand what each of them does. The median and mode are easy to comprehend; the mean is quite mysterious. That is why we now take a good deal of space to explain it.

The Mean as a Balance Point

Long before we knew anything about statistics, most of us learned something about teeter-totters (or seesaws.) In order to balance a big man like your father, you knew you had to sit further from the fulcrum

than he did. And by trial and error, you found just the right place to sit or, to put it another way, you found the spot between the two of you where the fulcrum belonged. Finding this place using mathematics is the subject we now investigate.

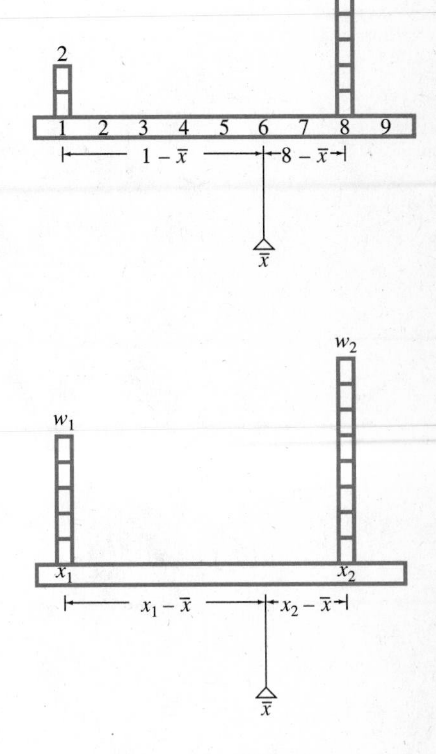

Consider a teeterboard of negligible weight. Along one edge, mark off a numerical scale using 1-foot intervals (see diagram in margin.) Suppose that a 2-pound weight rests at the 1-foot mark and a 5-pound weight at the 8-foot mark. Our question is: Where should the fulcrum be placed in order for the teeterboard to balance? Let's label this point $\bar{x}$ and try to determine its value on the scale we have marked off. Long ago, experimenters learned that a certain condition must be satisfied if the board is to balance; the *first moment* must be 0. What this means is that we must add together the product of each weight times its distance from the fulcrum (this distance being taken as negative to the left of the fulcrum and positive to the right) and get 0 as the result. Thus (see diagram in margin),

(Weight) times (distance) plus (weight) times (distance) is 0

$$2(1 - \bar{x}) + 5(8 - \bar{x}) = 0$$

$$2 \cdot 1 - 2 \cdot \bar{x} + 5 \cdot 8 - 5 \cdot \bar{x} = 0$$

$$2 \cdot 1 + 5 \cdot 8 = 2 \cdot \bar{x} + 5 \cdot \bar{x}$$

$$2 \cdot 1 + 5 \cdot 8 = (2 + 5)\bar{x}$$

$$\frac{2 \cdot 1 + 5 \cdot 8}{2 + 5} = \bar{x}$$

The answer is $\bar{x} = 6$, but what is more interesting to us is the form of the answer. If the first weight of size w_1 were placed at x_1, and the second weight of size w_2 placed at x_2, the same reasoning would lead us to

$$\frac{w_1 \cdot x_1 + w_2 \cdot x_2}{w_1 + w_2} = \overline{x}$$

The general situation is handled in exactly the same way. If there are m weights of size $w_1, w_2, \ldots, w_m$ spaced along the board at points $x_1, x_2, \ldots, x_m$,

$$\frac{w_1 \cdot x_1 + w_2 \cdot x_2 + \cdots + w_m \cdot x_m}{w_1 + w_2 + \cdots + w_m} = \overline{x}$$

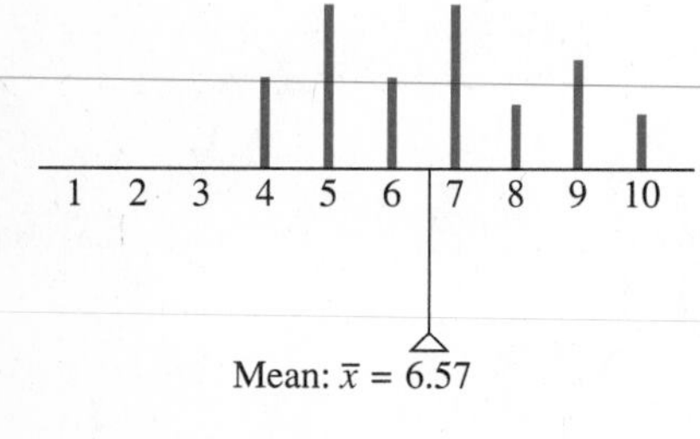

This is the result we were looking for. It tells us how to find the balance point $\overline{x}$ for a whole set of weight placed along a teeterboard.

The formula just displayed should look vaguely familiar. It is exactly the same formula we gave earlier in the section for the mean of a distribution of values, $x_1, x_2, \ldots, x_m$, provided we interpret w_i to be the frequency of x_i. We have arrived at a very important conclusion. *The mean of a distribution is the point at which the distribution balances when we interpret the frequency of a value as a weight at that value.*

Take the distribution of quiz scores (page 215) as an example. If we draw a bar graph for this distribution and think of the bars as weights, the distribution will balance at the mean, namely, at $\overline{x} = 6.57$.

Or refer back to the example of salaries at XYZ Enterprises. Now we can understand why there was a mean salary of $26,000 even though all but two employees got less than that amount. President Hornblower and his son with their huge salaries balance off the other 21 employees.

This example serves to emphasize one characteristic of the mean. It is strongly influenced by the presence of an extremely large (or small) value. The median, however, is not influenced at all by such values. Many people prefer to use the median as a measure of the center of a distribution in any situation, such as one involving salaries, where extreme values are likely to occur.

Grade-Point Averages

Grading schemes vary from college to college, but the basic principles are the same. At Podunk University, students take semester courses which vary from 1 to 5 credits. In each course a student receives a grade of A, B, C, D, or F. In order to treat grades numerically, A is given the value 4, B the value 3, and so on down to F which has the value 0. The basic unit is the credit; to figure the weighting factor for each grade, one must compute the number of credits at that grade. At the left is Susan Smart's transcript for her first year at Podunk University. She has 2 credits of F, 3 credits of D, 8 credits of C, 12 credits of B, and 8 credits of A. Thus her grade-point average is

$$\bar{x} = \frac{2 \cdot 0 + 3 \cdot 1 + 8 \cdot 2 + 12 \cdot 3 + 8 \cdot 4}{2 + 3 + 8 + 12 + 8} = \frac{87}{33} \approx 2.64$$

Table 7-2

Course	*Credits*	*Grade*
History 11	3	B
Economics 12	3	A
Math 13	5	B
Art 20	3	C
Pol. Sci. 12	3	B
Math 14	5	A
English 15	3	D
Physcis 17	5	C
Phys. Ed. 11	1	B
Music 40	2	F

Summary

For a distribution of values, we wanted to find a single number that was most representative of these values. Three candidates were offered: the mode (the most frequently occurring value), the median (the middle value), and the mean (the balance point). The last-mentioned is defined by the formula

$$\bar{x} = \frac{x_1 + x_2 + \cdots + x_n}{n}$$

and, even though it is strongly influenced by very large or very small values, it is the most commonly used.

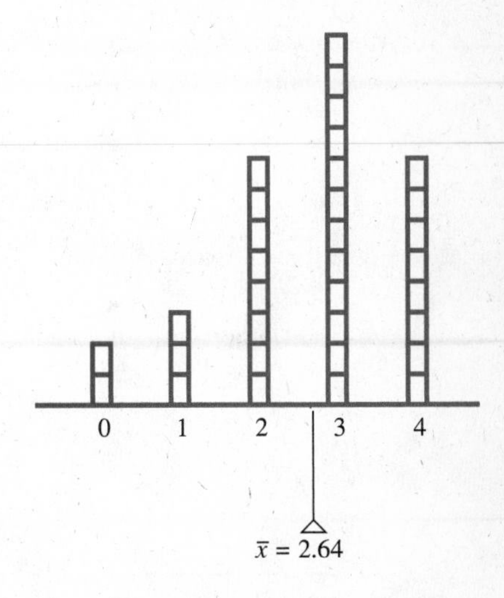

Problem Set 7.2

1. Calculate the mean, median, and mode for each of the following sets of data.
   (a) 2, 8, 7, 8, 6, 5, 10, 5
   (b) 10, 20, 18, 16, 20, 16, 13, 16, 17
2. Calculate the mean, median, and mode for each of the following sets of data.
   (a) −4, 8, −2, −1, 5, −1, 6, 2, 1
   (b) 51, 42, 36, 41, 50, 38, 43
3. Find the mean, median, and mode for the lengths (i.e., number of letters) of words in the first paragraph of this section. Begin by making a frequency chart.
4. Find the mean, median, and mode for the lengths of words in the quote from Galileo in the essay that opens Chapter 3. Begin by making a frequency chart.
5. A group of 100 women was weighed, with results as indicated in the margin. What is the mean weight? Use the midpoint of each class in making this computation. For example, assume that each individual in the first class weighs 110 pounds.

Table 7-3

Weight (pounds)	*Frequency*
105–115	10
115–125	12
125–135	15
135–145	20
145–155	14
155–165	15
165–175	10
175–185	3
185–195	1

Table 7-4	
Salary (dollars)	*Number*
Professor	
56,000–70,000	10
Associate professor	
38,000–56,000	20
Assistant professor	
30,000–38,000	40
Instructor	
22,000–30,000	30

6. Faculty salaries at Podunk University are as listed in the table in the margin. Compute the average salary, making the assumption indicated in Problem 5. However, explain why this assumption may cause a large error.

7. The incomes for the five employees of Friend's Furniture are

$18,000; $24,000; $24,000; $27,000; $79,000

Calculate the mean and median salaries. In your opinion, which number gives a better indication of salaries at Friend's?

8. Each of 10 students got 50 on a certain exam. What were the mean, median, and mode? Make up an example of 10 scores, not all equal, in which the mean, median, and mode are all 50.

9. A temperature of 25°C is considered ideal for human beings. In Boondocks, the mean temperature is 25°C. Does this mean that the temperature in Boondocks is ideal? Explain.

10. An instructor tells her class that she is going to disregard the lowest of the 10 quiz scores in Math 13 when she determines each student's final grade. Will this necessarily change a student's median score? Mean score? Explain.

11. Last term, Amy enrolled in Math, German, English, History, and Physical Education for 5, 4, 3, 3, and 1 credits, respectively. She got A in Math and German, B in English, and C in History and Physical Education. What was her grade-point average for the term? (A = 4, B = 3, C = 2.)

12. Homer took the same courses as Amy (Problem 11) and got C in Math, B in English, and A in the other three courses. Without calculating, decide whether Homer or Amy has the higher grade-point average. Now calculate Homer's average to make sure.

13. Five people weighing 120, 200, 150, 160, and 130 pounds are standing on a plank at distances 1, 2, 6, 9, and 10 feet, respectively, from one end of the plank. Assuming that the weight of the plank is negligible, where should the fulcrum be placed so that the system will balance?

x	y
1.375	1250
6.250	225
4.625	415
3.125	785
4.500	235
8.625	135
9.875	75
3.375	615
4.125	285
6.250	225
5.875	145
5.750	185
4.375	525

14. Horace has scores of 60, 75, 40, and 80 on the first four quizzes in Math 13. What must he average on the last two so that his overall average will be 75?

15.* Amy drove 90 miles at an average speed of 30 mph. What must she average on the return trip to have an overall average of 50 mph? Of 60 mph?

16.* Homer averaged 60 on the quizzes in Math 13 and 70 on the quizzes in Math 14. Yet the average of all his quiz scroes in Math 13 and Math 14 was not 65. How can this be?

17.* It is possible to have a class in which no student scores below the mean on the final exam. State precisely the conditions under which this can happen. Show that your answer is correct.

18. Find the mean of the X data shown in the margin. If your calculator has an $\bar{x}$ key, learn how to use it.

19. Find the mean of the Y data shown in the margin.

20. Suppose for the data in the margin that X represents the price of a share on the oca exchange and Y the number of shares sold at that price during a certain day. What was the mean price of a share that day?

21. At Hamline University, grades are assigned numerical values as follows. A(4.00), A^-(3.75), B^+(3.25), B(3.00), B^-(2.75), C^+(2.25), C(2.00), C^-(1.75), D^+(1.25), D(1.00), D^-(.75), F(.00). Anne has 4 credits of A^-, 6 credits of B^+, 8 credits of B^-, 9 credits of C, 5 credits of C^-. What is her grade-point average?

22. Dale has 3 credits of A^-, 6 credits of B^+, 9 credits of B, 3 credits of B^-, and 11 credits of C^+. Using the scale of Problem 21, what is his grade-point average?

For Research and Discussion

We said in this section that the median and the mode are easy to comprehend, but the mean is quite mysterious. That is due not only to the misleading influence that extreme values can exert, but also to properties of the mean that are counter-intuitive. For instance, consider the following calculation of baseball batting averages (the ratio, usually expressed to three decimals, of the number of times a batter gets a hit H divided by the number of times at bat AB). The annual All-Star game comes about midway in the season. Suppose two players, Willie Hitit and Claude Candue, have the records indicated before and after the All-Star game.

	Before			*After*		
	AB	H	Avg.	AB	H	Avg.
Hitit	240	80	.333	120	48	.400
Candue						

You should be able to make up some data so that Hitit will have a better average than Candue before the All-Star game and again afterwards, yet so that the season average of Candue will be better than that of Hitit. That's counterintuitive, but possible.

7.3 The Spread

An Adjustment

Professor Witquick tells classes that he designs tests of 100 points with the following scales in mind:

A: 100–88 B: 87–75 C: 74–60 D: 59–47

He further tells them that if he happens to make a test that the class finds so hard that raw scores fall badly out of this range, he scales them upward by adding points to the raw scores, but he doesn't specify his method for determining the points to be added. One day a class of 30 achieved the following scores:

56	62	66	68	72	78	48	52	58	62
74	69	62	58	52	46	70	75	80	42
33	48	54	60	64	65	70	70	61	64

Professor Witquick noted that he could bring the high score of 80 up to 100 with either of two schemes.

(a) Add 20 to all scores
(b) Multiply all scores by 1.25.

Discuss the effects of each plan. Which seems more fair to the students?

"How did you do on the test?"

"75? That's too bad. Oh . . . it was the 4th best in the class? Then you did pretty well."

The person trying to comment on a grade of 75 still needs more information. If more questioning reveals that there were only 5 in the class, the appraisal will be shifted again. A grade needs a context to be understood.

How many were in the class? What was the mean score? These will help, but one still wonders how the grades were spread out. How were they dispersed?

A picture or a graph gives some idea of the spread of a distribution. But once again, a statistician wants a number that measures this dispersion. One such measure is the **range,** which is defined as the difference between the largest and smallest values in the distribution. It is simple to calculate but has little else to recommend it. It's too coarse a measure; two distributions may have the same range but be very different in character. A much more discrimininating measure is the *standard deviation.*

The Standard Deviation

Before tackling the problem posed about a class of 30, let's look at a smaller class. Suppose that Professor Witquick has given two quizzes to his Advanced Calculus class of six students with the following results.

Quiz 1	*Quiz 2*
$x_1 = 10$	$x_1 = 7$
$x_2 = 8$	$x_2 = 7$
$x_3 = 7$	$x_3 = 6$
$x_4 = 5$	$x_4 = 6$
$x_5 = 5$	$x_5 = 6$
$x_6 = 1$	$x_6 = 4$

$$\bar{x} = \frac{10 + 8 + 7 + 5 + 5 + 1}{6} = 6 \qquad \bar{x} = \frac{7 + 7 + 6 + 6 + 6 + 4}{6} = 6$$

Note that both distributions have the same mean; their bar graphs balance at the same point. Yet the two distributions are very different. The first one spreads out much more than the second. We want to measure this spread.

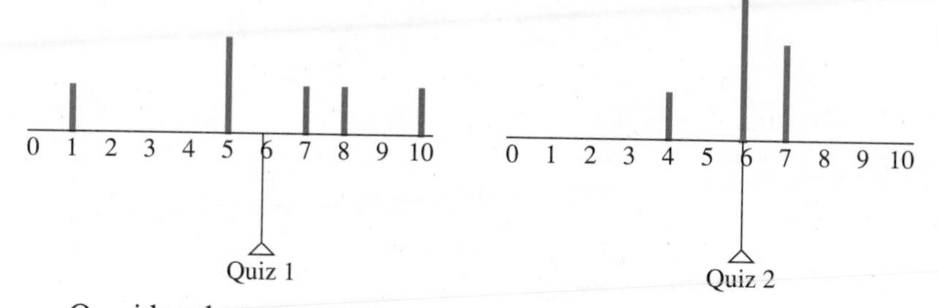

One idea that may occur to us is to look at the deviations of the scores from the mean. They are certainly much larger in the first case. Suppose we add up these deviations and use the sum as a measure:

Quiz 1		*Quiz 2*	
$x_1 - 6 =$	4	$x_1 - 6 =$	1
$x_2 - 6 =$	8	$x_2 - 6 =$	1
$x_3 - 6 =$	1	$x_3 - 6 =$	0
$x_4 - 6 =$	-1	$x_4 - 6 =$	0
$x_5 - 6 =$	-1	$x_5 - 6 =$	0
$x_6 - 6 =$	-5	$x_6 - 6 =$	-2
TOTAL	0	TOTAL	0

"Don't worry. That rope is one inch thick on the average."

That wasn't much help. In both cases the sum of the deviations is 0. (We should have known this would happen, since we really just calculated the first moment about the mean, which is 0 by definition.) Of course, we see the difficulty. Some deviations are positive and some are negative; in the sum, the deviations cancel each other out. What can we do to make all the deviations positive? How about squaring them?

Quiz 1		*Quiz 2*	
$(x_1 - 6)^2 =$	16	$(x_1 - 6)^2 =$	1
$(x_2 - 6)^2 =$	4	$(x_2 - 6)^2 =$	1
$(x_3 - 6)^2 =$	1	$(x_3 - 6)^2 =$	0
$(x_4 - 6)^2 =$	1	$(x_4 - 6)^2 =$	0
$(x_5 - 6)^2 =$	1	$(x_5 - 6)^2 =$	0
$(x_6 - 6)^2 =$	25	$(x_6 - 6)^2 =$	4
TOTAL	48	TOTAL	6

Now we have a reasonable measure of the spread. But there are still two things wrong. First, the sum may be large simply because there was a huge numer of values. We can take care of that by dividing by the number of values in each case (that is, by averaging the squared deviations). Second, the units of the original data (feet, pounds amperes, or whatever) have been squared. To get back to the original units, we take the square root. All of this discussion has led us to the following formula for the **standard deviation** s:

Think of the standard deviation as measuring the average distance the values lie from the mean.

$$s = \sqrt{\frac{(x_1 - \overline{x})^2 + (x_2 - \overline{x})^2 + \cdots + (x_n - \overline{x})^2}{n}}$$

Admittedly the formula is complicated; so we'll describe how to calculate it by the means of four rules:

1. Subtract the mean from each value (i.e., compute the deviation).
2. Square each deviation.
3. Average the squared deviations (add them up and divide by the number of values).
4. Take the square root of the result.

For quizzes 1 and 2, we get

Quiz 1

$$s = \sqrt{\frac{48}{6}} = \sqrt{8} \approx 2.83$$

Quiz 2

$$s = \sqrt{\frac{6}{6}} = 1$$

Note that the first value is much larger than the second, as expected. In the table below, we have worked out the standard deviation for a distribution of 15 values, using a convenient format.

Table 7-5			
Value, x_i	***Deviation,*** $x_i - \overline{x}$	***Squared Deviation,*** $(x_i - \overline{x})^2$	***Calculation of s***
10	3	9	$\overline{x} = \frac{105}{15} = 7$
9	2	4	
8	1	1	$s = \sqrt{\frac{36}{15}} \approx 1.55$
8	1	1	
8	1	1	
8	1	1	
7	0	0	
7	0	0	
7	0	0	
7	0	0	
6	−1	1	
6	−1	1	
5	−2	4	
5	−2	4	
4	−3	9	
105		36	

The Normal Distribution

To get a better idea of the significance of the standard deviation, we consider a curve that pervades all theoretical discussions in statistics, the **normal curve.** Its precise definition is complicated, involving ideas from calculus. For us, it is enough to say that its graph is a smooth, bell-shaped curve something like the curve shown below. Distributions that occur in real life are never exactly normal, but there is a remarkable tendency for naturally occurring distributions to be approximately normal. This is especially true when they arise from measurements of something, say weight, height, or scores on an intelligence test.

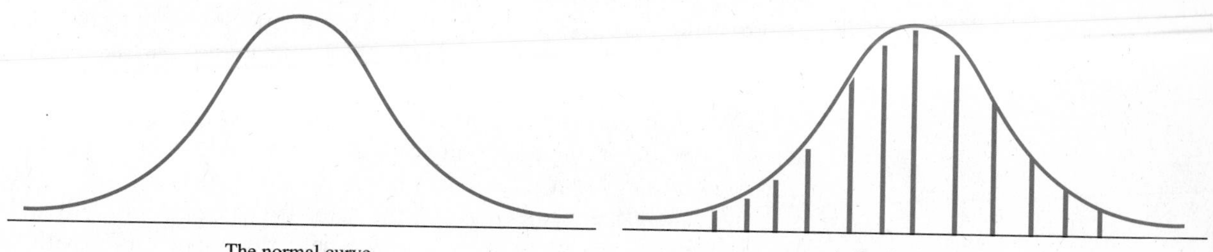

The normal curve

A distribution that is approximately normal

In the left-hand illustration below we show (approximately) how normally distributed values arrange themselves about the mean. For example, 68% of the values lie within 1 standard deviation of the mean, 96% lie within 2 standard deviations, and almost 100% are within 3 standard deviations. These percents are obtained by means of calculus; they correspond to areas under the normal curve.

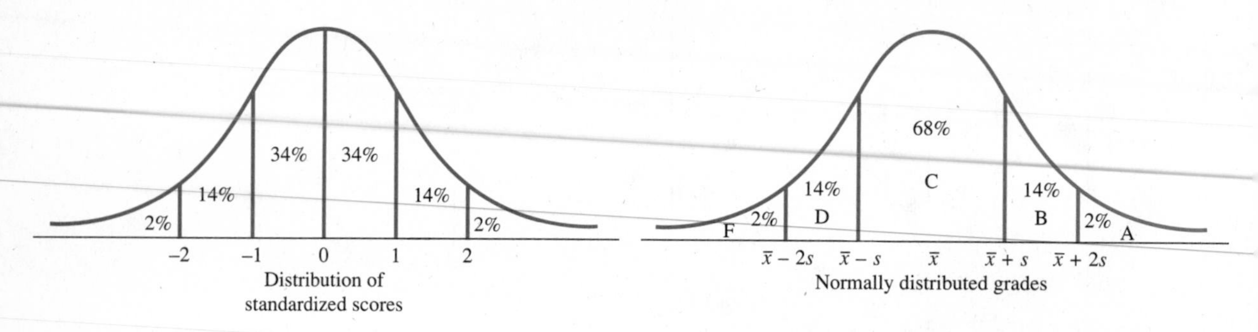

It is sometimes said that course grades ought to be normally distributed. In the right-hand figure above we show what some people mean by this. For example, to get an A a student has to score at least 2 standard deviations above the mean. This happens only about 2% of the time. Students who encourage their teachers to grade on a normal curve may want to rethink their position. And teachers who announce their intention to do so also should consider their position, making sure it can be defended. In fairness, it should be admitted that many teachers make grading on a normal curve to mean something less stringent than what a strict interpretation implies.

Standardized Scores

Facts by themselves sometimes convey little information. It is when facts are put in context that they begin to take on real meaning. To say that Susan Smart got an 8 on a quiz is a worthless piece of information. To say that she got an 8 on a quiz that had a mean of 6 and a standard deviation of 2.83 says a great deal more.

When it comes to comparing scores on tests in two different classes, the importance of context becomes especially significant. Let us suppose that Susan scored 8 in her Advanced Calculus class of six students where the mean was 6 with a standard deviation of 2.83, and that she also got 8 in British History where the mean was 7 with a standard deviation of 1.55, there being 15 students in that class. Did she do better in Advanced Calculus or in British History? How can we compare scores in two different classes? There is a way.

Our aim is to put all scores on a common scale. To do this, we compute the **standardized score** or **Z score,** defined by

$$Z = \frac{x - \overline{x}}{s}$$

Here x is a given score, and $\overline{x}$ and s are the mean and standard deviation of all the scores, respectively. Thus Susan's standardized scores in Advanced Calculus and British History are

$$Z = \frac{8 - 6}{2.83} \approx .71 \qquad Z = \frac{8 - 7}{1.55} \approx .65$$

Susan did slightly better in Advanced Calculus.

If you subtract $\overline{x}$ from each score, the resulting scores will have a mean of 0; if you then divide by s, the new scores will have a standard deviation of 1 (see Problems 9 and 10). Thus standardized scores are approximately normally distributed with mean 0 and standard deviation 1; that is, they are distributed as shown below. For example, 34% of the scores can be expected to fall between 0 and 1, and 14% between 1 and 2. Susan's scores of .71 and .65 are well above average but certainly are not exceptional.

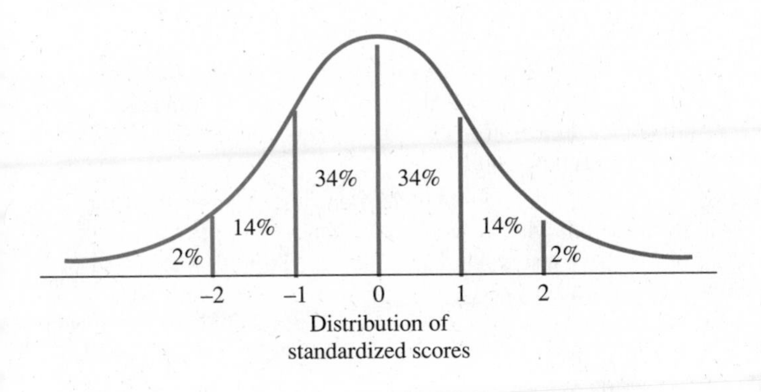

Distribution of standardized scores

Summary

The mean of a distribution is its balance point; it is a measure of the center of a distribution. Two distributions may have the same mean but be very different in character. For one, the values may cluster very closely around the mean; for another, they may be widely spread. To measure the spread, statisticians have introduced the standard deviation s defined by

$$s = \sqrt{\frac{(x_1 - \overline{x})^2 + (x_2 - \overline{x})^2 + \cdots + (x_n - \overline{x})^2}{n}}$$

It is always nonnegative. Roughly speaking, it measures the average distance the values lie from the mean. It is large when the values are widely dispersed and small (near 0) when they are tightly grouped.

Problem Set 7.3

1. Calculate the mean and standard deviation for the set {8, 6, 3, 5, 7, 6, 9, 8, 9, 9}.
2. Calculate the mean and standard deviation for the set {−3, 2, −4, 1, 0, 4, 6, −1, −5}.
3. Consider the two sets of data:

$$A = \{1, 5, 6, 9, 3, 5, 10, 5, 10, 6\}$$
$$B = \{5, 7, 6, 6, 5, 7, 6, 7, 5, 6\}$$

   (a) Make bar graphs for both sets.
   (b) Calculate the two means $\bar{x}_A$ and $\bar{x}_B$.
   (c) Calculate the two standard deviations s_A and s_B.
4. Ten students rated both Professor A and Professor B on a 1 (low) to 10 (high) rating scale. The results were the two sets of data in Problem 3. What conclusions can you draw about the two professors?
5. If the mean on a certain test is 75 with a standard deviation of 8, convert the scores 56, 70, 81, and 90 to standardized scores.
6. What is the meaning of a standardized score of 2? Of −1?
7. Suppose that sets A and B in Problem 3 are students' scores on a quiz in German and in Math, respectively. Amy got 5 on both quizzes. Compute her standardized score on both quizzes and decide whether she did better in German or in Math.
8. Rachel scored 456 on the national college entrance examination for which the mean is 500 with a standard deviation of 50. Raymond scored 25 on the BAT test which has a mean of 30 with a standard deviation of 6. If Hudson College weights these two examinations equally, which student has the best chance of being admitted to Hudson College?
9. Add 5 to each of the data in Problem 1. Compute the new mean and standard deviation. Make a conjecture about what happens to the mean and standard deviation when you add the same number to each data point.
10. Multiply each of the data in Problem 1 by 3. Compute the new mean and standard deviation. Make a conjecture about what happens to the standard deviation when you multiply each data point by a constant.
11. If a variable is normally distributed, approximately what percentage of its values will lie more than 2 standard deviations away from the mean? What percentage will lie more than 1 standard deviation above the mean?
12. Which of the following is more likely to be approximately normally distributed: the weights of all babies measured at birth, or the last digits of all the numbers in the Manhattan telephone directory? Explain.

13.* Consider the word lengths of all the words in the first paragraph of this section.
(a) Make a frequency chart.
(b) Calculate the mean.
(c) Calculate the standard deviation.

14.* Roll a pair of dice 50 times, keeping track of the total number of spots showing on each roll.
(a) Make a frequency chart.
(b) Compute the mean.
(c) Compute the standard deviation.

15.* If all the numbers in a set are equal, what is the standard deviation? Show that, if the standard deviation is 0, all the numbers in the set are equal.

16.* If the mean, median, mode, and standard deviation of two sets of five numbers are the same, are the two sets identical?

Table 7-6 contains data from the Metropolitan Transit Commission of Minneapolis and St. Paul. It gives the elapsed time between two specific bus stops for 17 different buses on three different days (X: April 22, Y: April 23, Z: April 28)

Table 7-6
Bus Time Data

X	Y	Z
10.1	13.3	13.2
12.9	12.5	12.3
15.1	13.3	14.1
12.4	14.3	13.4
13.1	14.3	12.3
13.4	12.2	11.4
10.5	14.7	11.2
13.4	10.5	11.2
14.1	11.5	21.2
12.4	12.4	12.4
13.7	13.1	12.4
11.2	11.4	10.3
11.1	12.2	14.2
12.5	12.5	9.6
13.1	12.5	12.3
11.2	14.4	11.2
13.1	14.3	15.2

17. For Table 7-6 calculate the mean and standard deviation for X. If your calculator has mean and standard deviation buttons, use them. However, be aware that some calculators use $n - 1$ rather than n in the definition of the standard deviation which may lead to slight variations in the answer.

18. For Table 7-6, calculate the mean and standard deviation for Y.

19. For Table 7-6, calculate the mean and standard deviation for Z.

For Research and Discussion

You are now equipped with some tools to analyze the two methods Professor Witquick was said to be considering to move the 30 grades up into the desired range. Find the following:

	Raw Score	*Raw Score +20*	*(Raw Score)(1.25)*
Mean			
Standard Deviation			

Draw a bar graph for the grade distribution that results from each of the two methods. What effect do the two methods have, if any, on the number of people receiving each grade? Does one method seem more fair to you? Now normalize the scores and consider the effect on grades if Professor Witquick was to grade strictly "on the curve."

7.4 Notation

Taxing Instructions

Form 1

1. List any n numbers.
2. Find the mean of the numbers listed in Step 1.
3. For each of the n numbers, find its difference from the mean found in Step 2.
4. Square each of the n numbers obtained in Step 3.
5. Find the sum of the n numbers obtained in Step 4.
6. Divide the total from Step 5 by n.
7. Go to step 8.
8. Take the square root of the number obtained in Step 6.

Form 2

1. List any n numbers.
2. Find the mean of the numbers listed in Step 1.
3. Go to Step 4.
4. Square each of the n numbers obtained in Step 1.
5. Find the sum of the n numbers obtained in Step 4.
6. Divide the total from Step 5 by n.
7. Subtract from the value obtained in Step 6 the square of the value obtained in Step 2.
8. Take the square root of the number obtained in Step 7.

If carried out correctly, the two sets of instructions above should, for the same set of n given numbers, both produce the same number (which is in fact the standard deviation of the original list). If you don't get the same number both times for an example of your own making, try again.

The instructions given are very similar, you will find, to those given on an income tax form. Income tax forms are not very popular. (Neither is income tax, but that's a whole different topic.) The forms could be simplified with the use of a little mathematical notation, if only the public was not so skittish about such matters.

We shall illustrate our meaning by showing how the two "forms" above can be simultaneously shortened and simplified by the use of concise notation that has been developed for such situations.

Writing Sums Compactly

$\sum_{i=1}^{n} x_i$ is read, "sum of x sub i as i runs from 1 to n."

"Sum" begins with "S"; and the Greek letter for "S" is Σ, pronounced "sigma." Think of Σ as standing for sum. In particular,

$$\sum_{i=1}^{n} x_i = x_1 + x_2 + x_3 + \cdots + x_n$$

Here Σ means add up; the symbol $i = 1$ below the sigma tells where to start adding; the n at the top tells where to stop. Try to follow these examples:

$$\sum_{i=1}^{5} a_i = a_1 + a_2 + a_3 + a_4 + a_5$$

$$\sum_{i=3}^{9} b_i = b_3 + b_4 + b_5 + b_6 + b_7 + b_8 + b_9$$

$$\sum_{i=1}^{5} i^2 = 1^2 + 2^2 + 3^2 + 4^2 + 5^2$$

$$\sum_{i=1}^{50} (2i - 1) = 1 + 3 + 5 + 7 + \cdots + 99$$

If you were able to follow them, you should be able to write each of these expressions in sigma notation:

$$c_1 + c_2 + c_3 + \cdots + c_{100}$$

$$1 + 2 + 3 + \cdots + 1000$$

$$1 + \frac{1}{2} + \frac{1}{3} + \frac{1}{4} + \frac{1}{5} + \frac{1}{6} + \frac{1}{7} + \frac{1}{8}$$

The answers appear in the summary.

Properties of Sums

Not only does sigma notation allow us to write sums in a very compact way; it also allows us to state certain properties of sums in an elegant fashion. For example, the three familiar facts:

$$\underbrace{c + c + c + \cdots + c}_{n \text{ terms}} = nc$$

$$ca_1 + ca_2 + ca_3 + \cdots + ca_n = c(a_1 + a_2 + a_3 + \cdots + a_n)$$

$$(a_1 + b_1) + (a_2 + b_2) + \cdots + (a_n + b_n) = (a_1 + a_2 + \cdots + a_n) + (b_1 + b_2 + \cdots + b_n)$$

become, when written in sigma notation,

$$\sum_{i=1}^{n} c = nc$$

$$\sum_{i=1}^{n} ca_i = c \sum_{i=1}^{n} a_i$$

$$\sum_{i=1}^{n} (a_i + b_i) = \sum_{i=1}^{n} a_i + \sum_{i=1}^{n} b_i$$

Sigma Notation in Statistics

Statisticians take great delight in sigmas; their books are full of them. This is because most statistical measures require that we add many terms. Here are the formulas we studied earlier for means and standard deviations, now written as statisticians prefer to write them:

$$\bar{x} = \frac{1}{n}\sum_{i=1}^{n} x_i \qquad s = \sqrt{\frac{1}{n}\sum_{i=1}^{n}(x_i - \bar{x})^2}$$

We have already hinted that s is messy to calculate; if you worked out the problems at the end of Section 7.3, you don't need to be told. This is particularly true when $\bar{x}$ is not an integer, which is usually the case. There is another formula for s, which is much nicer, and the properties of sigma mentioned above are very useful in deriving it. To avoid the square root we initially consider s^2 rather then s:

$$\begin{aligned}
s^2 &= \frac{1}{n}\sum_{i=1}^{n}(x_i - \bar{x})^2 \\
&= \frac{1}{n}\sum_{i=1}^{n}(x_i^2 - 2\bar{x}x_i + \bar{x}^2) \\
&= \frac{1}{n}\left[\sum_{i=1}^{n} x_i^2 + \sum_{i=1}^{n}(-2\bar{x}x_i) + \sum_{i=1}^{n}\bar{x}^2\right] \\
&= \frac{1}{n}\left(\sum_{i=1}^{n} x_i^2 - 2\bar{x}\sum_{i=1}^{n} x_i + \sum_{i=1}^{n}\bar{x}^2\right) \\
&= \frac{1}{n}\left(\sum_{i=1}^{n} x_i^2 - 2\bar{x}n\frac{1}{n}\sum_{i=1}^{n} x_i + n\bar{x}^2\right) \\
&= \frac{1}{n}\left(\sum_{i=1}^{n} x_i^2 - 2\bar{x}n\bar{x} + n\bar{x}^2\right) \\
&= \frac{1}{n}\sum_{i=1}^{n} x_i^2 - \bar{x}^2
\end{aligned}$$

Thus

$$s = \sqrt{\left(\frac{1}{n}\sum_{i=1}^{n} x_i^2\right) - \bar{x}^2}$$

a formula we claim is generally much easier to use than the original one.

Let's illustrate it first for Professor Witquick's first quiz in Section 7.3

Score	Score Squared
$x_1 = 10$	$x_1^2 = 100$
$x_2 = 8$	$x_2^2 = 64$
$x_3 = 7$	$x_3^2 = 49$
$x_4 = 5$	$x_4^2 = 25$
$x_5 = 5$	$x_5^2 = 25$
$x_6 = 1$	$x_6^2 = 1$
TOTAL 36	TOTAL 264

$$\bar{x} = \frac{36}{6} = 6$$

$$s = \sqrt{\frac{264}{6} - (6)^2}$$

$$= \sqrt{44 - 36}$$

$$\approx 2.83$$

Naturally, the result is the same as the one calculated in Section 7.3.

But the real merit of our new formula shows up when we have a large number of values and the mean is not integral. In the table below we have worked out an example in which there are 20 values.

Table 7-7

x_i	x_i^2	Calculation of $\bar{x}$ and s
10	100	
10	100	
9	81	$\bar{x} = 135/20 = 6.75$
9	81	
9	81	$s = \sqrt{(1027/20) - (6.75)^2}$
8	64	
8	64	$= \sqrt{51.35 - 45.56}$
8	64	
8	64	$= \sqrt{5.79}$
7	49	≈ 2.41
7	49	
7	49	
7	49	
6	36	
6	36	
5	25	
4	16	
3	9	
3	9	
1	1	
135	1027	

Summary

Two new ideas were introduced in this section. The first was simply a matter of notation, the sigma notation for sums. One could argue that the choice of notation is a trivial subject, hardly worthy of comment. But mathematicians have learned that the use of a carefully chosen, compact symbol can help them handle a complicated idea with ease and efficiency. This is particularly true of the symbol Σ which is used to denote sums. Its use is illustrated by

$$c_1 + c_2 + c_3 + \cdots + c_{100} = \sum_{i=1}^{100} c_i$$

$$1 + 2 + 3 + \cdots + 1000 = \sum_{i=1}^{1000} i$$

$$1 + \frac{1}{2} + \frac{1}{3} + \cdots + \frac{1}{8} = \sum_{k=1}^{8} \frac{1}{k}$$

The following problems give you an opportunity to confirm and improve your understanding of this symbol.

The second idea was a new formula for the standard deviation, which is generally easier to use in practice. It is conveniently written using the sigma symbol and was derived with its aid:

$$s = \sqrt{\left(\frac{1}{n}\sum_{i=1}^{n} x_i^2\right) - \bar{x}^2}$$

Problem Set 7.4

1. Calculate:

(a) $\sum_{i=1}^{8} 2i$ (b) $\sum_{i=1}^{8} (2i + 1)$ (c) $\sum_{i=1}^{5} i^2$ (d) $\sum_{i=1}^{4} \frac{1}{i}$

2. Calculate:

(a) $\sum_{i=1}^{5} 3i$ (b) $\sum_{i=1}^{5} (3i - 2)$ (c) $\sum_{i=1}^{5} (i - 1)^2$ (d) $\sum_{i=2}^{4} \frac{1}{i - 1}$

3. Rewrite using sigma notation.
(a) $1 + 2 + 3 + 4 + \cdots + 10$
(b) $1 + 4 + 9 + 16 + \cdots + 100$
(c) $y_1 + y_2 + y_3 + \cdots + y_{37}$
(d) $(y_1 - 3)^2 + (y_2 - 3)^2 + \cdots + (y_n - 3)^2$

4. Rewrite using sigma notation.
(a) $2 + 4 + 6 + 8 + \cdots + 30$
(b) $1 + \frac{1}{2} + \frac{1}{3} + \cdots + \frac{1}{75}$
(c) $w_3 + w_4 + w_5 + \cdots + w_{400}$
(d) $x_1y_1 + x_2y_2 + x_3y_3 + \cdots + x_ny_n$

5. Let $\sum_{i=1}^{100} x_1 = 431$. Calculate:

(a) $\overline{x}$

(b) $\sum_{i=1}^{100} 3x_i$

(c) $\sum_{i=1}^{100} (3x_i + 1)$

6. Let $\sum_{i=1}^{10} x_i = 20$; $\sum_{i=1}^{10} x_i^2 = 90$. Calculate:

(a) $\overline{x}$
(b) s

(c) $\sum_{i=1}^{10} (x_i^2 + 2x_i)$

7. Let $\sum_{i=1}^{100} x_i = 200$; $\sum_{i=1}^{100} x_i^2 = 4000$. Calculate $\overline{x}$ and s.

8. Let $\sum_{i=1}^{64} y_i = -32$ and $\sum_{i=1}^{64} y_i^2 = 1280$. Calculate $\overline{y}$ and s.

9. Redo Problem 1 in Section 7.3 using our new formula for s.

10. Find the mean and standard deviation of the set {25, 27, 30, 25, 32, 24, 27, 30, 28, 26}.

11.* Show, using sigma notation, that if $y_i = x_i + c$ where c is constant, $\overline{y} = \overline{x} + c$ and $s_y = s_x$.

12.* Show, using sigma notation, that if $y_i = cx_i$, where c is a positive constant, $\bar{y} = c\bar{x}$ and $s_y = cs_x$.

13. Show, using sigma notation, that $\sum_{i=1}^{n} (x_i - \bar{x}) = 0$.

14. Show that, if $z_i = x_i + y_i$, then $\bar{z} = \bar{x} + \bar{y}$.

For Research and Discussion

Progress in mathematics is tied more than you might think to the availability of good notation. Try the following.

(a) Multiply the Roman numerals (LXII)(CXVI) without converting the numbers to the base ten system.

(b) Read about attempts to solve the linear equation $ax = b$ before our current algebraic notation had been introduced.

(c) Without resort to letting letters represent numbers, that is using the "income tax" style, try to give instructions for finding the least common multiple of a set of three or four integers, such as 24, 42, and 56.

7.5 Correlation

Is One Guess as Good as Another?

We saw in Section 7.1 that Professor Witquick was trying to assess the abilities of incoming freshmen by analyzing their (fictitious) Standard Measure of Brainpower (SMOB) scores. His colleague, Professor Branebom, has become increasingly skeptical as to whether these scores are of any use in predicting student performance in college calculus. She conjectures that grades in high school geometry would be a better projector, and she has assembled the data in the table at the left for fifteen randomly chosen students. What do you think? Do the data support her conjecture?

Table 7-8
A Random Sample of 15 Podunk University Students

Geometry Grade, x	*Calculus Grade,* y	*SMOB Score,* z
4	3	48
4	4	60
4	3	78
4	4	62
4	3	70
3	4	80
3	3	60
3	3	72
3	2	64
3	2	62
2	2	52
2	1	56
2	2	60
1	0	58
1	1	40

Professor Branebom faces a problem which is very familiar to all scientists. What relationship (if any) exists between two variables? The great achievements in the physical sciences over the last 300 years are largely due to the discovery of almost exact relationships between important variables. For example, a physicist knows that $v = 32t$ and $s = 16t^2$ describe how velocity v and distance s are related to time t when an object falls under the influence of gravity.

The situation is much more complicated in the social sciences. Rarely do we find exact relationships. Yet there is great interest in establishing even the slightest tendency of one variable to depend on another and to measure the strength of that tendency. In particular, Professor Branebom would like to measure the dependence of y (grade in college calculus) on x (grade in high school geometry), and in turn the dependence of y on z (the SMOB score). But how?

Scatter Diagrams

Draw a picture.

By now we are conditioned to the notation that we should try to display data pictorially. The appropriate picture here is a **scatter diagram.** This is simply a plot of the (x, y) and (z, y) data on a standard coordinate plane (see diagrams below).

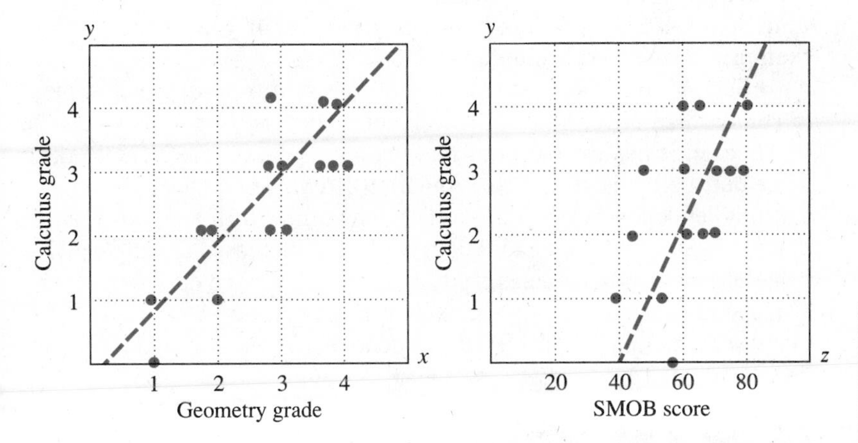

In each case, there is some indication of a trend, a trend that the dotted lines on the scatter diagrams are meant to suggest. Admittedly, neither scatter diagram comes close to the ideal situation in which the data points lie almost exactly on a straight line. That is too much to expect, for then we could predict with near certainty what the college calculus grade would be. Maybe some people can see from the scatter diagrams that high school geometry grades are a better predictor than SMOB scores, but we confess that this is far from obvious to us. We need stronger evidence than we have as yet.

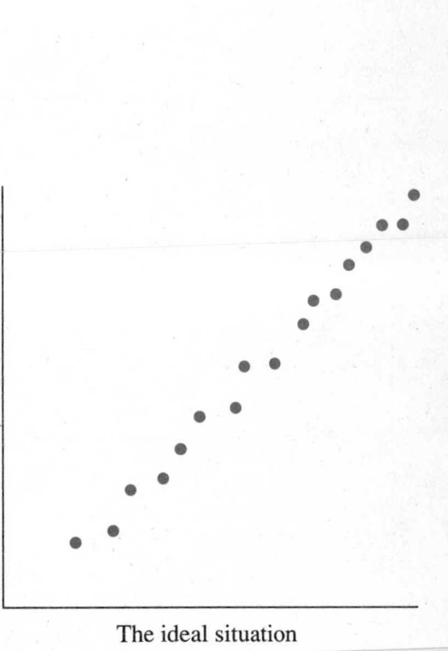

The ideal situation

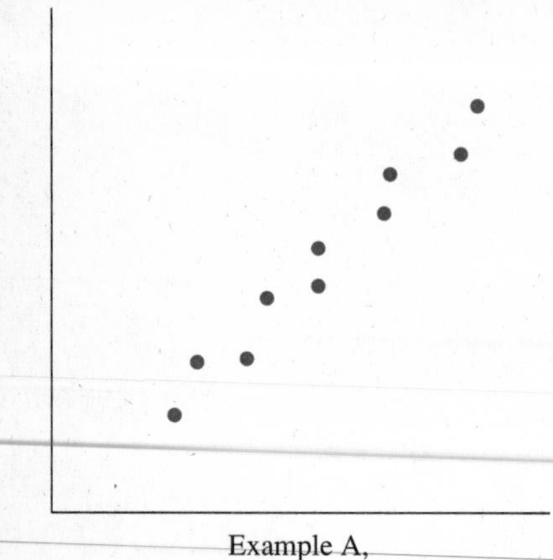

Example A,
r near 1

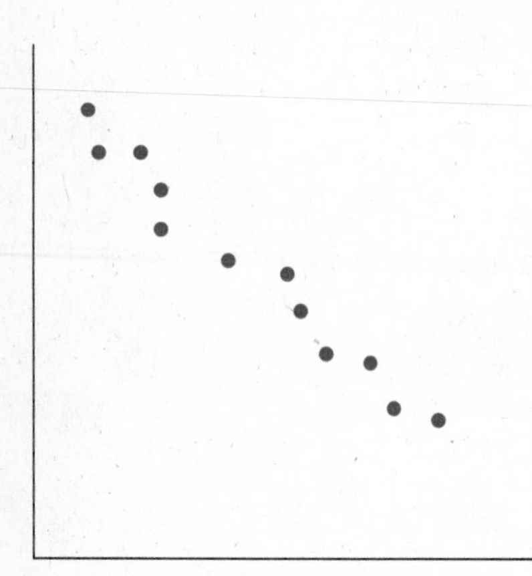

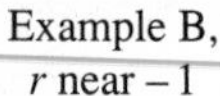

Example B,
r near -1

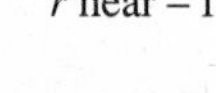

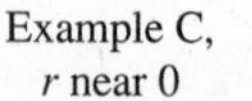

Example C,
r near 0

The Pearson Correlation Coefficient

To measure the tendency toward a linear (i.e., straight-line) relationship, Karl Pearson (1857–1936) introduced the sample correlation coefficient r. If $(x_1, y_1), (x_2, y_2), \ldots, (x_n, y_n)$ is a sample of paired values for the variables x and y, the **sample correlation coefficient** between x and y is defined by

$$r_{xy} = \frac{1}{ns_xs_y}\sum_{i=1}^{n}(x_i - \overline{x})(y_i - \overline{y})$$

Here $\overline{x}$, $\overline{y}$, s_x, and s_y are the respective means and standard deviations for the x and y data.

There are several reasons for Pearson's choice of this measure. First, $\Sigma\,(x_i - \overline{x})(y_i - \overline{y})$ is chosen because it is large and positive when the data closely approximate a line that slopes upward (example A), large but negative when the data approximate a line sloping downward (example B), and near 0 when the data spread out randomly (example C). Note, for instance, that when data follow a pattern like that in example A, x_i and y_i tend to be bigger or smaller than $\overline{x}$ and $\overline{y}$, respectively, at the same time. This means that the factors $x_i - \overline{x}$ and $y_i - \overline{y}$ tend to be positive or negative at the same time, thus making the product $(x_i - \overline{x})(y_i - \overline{y})$ consistently positive. This in turn makes the sum $\Sigma(x_i - \overline{x})(y_i - \overline{y})$ large and positive. See if you can reason out examples B and C in a similar fashion.

Second, the constant $1/ns_xs_y$ that stands in front of Σ is put there to standardize all correlation coefficients, that is, to put them on the same scale. It has the effect of making a correlation coefficient take a value between -1 and 1 (see Problem 9). A value of r near 1 indicates a strong tendency for the data to lie along a line with a positive slope, that is along a line on which y increases as x increases (example A). A value of r near -1 indicates a strong tendency for the data to lie along a line with a negative slope, that is a line on which y decreases as x increases (example B). Values of r close to 0 indicate that x and y are not linearly related.

A Simpler Formula for r

The formula as given is complicated to use. A somewhat better formula for computational work can be derived using the properties of Σ noted in Section 7.4. We include the derivation for algebraic experts and especially for wise but skeptical souls who won't believe anything unless they see it proved.

$$r = \frac{1}{ns_xs_y}\sum_{i=1}^{n}(x_i - \overline{x})(y_i - \overline{y})$$

$$= \frac{n\sum_{i=1}^{n}(x_iy_i - \overline{x}y_i - \overline{y}x_i + \overline{x}\,\overline{y})}{n\sqrt{\frac{1}{n}\sum_{i=1}^{n}x_i^2 - \overline{x}^2}\; n\sqrt{\frac{1}{n}\sum_{i=1}^{n}y_i^2 - \overline{y}^2}}$$

$$= \frac{n\left(\sum_{i=1}^{n} x_i y_i - \overline{x}\sum_{i=1}^{n} y_i - \overline{y}\sum_{i=1}^{n} x_i + \sum_{i=1}^{n} \overline{x}\,\overline{y}\right)}{\sqrt{n\sum_{i=1}^{n} x_i^2 - n^2\overline{x}^2}\sqrt{n\sum_{i=1}^{n} y_i^2 - n^2\overline{y}^2}}$$

$$= \frac{n\left(\sum_{i=1}^{n} x_i y_i - \overline{x}n\overline{y} - \overline{y}n\overline{x} + n\overline{x}\,\overline{y}\right)}{\sqrt{n\sum_{i=1}^{n} x_i^2 - (n\overline{x})^2}\sqrt{n\sum_{i=1}^{n} y_i^2 - (n\overline{y})^2}}$$

$$= \frac{n\sum_{i=1}^{n} x_i y_i - n\overline{x}n\overline{y}}{\sqrt{n\sum_{i=1}^{n} x_i^2 - (n\overline{x})^2}\sqrt{n\sum_{i=1}^{n} y_i^2 - (n\overline{y})^2}}$$

Thus

$$r = \frac{n\sum_{i=1}^{n} x_i y_i - \sum_{i=1}^{n} x_i \sum_{i=1}^{n} y_i}{\sqrt{n\sum_{i=1}^{n} x_i^2 - \left(\sum_{i=1}^{n} x_i\right)^2}\sqrt{n\sum_{i=1}^{n} y_i^2 - \left(\sum_{i=1}^{n} y_i\right)^2}}$$

With this formula for r in hand, we illustrate a calculation for the (x, y) data, that is the data that compares geometry and calculus grades, for the 15 students in Professor Branebom's sample.

Table 7-9

x_i	x_i^2	y_i	y_i^2	$x_i y_i$	***Calculation of r***
4	16	3	9	12	$r = \frac{(15)(121) - (43)(37)}{\sqrt{(15)(139) - (43)^2}\sqrt{(15)(111) - (37)^2}}$
4	16	4	16	16	
4	16	3	9	12	$= \frac{224}{\sqrt{236}\sqrt{296}}$
4	16	4	16	16	
4	16	3	9	12	$\approx .85$
3	9	4	16	12	
3	9	3	9	9	
3	9	3	9	9	
3	9	2	4	6	
3	9	2	4	6	
2	4	2	4	4	
2	4	1	1	2	
2	4	2	4	4	
1	1	0	0	0	
1	1	1	1	1	
43	139	37	111	121	

The calculation yields $r_{xy} = .85$. An identical calculation using the (z, y) data comparing SMOB scores and calculus grades gives $r_{zy} = .51$.

The Answer to the Original Question

Well, which is the better predictor of college calculus grades? High school geometry grades or SMOB scores? The calculations we have made (namely, $r_{xy} = .85$ and $r_{zy} = .51$) strongly support the view that high school geometry grades did a better job last year. As indicated earlier, we always hope for a value of r near 1 or near -1. A value of $r = .85$ is really quite good. One rarely finds correlation coefficients better than this arising from problems in the social sciences.

But now some words of caution. First, our data were completely artificial; we just made them up. But even if they were real data, we should not make exorbitant claims. Our sample size of 15 was rather small. Perhaps it was not representative of the total population of students who took calculus at Podunk University, and certainly we must be careful in claiming that it represents students who will take calculus in the future. It may be well to do the same experiment over, using a much larger sample. Then, too, we should avoid claiming that what is true at Podunk University is necessarily true at other colleges. Nevertheless, the hypothetical data give support to the conclusion that at Podunk University high school geometry grades are a better predictor of college calculus grades than SMOB scores. And, in fact, they suggest that geometry grades are a very good predictor.

Summary

We have introduced the correlation coefficient as a measure of the tendency for two variable data to lie along a straight line. The number r always lies between -1 and 1. A value of r near 1 or -1 indicates a strong linear relationship, near 0 none at all.

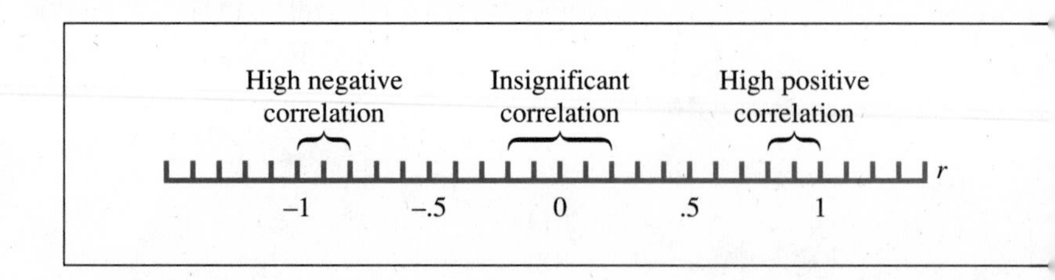

We emphasize that r is designed to measure the tendency toward a straight-line relationship. The sample data from two variables may lie along some other curve, indicating a more complicated but definite relationship between these variables, yet the value of r could be very small (see Problem 10).

Problem Set 7.5

Table 7-10 Data Set A

x	y
2	4
4	7
3	6
4	6
1	4
5	6
6	7
0	2
8	9
7	9

Table 7-11 Data Set B

x	y
10	2
9	3
9	4
8	4
6	4
5	6
5	7
3	8
3	9
2	13

1. Make a scatter diagram for data set A. Then calculate $\bar{x}$, $\bar{y}$, s_x, s_y, and r.

2. Make a scatter diagram for data set B. Then calculate $\bar{x}$, $\bar{y}$, s_x, s_y, and r.

3. How would you interpret each of the following values for a correlation coefficient (assuming a sample size of at least 25)?
(a) $r = -.9$ (b) $r = -.41$
(c) $r = .12$ (d) $r = .79$

4. Make a scatter diagram for the following xy data: $\{(1, 0); (1, 4); (5, 0); (5, 4)\}$. What do you estimate for a value of r? Now calculate r.

5. Would you expect a high positive, low positive, low negative, or high negative correlation between the following variables?
(a) The weight of a car and its gas consumption in miles per gallon.
(b) Median salary and gross national product.
(c) Athletic ability and mathematical talent.
(d) Average winter temperature and the cost of heating a home.
(e) The weight of a diamond ring and its cost.
(f) The amount of rain in the Midwest and the cost of beef in New York City.
(g) Intelligence and hat size.
(h) Weight and height for people.
(i) College grade-point average and median salary 20 years after graduation.

6. If $n = 20$, $\sum_{i=1}^{n} x_i y_i = 160$, $\sum_{i=1}^{n} x_i = 30$, $\sum_{i=1}^{n} y_i = 40$, $\sum_{i=1}^{n} x_i^2 = 180$, and $\sum_{i=1}^{n} y_i^2 = 200$, find r approximately.

7.* Obtain a set of two-variable data as follows. Roll a pair of dice (marked so that you can distinguish between them) 25 times. Let x_i be the number of spots on one die on the ith roll, and y_i the number of spots on the other. Guess the correlation coefficient. Now actually calculate r.

8.* Suppose $y_i = 3x_i + 1$ for $i = 1, 2, \ldots, n$. What does this imply about the scatter diagram for the xy data? Show, using properties of Σ, that r is 1 in this case.

9.* A famous inequality, due to A. Cauchy, states that, for any sets of numbers $a_1, a_2, \ldots, a_n$ and $b_1, b_2, \ldots, b_n$,

$$\left|\sum_{i=1}^{n} a_i b_i\right| \leq \sqrt{\sum_{i=1}^{n} a_i^2}\sqrt{\sum_{i=1}^{n} b_i^2}$$

Use this inequality to show that $|r| \leq 1$. Hint: Let $a_i = x_i - \bar{x}$ and $b_i = y_i - \bar{y}$.

10.* Suppose that all the xy data happen to lie scattered around a circle after they are plotted. You should expect to find a small value of r, but certainly x and y seem to be related very strongly. Explain this paradox.

11. Calculate r_{xy}, the correlation coefficient between X and Y, for the data of Table 7-7 of Problem Set 7.3. Interpret the result.
12. Calculate r_{xz} for the data of Table 7-7. Interpret the result.
13. Calculate the correlation coefficient between N and $\sqrt{N}$ using the data of Table 7-6. Interpret the result.

For Research and Discussion

There are two warnings that need to be sounded about possible misuse of the correlation coefficient.

1. High correlation between two sets of data does not imply that one phenomena is causing another. Both phenomena that generate highly correlated data may be caused by some third factor. For example, high correlation between years of formal education and income level is commonly cited by those who would encourage you to continue your education. We are in favor of higher education, particularly if it includes the study of mathematics, but it has to be acknowledged that higher education attracts people who are bright, and who have family emotional and financial support, two assets that auger well for future earnings, education or not.
2. It is quite possible (and there are numerous humorous examples in the literature) to find strong correlation between data that common sense dictates cannot possibly be related to each other in any way other than coincidental.

Find examples of both misuses of correlation.

chapter 8

Geometric Paths

A mathematician, like a painter or a poet, is a maker of patterns. If his patterns are more permanent than theirs, it is because they are made with ideas . . . . The mathematician's patterns, like the painter's or poet's, must be beautiful; the ideas, like the colours or the words, must fit together in a harmonious way. Beauty is the first test; there is no permanent place in the world for ugly mathematics.

G. H. HARDY

Impossible

The difficult we do right away. The impossible takes a little longer.

The can-do swagger that the United States Army Air Force tried to instill in its personnel with the World War II slogan quoted above still reflects the attitude with which many spirited people hear that a thing is impossible. That's simply a challenge for them to attempt something that no one up to that point has been able to do.

They have their reasons. There was a time when the building of a flying machine was thought to be impossible. Articles in sports publications in the late 1940s explained why it was impossible for a human to run a mile in 4 minutes. What one generation regards as impossible, the next generation accepts as necessary just to stay in the race. Nor is this simply a sign of our fast paced era; Pliny the Elder (23–79) observed, "How many things, too, are looked upon as impossible until they have been effected."

Mathematicians say it is impossible to trisect an angle using only a straight edge (unmarked ruler) and a compass. Since almost everyone knows (or knows that they once knew) how easy it is to bisect an angle, since the problem of the trisection is easily understood, and since that *impossible* word sits there as a defiant challenge, each generation produces a new crop of angle trisectors. They frequently send the product of their labors to the mathematics department of their state university

Since this work is usually very involved and hard to read, many departments have a form letter that they send out explaining that although they have not identified a specific error in the submitted work (because they have not read it), it is certain that it is wrong because it has been proved to be impossible.

That, a much badgered department chair once told me, is when the real trouble starts. The next letter will cite stories about the perseverance of Edison, berate the arrogance of the professional mathematician, or preach about the services that a state university owes the taxpayers

People do not understand that it is possible to *prove* something is impossible. One simply assumes the thing is possible, and shows that this assumption leads to something known to be wrong. In this way it can be proved that it is impossible to find more that the five regular solids known to the Greeks; it is impossible that the decimal expansion of $\sqrt{2}$ will ever fall into a repeating pattern, no matter how far it is computed; and many more examples exist.

The difficulty is that it sometimes requires a very involved line of reasoning from the assumption to the conclusion known to be wrong, and sometimes the wrong conclusion is only known to be wrong by experts in the field under consideration. And so it is that those who know a little history about things once thought impossible, and very little about mathematics, go right on trying to do the impossible. Someone could write a book; in fact, someone has [U. Dudley, *Mathematical Cranks,* Math. Assn. of Amer., 1992]. It doesn't stop them. As Thomas Jefferson once wrote of a wordy and illogical colleague in the nation's first Congress,

To refute was easy indeed, but to silence impossible.

8.1 Networks

The Seven Bridges of Königsberg

The old city of Königsberg (now Kalingrad) in East Prussia was located on the banks and two islands of the Pregel River. Seven bridges crossed the river approximately as shown in the diagram. On Sundays the citizens would walk around the town as is common in German cities. Here is a problem that occurred to them. Can a person find a route starting from his home which crosses every bridge exactly once and brings him home again?

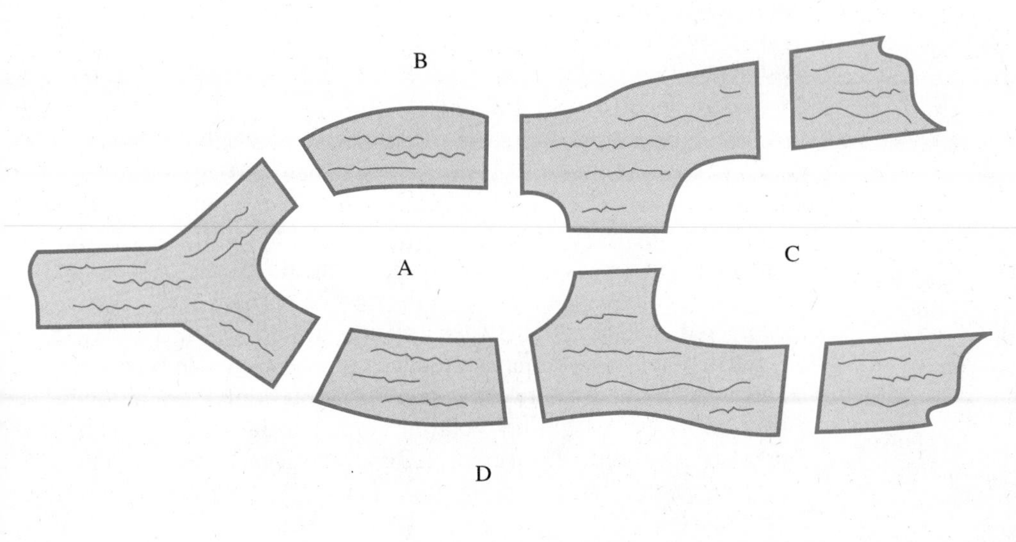

The problem of the seven bridges somehow came to the attention of the great Swiss mathematician, Leonhard Euler (1707–1783), who was residing in Russia as mathematician at the court of Catherine the Great. With customary acumen, he solved not only this problem but all others of the same type. His celebrated solution, here to be described, was presented to the Russian Academy at St. Petersburg in 1735.

Strip away irrelevant details.

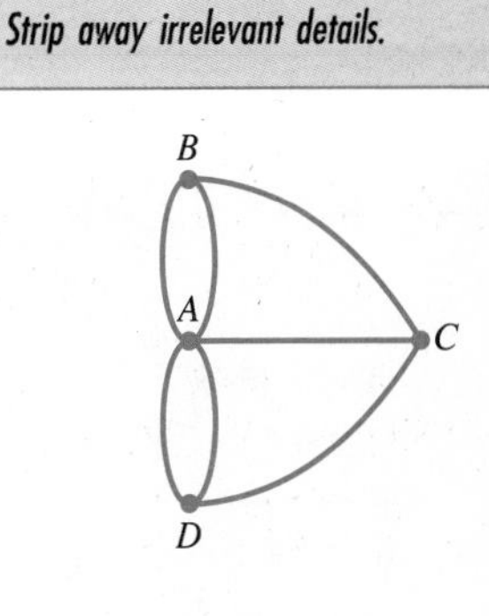

It is clear to anyone who tries to attack this problem with the head instead of the feet that the size of the islands, the length of the bridges, and the fact that these walks occurred on Sundays are irrelevant details. Euler saw that, if each land mass were represented by a point and each bridge by an arc joining two points, the problem would be one of moving around the network shown in the margin, traversing each arc exactly once, and returning to the starting point.

Before reading further, try your hand at traversing this network with a pencil, being sure not to lift it off the paper or retrace any arc. Try starting from different places. You will probably decide that it can't be done—but are you sure?

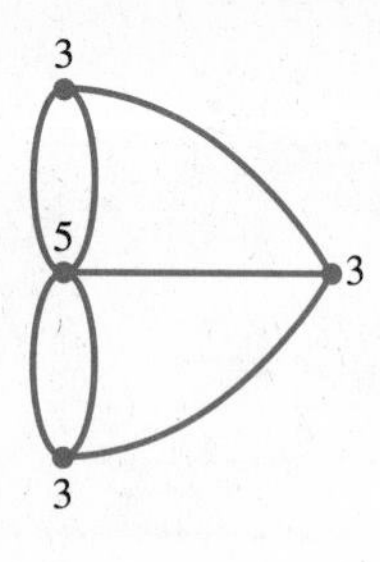

The next move seems simple enough, but it is the key to all problems of this type. It nicely illustrates the value of finding succinct notation that calls attention to the significant features of a problem. Simply attach to each of the four points (or **nodes,** as points are often called in network theory) a number that counts the number of arcs meeting at that node.

Now, any pass through a node contributes two to that number, one for the entrance and the other for the exit. Thus every node in a traversible network must have an even number attached to it. This includes the initial node, since the problem demands that the walker end where he or she started. The Königsberg bridges network doesn't have a ghost of a chance; all four of its nodes are odd. No one will ever find the required route across the seven bridges. It is impossible!

Euler Circuits

The seeds of a beautiful theory have already been planted. To see it in full bloom, we require precise terminology. The building blocks of a network are a set of points, from now on called **vertices** (or nodes), and a set of arcs, curved or straight, called **edges.** We require that each edge have two distinct vertices (its end points) and that each vertex be the end point of at least one edge. Two edges can meet only at a vertex. A **network** is a finite, connected collection of vertices and edges. Here "connected" means that between any two vertices of the network there is a **path,** that is, a continuous sequence of edges, joining them. All this is meant to say that a network is just what you think it ought to be; the examples in the margin help clarify some special points. Finally, the **degree** of a vertex is the number of edges that meet at that vertex.

Networks

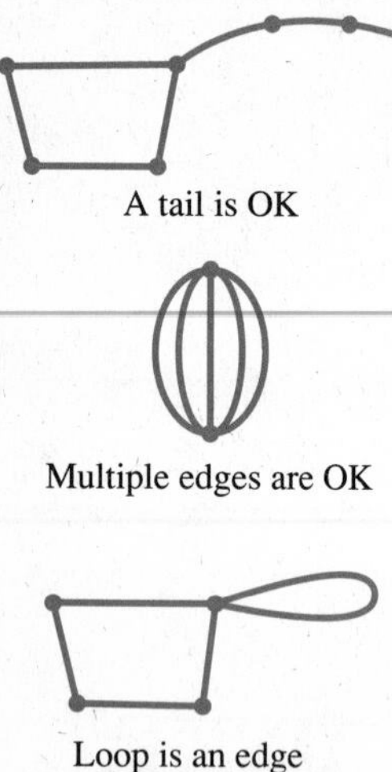

The general problem Euler attacked and solved was this: Is there a path around a network that traverses each edge exactly once and ends where it started? It is a matter of simple justice to call such a path an **Euler circuit.** Here is Euler's grand theorem on networks:

Theorem A network has an Euler circuit if and only if every vertex is of even degree.

Note that the theorem says two things. It says that, if a network has one or more odd vertices, there is no Euler circuit. We have already seen, in the case of the seven-bridges problem, an argument for this that suffices just as well in the general case. The theorem also says that a network having all even vertices has an Euler circuit. You may wish to enhance your appreciation for Euler's deceptively easy arguments by trying to prove this second part for yourself. Then read on to see how Euler leads us through such a network.

Not connected

Not networks

Take any network whatever that has all even vertices. Pick a vertex A, start a path there, and walk as far as you can, always leaving a vertex on an edge not previously used. Eventually, you will reach a vertex with no possible exit. But this dead-end vertex must be A, since at any other vertex there is an exit if there was an entrance, the degree being even.

But alas, the path P so obtained may leave out part of the network (see diagram). If so, there is a vertex B on path P where some unused edges meet. Start from B and obtain a path Q using previously unused edges and returning to B just as you did to A above. Now adjoin path Q to the original path P by following it until you get to B; then traverse path Q around its circuit back to B, finally completing path P back to A. If this combined path traverses the network, you are finished. If not, repeat the process. Eventually you will get an Euler circuit.

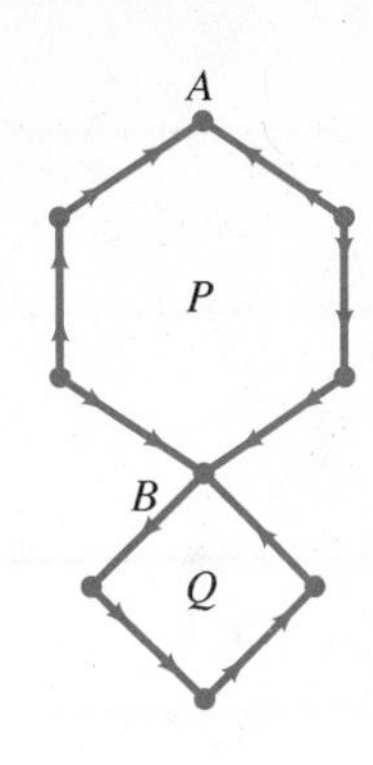

Hamiltonian Circuits

Horace and his good friend Hugo have discovered a strange coincidence. Horace is road inspector for the county highway department; Hugo is a salesman for a greeting card company. Both of them cover exactly the same territory; it is shown in the diagram below. Each Monday, Horace is required to make a complete inspection of the road system, checking for potholes and washouts. Each Tuesday Hugo calls at the drugstores in each of the towns shown. Naturally both men want to do their jobs as efficiently as possible.

Horace's job is easy. He needs an Euler circuit through the network. After checking that all the vertices are even, he draws a circuit on his map (see below, where successive steps are numbered).

Hugo's problem is quite different. He doesn't care a whit about the roads of the network. He simply wants a path that visits every vertex exactly once, returning him to his starting point. Such a path is called a **Hamiltonian circuit,** after another fine mathematician, William Rowan Hamilton. Fortunately there is one. Both men can do their jobs efficiently.

But now you can guess what the big question is. Is there a simple procedure for deciding whether a given network has a Hamiltonian circuit? No one knows, though many great mathematicians have looked long and hard. It is one of the famous unsolved problems of mathematics. A solution would be of considerable interest even to chemists who use networks to describe the structure of molecules—but this is a connection we don't have space to discuss.

W. R. Hamilton (1885–1865)
Known as the Royal Astronomer of Ireland, Hamilton was a great mathematician but an even greater physicist. Among his minor accomplishments was the invention of a game which required construction of the circuit now called by his name.

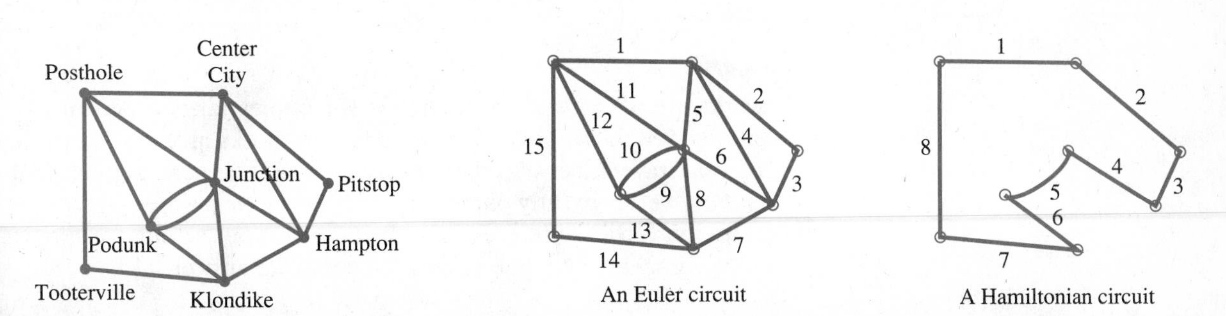

An Euler circuit

A Hamiltonian circuit

Spiderwebs and Subway Systems

Networks occur naturally all around us. One of the most interesting is the beautiful orbed web of the familiar garden spider. It is both an engineering marvel and a work of art. Here is an intriguing question for both people and spiders: What is the fewest number of continuous runs required for a spider to spin the web shown below, assuming it does not want to double any strands?

A Spiderweb

Surely the answer has something to do with the degrees of the vertices. A quick check reveals 14 odd vertices—12 around the rim and 1 at each end of the spiral. As Euler noted, odd vertices correspond to starting and ending points on noncircular paths. It therefore takes at least seven runs to spin the web; and seven will do. See if you can find them. Once again Euler's method gives us the complete solution for all problems of this kind.

Theorem If k is greater than or equal to 1, a network with $2k$ odd vertices contains a family of k distinct paths which together traverse each edge exactly once. Moreover, no family of fewer paths can do the job.

Summary

Networks are all around us, though it sometimes takes a person like Euler to notice them. Two problems about networks are: (1) Is there a circuit that traverses every edge exactly once? (2) Is there a circuit that visits every vertex exactly once? To the first we gave a simple answer—yes if all the vertices are even, no otherwise. To the second, we simply say—try your luck. No one knows the complete answer.

Problem Set 8.1

1. Determine which of the following has an Euler circuit. If there is one, draw it.

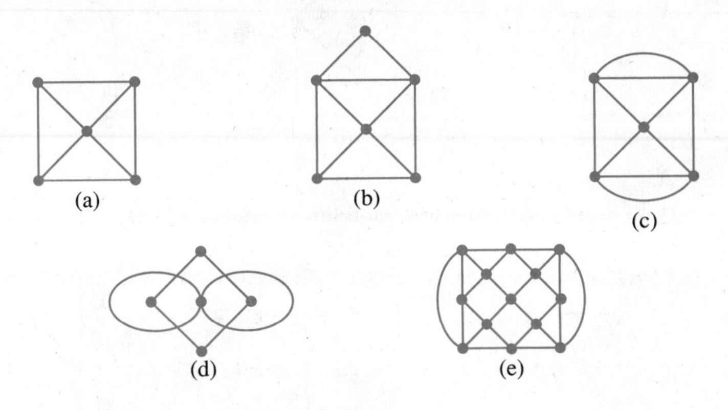

2. Follow the directions in Problem 1.

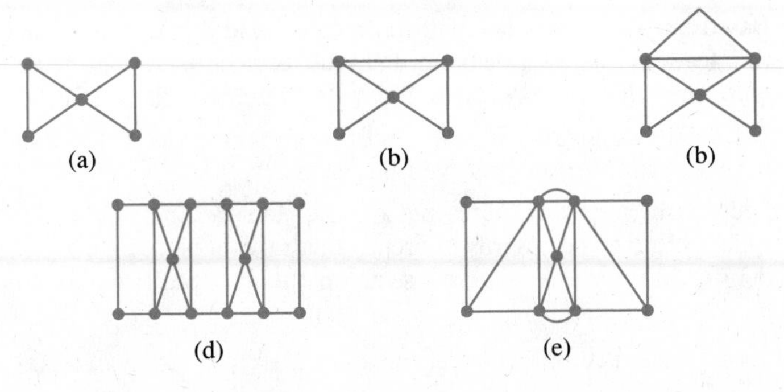

3. Try to draw a Hamiltonian circuit for each of the networks in Problem 1.
4. Try to draw a Hamiltonian circuit for each of the networks in Problem 2.
5. As in the Königsberg bridges problem, decide whether or not there is an Euler circuit for the following network of islands and bridges. If there is one, draw it.

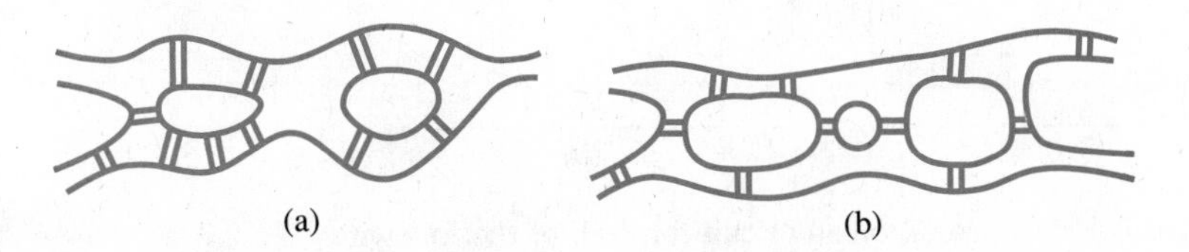

6. Can you add one more bridge to the Königsberg bridges network so that it will have an Euler circuit? If not, how about two bridges?
7. Call a path through a network that traverses each edge exactly once an **Euler path.** It differs from an Euler circuit in that it doesn't necessar-

ily begin and end at the same vertex. Try to draw an Euler path for each of the following.

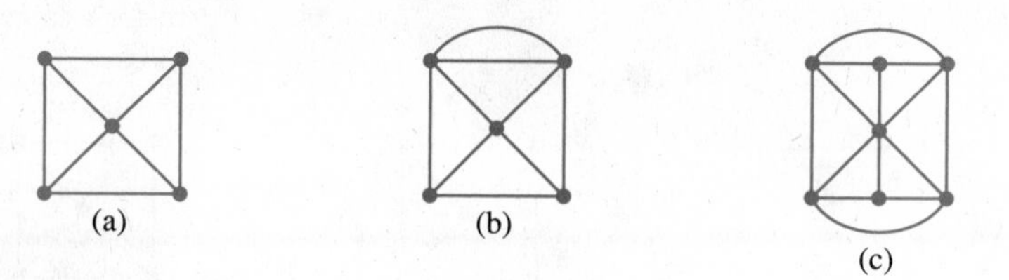
(a) (b) (c)

8. Try to draw Euler paths for the following networks.

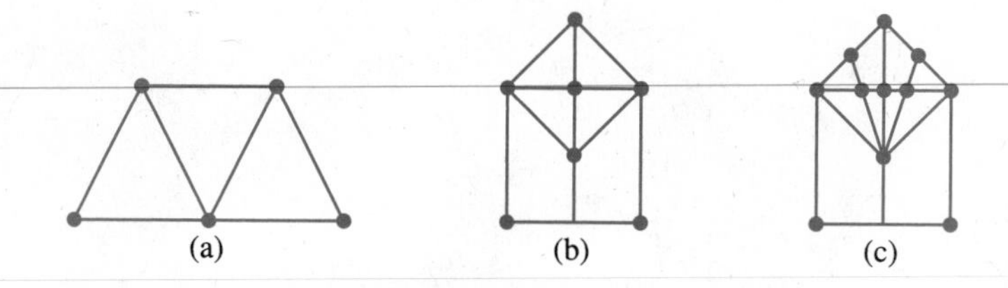
(a) (b) (c)

9. Based on your experience with Problems 7 and 8 and your reading of this section, suggest a theorem that tells us when there will be an Euler path through a network. Hint: Study the degrees of the vertices.

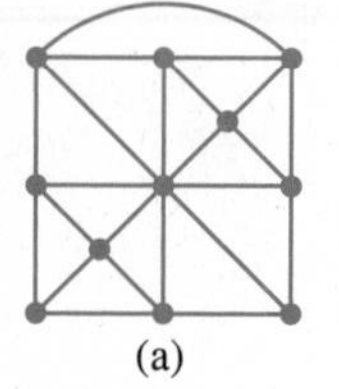
(a)

10. Use the theorem you suggested in Problem 9 to decide whether or not the figures in the margin have Euler paths.

11. Each of the following represents a house plan. Is it possible to find an Euler path (not necessarily a circuit) that takes a person through each door exactly once? If so, draw such a path. Hint: Superimpose a network as in part (a).

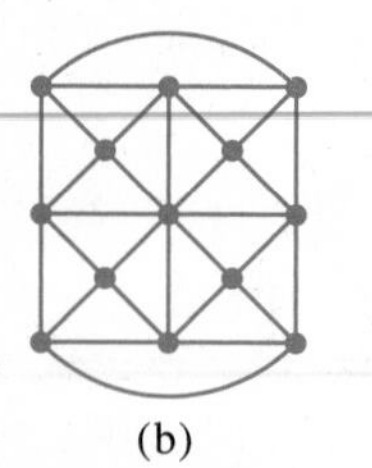
(b)

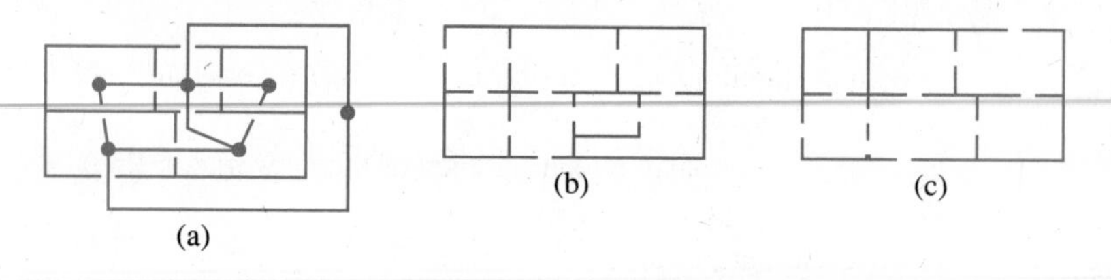
(a) (b) (c)

12. Follow the directions in Problem 11.

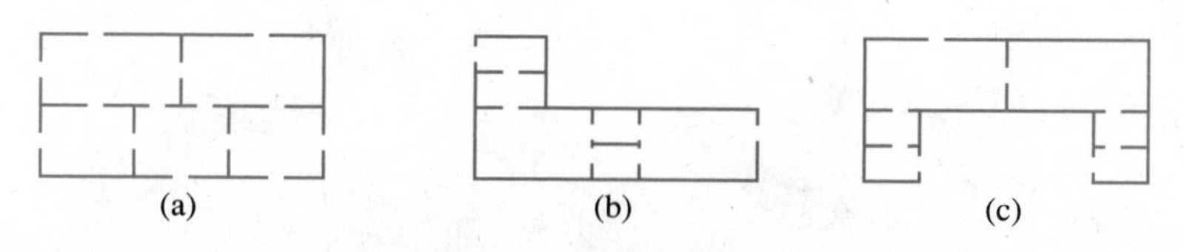
(a) (b) (c)

13. Find a Hamiltonian circuit for each of the following.

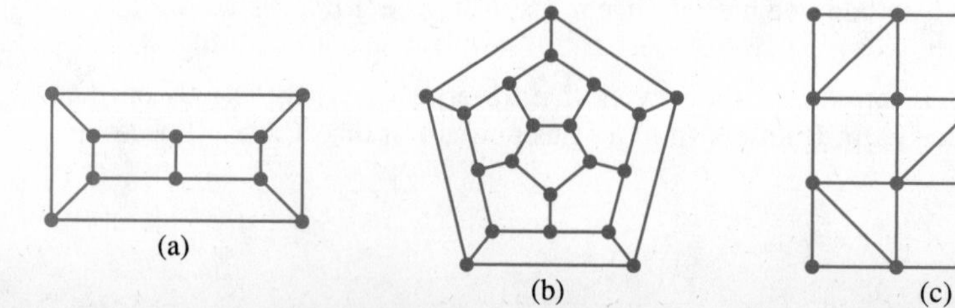
(a) (b) (c)

14. Assume that each of the networks in Problem 13 represents a spider web. How many runs are required to spin each web?

15. The diagram in the margin represents the subway system for a large city. How many separate lines are required when no two use the same track?

16.* Draw five fairly complicated networks. Label the degree of each vertex. Note for each network whether the total number of odd vertices is an even or an odd number. Make a conjecture. See if you can give an argument establishing your conjecture: Hint: Remove one edge after another from a network until you are down to one edge.

For Research and Discussion

When the first edition of this book appeared in 1978, we referred to plans for the Washington D.C. subway system that at one stage looked something like the diagram below. We noted that there were 10 odd vertices, and wrote, "Assuming the city wants the minimum number of lines and will not allow two different lines to use the same track, how many lines will be required?" Prudently, we also observed that the proposed network might be revised since "there are other factors to consider (e.g., cost, congestion, and congresspeople)." What in fact has happened since then? Which assumptions were right? which were wrong? why?

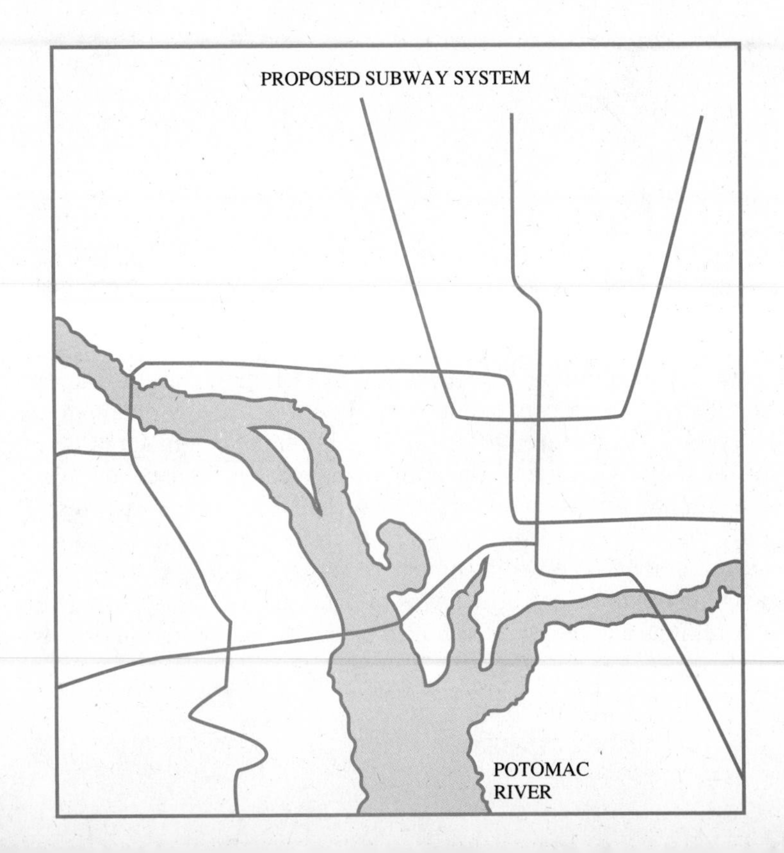

8.2 Trees

Degrees That Carry Weight

If, in the style of Euler, we label the degree of each vertex in the tree below, we will find 30 vertices labeled 1, 3 labeled 2, 18 labeled 3, and 5 labeled 4. Set

$$d_1 = 30$$
$$d_2 = 3$$
$$d_3 = 18$$
$$d_4 = 5$$

There is a formula that relates d_1, d_2, d_3, and d_4. Can you find it?

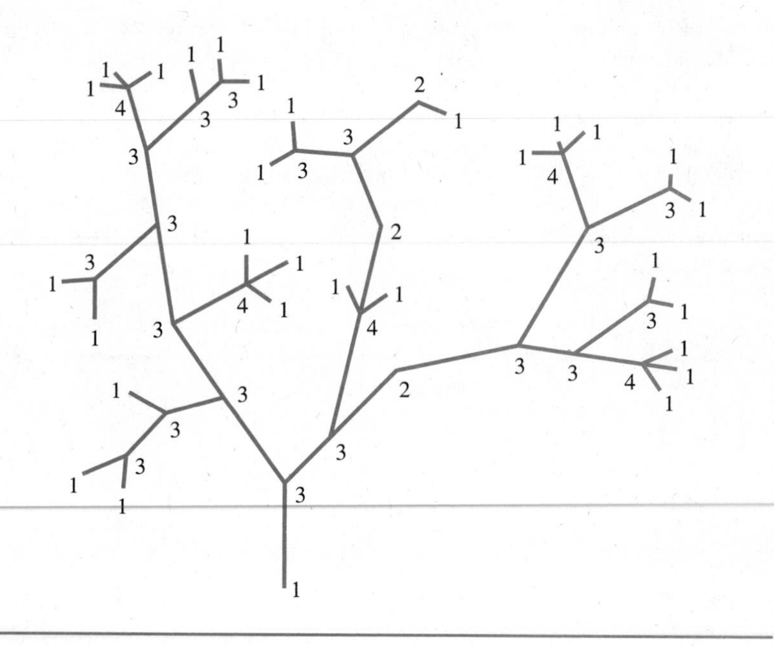

> A fool sees not the same tree that a wise man sees.
>
> *William Blake*

The reader will be forgiven for thinking the authors have no appreciation for what is beautiful about a tree. In our defense, we claim to have the imagination required to see in the diagram above much more than the skeleton of a fig tree shorn of that which makes for decency. We see, for example, rivulets flowing into creeks, thence into streams, and finally into a mighty river. Or, starting from the bottom, we can see a huge natural gas network coming from the oil fields of Texas and branching out to all the little towns of a midwestern state. We can even interpret it as a series of decisions to be made, each junction displaying the possible choices (with the three vertices of degree 2 indicating places where options have been cut off).

In any case, we have pictured a mathematical tree. For mathematicians, a **tree** is a (connected) network with no circuits or loops of any kind. Can something so simple be worth studying? We think so.

Start with a small example.

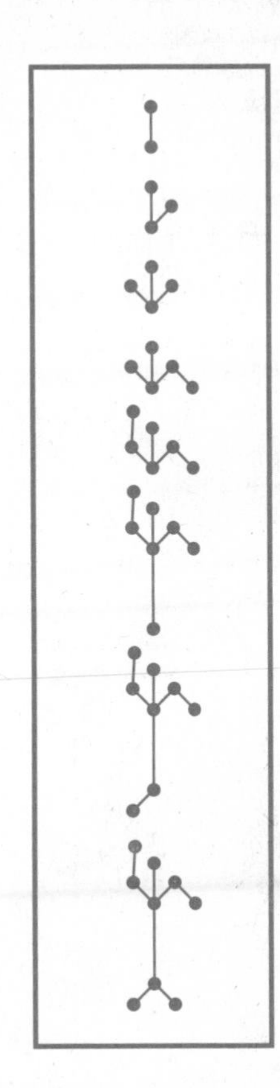

Edges and Vertices

As a warmup for the problem posed at the beginning of this section, we shall derive a formula that relates the number of edges and the number of vertices in any tree. We shall get to the root of the matter by starting with the smallest possible case of one edge and two vertices (a root and a leaf if you care for that terminology).

With our modest start, we have 1 edge, 2 vertices. Now start building from there to any complicated tree that you wish, one edge at a time. Each time you add an edge, you add a vertex. Thus you maintain the relationship; vertices always stay one ahead of edges. We have shown that

A tree with n edges has $n + 1$ vertices.

Degrees of Vertices

If you have given any thought at all to **Degrees That Carry Weight,** the problem that opened this section, you may have noted but been almost hesitant to write down the more or less obvious fact that $d_1 + d_2 + d_3 + \cdots = n$, where n is the total number of vertices. It does illustrate the kind of relationship we seek, but we agree that it is quite trivial, and we had something deeper in mind when posing the question.

We have in mind a formula that relates just the numbers d_i. See if you can find it. Don't forget the problem-solving techniques we have already found useful in this section: make up some examples for yourself, starting with small numbers of vertices and edges; draw pictures. Try this before you read on, lest you come to the answer that spoils the fun of solving this problem on your own.

Does the answer you are considering work for river systems? Such systems have no junctions of degree 2 and only rarely exhibit junctions of degree greater than 3. Thus most river systems obey the law $d_3 = d_1 - 2$. Now d_1, the number of first-degree vertices, is really just the number of sources plus the one exit. *Thus, for river systems, the number of triple junctions is equal to the number of sources minus 1.* For example, a system with 40 sources has 39 triple junctions.

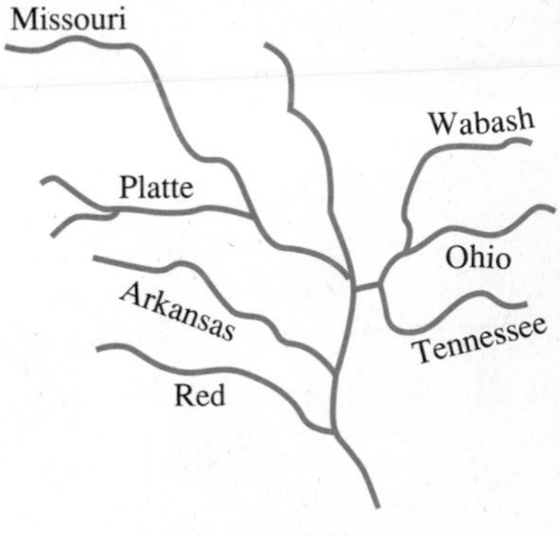

Mississippi River system

Have you guessed that $1d_3 + 2d_4 + 3d_5 + \cdots = d_1 - 2$? Try it on the examples you have looked at. Check it on **Degrees That Carry Weight.** Then try to prove it. We outline a proof in Problem 14.

> And this our life,
> Exempt from public haunt,
> Finds tongues in trees,
> Books in the running brooks,
> Sermons in stones,
> And good in everything.
>
> *William Shakespeare*

Minimizing the Cost of a Tree

Imagine that the 11 dots below represent buildings on a new college campus, which are to be connected by underground electric cables. The building code states that junctions can occur only within buildings. The contractor wishes to build the network so as to minimize its cost, that is, its total length. Clearly the network should not contain any circuits, for one link of a circuit could be left out and the buildings would still be connected. The contractor wants an **economy tree.** Here are the rules for building one:

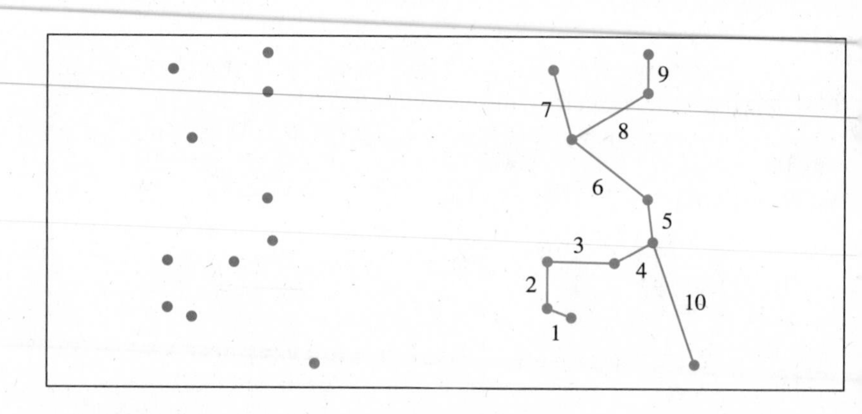

1. First build the shortest edge.
2. Thereafter connect the vertex closest to the part already constructed.
3. Break any ties in an arbitrary manner.

For our example, the result of following the rules gives the tree shown above (with the steps numbered).

There remains the essential point: to show that these rules produce a tree of minimum length. That of course is the rub; we know of no easy demonstration (see Problem 15).

Spirals, Meanders, and Explosions

Sometimes there are factors other than total length to be considered in building a tree. If edges represent steam heating lines, one should make sure no building is too far from the central heating plant and that no one line services too many buildings. For similar reasons, no branch of an oak tree can be too long; otherwise the trunk will not be able to force nutrients out to the farthest twig.

Suppose, for example, that a heating plant is located at H in the accompanying diagrams, where for simplicity the 25 buildings are at the intersection points of a regular grid. The total lengths of the spiral and the meander are 24 units, the least possible (they are economy trees); but we can guarantee that the dorm residents at the end of the line will complain on a cold February morning.

The explosive pattern in the middle excels in directness, but at the expense of adding length; it measures 37 units overall. Now the contractor will complain. To find some middle ground takes us back to Section 1.3; experiment, guess, demonstrate. The branching pattern is an economy tree (of length 24) and is quite good on directness. But the compromise shown at the bottom achieves almost the best of both worlds. It measures 27.3 units, and yet each building is within 2.8 units of H.

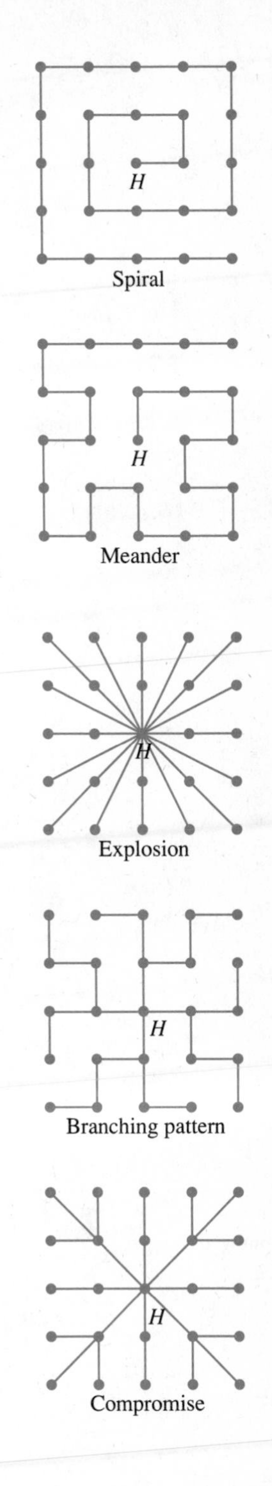

Summary

We have discovered two important laws that trees obey. Vertices always exceed edges by exactly one, and their degrees inevitably conform to the formula

$$1 \cdot d_3 + 2 \cdot d_4 + 3 \cdot d_5 + \cdots = d_1 - 2$$

Moreover, for a given set of vertices, we gave a clearly prescribed set of rules that will yield a tree of shortest length (an economy tree).

Problem Set 8.2

1. Label the degree of each vertex in tree A. Then calculate d_1, d_2, d_3, . . . , and verify the formula displayed in the summary above.

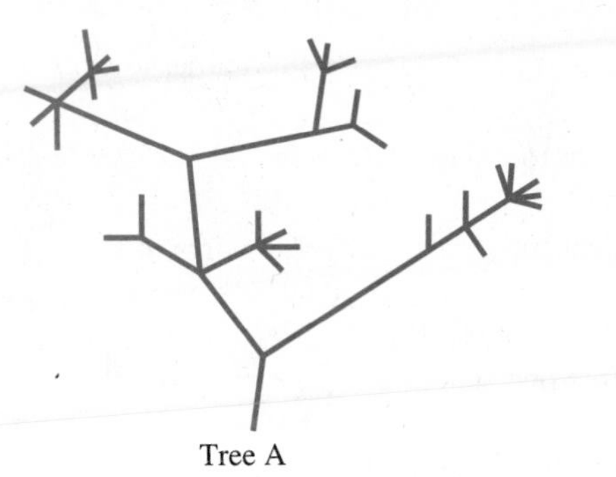

Tree A

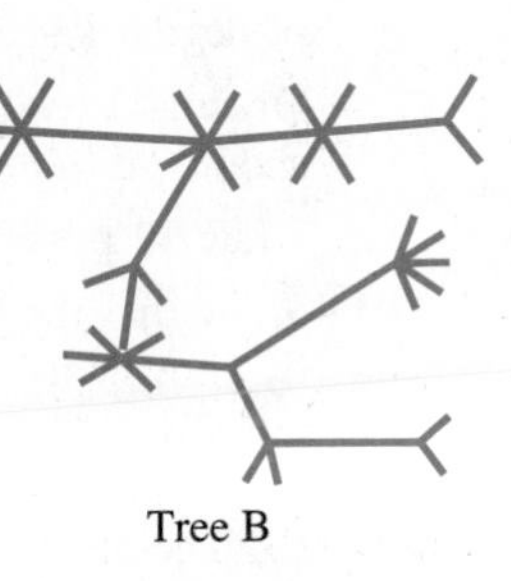

Tree B

2. Follow the instructions in Problem 1 for tree B.
3. Count the number of vertices in tree A. Without counting, calculate the number of edges. Hint: We derived a formula for this.
4. Count the number of the edges in tree B. Without counting, calculate the number of vertices.
5. Draw (if possible) a tree with $d_1 = 17$, $d_2 = 5$, $d_3 = 3$, $d_4 = 3$, and $d_5 = 2$.

6. Draw (if possible) a tree with $d_1 = 20$, $d_2 = 1$, $d_3 = 2$, $d_4 = 4$, $d_5 = 4$, and $d_6 = 1$.

7. Assuming that all junctions in a certain river system are triple junctions and that there are 49 of them, how many sources does the system have?

8. Tree C represents the pairings in a tournament. Note that there are no vertices of degree greater than 3, so $d_3 = d_1 - 2$. A vertex of degree 3 corresponds to a game. Derive a rule for the number of games if there are n contestants. Compare with Section 1.4.

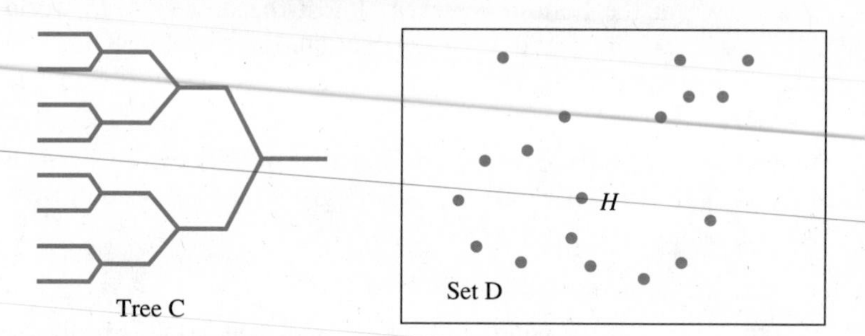

Tree C

9. Make two copies of set D, shown above (do this by placing a sheet of onionskin paper over the figure and marking the dots).
   (a) Use one copy to draw an economy tree.
   (b) Use the other copy to draw a good distribution system for steam pipes, assuming H is the heating plant and that connections can occur only at dots.

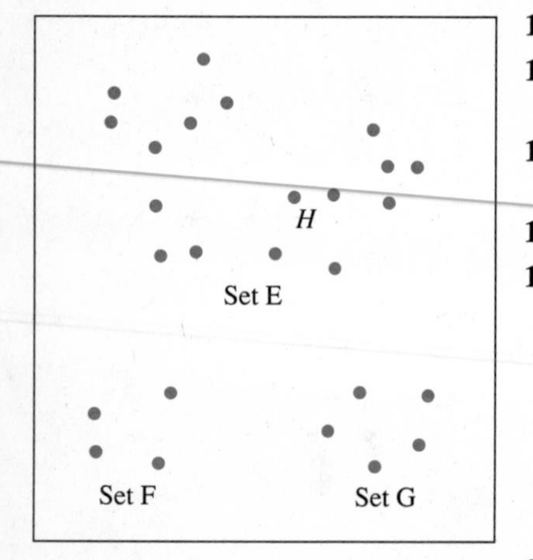

10. Follow the directions in Problem 9 for set E.

11. Show that every tree (with at least one edge) has at least two vertices of degree 1 (i.e., $d_1 \geq 2$).

12.* Draw all possible different trees connecting the points in set F. (What should "different" mean?)

13.* Draw all possible different trees connecting the points in set G.

14.* Show that the formula $1 \cdot d_3 + 2 \cdot d_4 + 3 \cdot d_5 + \cdots = d_1 - 2$ holds by convincing yourself of the following.
   (a) We can build any tree one step at a time, each step (after the first) consisting of adding one edge and one vertex.
   (b) The formula holds for the simplest tree, one edge with two vertices.
   (c) Each step does two things. It replaces a vertex of degree j by a vertex of degree $j + 1$, and it adds a vertex of degree 1. The result is simply to add 1 to both sides of the formula.

15.* Show that the rules given in the text for producing an economy tree actually work by filling in the details of the following argument.
   (a) Let E be a tree constructed according to the rules and suppose there is a tree of shorter length reaching all the given vertices. Let F be a shortest such tree.

(b) Let $E_1, E_2, E_3, \ldots, E_n$ be the edges of E listed in the order in which they were adjoined to make E. There is a first edge, say E_i, in E that is not in F.

(c) Adjoin E_i to F. The result, $F \cup E_i$, must have a circuit, since there are two paths connecting the vertices joined by E_i. This circuit must have an edge, say E_j, not in E.

(d) Create a new tree G from F by removing E_j and replacing it with E_i:

$$G = (F - E_j) \cup E_i$$

(e) If L denotes length,

$$L(G) = L(F) - L(E_j) + L(E_i)$$

Since $L(F)$ is as small as is possible for a tree, $L(E_i) \geq L(E_j)$.

(f) But E_i was (according to the rules of construction) the shortest edge that could be connected to $E_1, E_2, \ldots, E_{i-1}$ without making a circuit. Since E_j can also be joined to $E_1, E_2, \ldots, E_{i-1}$ without making a circuit, $L(E_j) \geq L(E_i)$ and $L(E_j) = L(E_i)$. Thus G and F have the same length.

(g) Note that G has one more edge, namely, E_i, in common with E than did F. Now repeat the above operations on G. We eventually obtain a tree coinciding with E, which has the same cost as F, contrary to our supposition.

16.* A, B, C, and D are the four corners of a square of side length 1 unit. A tree of minimum length is to be constructed connecting these four points but for this problem we will allow the introduction of new vertices. For example, we will allow the tree shown in the margin which has length $2\sqrt{2} \approx 2.8284$. It can be shown that the shortest tree connecting A, B, C, and D (allowing new vertices) has length $1 + \sqrt{3} \approx 2.7321$. Try to find the construction that leads to this answer.

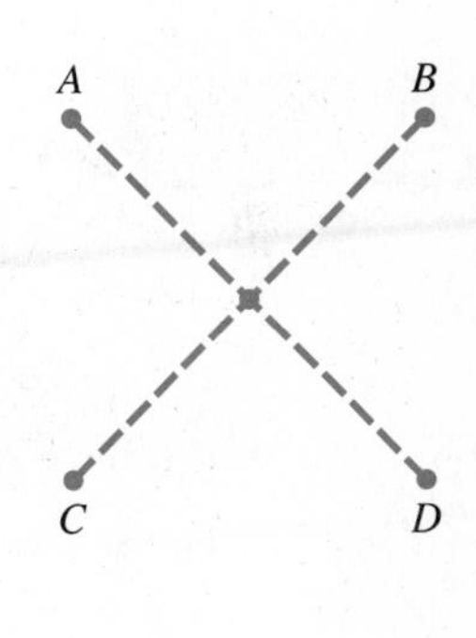

For Research and Discussion

The subject matter of this chapter, and particularly of this section, is at the heart of discrete mathematics, a subject that has become increasingly important in the last 25 years. Read the introductions to several books on discrete mathematics to get some understanding of what it is, and how it differs from continuous mathematics. Write a report that includes

a. a description of what you understand discrete mathematics to be.

b. several typical textbook problems from the subject.

c. a statement of several of the classic problems of discrete mathematics (traveling salesman, Chinese postman, bin packing, the knapsack problem, . . .).

8.3 The Platonic Solids

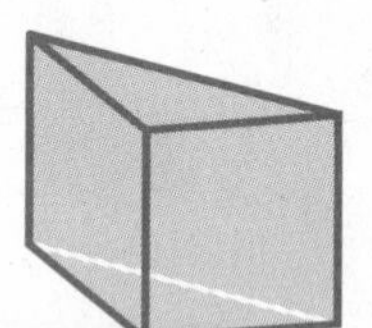

Triangular prism

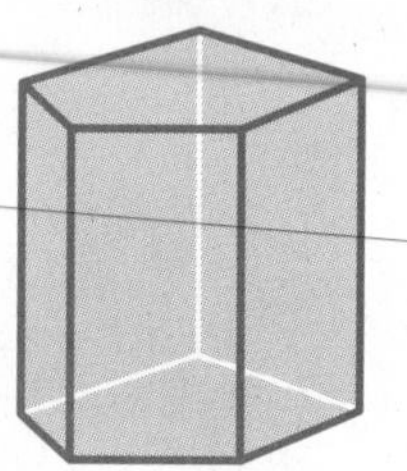

Pentagonal prism

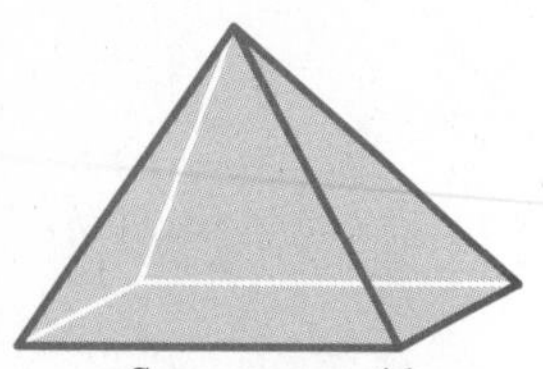

Square pyramid

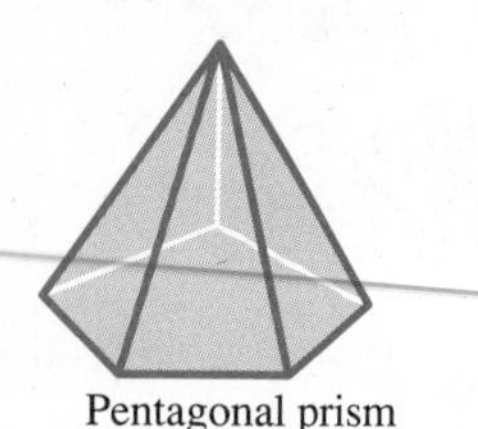

Pentagonal prism

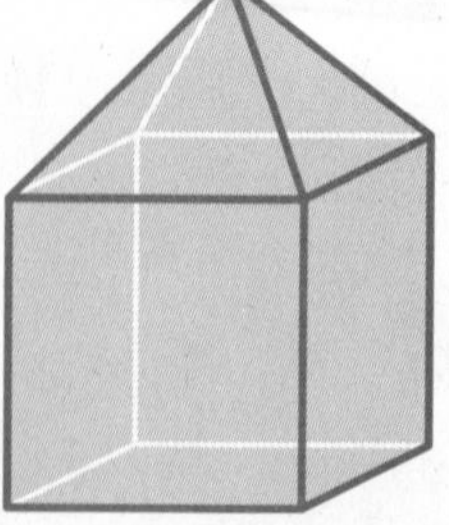

Roofed box

A Solid Problem

A polygon is a planar figure without holes that is bounded by line segments (e.g. a triangle, a quadrilateral, etc). A regular polyhedron is a three dimensional solid having identical regular polygons as faces and the same configuration at the corners. The tetrahedron and cube in the adjacent picture are examples of regular polyhedra. Can you find other regular polyhedra.

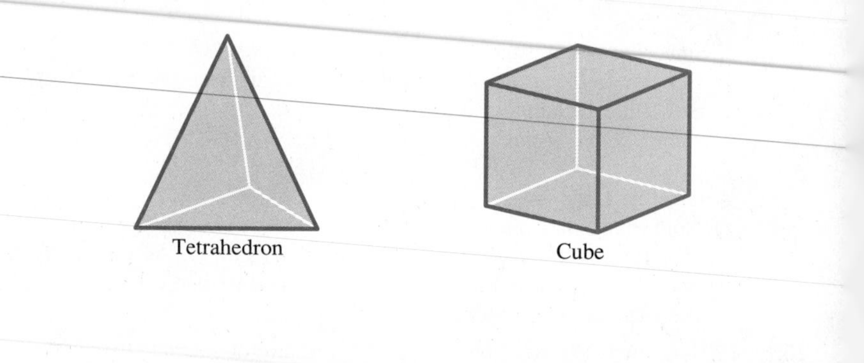

Tetrahedron Cube

Regularity! Symmetry! Already identified as touchstones for problem solving, these are ideas that have been central considerations in art forms, in scientific exploration, and even in religious philosophy throughout history. We would expect then that solid figures having all their faces as identical polygons would figure prominently in all of these fields, and in this expectation we are not to be disappointed.

Guided by Max Friedländer's observation, "A fool he who gives more than he has," we shall not venture into cubism as an art form, but shall rather exhort the reader to experience the satisfaction of discovering something "really neat" by putting some genuine effort into the question that opened this section. You should be able to find at least one more regular polyhedron for yourself before having the joy of personal discovery taken from you by reading on.

Euler's Formula for Polyhedra

It may seem to you that regular polyhedra have little to do with networks, but appearances are deceptive, as we shall see by once again following the insights of Euler. Since you should have discovered by now that it is not easy to find examples of regular polyhedra, let us invoke the principle of beginning with a related problem, simplifying things by dropping the restriction that the polyhedra we consider be regular.

A **polyhedron** is a solid figure, all of whose faces are polygons. A few are shown in the margin. We are primarily interested in convex polyhedra, that is, those without holes or dents. (In technical language, a figure is convex if any two of its points can be joined by a line segment within the figure.) Each convex polyhedron has a certain number of faces F, vertices V, and edges E. Is there a relationship between these three numbers that is always valid? Let's make a table for the solids we have pictured and see if any pattern appears.

Table 8-1

Polyhedron	*F*	*V*	*E*
Triangular prism	5	6	9
Pentagonal prism	7	10	15
Square pyramid	5	5	8
Pentagonal pyramid	6	6	10
Roofed box	9	9	16
Tetrahedron	4	4	6
Cube	6	8	12
Octahedron	8	6	12
Dodecahedron	12	20	30
Icosahedron	20	12	30

The formula that relates F, V, and E is not obvious, but it is not so difficult as the one we discussed in the last section relating the degrees of vertices in a tree. Again we urge you not to be too hasty in reading the solution before you give yourself the joy of discovering a relationship for yourself.

Experiment with different representation of the information.

If you have put some genuine effort into the problem at hand, you are allowed to read on for either confirmation of your ideas, or for some hints. Take any convex polyhedron and suppose that its insides have been removed, leaving only its outer surface which we imagine to be made of thin, flexible rubber. Cut out one face and then stretch out the part that's left so that it lies flat in the plane (see marginal illustration for the case of a cube). This transforms the surface of the polyhedron into a network in the plane and in the spirit of Section 2.4, transforms our problem as

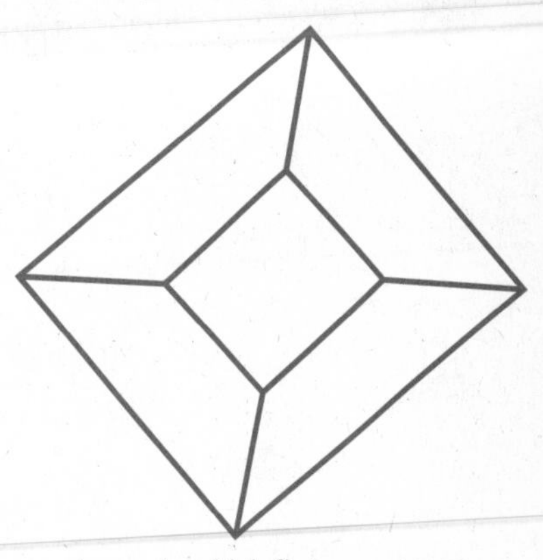

A case in which flat-ery gets you somewhere

Transform the problem.

well. Imagine the region around the outside of the network as corresponding to the removed face. Then the network will have exactly the same number of faces, vertices, and edges as the polyhedron. If we can prove Euler's Formula for networks in the plane, we will get it free for convex polyhedra.

Start with a small example.

Having reduced our question to one about networks in the plane, we can again try the trick that served us well in the last section. Start with the smallest possible network, and then watch carefully what happens as we build it up to something more complex.

Begin with a single edge ($E = 1$) joining two vertices ($V = 2$). Remember that the region outside the network is counted as a face ($F = 1$). There are many formulas that relate E, F, and V for this example. We have a start, but we need more evidence.

What happens if we add one edge? Since we are not stydying trees as we were in the last section, there are two different ways in which the new edge may form a new network.

(1) The new edge joins the two existing vertices, thereby adding one to F and one to E in the formula that relates E, F, and V.
(2) The new edge joins one of the existing vertices to a new vertex, thereby adding one to E and one to V.

No matter how complicated a network becomes, the next step to building it up involves adding one more edge, and the addition of one edge will necessarily have one of the two effects noted above. That is, with each addition, E increases by one, and either V or F is increased by one. A formula that has E on one side of the equal sign and both V and F on the other will, if in balance before we add an edge, remain in balance after a new edge is added.

Does it work for known special cases?

Everything we need is now in hand. Go back to the first data we collected for this problem, the single edge tree where we saw that $E = 1, F = 1$, and $V = 2$. Taken together with the last statement of the last paragraph, you should try guessing a formula. Don't forget that such a formula should work for the values listed in Table 8-1. In order not to spoil things for you, we refrain from stating the formula here; it appears where we need it below.

Regular Polyhedra

A polygon is regular if it has equal sides and equal angles. Examples are equilateral triangles, squares, regular pentagons, etc. A convex polyhedron is called **regular** if

1. Each face is a regular p-gon (a polygon with p edges).
2. Each vertex is surrounded by the same number, say q, of these faces (equivalently, q edges meet at each vertex).

Regular convex polyhedra are also called Platonic solids. There are only five of them. Let's see why.

Notice that each edge is shared by two faces. Thus pF counts each edge twice; i.e.,

$$pF = 2E \qquad \text{or} \qquad \frac{F}{E} = \frac{2}{p}$$

However, q edges meet at each vertex. So qV counts the number of edge ends, which again is twice the number of edges.

$$qV = 2E \qquad \text{or} \qquad \frac{V}{E} = \frac{2}{q}$$

Now we need Euler's formula for polyhedra. We hope you discovered it by now, but if you didn't, reread this section now, noting how nicely all the evidence fits for

$$F + V = E + 2$$

If you multiply both sides of this equation by $\frac{1}{E}$, you will get

$$\frac{F}{E} + \frac{V}{E} = 1 + \frac{2}{E}$$

The formula written in this form enables us to use the highlighted information above. Substituting,

$$\frac{2}{p} + \frac{2}{q} = 1 + \frac{2}{E} > 1$$

The last inequality follows from the fact that addition of the positive number $\frac{2}{E}$ to 1 certainly gives us a number greater than 1. Multiplication by $\frac{1}{2}$ gives us

$$\frac{1}{p} + \frac{1}{q} > \frac{1}{2}$$

and this formula enables us to answer the question of how many regular polyhedra there can be.

What integers p and q can possibly satisfy this last formula? Keep in mind that both p and q must be at least 3. Observe that, if $p \geq 6$, then $q < 3$, which we have just ruled out. We quickly conclude that both p and q are between 3 and 5. On checking we find exactly five solutions, and each of them corresponds to a Platonic solid. The possible values for p and q and then for E, F, and V are shown below.

Table 8-2

Solid	p	q	E	F	V
Tetrahedron	3	3	6	4	4
Cube	4	3	12	6	8
Octahedron	3	4	12	8	6
Dodecahedron	5	3	30	12	20
Icosahedron	3	5	30	20	12

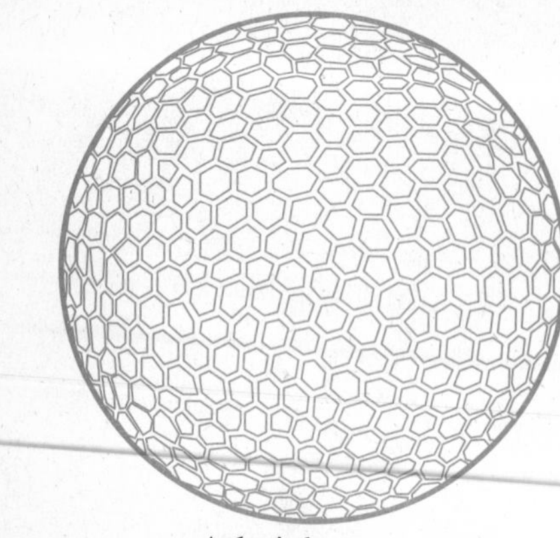
Aulonia hexagona

No one will ever find another regular polyhedron. It's impossible.

Actually we have proved considerably more than we have claimed. Nowhere did we use the assumption that the faces are regular, but only that they have the same number of edges. We have proved in fact that any convex polyhedron with p-gons as faces, q of them meeting at each vertex, fits into one of five classes. A class is determined by one of the five possible values for p and q. The skeleton of a radiolarian, called *Aulonia hexagona,* appears to defy this law. It seems to be made of a huge number of hexagons ($p = 6, q = 3$). But a close check reveals that some of the faces are pentagons, this being the only way that Euler's Formula can survive.

The Platonic Solids in History

The Greeks discovered the five regular solids shown here and attached a certain mystical significance to them. Why, they asked, could they find only five of these solids with identical regular polygons as faces and the same configuration at the corners. Both they and those who followed found ingenious philosophical and religious reasons for this strange fact of nature.

Kepler, the great sixteenth-century astronomer, worked at a time when just six planets (Mercury, Venus, Earth, Mars, Jupiter, and Saturn) were known. The methods of the time called for explanations as to why there would be six, as well as for explanations of why their speeds and distances were related as they were.

Kepler, knowing that Euclid had demonstrated that there could only be five regular solids, concluded that the five solids corresponded to gaps between the six planets. He represented the orbit of the earth by a sphere. The relative distances of the other planets could then be explained roughly (actually, it was a bit lumpy) by inscribing and circumscribing the platonic solids in a prescribed order.

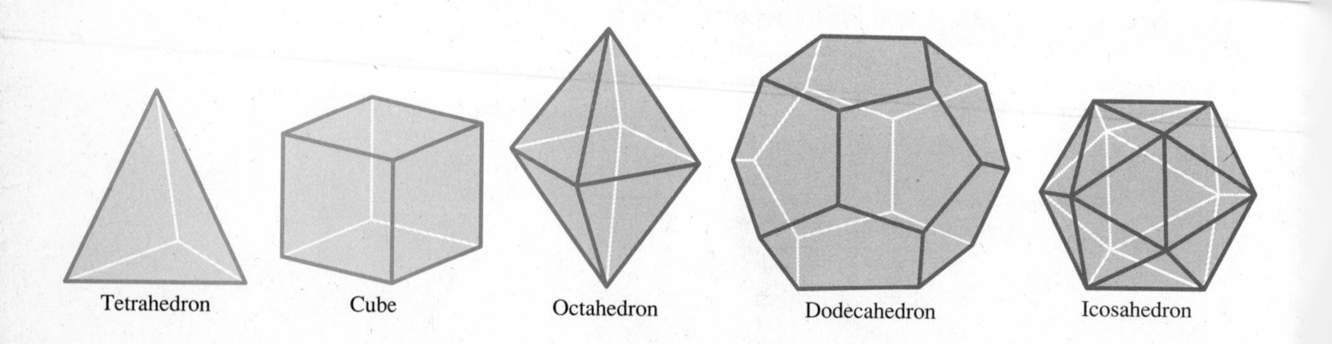
Tetrahedron Cube Octahedron Dodecahedron Icosahedron

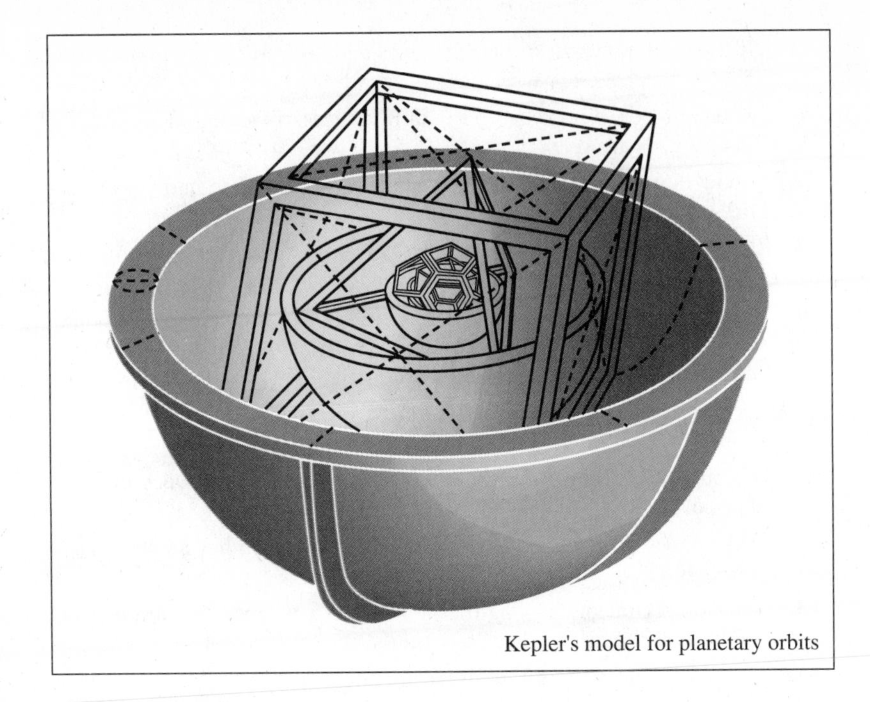

Kepler's model for planetary orbits

Network A

Summary

Certainly the most important result in this section is Euler's Formula, $F + V = E + 2$, which relates the number of faces, vertices, and edges of a convex polyhedron or of a planar network. It is so simple that anyone can understand it; yet its consequences are far-reaching. We saw one in this section: there are only five Platonic solids. We will see another in Section 8.4.

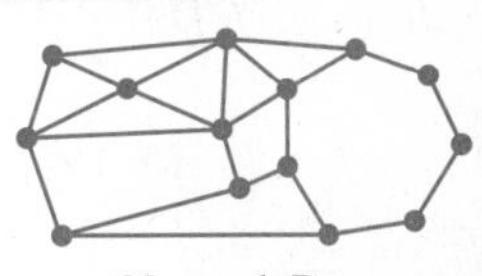

Network B

Problem Set 8.3

1. Calculate F, V, and E for network A shown in the margin. Verify Euler's Formula. Remember that the region outside the network counts as a face.
2. Follow the directions in Problem 1 for network B.
3. Calculate F, V, and E for polyhedron C and verify Euler's Formula. Note that polyhedron C, which is formed by pasting two regular tetrahedra together, has six equilateral triangles as faces; yet it is not a regular solid. Why?
4. Calculate F, V, and E for polyhedron D and verify Euler's Formula.

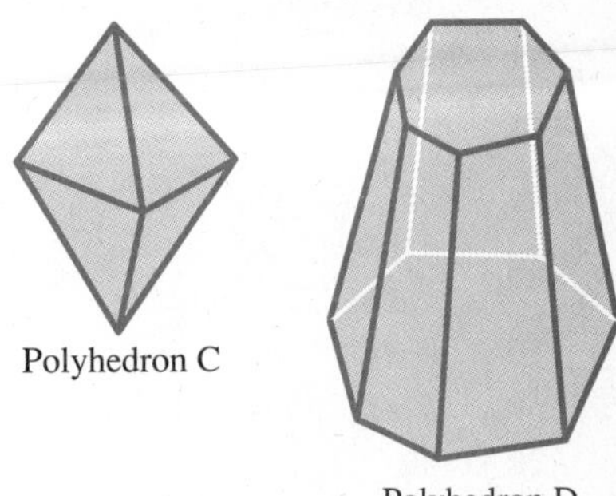

Polyhedron C

Polyhedron D

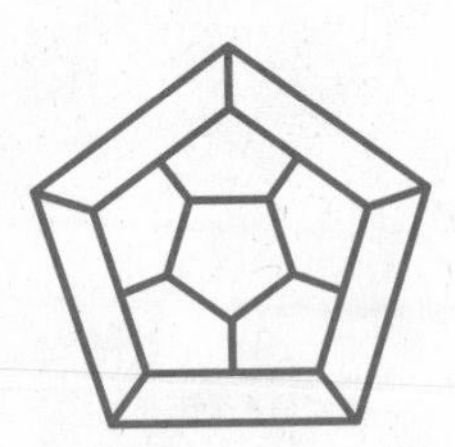
Network for dodecahedron

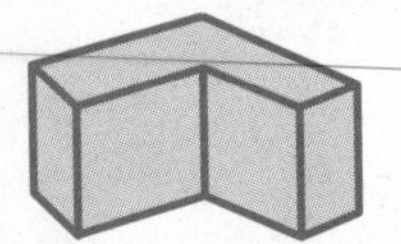
Polyhedron E

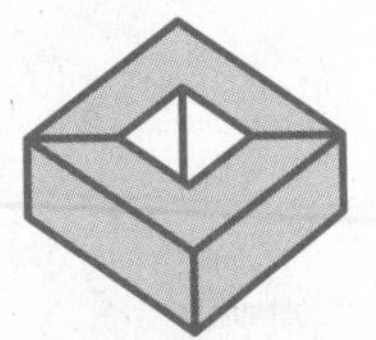
Polyhedron F

5. If a convex polyhedron has five faces and six vertices, how many edges does it have? Sketch such a polyhedron.
6. Is there a convex polyhedron with 8 vertices, 5 faces, and 13 edges? Justify your answer.
7. Show that a convex polyhedron cannot have the same number of vertices as edges.
8. Show that, if a convex polyhedron has an even number of vertices and faces, it also has an even number of edges.
9. Which of the five regular polyhedra have Euler circuits (see Section 8.1)? Hint: Draw the corresponding networks in the plane. For example, the dodecahedron network is shown at left.
10. Which of the five regular polyhedra have Hamiltonian circuits (see Section 8.1)? Follow the hint in Problem 9.
11. Calculate F, V, and E for the nonconvex polyhedron E. Does Euler's Formula hold?
12. Does Euler's Formula hold for polyhedron F which has a hole through it?

13.* Give an argument showing that a nonconvex polyhedron with dents but no holes still satisfies Euler's Formula (see Problem 11).

14.* Draw several polyhedra, each having one hole (see Problem 12). Make sure all their faces are polygons. Calculate $F + V - E$ for each of them. Make a conjecture.

15.* Draw several polyhedra, each having two holes. Calculate $F + V - E$ and make a conjecture.

16.* Conjecture a formula for $F + V - E$ for a polyhedron with n holes.

17.* Without cutting it into more than one piece, determine how to fold and cut a piece of construction paper to make models of a tetrahedron and an octahedron (see below for a cube). Make these models.

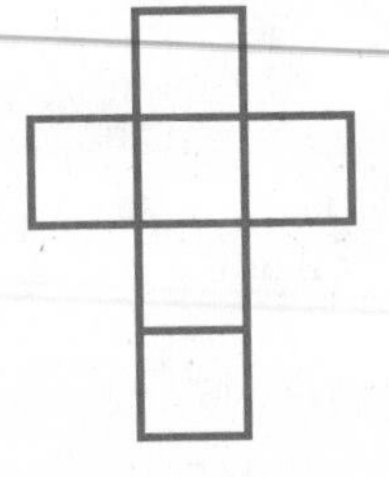

18.* Follow the directions in Problem 17 for a dodecahedron and an icosahedron.

19.* Show that every convex polyhedron satisfied $V \geq 2 + \frac{1}{2}F$.

20.* Show that every convex polyhedron satisfies $F \geq 2 + \frac{1}{2}V$.

21.* Imagine a cube made of clear plastic sheets. Suppose that a hole is drilled at the center of each face and stiff wires are inserted through these holes. The outline of another regular solid, an octahedron, is obtained (see marginal illustration). A similar phenomenon occurs if we start with any of the other regular solids. Figure out exactly what happens in each case. Hint: The centers of the faces of the original solid become the vertices of the new solid.

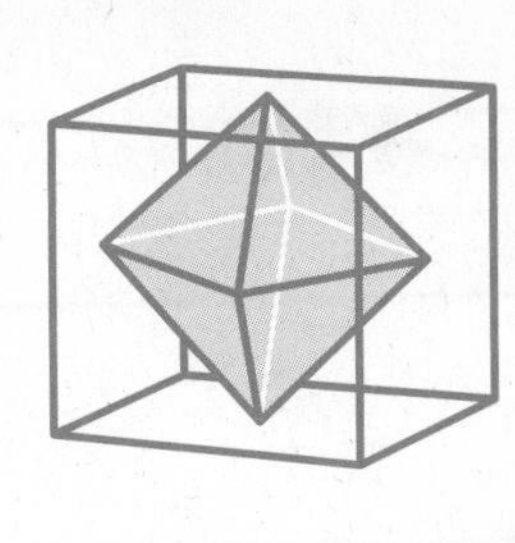

For Research and Discussion

By truncating (slicing off) the corners of a cube we obtain a semiregular polyhedron—a truncated cube. A convex polyhedron is semiregular if its faces are regular polygons of two or more types and if each vertex of the solid is surrounded by these polygons in the same way. There are exactly 14 semiregular polyhedra. See how many you can find.

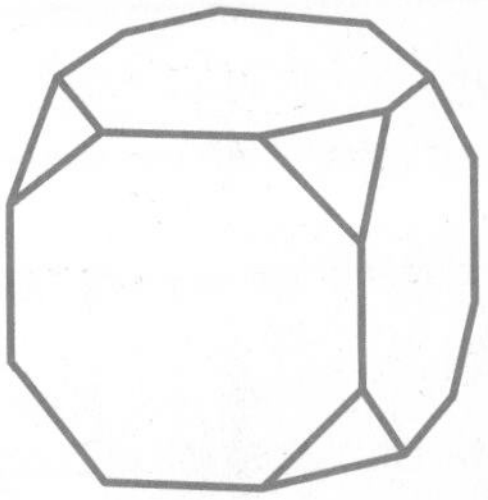

8.4 Mosaics

A Puzzle That Fits

Two identical squares joined to have a common edge from a domino. Three squares joined together form a tromino, and they come in 2 shapes. Pentominoes, composed of five squares, come in 12 different shapes, 3 of which are shown below. We have illustrated a tiling of the plane using one of them. How many of the 12 different pantominoes can be used to tile the plane (that is, fitted together so as to cover the whole plane without gaps or overlap)?

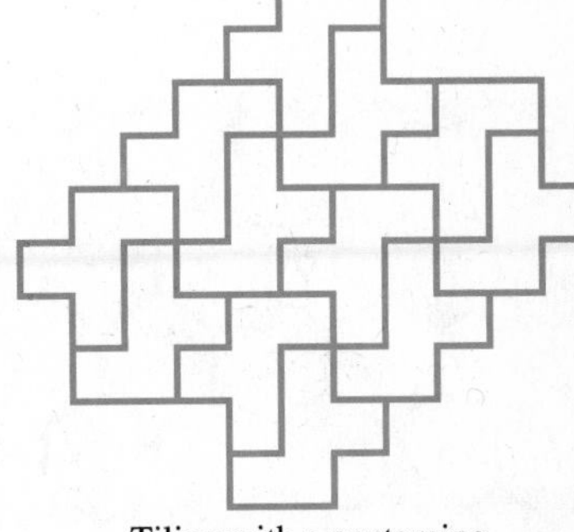

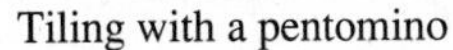
Tiling with a pentomino

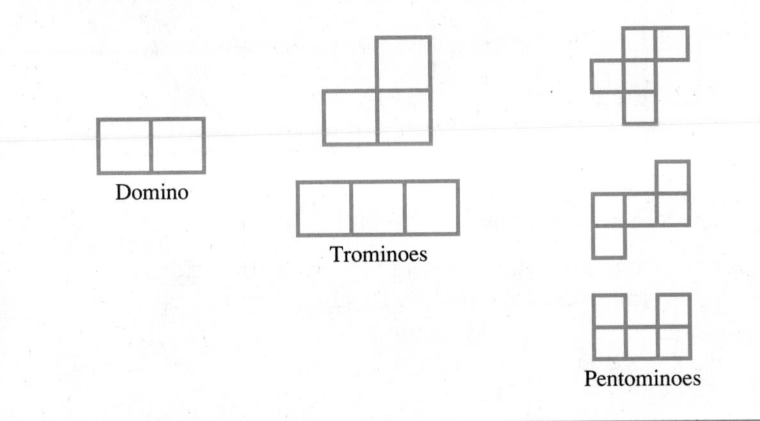

Take a jigsaw puzzle piece (for example, one shaped like the bird in the Escher picture at the end of this section on page 270) and reproduce it *ad infinitum.* Then try to fit the resulting identical pieces together so they fill the whole plane without gaps or overlap. If you succeed, you will have made a pattern called a **mosaic** (or tessellation), and the basic piece you started with is said to **tile** the plane. The great Dutch artist, M. C. Escher, created many beautiful mosaics using a bird or an animal as the basic piece. Lest you think it is easy, try to create a mosaic of your own. You will soon discover that not just any old piece will work. In fact, this is the question we want to consider: What shape pieces will tile the plane?

Regular Mosaics

Among all mosaics, certainly the simplest and in some ways the most elegant are those having regular polygons as faces. Three come immediately to mind: There is the familiar hexagonal pattern seen in honeycombs and on many bathroom floors; there is the pattern of squares that appears on chessboards and maps of city streets; there is the pattern of equilateral triangles occasionally found in American Indian arts and crafts. But there are no others. This will take some explaining.

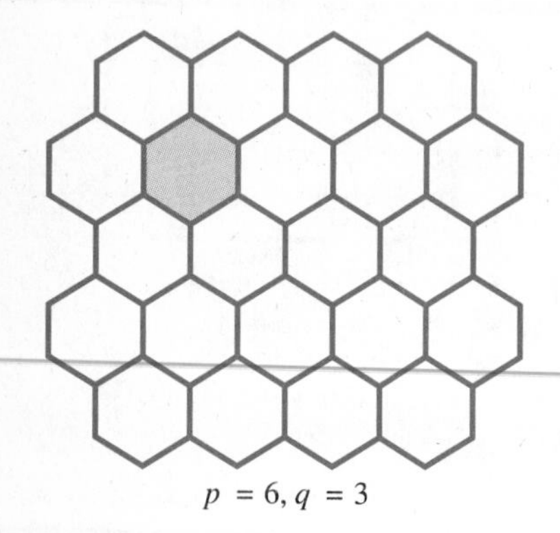

$p = 6, q = 3$

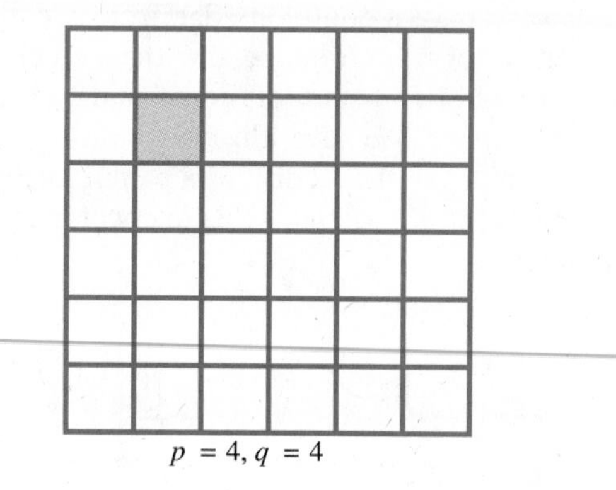

$p = 4, q = 4$

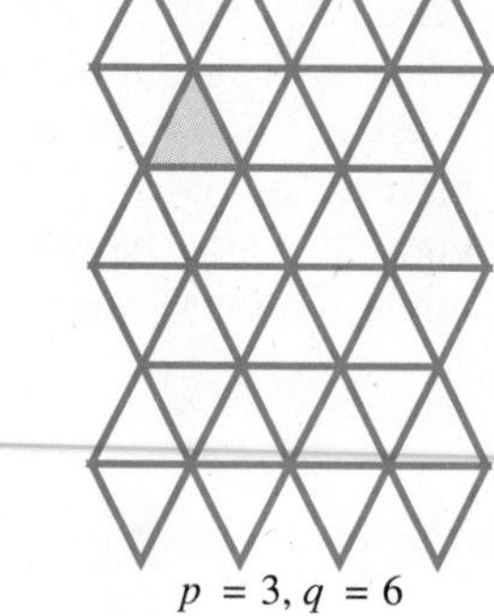

$p = 3, q = 6$

Recall that a regular polygon is a polygon with equal sides and equal angles. Now call a mosaic a **regular mosaic** if

1. Each face is a regular p-gon (a polygon with p sides).
2. Each vertex is surrounded by the same number, say q, of these faces (equivalently, q edges meet at each vertex).
3. Two faces abut only along a common edge.

Condition 3 may seem superficial. It is included to eliminate certain modifications of regular mosaics, which can be obtained by sliding a whole row of polygons one way or the other along a line (see margin). With this understanding, we assert that there are only three regular mosaics.

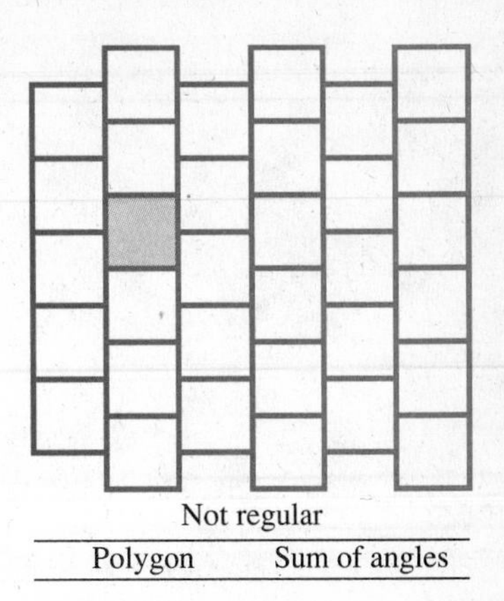

Not regular

We are going to resurrect a fact from high school geometry. Almost everyone remembers that the angles of a triangle add up to 180°; some may remember that the angles of a quadrilateral (a four-sided polygon) add to 360° or twice 180°. The general result is that the angles of a p-gon add up to $(p - 2)$ 180 degrees, a result you can see by decomposing the polygon into triangles as indicated in the margin. Now if a p-gon is regular, so that all its angles are equal, each of them will measure

$$\frac{(p-2)\,180^\circ}{p}$$

Polygon	Sum of angles
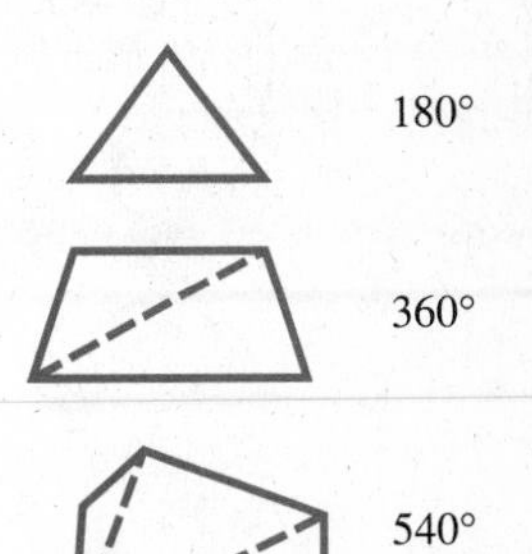	180°
	360°
	540°

Now in a regular mosaic, q of these angles surround each vertex. This implies that each angle measures $360/q$ degrees. If we equate these two expressions and do a bit of algebra, an interesting formula appears (compare it with a similar formula in Section 8.3):

$$\frac{(p-2)\,180}{p} = \frac{(2)\,180}{q}$$

$$\frac{p-2}{p} = \frac{2}{q}$$

$$1 - \frac{2}{p} = \frac{2}{q}$$

$$1 = \frac{2}{p} + \frac{2}{q}$$

$$\boxed{\frac{1}{2} = \frac{1}{p} + \frac{1}{q}}$$

Take a look at the last formula which we remember holds for any regular mosaic. What positive integers p and q can satisfy this equation? It takes only the least bit of experimenting to discover that there are only three solutions:

$$p = 6 \quad p = 4 \quad p = 3$$
$$q = 3 \quad q = 4 \quad q = 6$$

These correspond to the three regular mosaics we have already discussed. That's all there are.

My house is constructed according to the laws of a most severe architecture; and Euclid himself could learn from studying the geometry of my cells.

The bee in Arabian Nights

For untold ages, bees have employed a hexagonal mosaic as the basic pattern in the construction of their honeycombs. This may be dictated by another fact about this particular arrangement of cells. Among all divisions of the plane into parts of equal area, this is the one for which the network of edges has the least length. The bee has learned to economize on the use of wax for cell partitions. However, it would take us too far afield to try to demonstrate this.

Quasi-Regular Mosaics

Let's relax our requirements. Call a mosaic **quasi-regular** if

1. Each face is an identically shaped convex p-gon (convex means without dents or holes).
2. Two faces abut only along a common edge.

This allows for great variety but maybe not as much as you would expect. We now show that no convex polygon of more than six sides works.

First note that any triangle works. Simply take two copies of it, rotate one through 180° and fit them together as a parallelogram. The resulting parallelogram clearly tiles the plane. A similar argument applies to any convex quadrilateral. Fit two copies together, one flipped over, and the resulting hexagon tiles the plane (see Problem 10).

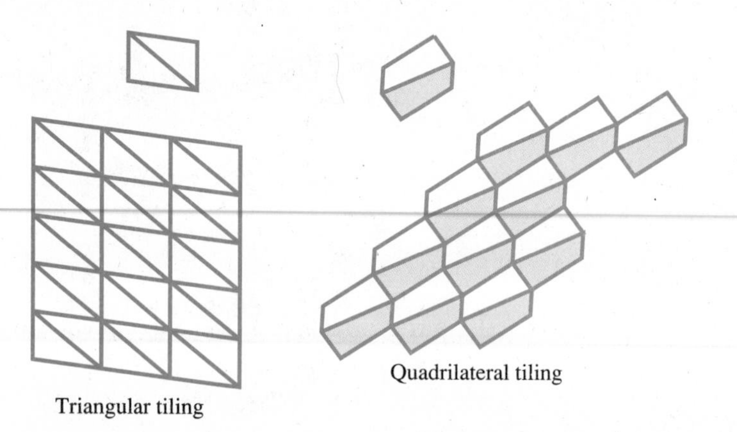

Triangular tiling

Quadrilateral tiling

Things get more complicated and more interesting when we move to pentagons. Not all pentagons work; we already know that a regular pentagon fails. Several authors have at one time claimed to have found all convex pentagons that will tile the plane only to have someone else find a new example. We still do not know whether the list is complete. For an interesting discussion of this problem, see Martin Gardner, "Mathematical Games," *Scientific American,* July 1975, pp. 112–117, and December 1975, pp. 116–119, or D. Schattschneider, "Tiling the Plane with Congruent Pentagons," *Math. Magazine* 51, 1978, pp. 29–44.

Finally, we mention that there are three different types of hexagons that can tile the plane. An example of a pentagonal and a hexagonal tiling appears below.

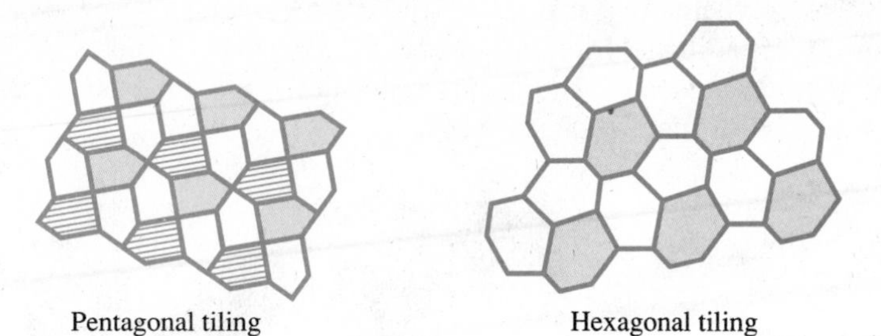

Pentagonal tiling   Hexagonal tiling

Why is it that a convex polygon that tiles the plane can have at most six sides? Once again, it is Euler's Formula that dictates this fact. Consider any mosaic formed by using copies of a convex p-gon. Look at some bounded portion of it, say, the part inside a big circle. Note that each edge is shared by two faces; so

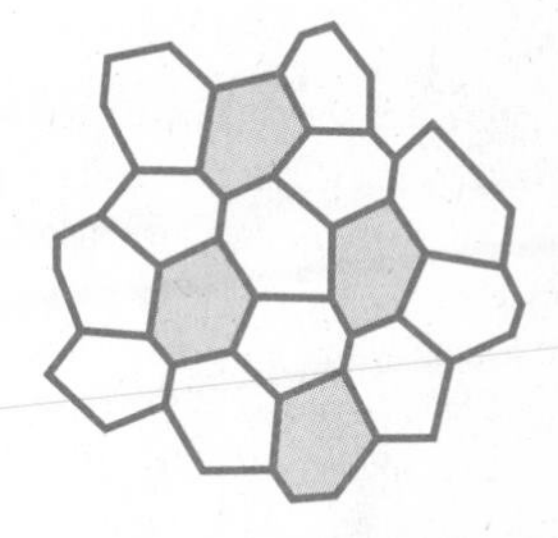

$$pF = 2E \qquad \text{or} \qquad \frac{F}{E} = \frac{2}{p}$$

Furthermore, at least three edges enter each vertex; that is, there are at least $3V$ edge ends. Since each edge has two ends, we see that

$$3V \leq 2E \qquad \text{or} \qquad \frac{V}{E} \leq \frac{2}{3}$$

Oh, we admit to having cheated slightly. Our analysis is not quite correct for the part around the outer boundary. But what happens on the boundary becomes less and less significant as the circle gets bigger and bigger; and the shaded formulas get better and better. Now proceed as we did in Section 8.3. Take Euler's Formula, divide by E, and substitute the above results:

$$E + 2 = F + V$$

$$1 + \frac{2}{E} = \frac{F}{E} + \frac{V}{E}$$

$$1 + \frac{2}{E} \leq \frac{2}{p} + \frac{2}{3}$$

$$\frac{1}{3} + \frac{2}{E} \leq \frac{2}{p}$$

Note that, as the circle gets bigger, E gets larger and $2/E$ fades to 0. We are left with the inequality

$$\frac{1}{3} \leq \frac{2}{p}$$

which implies that $p \leq 6$.

Summary

The repetitive pattern one obtains when tiling a floor with identically shaped pieces is called a mosaic. While great variety is possible, the laws of mathematics, especially Euler's Formula, $F + V = E + 2$, impose surprising restrictions. There are, for example, only three possible regular mosaics. And no mosaic can be composed of convex polygons with more than six edges.

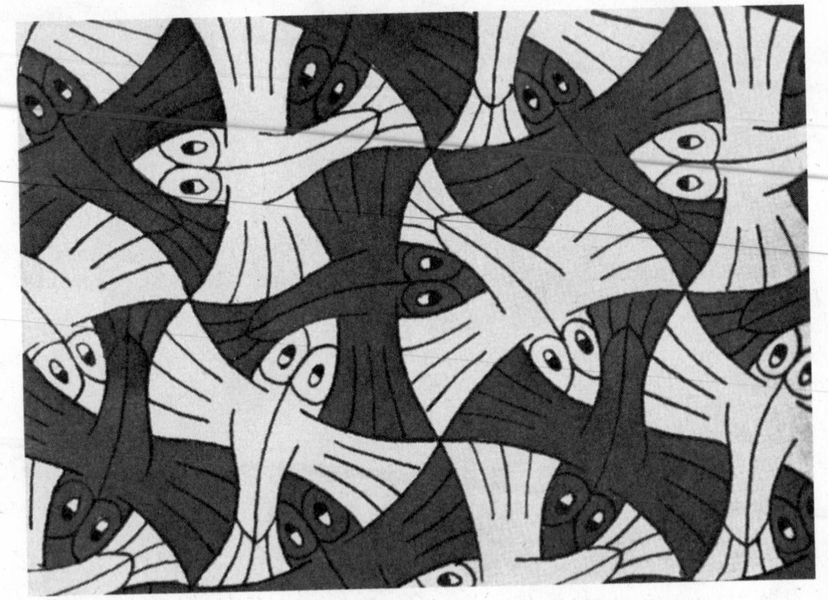

This is the richest source of inspiration I have ever struck; nor has it dried up yet. The symmetry drawings. . . . show how a surface can be regularly divided into, or filled up with, similar-shaped figures which are contiguous to one another, without leaving any open spaces. The Moors were past masters at this. They decorated walls and floors, particularly in the Alhambra in Spain, by placing congruent, multi-coloured pieces of majolica together without leaving any spaces between.

M. C. Escher

Problem Set 8.4

1. Make a quasi-regular mosaic using triangle A.
2. Make a quasi-regular mosaic using quardilateral B.
3. Make a quasi-regular moasic using quadrilateral C.

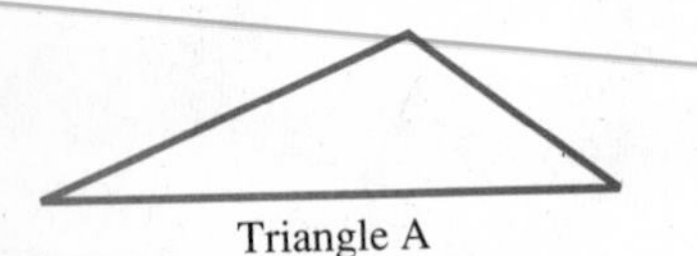

Triangle A

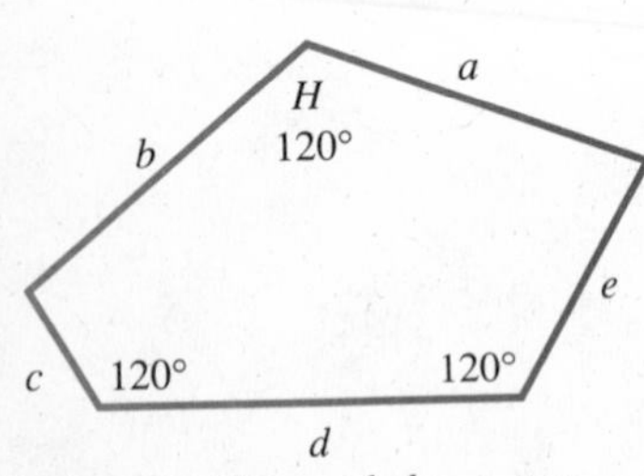

Pentagon D, $a = b\ d = c + e$

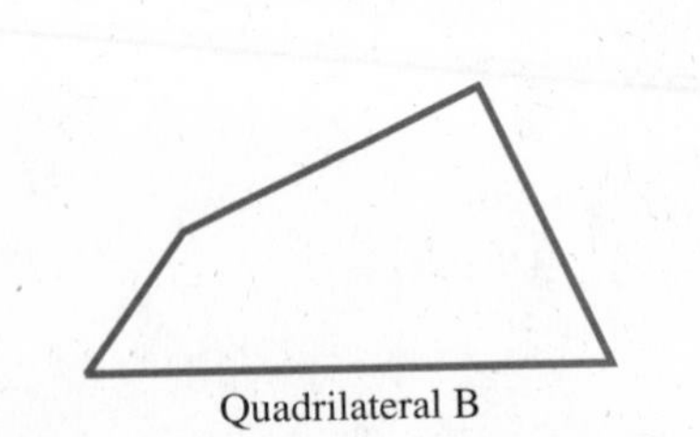

Quadrilateral B

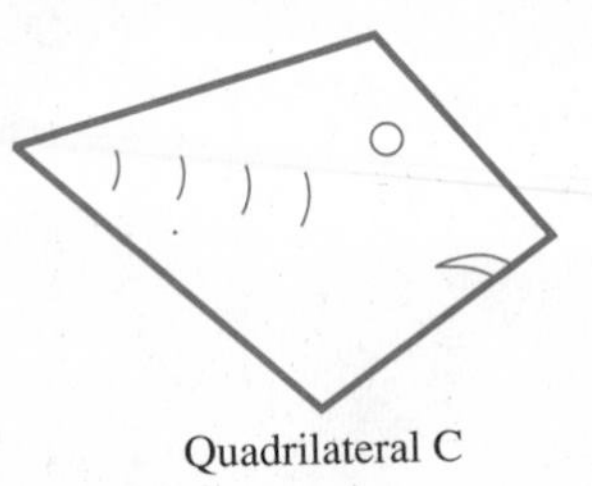

Quadrilateral C

4. Make a quasi-regular mosaic using pentagon D. Hint: Fit three pentagons around point H and then reproduce the resulting figure.

5. Make a quasi-regular mosaic using pentagon E. Hint: Fit six pentagons around point H and then reproduce the resulting figure.

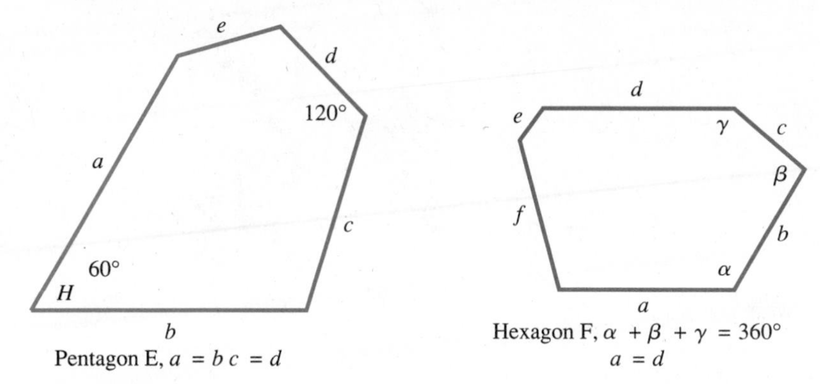

Pentagon E, $a = b$ $c = d$

Hexagon F, $\alpha + \beta + \gamma = 360°$
$a = d$

6. Make a quasi-regular mosaic using hexagon F. Hint: Fit two hexagons together and then reproduce the resulting figure.
7. Find some hexaminoes (six squares) that can tile the plane.
8. Show how to tile the plane with the heptomino shown at right.
9. Show how to tile the plane with the heptomino shown at far right.
10.* Show that any convex quadrilateral can tile the plane. Hint: Fit four copies together around a vertex as shown below. Note that the angles add up to 360°. This can be done at each vertex.

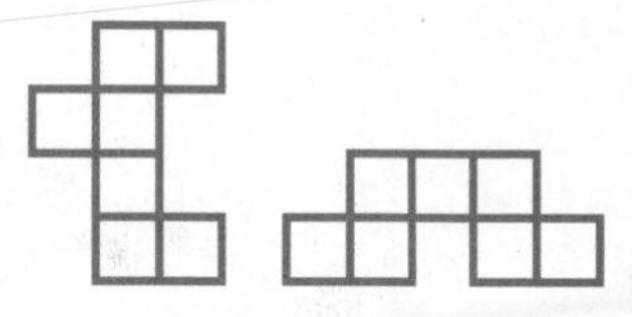

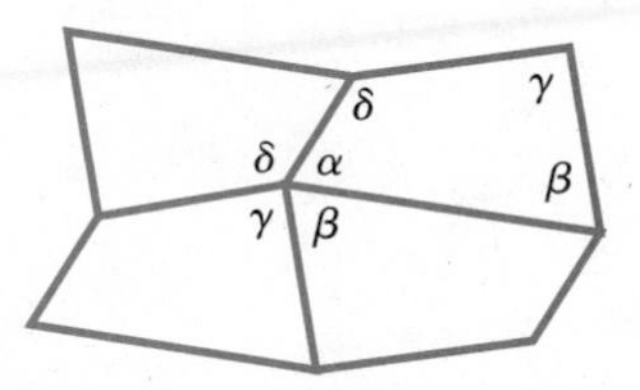

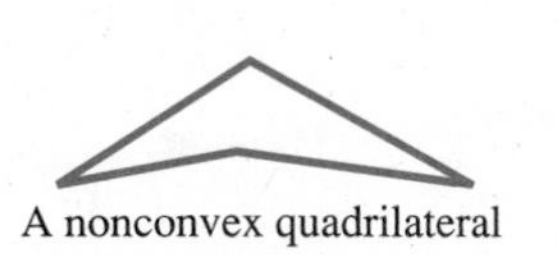

A nonconvex quadrilateral

11.* Show that the plane can be tiled with the nonconvex quadrilateral shown above. Does any nonconvex quadrilateral work?
12.* A mosaic is called semiregular if it is composed of regular polygons of two or more types so that each vertex is surrounded by these polygons in the same way. It can be shown that there are only eight semiregular mosaics. One is shown on page 272. See how many others you can find.

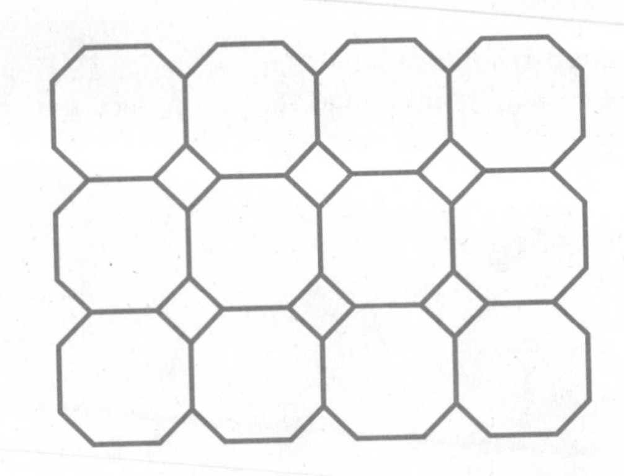

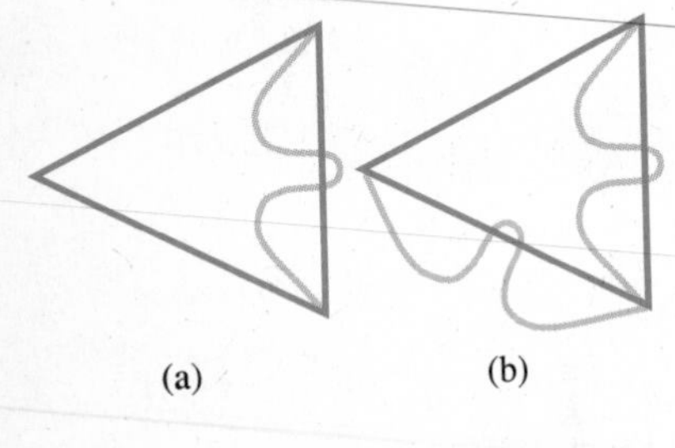

(a) (b)

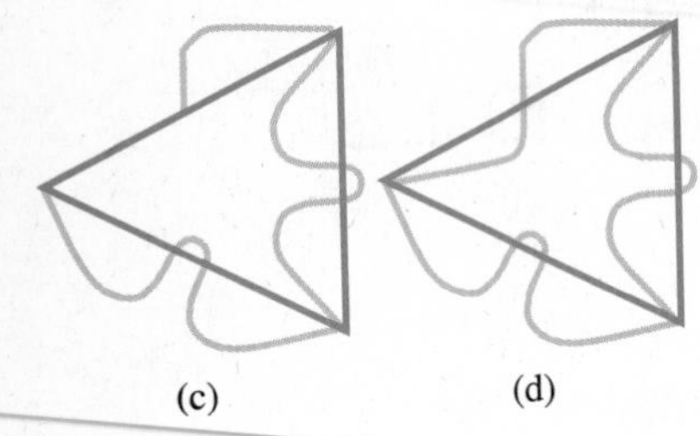

(c) (d)

For Research and Discussion

Many beautiful mosaics (including the one by Escher shown with the summary of this section) can be created using an equilateral triangle as the basic starting piece. The steps are illustrated in the margin.

a. Draw a curve along one side of the triangle.

b. Reproduce this curve along a second side but in a reflected position.

c. Draw a curve along one-half of the third side.

d. Reflect this curve in the midpoint of the third side.

Make a mosaic of your own following this procedure.

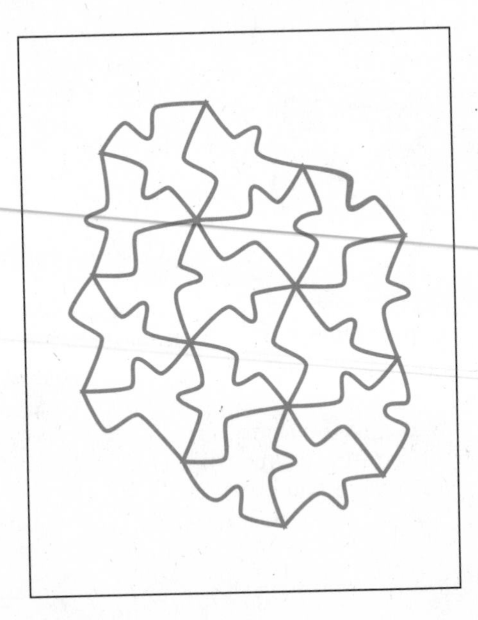

8.5 Map Coloring

A Coloring Problem

A student of mine asked me today to give him a reason for a fact which I did not know was a fact, and do not yet. He says that if a figure be anyhow divided, and the compartments differently coloured, so that figures with any portion of common boundary line are differently coloured—four colours may be wanted, but no more. . . .

What do you say? And has it, if true, been noticed? My pupil says he guessed it in colouring a map of England. The more I think of it, the more evident it seems.

Augustus DeMorgan to W. R. Hamilton, October 23, 1852

What do you say? Is there a map that requires five colors in order that regions with a common boundary will be differently colored? This innocent-sounding question, posed by an obscure student, defied the best efforts of the world's best mathematicians for well over 100 years. As late as 1975, the complete solution seemed farther away than it must have to DeMorgan in 1852. Of course some progress had been made. For instance, it had been shown that five colors would always suffice, and that any map with less than 39 countries (or compartments or regions) could be colored with four colors. But the answer to the question, Can any map whatever be colored with four or less colors? seemed

A conversation between Huck Finn and Tom Sawyer in their flying boat:
"We're right over Illinois yet. And you can see for yourself that Indiana ain't in sight. . . . Illinois is green, Indiana is pink. You show me any pink down there, if you can. No sir, it's green."
"Indiana pink? Why, what a lie!"
"It ain't no lie; I've seen it on the map, and it's pink."

Mark Twain

as elusive as ever and carried with it the intimidating reputation of having gone unanswered for over 120 years. In late 1976, K. Appel and W. Haken reported ["Every Planar Map is Four Colorable", *Bulletin Amer. Mathematical Society,* Vol. 82 (1976), 711–712] that they had demonstrated an affirmative answer by making extensive use of a large computer.

Three observations are in order. (1) A good teacher listens to a good question. (2) There are easily stated mathematical problems that even the best mathematicians find terribly difficult. (3) It pays to study a hard problem even if you don't solve it. It is to the last-mentioned point that we turn our attention.

Two-Color Maps

Chessboard

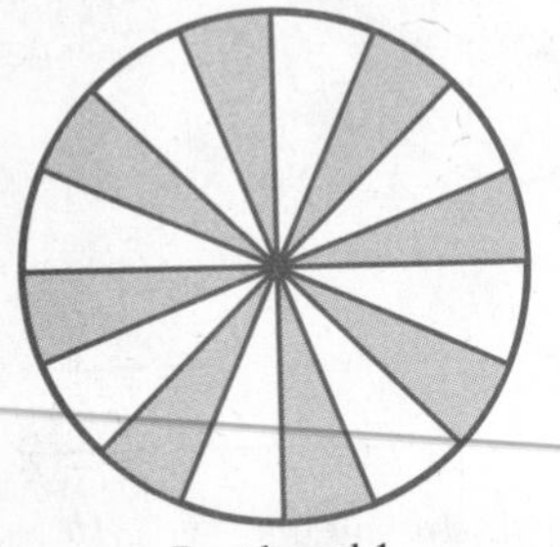
Dart board 1

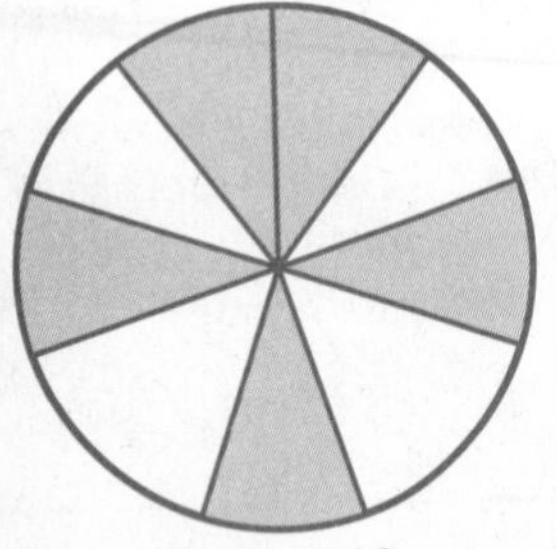
Dart board 2

If the four-color problem is too hard for us, maybe we can solve a simpler one. What kind of maps can be colored with one color? Clearly, maps with just one country. That was easy enough. Well, what kinds of maps can be colored with just two colors? Think of some two-color maps with which you are familiar, such as a chess-board or a dart board of the kind pictured. What is special about them?

The dart board is a good example to think about. Not every dart board can be colored with two colors. Try as you will, you can never make two colors do for the second dart board. That's because the colors must alternate as you go around a vertex; and that's possible only if that vertex is of even degree.

We must stop and point out a temporary assumption being made. Each of our maps (the chessboard and the two dart boards) are to be thought of as islands imbedded in an infinite ocean requiring no color. With this in mind, call any vertex not on the shore of the island a dry vertex. Then one conclusion is already obvious. If the regions of an island can be colored with two colors, then each dry vertex is of even degree.

The truly remarkable fact is that the converse also holds. If each dry vertex is of even degree, then the regions of the island can be colored using two colors.

Any proof of the latter statement requires some ingenuity. Ours is borrowed from Sherman Stein, *Mathematics, the Man-made Universe,* 3rd ed., (San Francisco: Freeman, 1976), pp. 336–340. Consider an arbitrary island with dry vertices of even degree, for example, one like the island shown on page 275. Stick a tack somewhere in the center of each region, making sure that the direct line between any two tacks misses all the vertices. Pick two of the tacks; label one London and the other Bath.

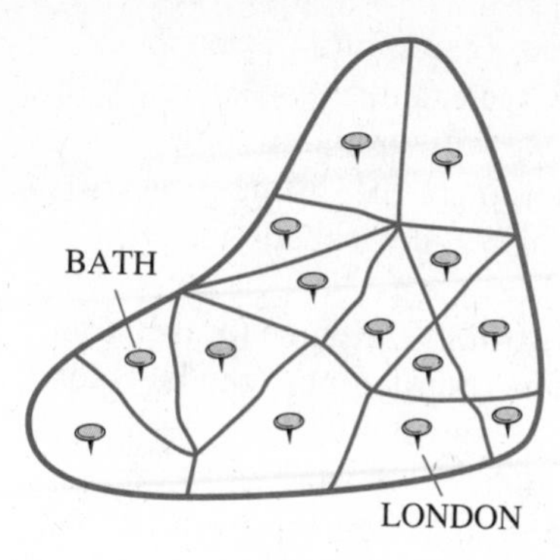

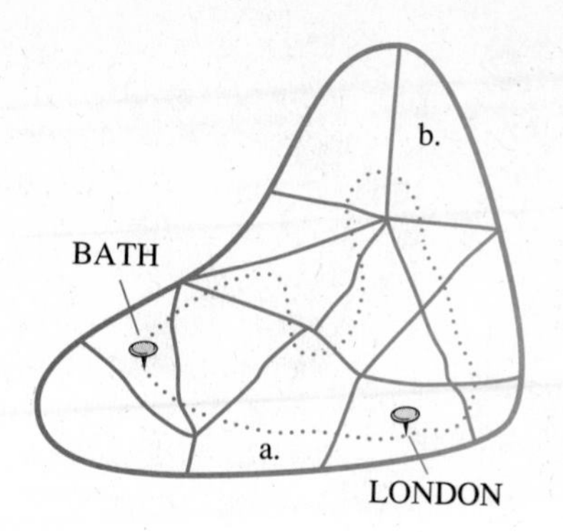

Now any traveler from London to Bath has her choice of many different paths. She can take the direct route we have labeled a; she can take a meandering path such as b. The astute traveler notices a very strange fact. If the direct path has an odd number of border crossings (as it does in our example), so does every other path. If it has an even number of border crossings, so does every other path. Why is this so?

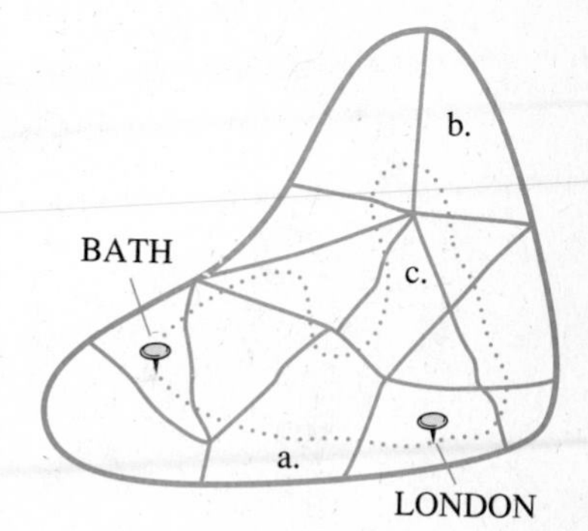

Imagine any meandering path from London to Bath to be an elastic string which is steadily contracting toward the direct path. As it passes through a vertex (moves from position b to position c), it loses some border crossings and perhaps picks up some brand new ones. The number lost plus the number gained is the degree of the vertex, an even number. Now, when the sum of two numbers is even, so is their difference. Thus the change in the number of crossings is even.

The same is true when the string frees itself of a border, as in going from position d to position e. The change is even. It follows that the *parity* (oddness or evenness) of the path is unchanged as it shrinks. We conclude that any path from London to Bath has an odd or even number of border crossings according as the most direct path does so. And keep in mind that Bath could have been located in any of the regions.

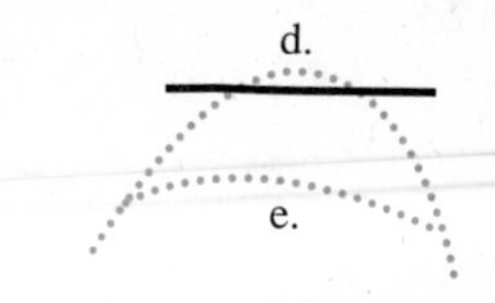

That was the tricky part. Now we need only to give our imaginary traveler some red and blue paint and some instructions. Tell her to paint London and its region red. Tell her to change colors whenever she crosses a border. Tell her to keep going until she has painted every region. Our argument shows that no matter what path she takes she can't possibly get mixed up.

That takes care of the island. What if we want to color the ocean too? We won't get by with two colors in the example just studied but, if all the wet vertices are of even degree, we will have no trouble. We just treat the ocean like another country and forget the distinction between dry and wet vertices. Here is the general two-color theorem:

Theorem 1 Suppose that a network partitions the plane (or the surface of a sphere) into two or more regions. The resulting map can be colored in two colors if and only if each vertex is of even degree.

We have glossed over one little point. What is a region (or country)? We wish to rule out certain kinds of countries, for example, those shaped like a figure eight or even those like the United States which consists of three separate parts. To include such countries complicates the coloring problem still further. To avoid these difficulties, we insist on talking about regions that can be thought of as having a single loop of string as their boundary, a loop that neither meets nor crosses itself.

Three-Color Maps

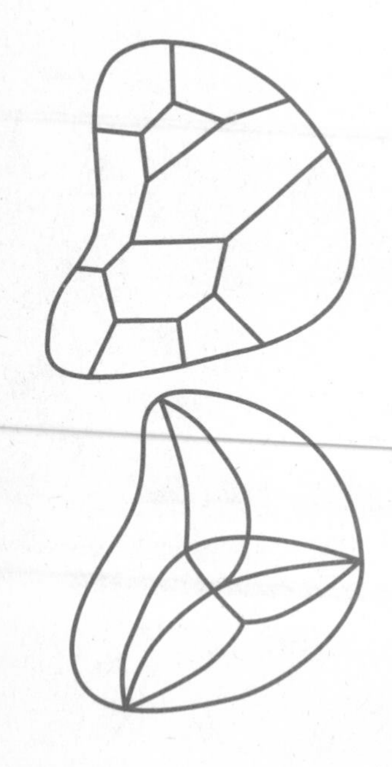

What kind of maps can be colored with just three colors? No one knows a very good answer to this question. But there are two partial results worth stating.

Theorem 2 If the plane (or surface of a sphere) is partitioned into regions, each with an even number of edges, and if each vertex is of degree 3, the resulting map can be colored with exactly three colors.

Theorem 3 If the plane (or surface of a sphere) is partitioned into at least five regions, each sharing its borders with exactly three neighboring regions, the resulting map can be colored in three colors.

Coloring Spheres and Doughnuts

We have already hinted that coloring spheres and coloring planes are equivalent problems. To see why, take a sphere and place it on a plane so that it rests on its south pole. Poke a long straight hatpin through its north pole and on through point P until you hit the plane at P'. This process, since it can be done for any point P, transfers a map on the sphere to a map on the plane. Conversely, it can be used to transfer a map of the plane back onto the sphere. It follows that, if you can color one, you can color the other.

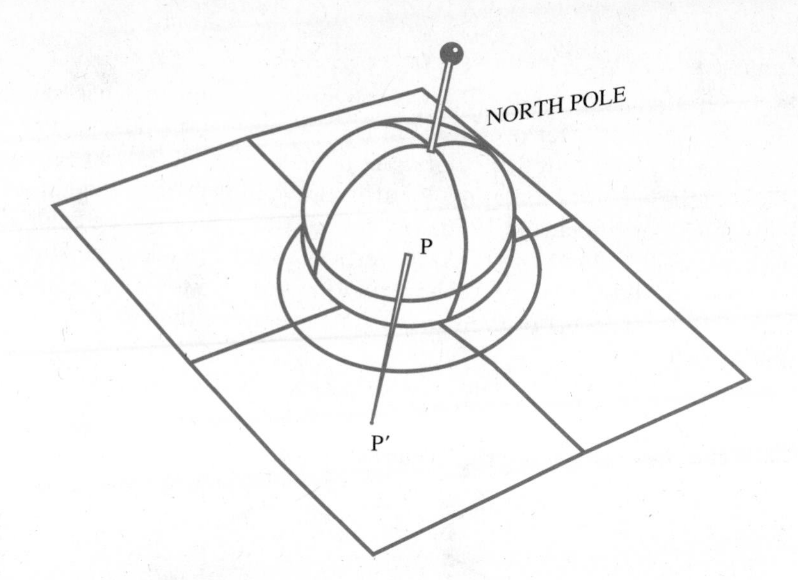

The surface of an inner tube (or doughnut) is an entirely different matter. Take a rectangular sheet of rubber and color it with seven colors as indicated. Then roll it up into a cyliner (like the paper on a cigarette). Finally, bring the two ends around and stick them together, forming a surface like the inner tube in a bicycle tire. Note that each of the seven regions touches the other six. It takes seven colors to color them. Now the clincher. Mathematicians have shown that any map whatever on a doughnut can be colored with seven colors.

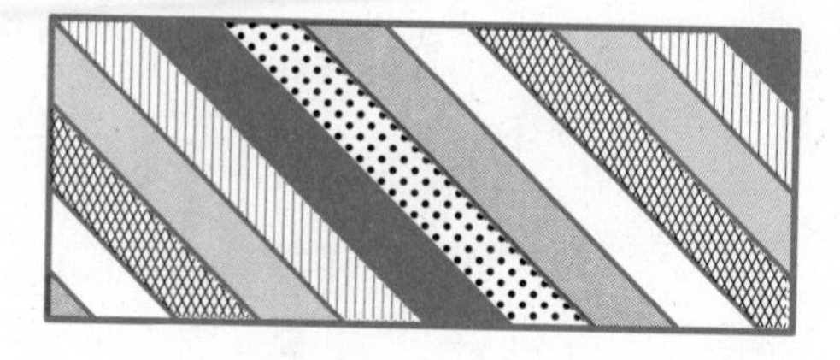

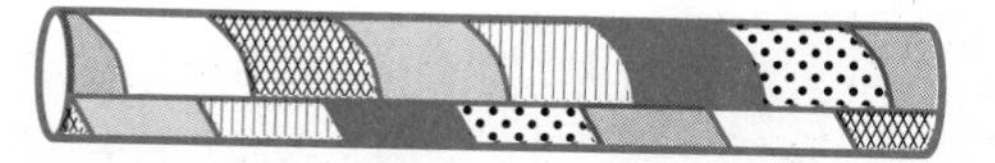

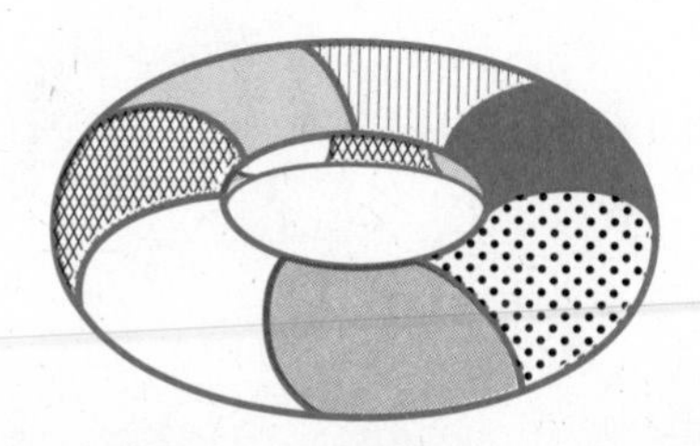

Summary

In 1852, a student posed a simple-sounding question that has given mathematicians fits for over 100 years. Can every map on a sphere be colored with four or fewer colors? As late as 1975, no one knew the answer to this question, though an affirmative answer was announced by two mathematicians in 1976. We showed that a map can be colored with two colors if and only if all its vertices are of even degree. And to top that, we stated that the general coloring problem for a doughnut is solved. It may take up to seven colors, but never more than that, to color a doughnut.

Problem Set 8.5

1. Color each of the following islands using the smallest number of colors.

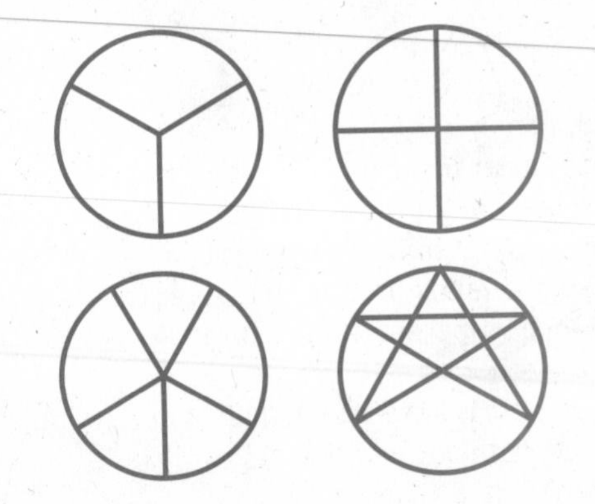

2. Color each of the following islands using the smallest number of colors.

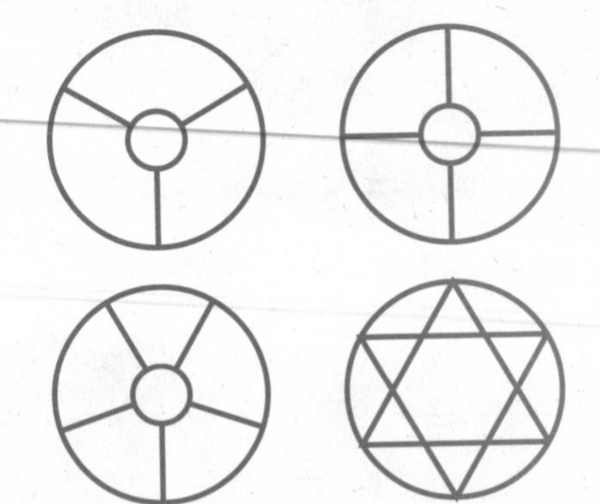

3. Color map A with the fewest number of colors.

4. Color map B with the fewest number of colors.

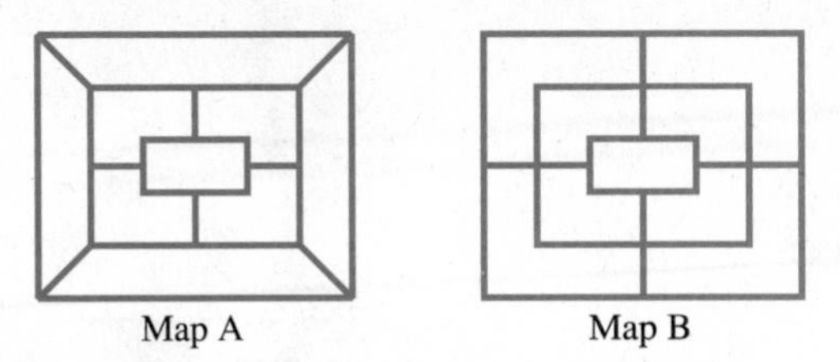

Map A Map B

5. Draw a map of your own that illustrates Theorem 2.
6. Draw a map of your own that illustrates Theorem 3.
7. A map covering a sphere has 50 regions, each with five edges. How many vertices does it have? Hint: Euler's Formula should help.
8. Fifty dots are placed on a sphere. Homer and Hilda are each asked to draw maps so that each region has three edges, using the dots as vertices. They inevitably wind up with the same number of regions. Why?
9. What is the minimum number of colors required to color a
   (a) Tetrahedron?
   (b) Cube?
   (c) Octahedron?
   (d) Dodecahedron?
   (e) Icosahedron?
   See Section 8.3 for pictures of these solids.
10.* What is the least number of colors required for a map of the United States if no adjoining states get the same color? Color a map of the United States to make sure.
11.* Show that any map on a sphere must have at least one region with five or fewer edges.
12.* In how many essentially different ways can cube be colored with three given colors?

For Research and Discussion

The proof by Appel and Haken that every map is four-colorable remains a matter of discussion in mathematical circles because of its heavy reliance on computer computations so numerous as to defy human checking. It is not uncommon for users of commercially marketed software to find errors that the developers did not detect, so how can we be sure that the computations on which the proof relies do not have an error? And what constitutes a proof anyhow? Doesn't a proof have to be understandable by human beings experienced in reading and analyzing a deductive argument? Read T. Tymoczko, "The Four-color theorem and its philosophical significance," *Journal of Philosophy* 76 (1979) 57–83. What other questions are raised? How do you feel about them?

part III
Reasoning and Modeling

A scientist has been compared to a person trying to understand the mechanism of a watch from observations he can make without removing the case. He naturally looks around for some kind of model that can be used to describe what makes the thing tick. The difference is of course that a watch is a human creation, and so the possibility of duplicating it actually exists. But most scientists do not expect to duplicate the world of their experience, so they rely on being able to form models.

The models of which we speak are not made with plastic and balsa wood. Rather, they consist of ideas. They are assumptions, sometimes (misleadingly) called laws, which together with their consequences seem to explain some phenomena we observe in the world. The entire enterprise, dare we admit it, is an activity of the mind in which one begins not with something proved beyond doubt, but with something tentatively set forth as a hypothesis. Proof enters the picture only when we ask what certain conclusions must follow if the hypotheses are accepted. Then, armed with these conclusions, we go back to the world of our experience, asking whether or not the conclusions proved are consistent with the facts observed. If they are, the assumptions together with their implications become our model.

It is this view of science that explains not only why mathematics has played a central role in the natural sciences, but why it is playing an increasingly important role in the social sciences. When reduced to its logical foundations, mathematics is a subject that clearly states its assumptions (axioms) and, without asking whether they are true or false, sets about deducing the consequences (theorems). This is the axiomatic method. It lies at the heart of modeling.

A complete system of theoretical physics is made up of concepts, fundamental laws which are supposed to be valid for these concepts, and conclusions to be reached by logical deduction. It is these conclusions which must correspond with our separate experiences; in any theoretical treatise their logical deduction occupies almost the whole book.

A. EINSTEIN

Albert Einstein
(1879-1955)

All that we create is false.

A. Einstein

If asked to name a famous scientist, many if not most people may well respond with Einstein's name. Born in Germany to a family owing its circumstances to modest beginnings and the happy-go-lucky character of his father, Albert showed little early evidence of what he was to become. A teacher, when asked what profession Albert should adopt, reportedly answered, "It doesn't matter; he'll never make a success of anything." Unable to enter a German university because he never finished his education at the gymnasium (high school) he went to Zürich to attend the Swiss Federal Institute of Technology. Thus began the chain of events in which wars, his Jewish background and Zionist convictions, and his professional affiliations combined to make him in one sense an international citizen and, in another, a man without a country. On graduation from the institute, he needed the influence of friends to obtain a post as Technical Expert, Third Class (he applied for Second Class, but lacked qualifications) in the Swiss Patent Office. It was in this position, not from a post in a major research center, that he used the tools of mathematical reasoning, not the instruments of a scientific laboratory, to fashion a model that was to change the way in which we understand our universe.

chapter 9

Methods of Proof

Considering that among all those who have previously sought truth in the sciences, mathematicians alone have been able to find some demonstrations, some certain and evident reasons, I had no doubt that I should begin where they did, although I expected no advantage except to accustom my mind to work with truths and not to be satisfied with bad reasoning.

RENÉ DESCARTES

So You Want Certainty

An old academic joke, which can be modified to poke fun at a discipline of your choice, tells of a professor who gives the same questions each year on the final exam. When asked if this doesn't make it too easy for the students, the cheerful reply is, "Oh no. In our discipline we change the answers every year."

There is just enough truth in the answer that we are amused. We know that economists do change their thinking on the likely effects of changing interest rates; physicians do change their ideas about what is the proper treatment for our woes, sometimes returning us to remedies they told us to abandon a generation earlier. Students of religion seem never to tire of proposing new answers to the question, "What is the nature of God?" Even in the so-called hard sciences we see chemists change their answers to "What causes fire?" or physicists to "Why does mercury rise in a vacuum?"

The joke does not get told about mathematics, however, because this is a discipline in which the answers never change. They are absolutely certain, and are the same in different cultures as well as for different generations. When asked "What are the rational roots of $2x^3 - x^2 + 2x - 1 = 0$?" the person questioned may or may not know how to find such roots, but there is only one right answer. It is 1/2. You don't need to know much at all to see that setting $x = 1/2$ would have satisfied the question in the time of Methuselah; it will do so during the millennium to come.

I am reminded of the student who asked me outside of our classroom one day to sign a drop card. I asked why so wonderful a course, the flowering of humankind's most certain achievements, was being dropped. "Because," said the student, "I hate answers in the back of the book; I want to take a course where my opinion counts for something."

There are in fact good pedagogical reasons for mathematicians to rephrase the questions in their books. People trying to apply mathematics to problems that occur in science and engineering do not encounter questions phrased so crisply, and fresh mathematical insight frequently comes only when one thinks of a new way to ask a question. It nevertheless remains the case that when a mathematical problem is clearly posed, there is a single answer that, when found, will always be correct.

It has been said that mathematics could be defined as the one academic subject in which the questions that are considered have answers that are absolutely certain. Remarkably, however, even more is true. Not only are we certain of answers to questions having numeric answers, but there is agreement among philosophers that mathematicians have achieved a level of certainty in their proofs of theorems that is not attained in other branches of human endeavor.

> It has been generally taken for granted, that mathematics alone are capable of demonstrative certainty.
>
> John Locke

> . . . it is a common opinion that only the mathematical sciences are capable of demonstrated certainty.
>
> Gottfried Wilhelm Leibnitz

We may now smile to think that Aristotle identified fire as one of the four elements of the world, or was satisfied to explain certain phenomena with the pronouncement that nature abhors a vacuum, but we should reflect on the fact that every schoolchild still studies not only the results but the proofs of the results in Euclid's geometry. And some schoolchildren find this kind of certainty to be very heady stuff.

> At the age of 12 I experienced a second wonder . . . in a little book dealing with Euclidean plane geometry, which came into my hands at the beginning of a school year. Here were assertions, as for example the intersection of the three altitudes of a triangle in one point, which—though by no means evident—could nevertheless be proved with such certainty that any doubt appeared to be out of the question. This lucidity and certainty made an indescribable impression upon me . . . .
>
> Albert Einstein

9.1 Evidence But Not Proof

$p(n) = n^2 + n + 41$
$p(0) = 0^2 + 0 + 41 = 41$
$p(1) = 1^2 + 1 + 41 = 43$
$p(2) = 2^2 + 2 + 41 = 47$
$p(3) = 3^2 + 3 + 41 = 53$
$p(4) = 4^2 + 4 + 41 = 61$
$p(5) = 5^2 + 5 + 41 = 71$
$p(6) = 6^2 + 6 + 41 = 83$
$p(7) = 7^2 + 7 + 41 = 97$
$p(8) = 8^2 + 8 + 41 = 113$
$p(9) = 9^2 + 9 + 41 = 131$

A Prime Lesson

The first ten integers n, when substituted into the formula $p(n) = n^2 + n + 41$, produce primes, that is, integers having no smaller integers (other than 1) as factors. Do these examples mean that we will always get primes? Do we in fact always get primes? or is there some n for which $p(n)$ will be non-prime? How many examples does it take before you can be sure that something will always work?

We seldom think about the processes by which we come to believe something, and some people may be happier not thinking about them. We believe, however, that it is useful to examine a few of these processes so that we can distinguish between evidence for believing something and an argument that constitutes a "proof."

Authority

We believe many things, whether or not we recognize it, because we learned them from a source we regard as authoritative. To be sure, we are selective in deciding what constitutes an authoritative source, and this decision varies with topic and time. A child believes his parents when they tell him about their family tree, but he is likely to check with his friends any information his parents give on how new twigs come into being.

As we grow older, our sources of information increase. We choose and then listen to certain people we believe (or fervently hope) to be competent: a dentist, a lawyer, a mechanic, perhaps (if we may say so) a teacher. We turn to the written word and, after making the proper disclaimers about not believing everything in print, we decide which book on gardening, which history book, and which newspaper or magazine we will quote as the gospel truth. It is surprising, sometimes even alarming, to realize how many things we believe on the authority of a spoken or written word.

No proposition should ever be considered as proved, however, just because some authority says it is so. Never! This is certainly a universal rule of thinking, but mathematics is a particularly nice place to illustrate it. The 350 years of effort to establish Fermat's Last Theorem (discussed on page 5) dramatically demonstrate that the assertion of even a towering figure like Fermat is not accepted without proof. Neither did the failure of generation after generation of mathematicians to find a single counter example suffice. A proof was what was wanting and even as this is written, mathematicians are waiting to see if Andrew Wiles can fill in the gaps in what he recently announced was a proof.

> In questions of science the authority of a thousand is not worth the humble reasoning of a single individual.
>
> *Galileo*

Induction

It is no problem in elementary science classes to convince youngsters that dark clouds carry water. They already believe it; they have probably been baptized into faith by being sprinkled. This process of drawing conclusions on the basis of many observations is called **induction.** We expect, consciously or unconsciously, that what has happened before will happen again. If it doesn't, we feel betrayed.

Once again, however, we find in mathematics some forceful reminders that induction does not constitute proof.

Mathematicians tenaciously cling to the principle that no number of examples suffices to prove an assertion. Again we draw attention to Fermat's Last Theorem. For years mathematicians struggled to prove or disprove this result. Following the advice offered in Section 1.1, they worked on special cases, considering particular values of n. Their combined efforts, augmented in recent years by computers, established that the theorem was true for all n through 600, and for every prime n less than 125,000. Yet, lacking a general proof for arbitrary n, it was universally held by mathematicians that Fermat's Last Theorem was still unproved until Wiles presented his work, and the matter is still in doubt.

> The mathematician as the naturalist, in testing the consequences of a conjectured law by a new observation, addresses a question to nature. "I suspect this law is true. Is it true?" . . . Nature may answer YES or NO, but it whispers one answer and thunders the other; its YES is provisional, its NO is definitive.
>
> *George Polya*

How did you answer the question raised in the problem called A Prime Lesson that opened this section? If you tried the next 20, or the next 30 integers, you still got a prime every time. The fact is that you get a prime for $n = 0, 1, \ldots, 39, 40$, but not for 41; $p(41) = 1763 = (41)(43)$, not a prime. Just because something works the first 41 times you try it doesn't mean it will always work, and this example is nothing compared to the one cited in the note titled, *How Many Are Enough?*

How Many Are Enough?

"How do you know it won't bring you bad luck?"

"Because I've done it hundreds of times."

"Maybe you've just been lucky."

An assertion is not to be believed simply because one observation seems to confirm it. Neither is it to be believed on the basis of two or three examples. Well, okay. How about 1500 examples? 600,000 examples? Consider the following. Let us call an integer $k > 1$ an even type if, when factored into primes, the number of prime factors is even. (A prime is a whole number greater than 1 with no factors other than itself and 1.) And call it an odd type if the number of prime factors is odd. Examples: $4 = 2 \cdot 2$, $6 = 3 \cdot 2$, and $9 = 3 \cdot 3$ are even types; $8 = 2 \cdot 2 \cdot 2$, $12 = 2 \cdot 2 \cdot 3$, and $18 = 3 \cdot 3 \cdot 2$ are odd types; any prime number is an odd type.

Considering all the numbers from 2 through 12, we find

Odd types: 2, 3, 5, 7, 8, 11, 12
Even types: 4, 6, 9, 10

Counting the odd types and the even types, we find that through 12 the odds lead the evens seven to four. The mathematician George Polya found in 1919 that, for all k up to 1500, the odds led the evens. He then made the following conjecture.

For any integer k, the number of integers from 2 through k of odd type exceeds the number of integers from 2 through k of even type. Poyla knew it was true for $k = 1, 2, \ldots, 1500$. It was later proved for all k through 600,000. But in 1960, R. S. Lehman showed that, for $k = 906{,}180{,}359$, from 2 through k there are exactly as many integers of even type as there are of odd type.

This is a remarkable example. An assertion, known to be true in 600,000 cases (and probably more), fails to be true in general. Mark it! No particular number of examples proves an assertion.

Is this peculiar to mathematics? Are there areas of human endeavor in which a given number of examples constitutes final proof? What things do you believe on the basis of a certain number of examples?

Experimentation

Sometimes, as in the controlled procedures of the laboratory or in the informal testing of a new product in our home, we deliberately put a proposition to the test. At other times, as when we hit a nonfunctioning television set—only to discover that it then works—we come upon learning experiences quite by chance. But however it happens, much of what we believe is the result of something we have tried.

Learning from an experiment is of course just a form of induction in which we substitute for many observations a few test runs we hope can be repeated. As such, the experiment shares the limits of any inductive process as a means of proof. But there is an additional problem with experimentation, which is once again conveniently illustrated in mathematics. Consider the Theorem of Pythagoras, which says that, in every right triangle,

$$x^2 + y^2 = z^2$$

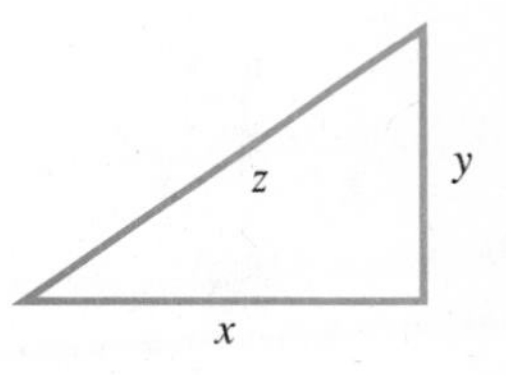

Suppose we wish to verify this experimentally in the case of the triangle pictured. The sides, measured in centimeters, are

$$x = 2.7 \qquad y = 1.8 \qquad z = 3.2$$

Now $x^2 + y^2 = (2.7)^2 + (1.8)^2 = 10.53$ and $z^2 = 10.24$. Does $x^2 + y^2 = z^2$? How much deviation do you allow for experimental and/or roundoff error? This is of course a problem encountered in all quantitative sciences. In mathematics, the answer is easy. Mathematical propositions are not established by physical experiments.

Can We Forget about Small Errors?

In 1840, a young French astronomer named Urban Jean Leverrier showed that the observed path of the planet Mercury deviated from the path predicted by Newton's theory. The deviation amounted to about 42 inches each century! The layman may wonder, in view of the astronomical distances involved, how such a deviation could be detected, much less worried about. Nevertheless, astronomers did worry about it. Several possible explanations were offered, all of them so complicated as to seem unlikely. Finally, in 1915, Einstein demonstrated that his new theory offered an explanation. This demonstration was in fact one of the few actual pieces of experimental evidence Einstein was able to cite as support for his theory of relativity.

Summary

We have examined several of the rationales we commonly offer as evidence for believing something: authority, induction, and experimentation. We have illustrated each of these with examples of things people commonly learn in these ways, and we have shown, using examples from mathematics, that none of these methods can be used as proofs.

Ideas from mathematics that were used included Fermat's Last Theorem. Polya's classification of numbers into even and odd types, and the Theorem of Pythagoras.

Self-Test

Choose the completion that seems best to you. If none of them seem quite correct to you, be prepared to tell why and to supply an ending of your own in class discussion.

1. Experimentation
(a) has no place in mathematics.
(b) never really proves anything, in mathematics or anywhere else, because it is really a form of induction.
(c) is the only certain way to prove anything.

2. Polya's classification of the integers into even and odd types was mentioned
(a) because it illustrates how much a great mathematician can see in so familiar an idea as the classification of integers into evens and odds.
(b) because it cautions us not to believe that something is sure to happen every time just because it happened the first 600,000 times we tried it.
(c) because it illustrates that even in mathematics, people tend to accept a thing as true if a well-known mathematician says he thinks it is true.

3. Fermat's Last Theorem
(a) would be considered proved by almost anyone (except for a professional mathematician) since it has been shown to be true for every one of the more than 600 cases that have been investigated.
(b) cannot be verified experimentally because, as is the case with the Theorem of Pythagoras, one cannot distinguish between a genuine error and an error that is introduced by rounding off decimals.
(c) is an example of a theorem that cannot ever be proved for certain because it asserts that something is true for all numbers n, and we cannot possibly know that something will be true for all n.

Problem Set 9.1

1. Make a list of four things you believe because
(a) You read or heard them from a source you consider reliable (authority).
(b) You have seen them so many times that you have come to expect them (induction).
(c) You have personally put them to a test (experimentation).

2. Think of at least one example of something you once believed on the basis of authority, which you no longer believe. Do the same for induction and experimentation. Then identify the reasons that caused you to change your mind (different authority, a chance to make observations under new circumstances, contradictory experiences, etc.).

3. Do you believe or disbelieve the following statements? On what grounds?
   (a) George Washington and his troops camped at Valley Forge.
   (b) The moon gives off no light of its own.
   (c) When water freezes, it expands.
   (d) Everyone should be vaccinated against smallpox.
   (e) Air travel is safer than automobile travel.
   (f) Each January, there are some days when the temperature in St. Paul, Minnesota, drops below 0°F.

4. Do you believe or disbelieve the following statements? On what grounds?
   (a) Abraham Lincoln debated Stephen Douglas.
   (b) An American has stood on the moon.
   (c) When chilled, iron contracts.
   (d) Children should drink lots of milk.
   (e) Americans spend more money on commercially prepared dog food than they contribute to cancer research.
   (f) High humidity increases discomfort on a hot day.

5. Verify Polya's conjecture about numbers of odd type and even type for $k = 27$.

6. Let $k \geq 5$ be fixed. For each even number n less than k, there is another number m less than n such that m and n are of different type. Explain.

Mathematicians distinguish sharply between a conjecture (for which experimental evidence and induction are very much in vogue) and a proof (for which these methods are out of style). In Problems 7 through 10, you are given some examples from which you are to make a conjecture. How can your conjectures be proved or disproved?

7. (a)
$$1 = 1^2$$
$$1 + 2 + 1 = 2^2$$
$$1 + 2 + 3 + 2 + 1 = 3^2$$
   (b) We have, in the following array of numbers, circled the prime numbers

(7)	8	9	10	(11)	12
(13)	14	15	16	(17)	18
(19)	20	21	22	(23)	24

8. (a)
$$(1)(1) = 1$$
$$(11)(11) = 121$$
$$(111)(111) = 12321$$
   (b) $2^2 - 2 = 2$, $3^2 - 3 = 6$, $4^2 - 4 = 12$, . . . are all divisible by 2.
   $2^3 - 2 = 6$, $3^3 - 3 = 24$, $4^3 - 4 = 60$, . . . are all divisible by 3.

9. (a) Draw three line segments Aa, Bb, and Cc so that they intersect at a common point. Then draw triangles ABC and abc, extending the sides as necessary to form the intersections of AB with ab, AC with ac, and BC with bc.

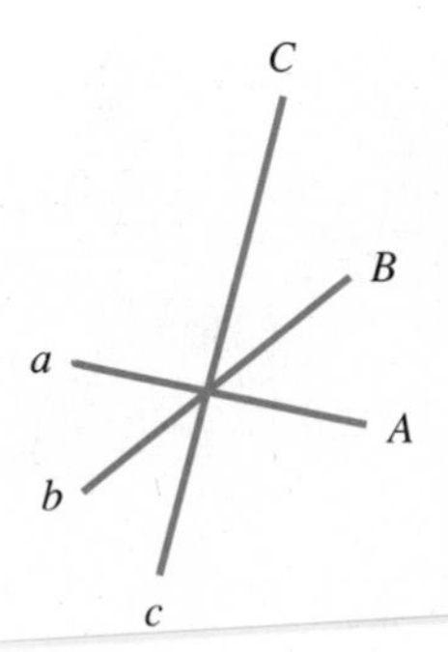

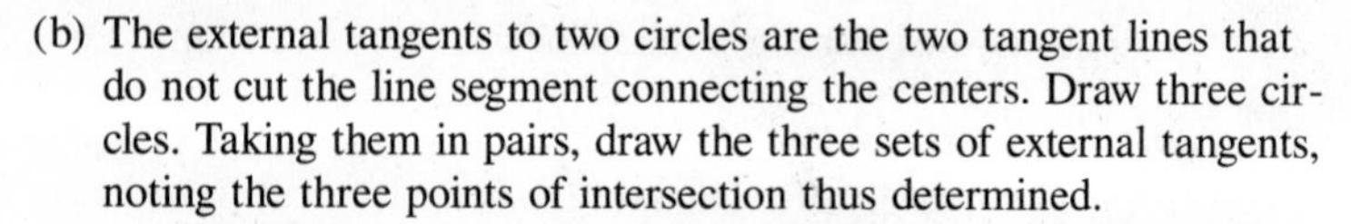

(b) The external tangents to two circles are the two tangent lines that do not cut the line segment connecting the centers. Draw three circles. Taking them in pairs, draw the three sets of external tangents, noting the three points of intersection thus determined.

10. (a) Locate points A, B, and C on one line, and points a, b, and c on another. Consider the intersections of Ab with aB, Ac with ab, AC with ac, and Bc with bc.

(b) Draw a circle and circumscribe an arbitrary quadrilateral, labeling the points of tangency in clockwise order A, B, C, and D. Draw segments AC and BD. Draw the two diagonals of the quadrilateral.

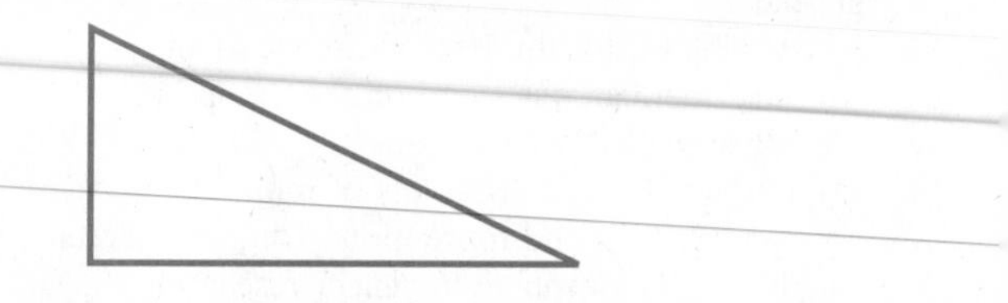

11. Measure the sides of the right triangle above. How close can you come to verifying the Theorem of Pythagoras?

12. Measure the angles in the right triangle above. How close do you come to verifying the theorem that says that the sum of the interior angles is 180°?

13.* Draw a circle of radius 3 centimeters on onionskin paper. Lay it over a grid of 1-millimeter squares. By counting the squares included in the circle, see how close you come to verifying that the area $A = \pi r^2$. (Use $\pi = 3.1416$.)

14.* Lay a piece of string along the circumference of the circle drawn in Problem 13. Try to verify that the circumference $C = 2\pi r$.

15.* Summarize the reasons for rejecting authority, induction, and experience or experimentation as methods of proof in mathematics.

For Research and Discussion

The availability of computers has opened up new possibilities for having students discover for themselves theorems that in former days could only be introduced in the context of a coherent theoretical development. This has given rise to laboratory manuals and guides to self-discovery in an area of study that was once described as the discipline that knows nothing of experiment. It has also led to fears on the part of some mathematicians that the teaching of proof might get lost in the rush to experiment. These developments suggest two projects.

(a) Look for some of the self-discovery lab manuals now available, probably in your computer lab. Try an exercise or two. Report on any "discovery" you make.

(b) Read some of the arguments now appearing in publications aimed at teachers about the balance that should exist between learning by experiment and demanding that some things be proved. You might start by reading the summary in *Focus* (the newsletter of the Mathematical Association of America), February, 1994 of the Presidential address given by Deborah Tepper Haimo at the Association's annual meeting that year. Her address, "Experimentation and Conjecture Are Not Enough," makes reference to the Polya conjecture described in this section.

9.2 Deduction

Is My Hat Black?

The black and white hat game is played by seating contestants in a circle so they can see each other. A referee places on each head a hat which may be black or white. No player sees his own hat. The rule is that a player who sees a black hat on any opponent must raise his hand. The first player to deduce (not guess) the color of his own hat wins.

At a certain party, Simple, Simon, and Professor Witquick were all given black hats. Of course all three raised their hands immediately. After a few moments of silence, the professor confidently announced that his hat was black. How did he deduce this?

If you believe that Jack, because of his conviction that the national government should own the railroads, is a communist, and if you also believe that all communists are atheists, it follows that you believe that Jack is an atheist. This process, in which we infer a further proposition on the basis of some principle or principles already accepted, is called **deduction.** It tells us, on the basis of what we say we believe, what else we must believe. And the process works in reverse, telling us what not to believe. Having accepted the proposition that rabbits can be produced (and probably will be produced) only by other rabbits, we smile when the magician produces a rabbit out of the air, ignore the apparent evidence, and assume that our powers of observation have somehow betrayed us.

As a means for deciding whether or not to accept a proposition, the method of deduction depends on two things. In the first place, we must have some previously accepted proposition with which to start. In the second place, we must be able to reason from premise to conclusion in a way that we (to say nothing of others) believe to be reliable.

Books on game theory sometimes point out that, in planning a strategy, you must assume that your opponent will do his best, and that his best includes intelligent reaction to your moves.

In winning the black and white hat game described above, Professor Witquick used both features of deductive thinking. The professor reasoned that, if his hat were white, then Simon, seeing Simple's hand up, would know that his (Simon's) own hat was black. He realized from the silence that Simon was momentarily confused by seeing two black hats and so concluded that his own hat was black. Thus, on the basis of an assumption (Simon wants to win), confidence in Simon's ability to think (if he saw a white hat on me, he'd figure out that his was black), and the fact of Simon's silence, the professor was able to deduce the color of his own hat.

The truth of the conclusion of a deductive argument depends of course on the truth of the proposition accepted to start with. If the initial proposition is wrong, the best reasoning in the world cannot give us confidence in the conclusion. But if we believe the initial proposition to be true, and if the reasoning is sound, we are compelled to believe the conclusion.

The nature of the initial propositions is usually emphasized by referring to them as the assumptions, the premises, or the hypotheses. In assessing the soundness of an argument, we do not allow ourselves to be drawn into a discussion of the truth of the assumptions. Neither therefore are we in a position to discuss the truth of the conclusion. We ask rather if the conclusion is valid. Is it the inescapable consequence of the assumptions?

Venn Diagrams

One picture, it is said, is worth a thousand words. And it is true that a picture called a **Venn Diagram** is sometimes of help in analyzing an argument. The technique calls for drawing ovals to represent the sets under consideration and then asking whether or not the hypotheses, the assumed part of the argument, force you to draw a picture that corresponds to the desired conclusion. Consider the following argument.

Hypotheses: All tubs float.
Some tubs are gray.
All battleships are gray.
She is a tub.

Conclusions: She is a battleship.
She is gray.
She floats.
All battleships float.

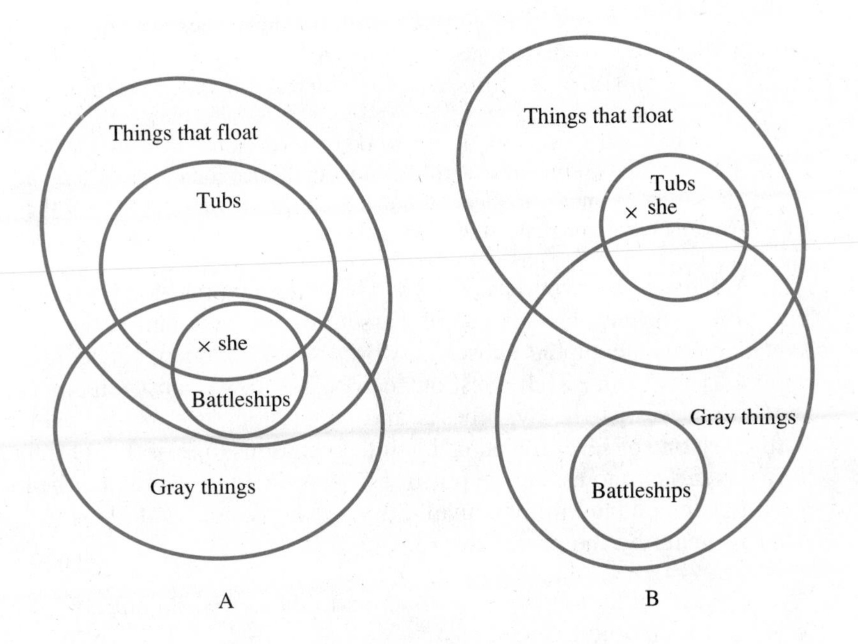

We have drawn, in Fig. A, a picture in which all the hypotheses are satisfied. That is, tubs are completely contained in the set of things that float, and at least some tubs are gray things. Similarly, battleships are completely contained in the set of gray things, and she is indicated as being a member of the set of tubs. At the same time, as Fig. A is drawn, all the conclusions are satisfied. She is pictured in the set of battleships, in the set of gray things, and in the set of things that float; and the set of battleships lies entirely in the set of things that float.

The situation is quite different in Fig. B, however. We again have a picture in which all the hypotheses are satisfied, but this time only one of the conclusions is satisfied.

The one conclusion that is satisfied, "She floats," is inescapable. Logic drives us to it. This is what is meant by a valid conclusion; it is a conclusion from which we cannot escape if we accept the hypotheses. In terms of Venn Diagrams, we are driven to it in the sense that any picture that satisfies the hypotheses must also satisfy a valid conclusion.

This example gives us one more opportunity to comment on the distinction between validity and truth. It may be true that all battleships float, but we are not driven to this conclusion on the basis of the stated hypotheses. We did not reject as invalid the statement, "All battleships float," because we were thinking about those that have sunk. We rejected the statement because it was possible to satisfy all the hypotheses without satisfying that conclusion.

When the hypotheses and conclusions are stated as clearly as they are in the example above, it is generally easy to draw a Venn Diagram showing which conclusions are valid and which are not. When one is confronted with a paragraph in a written argument (editorial, political commentary, etc.) or an oral presentation, however, it is more difficult to determine the validity of a conclusion, because one must first state clearly what seem to be the assumptions (hypotheses) being made and what conclusions are being drawn. Consider the following argument.

> Inflation is very damaging to the worker who does not hold capital assets that increase in value with the inflation. Since inflation can only be curbed by balancing the federal budget, it is in the best interests of the working class to balance the federal budget.

What are the hypotheses? Does the writer assume that no members of the working class hold capital assets, or are we talking in the first sentence only about the subset of workers who happen not to hold capital assets? Since such questions can be asked, we must remember that, when we list the hypotheses and conclusions, we are probably listing just one of several reasonable interpretations (one of the reasons why it is difficult to assess the position held even after you read or hear a political candidate, for example). Suppose we decide the following is what is meant by the paragraph above.

Hypotheses: Inflation is very damaging to those who do not hold capital assets.
Some workers do not hold capital assets.
Inflation can be curbed only by balancing the federal budget.

Conclusion: It is in the best interest of the working class to balance the federal budget.

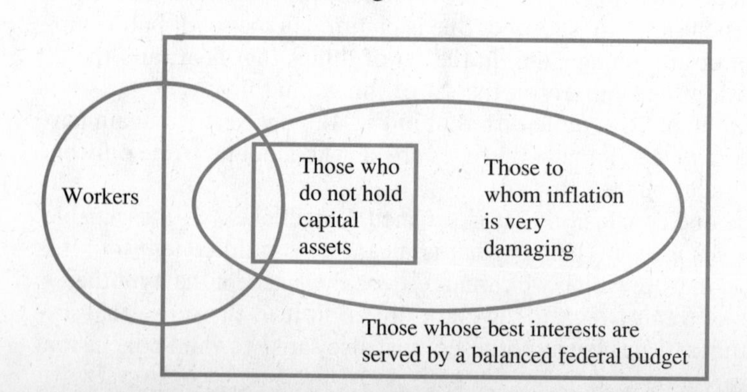

Our Venn Diagram can now be drawn as indicated. We see that the conclusion is not valid. We also see that it can be made valid by changing our interpretation of the second hypotheses to:

No workers hold capital assets.

or by changing the conclusion to:

It is in the best interests of some workers to balance the federal budget.

The first change has the effect of significantly decreasing the number of people willing to accept our hypothesis; the second change leads to a conclusion not nearly so dramatic. A writer wishing to make a strong case that appeals to a large number of people is best served by the rather imprecise prose so familiar to us all.

Summary

Deduction is a method of arguing in which we show on the basis of a compelling argument that, if we accept certain propositions (the hypotheses of the argument), we are forced to accept some other proposition (the conclusion). Since no judgment is made about the truth or falsity of the hypotheses, none can be made about the conclusion. We only ask whether the argument is valid and, if it is, we say the conclusion has been proved.

> By deductive reasoning, we are enabled only to reveal to ourselves implications already included in our assumptions.
>
> *Paul Samuelson*

Self-Test

Choose the completion that seems best to you. If none of them seem quite correct to you, be prepared to tell why and to supply an ending of your own in class discussion.

1. If a statement has been proved by deduction, then
(a) it certainly is true.
(b) it will be true if and only if the hypothesis on which it is based is true.
(c) it is properly said to be valid.

2. A statement obtained by valid deduction
(a) must be true.
(b) may be true.
(c) will be false if it is based on a false hypothesis.

3. Suppose we have drawn a Venn Diagram in which all of the hypotheses are satisfied.
(a) If the conclusion is also satisfied, then we may be sure that the conclusion is valid.
(b) If the conclusion is not satisfied, then we may be sure that the conclusion is not valid.

Problem Set 9.2

In Problems 1–8, decide which (if any) of the conclusions follow from the hypotheses by a valid argument.

1. *Hypotheses:* Some x are y.
All y are z.
Some y are w.
Conclusions: (a) Some x are not z.
(b) Some x are z.
(c) Some x are w.

2. *Hypotheses:* All x are y.
All x are z.
Some w are y.
Conclusions: (a) Some z are y.
(b) Some y are z but not x.
(c) Some x are not w.
(d) Some x are not w.
(e) Some x are w.

3. *Hypotheses:* Some humans are hairy animals.
All dogs are hairy animals.
All dogs should be kept on a leash.
Conclusions: (a) Some dogs are not humans.
(b) Some dogs are human.
(c) Some hairy animals that are not dogs should be kept on a leash.
(d) Some animals that should be kept on a leash are hairy.

4. *Hypotheses:* Everyone who works hard is well educated.
Some who work hard are rich.
Some professors work hard.
Conclusions: (a) Some professors are not well educated.
(b) Some professors are rich.
(c) Some professors are well educated.

5. *Hypotheses:* Full professors are old creatures.
No goats receive social security checks.
Some old creatures receive social security checks.
Some full professors are good teachers.
Conclusions: (a) Some good teachers are old creatures.
(b) Some good teachers receive social security checks.
(c) No full professors are old goats.
(d) Some professors are old goats.

6. *Hypotheses:* Some hot things are very colorful.
All red things are very colorful.
Red things make us cautious.
Some dogs are red.

Conclusions: (a) Some dogs are colorful.
(b) Anything that makes us cautious is colorful.
(c) There are hot dogs.
(d) There are red hot dogs.
(e) Some hot things are red.

7. *Hypotheses:* All capable people are cantankerous.
No conceited people are capable.
Some clever people are conceited.
Some clever people are capable.

Conclusions: (a) There are no clever people who are not cantankerous.
(b) All clever conceited people are cantankerous.
(c) No cantankerous people are conceited.

8. *Hypotheses:* All timid people are followers.
All timid people are quiet.
All followers contribute to the success of the leader.
Some timid people are irresponsible.

Conclusions: (a) All quiet people contribute to the success of the leader.
(b) Some people who contribute to the success of the leader are irresponsible.
(c) Some followers are quiet.

9. Professor Witquick classifies people as conversationalists as follows:
Some informed people are provocative.
Some, but not all, overbearing people are provocative.
All overbearing people, by definition, at least have the virtue of being informed.
Scintillating people are always provocative.

Can we conclude that the Professor believes the following?
(a) Some informed people are not provocative.
(b) Scintillating people are sometimes overbearing.
(c) People may be scintillating without being informed.

10. John Q. Public classifies politicians as follows:
Some are crooks who are getting rich at public expense.
Many are do-gooders, but all politicians are incompetent.
There are a few sincere ones who are not getting rich at public expense.

From these opinions, is it clear that John Q. Public also believes the following?
(a) The do-gooders are all crooks.
(b) The sincere politicians are all competent.
(c) Some crooks are competent.

In Problems 11 through 14, we cite a quotation which is, for the purpose of the problem, accepted. You are then to decide whether or not the statements that follow are logical deductions from the given information.

11. "Silence is the best tactic for him who distrusts himself" (F. de La Rochefoucauld).
(a) Hugo distrusts himself, so he is silent.
(b) Hugo is silent, so Horace concludes that Hugo distrusts himself.
(c) Hugo distrusts himself, so silence is his best tactic.

12. "We always like those who admire us" (F. de La Rochefoucald).
(a) Professor Witquick admires Librarian Hardback, so we can be sure that Witquick likes Hardback.
(b) Professor Witquick admires Librarian Hardback, so we can be sure that Hardback likes Witquick.
(c) Professor Witquick likes Hardback, so we can be sure that Witquick admires Hardback.
(d) Professor Witquick likes Hardback, so we can be sure that Hardback admires Witquick.

13. "A really busy person never knows how much he weighs" (Edgar Watson Howe).
(a) A person who never knows how much he weighs is really busy.
(b) A person who is not really busy sometimes knows how much he weighs.
(c) A person who is not really busy always knows how much he weighs.
(d) Since Professor Witquick is a really busy person, there are times when he does not know what he weighs.

14. "The greatest minds are capable of the greatest vices as well as the greatest virtues" (René Descartes).
(a) Professor Witquick has a great mind; his vices are therefore likely to be great.
(b) Those who exhibit the greatest virtue are likely also to exhibit the greatest vice.
(c) Those capable of the greatest vices have the greatest minds.
(d) One not capable of the greatest vices and the greatest virtues is not numbered among those having great minds.

15. For each of the following paragraphs, identify the hypotheses and the conclusion.
(a) Since my home state is certain to collect income taxes on any money made in the state, I was careful to conduct all my business in another state. That way, I won't have to pay income tax in my state.
(b) Dogs can't pull a sled that is too heavy. If you insist on taking all that gear, the sled will be too heavy. The dogs won't be able to pull it.
(c) A sensitive person understands that he or she cannot continually refuse offers of help. Homer accepts any offer of help that comes his way. He must be a very sensitive person.

16. For each of the following paragraphs, identify the hypotheses and the conclusion.
(a) Every school in the state is trying to terminate several positions on its professional staff. I have decided to accept a position in another

state so that I can avoid the anxiety of wondering if my position will be terminated.
(b) If a person combines hard work with ability in this business, financial reward is certain. Jones has received great financial reward from the business. He must have combined hard word with ability.
(c) There is no way a person with ability can fail if he works hard. Smith has failed, in spite of hard work. Poor Smith just doesn't have the ability.

17. Suppose that you, as a student government leader, employ someone to survey students standing in line outside the dining hall. You are told the next day that, of the 141 students interviewed, 72 were satisfied with the food service; 97 were satisfied with their dormitory room, 61 were satisfied with both, and 39 were dissatisfied with both. Deduction should serve to warn you not to trust the work of the one who took the survey. Why?

18. Hugo asked Horace if he had any change, and when Horace reported that he had $1.15 in change, Hugo asked for change for his $1.00 bill. Horace responded that he could neither change a $1.00 bill nor a half-dollar. What coins did Horace have?

19. Can you give examples of valid arguments in which you would say
(a) The hypotheses are false and the conclusion is false?
(b) The hypotheses are false and the conclusion is true?
(c) The hypotheses are true and the conclusion is false?
(d) The hypotheses are true and the conclusion is true?

20. Can you give examples of arguments that are not valid in which you would say
(a) The hypotheses are false and the conclusion is false?
(b) The hypotheses are false and the conclusion is true?
(c) The hypotheses are true and the conclusion is false?
(d) The hypotheses are true and the conclusion is true?

21.* Analyze the black and white hat game if the professor plays with three others. In what situations can he determine the color of his hat? Are there any situations in which he cannot determine it?

For Research and Discussion

In our analysis of the paragraph on the effect of inflation on workers, we

1. expressed in a series of crisp, declarative sentences what might have been inferred to be the writer's hypothesis.

2. similarly wrote a conclusion the writer may very well have wanted us to accept.

3. showed why the argument as interpreted was not valid.
4. noted several ways of changing either the hypothesis or the conclusion so as to make the argument valid.
5. noted that the changes we considered all seemed to leave us with a less forceful article.

Try to find an article (editorial, syndicated column, etc.) that you can similarly analyze.

9.3 Difficulties in Deductive Thinking

Where's the Dollar?

Three men register for a hotel room and are told that the charge for the room is $90. They pay $30 apiece and have gone to their room when the clerk realizes that she has made an error; the charge should have been only $85. She sends a bellhop up with the $5, but the bellhop, anticipating the difficulty of dividing $5 among three men, decides they'll be pleased to get back $3 which they can easily split up. For this thoughtfulness, he rewards himself with the $2. Then the men will have paid just $87 for the room, and the bellhop will have $2. Your problem: Shouldn't there be another dollar somewhere?

Since we have claimed that deduction is the only way to prove something conclusively, it is perhaps time to acknowledge that such a proof is only as good as the logic on which it depends. Let us look therefore at some of the common errors that can mar a deductive argument.

In the first place, we often get into trouble because of the language we use. Words have multiple meanings, a fact often exploited in jokes and quips. More seriously, most of us have been in arguments in which the meaning of a term was shifted in midstream. There are also the more subtle problems that come up when we confuse the name of a thing and the thing itself.

Mouse is a syllable.
A mouse eats cheese.
So some syllables eat cheese.

In this case the name of the thing is "mouse" and the thing itself is the rodent that eats cheese.

Another kind of difficulty arises when we try to use language to make statements that are supposed to apply to the language itself, such as:

Every statement on this line is false.

More common than troubles with language are errors often made in connection with the simple statement that A implies B, written $A \Rightarrow B$, or expressed in the form, "If A, then B."

Then statement, "If B, then A," is the **converse** of "If A, then B." When expressed so succinctly, there is little difficulty in getting people to see that a statement can be true while its converse is false. But in the heat of an argument or even in ordinary discourse, this distinction is often blurred over. Tell a child, "If I buy you some candy, then I will not buy you ice cream." Then drive past the ice cream shop. No logic compels you to buy candy, but the child may.

The converse, which the child not only supplies, but believes is, "If I do not buy you ice cream, then I will buy you some candy."

Or, to illustrate the same idea with an example closer to the reader's heart, consider the college president who is quoted in the student paper, "If we provide more scholarship aid for those needing it, then we will have to have a general raise in tuition." Suppose this is followed several weeks later with the announcement of a tuition increase. What will be the "logical" expectation of the student body?

A second problem with $A \Rightarrow B$ is that there are those who draw conclusions from "not A." A father tells his son, "If it is a nice day tomorrow, then I want you to help me with some yard work." The next day turns out to be, by mutual agreement, not nice. If the father decides to go ahead with the yard work anyhow, he may have to listen to the complaint, "You said I'd have to work if it was a *nice* day." The father of course had said nothing at all about what he would do if it were not a nice day. It's a point of logic. It may be the only point he gets in the ensuing discussion.

Finally, let us turn to what is potentially the most damaging criticism of deduction as a method of proof. We have pointed it out before. Deduction is absolutely dependent on the idea that we are able to reason from hypotheses to conclusion in a reliable manner. We bank on the idea that what appears to be a compelling argument to one person is just as compelling to another, that we all share something so universal that we call it "common sense." The great thinkers have commented on this thread of good sense running through mankind, and it does seem that most people, even those with appetites almost insatiable in other respects, feel that they are abundantly supplied with common sense.

The power of forming a good judgment and of distinguishing the true from the false which is properly speaking what is called Good Sense or Reason, is by nature equal in all men.

René Descartes

> Everyone complains of his memory and no one complains of his judgement.
>
> *F. de La Rochefoucauld*

A **paradox** is an argument which, at each step, looks correct. Yet it leads to an absurd conclusion. One might expect that encounters with a few paradoxes would cause us to have doubts about how much confidence we should place in our sense of reason. In fact, however, the reaction of most people to a paradox seems only to confirm that our confidence in our power of reason is deep-seated indeed.

> The hopelessness of conversation between two people not sharing a common sense of logic is illustrated very cleverly in the story, "What the Tortoise Said to Achilles" (J. R. Newman, *The World of Mathematics,* Simon and Schuster, 1956, pp. 2402–2405) by Lewis Carroll. Lewis Carroll, the children's story teller (*Alice in Wonderland,* etc.) was in fact Charles Dodgson, a professor of mathematics with a lively interest in logic.

We see this in a reaction typical of many people when they first encounter Zeno's Paradox. They follow the argument a step or two (or three) until they see where it is leading. Then they break into a grin and try to offer some objection to the argument that seemed so logical to them a moment before. Finally, (after failing to identify any real error in the argument), they exhibit the confidence in their good sense that we referred to above. Amused, and perhaps a little perplexed, they shrug off the arguments of the learned Zeno and walk away no less certain that Achilles will surely pass the tortoise.

Zeno's Paradox

Achilles and a tortoise are to have a race. Achilles, being twice as fast, gives the tortoise a head start of 1 mile. The race is begun, and in due time, Achilles reaches the spot T_1 where the tortoise had been when the race started. The tortoise of course has moved on to a point T_2, $\frac{1}{2}$ mile up the road.

The race is not over and in due time Achilles reaches T_2. By then the tortoise is at T_3, $\frac{1}{4}$ mile ahead of Achilles. Again we observe that the race isn't over, that Achilles will eventually reach T_3. And so on. Is it now clear that Achilles will never catch the tortoise?

It is fortunate of course that our thinking processes are similar, for if it were not so, we would have great difficulty in communicating with each other. What, after all, remains to be said to someone who completely understands what you said, but fails to see that it makes any sense. Even the technique of raising one's voice under such circumstances seems not to be terribly effective.

Summary

We have identified several difficulties commonly encountered in deductive arguments: words can carry a multiplicity of meanings, statements can appear self-contradictory, the converse of a true statement is sometimes mistakenly taken to be true, and we are tempted to draw conclusions from $A \Rightarrow B$ when we know that A is false.

We have also noted that deduction rests squarely on the idea of common sense. The great thinkers have generally supported this notion, and most people trust their ability to think in spite of paradoxes in which we seem to get misled.

A Visual Paradox

We have described in this section an example which raises questions about our ability to reason correctly. Perhaps it is not entirely inappropriate to remind the reader that one can't always trust one's eyes either.

A B

Which segment is larger, A *or* B*?*

Self-Test

Choose the completion that seems best to you. If none of them seem quite correct to you, be prepared to tell why and to supply an ending of your own in class discussion.

1. This section begins with a "logical" syllogism which concludes that some syllables eat cheese. This absurdity occurs because
   (a) we have played on the fact that words often have multiple meanings.
   (b) we have constructed a paradox.
   (c) we are trying to use language to make a statement about the language itself.
   (d) we are confusing a symbol with the actual thing represented by the symbol.

2. A valid argument that reasons from a true hypothesis to a false conclusion.
   (a) illustrates one of the pitfalls of deductive thinking.
   (b) is a contradiction in terms; it simply can't happen.
   (c) is called a paradox.

3. Zeno's paradox is used in this section to illustrate
   (a) that arguments that seem correct at each step can lead to conclusions that are not correct.
   (b) that people can be made to lose confidence in their "common sense."
   (c) that the people of ancient Greece reasoned in ways that seem wrong to us today.

Problem Set 9.3

1. State the converse of each of the following ideas of Henry Thoreau, rewritten here in if-then style.
   (a) If a government imprisons any person unjustly, then the true place for a just man is in prison.
   (b) If a man is thinking or working, then he is alone.
   (c) If one is truly rich, there are many things he can afford to leave alone.
   (d) If a man does not keep pace with his companions, it is because he hears a different drummer.

(e) If a man must earn his living by the sweat of his brow, he sweats easier than I do.

2. State the converse of each of the following proverbs of Benjamin Franklin, rewritten here in if-then style.
(a) If there's marriage without love, then there will be love without marriage.
(b) If two are dead, then three can keep a secret.
(c) If the well runs dry, we learn the worth of water.
(d) If you are willing to give up essential liberty to obtain a little temporary safety, then you deserve neither liberty nor safety.
(e) If you would not be forgotten as soon as you're dead and rotten, then either write things worth reading or do things worth the writing.

The **contrapositive** *of the statement, "If A , then B," is, "If not B, then not A." A statement and its contrapositive are both true or both false.*

3. State the contrapositive of each of the statements in Problem 1.

4. State the contrapositive of each of the statements in Problem 2.

5. In each instance, identify the error in the reasoning.
(a) When the Republicans are in office, the country always heads for a depression. My doctor says I am showing signs of depression. If the Republicans win, I will be in a depression.
(b) You said that, if I didn't turn in this paper, I'd flunk. But I did turn it in, so I don't see how you can flunk me.
(c) It is clear that, if the courts are too lenient, crime in the streets will increase. Moreover, statistics make it evident that crime in the streets is increasing. The conclusion is inescapable; the courts must be too lenient.

6. In each instance, identify the error in the reasoning.
(a) Jake (the Brake) Stout is no mean linebacker. Unless a linebacker is mean, he cannot succeed at his position. Jake is not a success as a linebacker.
(b) He said that, if it rains, he will be in a bad mood. I'm happy the sun is shining so that he'll in a good mood.
(c) Senator Hornblower said that, unless the Senate adopted her plan, the country would be in a terrible mess. Considering the mess we're in, I take it that the Senate did not adopt her plan.

There are numerous popular puzzles in which logical analysis seems to defy good sense. Problems 7 through 9 are illustrations.

7. Two exhibitors are selling identical wooden gizmos at a sidewalk art sale. One is selling them at three for \$1, and the other at two for \$1. Both have 30 items left late in the afternoon, and they decide to leave for the day, asking a third party if he will see them at five for \$2. If they had each sold their items separately, they plainly would have netted \$25; so they were surprised when their friend gave them \$24, reporting that all 60 had been sold at the requested five for \$2. What happened to the other \$1?

8. Melinda had two friends, Esther and Wanda, each of whom was jealous of the time Melinda spent with the other. Since Esther lived east, and Wanda lived west of Melinda along the same subway line, and since trains ran in both directions at ten-minute intervals, Melinda decided to solve her problem by entering the station at random times and taking the first train to arrive. She was dismayed to find after trying this random process for a month that she was seeing Esther eight times for every two visits to Wanda. What could the explanation be?

9. A farmer decides that, on his death, his oldest son should have one-half of his possessions, the next son should have one-third, and the youngest should have one-ninth. He does die of couse, and when he goes he has 17 horses. With the market for butchered horses depressed at the time, the sons are in a dilemma as to how to split up the legacy. A neighbor with good horse sense comes to the rescue, giving them a horse of his own. They then take, respectively, 9, 6, and 2 horses. To everyone's surprise, they have a horse left over to return to the neighbor. How come?

10. Here is another question to which an incorrect answer seems plausible. The large wheel rolls along the lower track, so the distance AA' is the circumference of the large circle. The small wheel, physically fastened to the large one, moves along the upper track. Does BB' equal the circumference of the small circle?

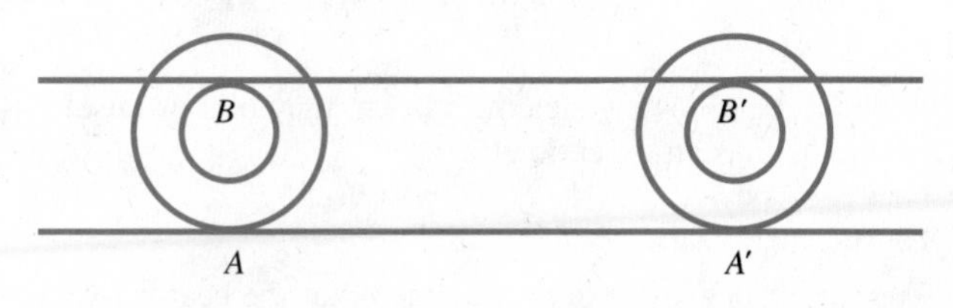

11. Some writers include as paradoxes any question to which an incorrect answer seems immediately obvious. Some questions of this sort were considered in Chapter 1. Here are a few more.
   (a) Suppose a 5-pound ball and a 10-pound ball are dropped simultaneously from a building rooftop. Will the 10-pound ball reach the ground faster?
   (b) Two identical coins are laid adjacent to each other, and the one on the left is rolled along the half-circumference of the other. Will the head that started right side up again be right side up or upside down?

12.* Zeno's Paradox about Achilles is nicely explained by making use of what we learned in Section 4.3 about the sum of a geometric series. Suppose Achilles runs 2 mph. Then in going from A to T_1, he uses up $\frac{1}{2}$ hour; in going from T_1 to T_2, he uses up $(\frac{1}{2})(\frac{1}{2}) = (\frac{1}{2})^2$ hour; etc. He will therefore reach T_n in the total elapsed time of

$$\frac{1}{2} + \left(\frac{1}{2}\right)^2 + \cdots + \left(\frac{1}{2}\right)^n$$

That he will catch the tortoise follows from the fact that an infinite number of terms can have a finite sum.

(a) Use the formula in Section 4.3 for the sum of the terms indicated above.
(b) Compute this for $n = 4, 5, 6$.
(c) What happens as n gets very large? How long, then, before Achilles catches the tortoise?
(d) Does your answer to part (c) correspond to your common sense answer to the question, If Achilles, running at 2 mph, gives the tortoise, running at 1 mph, a 1-mile head start, how long will it be until Achilles catches the tortoise?

There are some deep, difficult paradoxes arising from the way we use language, especially when we talk about classes of all objects of a certain kind. The reader who finds the following examples engaging should consult the article, "Paradox," by W. V. Quine in the April 1962 issue of Scientific American.

13. Think about the following statements. Are they true or false?
(a) Every rule has an exception.
(b) Never say never.
(c) Every generalization is false.

14. (a) The sheriff of a western town, angered at the bearded youths hanging around town, instructs the town's only barber that he is to shave all those, and only those, who don't shave themselves. What happens to the barber's beard?
(b) The same hapless sheriff decides to rid the town of long-haired men. He summarily cuts hair he considers too long. He asks people whom he picks up one question and decides whether or not they are telling him the truth. Those he judges to be telling the truth get a very short haircut; those whom he believes to be lying are shaved bald. Hearing that this was his policy, one young man—logically inclined—answered the question, "Do you know what is going to happen to you?" by saying, "I am going to be shaved bald." What should the sheriff do?

For Research and Discussion

Zeno was responsible for more than the one paradox mentioned in this section. What others can you learn about? Find as many examples of paradoxes as you can, adhering to the notion that a paradox is a line of reasoning that seems correct at each step, except for the misfortune of a spurious conclusion.

9.4 Deduction in Mathematics

Ten in a Row

In a railroad switchyard, 15 boxcars and 15 flatcars sit coupled together in random order on a siding. There is a call for 5 boxcars and 5 flatcars at another location. Is it always possible to find a string of 10 contiguous cars, 5 boxcars and 5 flatcars, that can be uncoupled and pushed as a unit through a switch onto the main track?

A mathematical theorem must be true without exception. If there is an exception, someone is sure to find a counterexample that makes it embarrassingly obvious that the theorem does not always hold. In this case, there is no way to hedge; the theorem is false. The terms used in mathematics do not carry the multiple meanings that create (or sometimes, we suspect, allow) the ambiguity often encountered in other disciplines. And while the paradoxes encountered in Section 9.3 do not shake the general faith we have in common sense, mathematicians are very much aware that they often deal with complicated relationships in which one cannot rely on visual pictures or intuition to correct lapses in step-by-step arguments. For this reason, they have given considerable thought to the kinds of arguments they allow in their proofs.

It is often alleged that studying the methods of reasoning allowed in mathematics has beneficial effects on one's reasoning in general. While we hesitate to venture an opinion as to whether this is so, we do believe that an introduction to these techniques is essential to understanding what mathematics is about. We therefore turn now to a discussion of some of these methods.

> In every man there is an eye of the soul which, when by other pursuits is lost and dimmed, is by these [arithmetic, geometry] purified and re-illuminated; and is more precious far than ten thousand bodily eyes, for by it alone truth is seen.
>
> *Plato*

The Arbitrarily Chosen Element

A theorem of elementary plane geometry asserts that all triangles inscribed in a semicircle are right triangles. Suppose the following proof is offered.

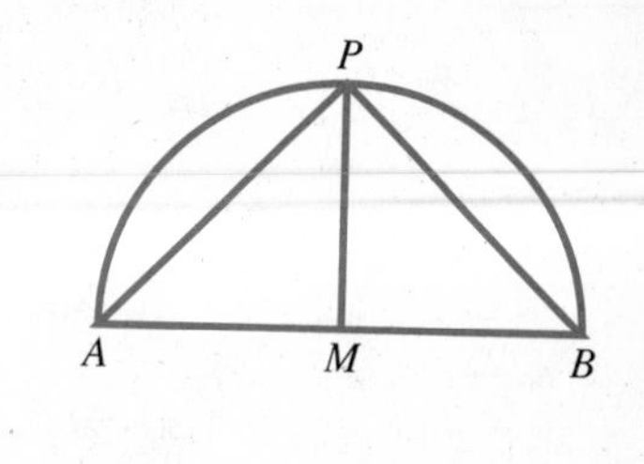

> Erect at the center M the perpendicular bisector of AB, letting P be the point of intersection with the semicircle. Since $AM = MP$, ΔAPM is an isosceles right triangle; thus $APM = 45°$. Similarly, $MPB = 45°$. This means that $APB = 90°$; the inscribed ΔAPB is a right triangle.

Someone will surely point out that this proof works for only a very special inscribed triangle. If the theorem is to be proved for all inscribed triangles, we must begin by considering an *arbitrary* inscribed triangle. The proof must begin as follows.

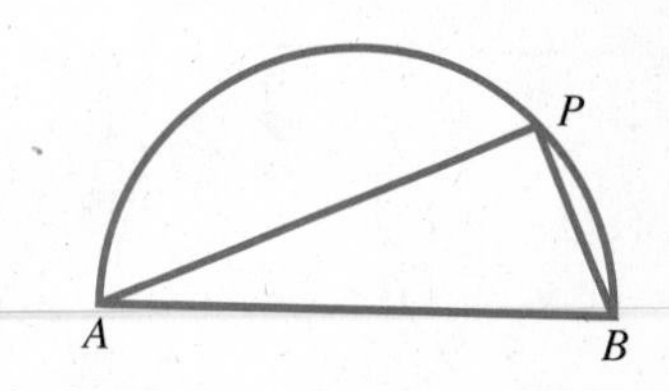

> Let P be a point chosen arbitrarily on the semicircle; draw the inscribed ΔAPB.

The argument that follows must then show that triangle APB is a right triangle. See Problem 18 for the complete argument. The principle we are trying to illustrate is the following.

> *To establish an assertion about all members of a certain set, choose an arbitrary member; that is, choose a member having no properties not shared by all members of the set under consideration. Show the assertion to be true for this arbitrary member.*

A professor who taught a large section of an introductory course was annoyed at the seeming indifference of students to his lectures. He went so far as to claim that not one student in the class brought anything to write with so as to be able to take notes. When challenged, he offered the following proof. He arbitrarily selected a student seated about halfway back in the lecture hall and asked this unfortunate if she had a pencil or a pen with her. The student admitted that she had neither. Since the student had been chosen arbitrarily, the professor claimed that he had proved his point. What is wrong with his argument?

The idea is of course that the argument given for the arbitrary element can be given for any other element just as well.

Contradiction

One commonly hears the phrase, "Suppose for the sake of the argument that. . . ." What is to follow can be anticipated. The speaker intends to show that this supposition leads to something undesirable if not impossible, hoping then that some interior linkage in the thinking apparatus of the listener will cause him or her to reject the supposition.

This is the idea of the form of argumentation known as *reductio ad absurdum* (reduction to the absurd) or proof by contradiction. The principle is as follows.

> The first person he met was Rabbit. "Hello, Rabbit," he said. "Is that you?"
>
> "Let's pretend it isn't" said Rabbit, "and see what happens."
>
> A. A. Milne

> *To prove that A is the case, assume the negation of A (not A). Show that the assumption of not A , logically developed, leads to something absurd.*

We illustrate this by proving a very simple fact about even and odd numbers.

Let us recall several definitions. A positive integer is even if it has a factor of 2; that is, n is even if it can be written in the form $n = 2k$ for some integer k. And n is odd if it is 1 more than some even number; that is, n is odd if it can be written in the form $n = 2k + 1$. Now we are ready to prove a theorem by contradiction.

Theorem If n^2 is even, n is even.

We prove this by assuming that it is false, that is, we assume that n^2 is even but n is odd. Then

$$n = 2k + 1$$

$$\begin{array}{r} 2k + 1 \\ \underline{2k + 1} \\ 2k + 1 \\ \underline{4k^2 + 2k } \\ 4k^2 + 4k + 1 \end{array}$$

from which it follows that

$$\begin{aligned} n^2 &= 4k^2 + 4k + 1 \\ &= 2(2k^2 + 2k) + 1 \end{aligned}$$

It is now clear, however, that n^2 is odd, contradicting the given fact that it is even. This shows that n cannot be odd; it must be even.

The Pigeonhole Principle

Suppose a mail carrier has five letters, all to be delivered to a building having just four apartments, hence four mailboxes. What conclusion can you draw? (We once put this question to a student who, after a period of nervous silence, responded, "It's been several years since I've had any mathematics.") We hope that, independent of the reader's training in mathematics, it is obvious that at least one mailbox will get more than one letter.

The principle illustrated by this example is called the **pigeonhole principle:**

If n objects are to be placed in m slots with $n > m$, at least one slot will get more than one object.

A similar principle is also referred to as the pigeonhole principle:

If n objects are to be placed in n slots with no more than one object in each slot, each slot will get exactly one object.

There is a well-known problem often used to illustrate the pigeonhole principle. The idea is to show that, in any sufficiently large city, at least two people have exactly the same number of hairs on their head. We quote one solution to this problem (M. Kac and S. M. Ulam, *Mathematics and Logic,* Praeger, New York, 1968, p.11) and then comment on it.

The Solution to a Hairy Problem

In the case of New York City all one needs to know is that the number of hairs on any head is less than the city's population of roughly 8,000,000. (A person would collapse under the weight of 8,000,000 hairs.) If each person is tagged by his specific number of hairs, at least two people must be tagged by the same number (i.e., have the identical number of hairs).

Before including the solution to this problem, we decided to check the assertion that a person could not stand up under 8 million hairs. Accordingly, we went to the chemistry department and had the chemists weigh a hair about $2\frac{1}{2}$ inches in length. The announced result was 0.0004 gram. A little computation (the chemists did it for us) showed that 8 million such hairs would weigh about 7 pounds. Few people would collapse under such a burden, but it is also true that few people are likely to buy a 7-pound hat; and in several class discussions we have gotten general agreement that even the hairiest people on campus probably do not carry 7 pounds of hair. In any case, the solution gives us an opportunity to identify the several stages of an attempt to solve a problem.

Observation Some hair was weighed. Some computations were made. Varying hair styles were observed.

Premises No human being has as many as 8 million hairs on his or her head. New York City has a population of more than 8 million.

Mathematical Argument Consider the residents of New York City. Put persons with no hairs on their heads in group 0, one hair on their heads in group 1, those with two hairs in group 2, etc. By hypothesis, we will need at most 8 million groups to accommodate everyone. But there are more than 8 million people to be put into these groups. By the pigeonhole principle, at least one group must contain more than one person.

Conclusion At least two people in New York City have the same number of hairs.

Now notice something. If you wish to dispute our conclusions, you will need to question the initial assumptions (premises). You may convince someone (in fact, you may be right) that, if one takes careful account of the short hairs on the nape of the neck, the average weight of a hair will be much less than 0.0004 gram. Or you may question whether or not New York really has 8 million people; this requires census information which is, of course, subject to change. All kinds of questions can be raised. But our point is that they all have to do with the premises. Once they are granted, the conclusion is inescapable. A mathematician need not worry that his or her part of the job will be spoiled by future investigation, or even the news that the analytical balance was out of whack on the day of the observations. A mathematician's satisfaction derives from having showed that those who accept the premises must accept the conclusion.

Ten in a Row

We do not claim that the three methods described so far exhaust all the methods ever used in mathematics, but the methods used really are not very numerous, nor are they when considered individually, terribly complex. The complexity of mathematical arguments comes when several of the basic ideas are varied a bit, and then used in conjuction with one another. A solution to the **Ten in a Row** problem illustrates the interplay of the ideas above, each used with a slight variation.

Begin by thinking of the 30 cars partitioned into three sets; the first 10 cars, the middle 10, and the last 10.

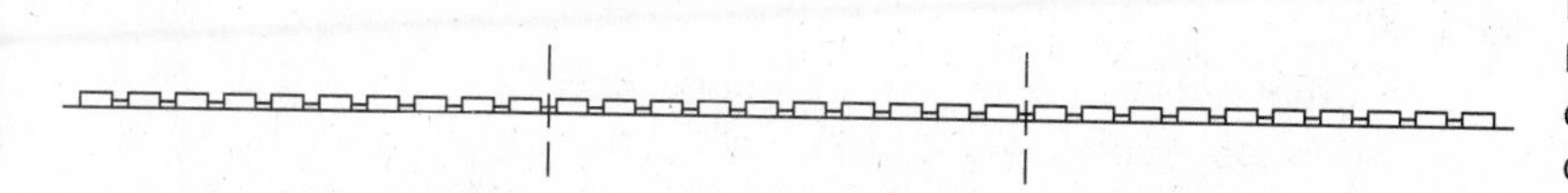

Proof by contradiction! Suppose all three sets had more boxcars than flatcars. At least 6 boxcars in each set would give us at least 18 boxcars, whereas we have only 15.

If one of these sets contains 5 boxcars and 5 flatcars, we are through. If not, observe that not all three sets can have an excess of boxcars, nor can all have an excess of flatcars. Evidently two of the sets will have an excess of one type of car, while the third will have a deficiency of these cars. Moreover, one of the sets having an excess of these cars will be adjacent on the track to the set having a deficiency. Number the cars of these two sets 1–20.

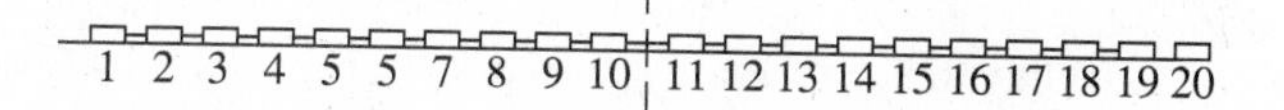

What we say about one set must be something we could say about any set.

To avoid cumbersome language or notation, let's assume that the first set of 10 cars has an excess of boxcars, the last 10 cars have a deficiency of boxcars. (You will see that the argument would work just as well for flatcars, and that the excess could just as easily be in the set to the right.)

Here our goal will be to show that some n_k must go into the pigeonhole labeled 5.

We next consider the eleven sets S_k of cars, each labeled by an integer n_k that records the number of boxcars in the set.

$$\begin{array}{ll} S_1 = \{1, \ldots, 10\} & n_1 \text{ boxcars in } S_1 \\ S_2 = \{2, \ldots, 11\} & n_2 \text{ boxcars in } S_2 \\ \vdots & \\ S_{10} = \{10, \ldots, 19\} & n_{10} \text{ boxcars in } S_{10} \\ S_{11} = \{11, \ldots, 20\} & n_{11} \text{ boxcars in } S_{11} \end{array}$$

Each of the integers goes into the appropriately numbered pigeonhole.

10	9	8	7	6	5	4	3	2	1	0

We know that n_1 goes into one of the pigeonholes to the left of 5 (because S_1, one of our original sets, was selected to have an excess of boxcars). Similarly, n_{11} goes in a pigeonhole to the right of 5.

We are ready to complete our argument. Note that no integer n_k is ever more than one pigeonhole to the left or the right of n_{k-1}, since omitting one car at the front of the line and adding one at the end cannot change the number of boxcars by more than 1. Thus, if we start to the left of pigeonhole 5, move one step at a time, and end up in a pigeonhole to the right of 5, we cannot have skipped over 5. There must be some n_k in pigeonhole 5. That is, some set S_k contains 5 boxcars. This is the set of cars to be pushed as a unit onto the main track.

Galileo (1564–1642)

For just as in nature itself there is no middle ground between truth and falsehood, so in rigorous proofs one must either establish his point beyond doubt, or else beg the question inexcusably. There is no chance of keeping one's feet by invoking limitations, distinctions, verbal distortions, or other mental acrobatics. One must with a few words and at the first assault become Caesar or nothing at all.

Summary

Certain methods of deductive argument have proved themselves very useful in mathematics. We have described three of them in this section. In our discussion, we have given (or at least started) three proofs the reader should not only understand, but also see as illustrations of the corresponding methods.

Methods
1. Use of an arbitrary element.
2. Contradiction.
3. Pigeonhole principle.

Illustrations
1. Triangles inscribed in semicircles.
2. If n^2 is even, n is even.
3. Hair counts in New York.

Self-Test

Choose the completion that seems best to you. If none of them seem quite correct to you, be prepared to tell why and to supply an ending of your own in class discussion.

1. We quoted Galileo's assertion that in a rigorous proof, one must either establish his point beyond doubt, or else beg the question inexcusably. In this quote, we see
(a) that Galileo failed to grasp the idea that in the final analysis, nothing can be proved.
(b) an example of the arrogance that ultimately brought Galileo into conflict with the church.
(c) that Galileo identified a rigorous proof with what we have called a deductive argument.

2. In our proof that there are two individuals in New York City with the same number of hairs on their heads,
   (a) we see that the role of experiment is to help us decide whether the pigeonhole principle is an appropriate argument to use.
   (b) we see that an experiment is useful in helping us decide on the truth of the hypothesis, but is of no help at all in establishing the validity of the argument.
   (c) we see that one can give perfectly good mathematical proofs of statements that no one would really believe.

3. In our proof that there are 5 boxcars and 5 flatcars in a contiguous string of 10 cars, we assumed that the first 10 cars had more than 5 boxcars, the second 10 had less than 5 boxcars.
   (a) We did this because this problem could not be solved without making additional assumptions.
   (b) Our argument would clearly have worked just as well if the first 10 cars had more than 5 flatcars, and the second 10 had less than 5 flatcars. The only problem would have come if the first 10 cars and the second 10 cars had both had an excess of the same type of cars.
   (c) The reader was expected to understand that any other arrangements of the first, second, and third groups of 10 could be handled with the same argument.

Problem Set 9.4

1. Using the notion of an arbitrarily chosen element, show that the square of an even number (a number that can always be written in the form $2n$) is again an even number.
2. Show that the square of an odd number (a number that can always be written in the form $2n + 1$) is again an odd number.
3. Show that the square of a number not divisible by 3 (a number of the form $3n + 1$ or $3n + 2$) is not divisible by 3.
4. Show that the square of a number divisible by 3 is again divisible by 3.
5. Show that the product of an odd number and an even number is even.
6. Show that the product of two odd numbers is odd.
7. Show that the square of an odd number can be written in the form $8n + 1$.
8. Show that the product of three consecutive positive integers is divisible by 6.
9. It is a fact that one cannot construct the trisectors of an arbitrary angle using only a straightedge and compass. Can you construct an angle equal to one-sixth of an arbitrary angle? Hint: Assume you can; then use contradiction.
10. In a certain city, while you can get from A to B by bus, you cannot get from A to C by bus. Prove that you cannot get from B to C by bus.
11. Prove that, if n^2 is odd, n is odd. Hint: Assume n is not odd; note Problem 1.
12. Prove that, if n^2 is divisible by 3, then n is divisible by 3. Hint: Assume n is not divisible by 3; note Problem 3.

13. Assume there are more maple trees in the world than there are leaves on any single maple tree. Show that it follows that there must be at least two maple trees having exactly the same number of leaves.

14. Assume there are more chickens in the world than there are feathers on any one chicken. Without any chicken plucking, show that there must be at least two chickens having the same number of feathers.

15. If all the sand on the shore of Lake Michigan were scooped into thimbles, would two thimbles necessarily contain exactly the same number of grains of sand?

16. Prove that if 5 points are placed anywhere within or on the boundary of an equilateral triangle of side 1, then at least two of the points must be within $\frac{1}{2}$ of each other.

17. Let P be an arbitrary polyhedron. Show that at least two faces have the same number of edges.

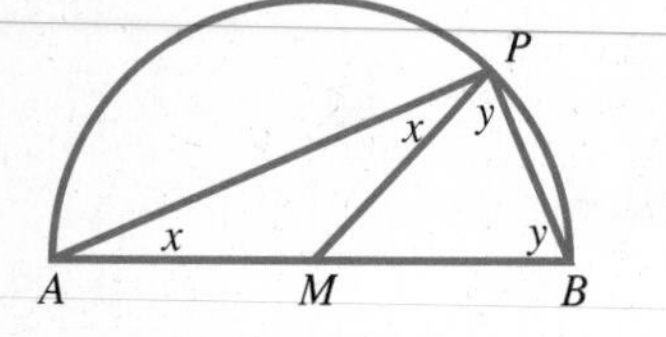

18. Complete the proof that a triangle inscribed in a semicircle is a right triangle, using the following hints. Draw the radius MP, forming two isosceles triangles, ΔAMP and ΔPMB. Recall that the base angles of isosceles triangles are equal. Recall also that the sum of the interior angles of any triangle is 180°.

For Research and Discussion

Numerous mathematical puzzles can be solved using some variation of the pigeonhole principle. Among our favorites:

1. A child scribbles on an $8\frac{1}{2}$ in $\times$ 11 in sheet of white paper with a red crayon. Prove that no matter how random (or how carefully contrived) the coloring, when it is finished, it will be possible to find two points at a distance one inch from each other that are the same color.

2. List 7 integers $a_1, a_2, \ldots, a_6, a_7$. It will always be the case that one of the numbers listed will be a multiple of 7, or some subset of the numbers will have a sum that is a multiple of 7. Moreover, this same statement will be true if 7 is everywhere replaced by any integer n.

See A. Soifer, *Mathematics as Problem Solving,* (Colorado Springs: Center for Excellence in Mathematical Education, 1987, pp. 10–15) and other books in the problem-solving genre. Find a problem solvable by the pigeonhole principle that you find particularly attractive.

chapter 10

From Rules to Models

I think that everything that can be an object of scientific thought at all, as soon as it is ripe for the formation of a theory, falls into the lap of the axiomatic method and thereby indirectly of mathematics. Under the flag of the axiomatic method mathematics seems to be destined for a leading role in science.

DAVID HILBERT

Intellectual Honesty

Students learn very early in school geometry that a proof begins with a clear statement of what is to be assumed. That characterizes all mathematical proof; the hypothesis of a theorem must be made clear. In other pursuits, we may be unaware, or even be deliberately evasive about some of our initial assumptions. In mathematics, it is a professional embarrassment to be caught, consciously or not, making an assumption not stated in the beginning.

More mature students of mathematics are sooner or later impressed with the fact that not all of their terms can be defined. Triangles can be described as line segments connecting points, and lines may be said to be collections of points, but having done so, defining a point as the intersection of two lines, even of two straight lines, will be seen to be circular. Mathematicians come pretty quickly to the realization that they cannot, and had better be very open about the fact that they cannot define all their terms.

Building on clear assumptions, the mathematician's next task is to give an argument, the purpose of which is not to persuade, but to compel any straight-thinking person to accept the conclusion. A lapse in one's logic is not something that can be charitably dismissed because the result is so nobel, so beautiful, or so useful. The argument itself is the centerpiece, the essence of what a mathematician creates. One does not knowingly gloss over the very thing upon which one's work is judged.

A mathematician's presentation, whether written or oral, is therefore shorn of hyperbole or passion. Knowing that the proof at hand is being carefully scrutinized by an audience that holds to a high degree of logical rigor, one is drawn more to a cold rationality than to a fiery oratorical style which, in the end, can do nothing to overcome a hole in the argument. Understatement rather than exaggeration is the preferred style of the person who knows it will be taken as an indication of incompetence if more is claimed than has been established beyond all doubt.

Finally, it is to be noted that a person who knowingly works with undefined terms, and makes it conspicuously clear that every argument ultimately rests on assumptions will not claim to have said something true about the world of experience. Whether or not the assumptions say something true about the world is left for others to judge, so in this way too the mathematician is careful not to claim to have done more than has in fact been done.

On these grounds, the recognition that not all the terms can be defined, the clear identification of all of assumptions, and the supreme effort to reason in a dispassionate, clear, and logical fashion, it has been argued that mathematics is the ideal place to teach the meaning of intellectual honesty. This is not a claim that mathematicians have as individuals achieved a higher level than other humans, but it is a suggestion that the mathematics classroom, shorn as it usually is of highly controversial and emotional issues, is an ideal place to discuss the features that characterize intellectual honesty. Howard Fehr said it this way:

> Mathematics is the one area of human enterprise where the motivation to deceive has practically been eliminated. Not because mathematicians are necessarily virtuous people, but because the nature of mathematical ability is such that deception can be immediately determined by other mathematicians. This requirement of honesty soon affects the character of the continuous student of mathematics.

In 1974, the Notices of the American Mathematical Society carried an exchange of letters about the nature and purposes of the thesis required for the doctorate in mathematics. The theme of intellectual honesty as we have described it was nicely summarized in a letter from Leo Zippin:

> The recently published exchange of views on the issue: "The Ph.D. Thesis" omits insistence on the most sharply distinctive aspect of mathematics, the single respect in which it is, perhaps, like no other discipline on earth: its practitioners must accept differences as to whether a particular mathematical result was worth proving—but they cannot tolerate differences as to whether it was indeed proved.
>
> It seems to me that many conclusions can be drawn from this, but the most notable might well be that for a brief time at least while he is busy thinking through the rigor of his work, your author has to be a modest, even an humble fellow eagerly dedicated to truth. All of the graduate students I have known, and I myself, have found this a refreshing—even an exhilarating experience.

10.1 The Consequences of Given Rules

A Colorful College

A college president, a professor, an instructor, and a janitor are named Mr. Brown, Mr. Green, Mr. White, and Mr. Black, but not respectively. Four students with the same names will be designated here as Brown, Green, White, and Black. The student with the same name as the professor belongs to Black's fraternity. Mr. Green's daughter-in-law lives in Philadelphia. The father of one of the students always confuses White and Green in class, but is not absent-minded. The janitor's wife has never seen Mr. Black. Mr. White is the instructor's father-in-law and has no grandchildren. The president's oldest son is seven. What are the names of the president, professor, instructor, and janitor?

Litton's *Problematical Recreations* compiled by James F. Hurley (New York: Van Nostrand Reinhold, 1971).

A good first step in a problem like the one above is to rewrite the given information in a series of clear, crisp sentences, each stating one bit of information in an unambiguous way. One student began the problem above by listing the given information as follows:

1. The student with the same name as the professor belongs to Mr. Black's fraternity (who, from the common use of language, we therefore take to be someone other than Mr. Black).
2. Mr. Green has a daughter-in-law living in Philadelphia.
3. One of the teachers, not named Mr. White or Mr. Green, is the father of one of the students (who therefore has the same name as that teacher).
4. Mr. White is the instructor's father-in-law and has no grandchildren.
5. The president's oldest son is 7 (and so, we presume, is unmarried).

She then identified the people as follows:

6. *The professor is Mr. Brown.* The father of one of the students, named either Mr. Black or Mr. Brown, teaches (item 3); so he is either the professor or the instructor. He must be the professor, since the instructor has no children (item 4). The professor is not named Mr. Black (item 1), so he must be Mr. Brown.
7. *The instructor is Mr. Black.* The instructor has no children (item 4), so he cannot be Mr. Green (item 2). He is not Mr. White (item 4) and he is not Mr. Brown (item 6).

8. *The president is Mr. White.* We know that neither Mr. Brown (item 6) nor Mr. Black (item 7) is the president. The president does not have a married son (item 5), so Mr. Green is not the president (item 2). That leaves Mr. White.

The reader will note, among other things, that we are making use of the pigeonhole principle in this argument. The final identification follows directly from this principle:

9. *The janitor is Mr. Green.*

Several instructive observations can be made about this student's solution. In the first place, we call attention to the parts of items 1 and 5 that are enclosed in parentheses. It can be argued that one or both of these bits of information go beyond what is given in the statement of the problem. The only comeback to this is that she at least placed the argument where it belongs; that is, it is clear that the argument concerns what information is given to start with. Her solution makes use of the information as stated in items 1 and 5. Anyone who refuses to accept this information must either find another solution or argue that the problem as stated cannot be solved.

Second, we note that she never used certain information that was given. It is always a good exercise to review unused information. Did we use it unconsciously? If we argued differently so as to use it, could we have avoided some of the questionable assumptions we put in parentheses? Or is the information actually irrelevant to solving the problem?

Finally, we note that several times she made use of the idea that a father and his son have the same last name. This is generally true of course, but not always. Hence it should be stated as one of the things taken to be understood at the beginning. We have more to say later about making certain that all initial assumptions have been identified.

In any argument in which inferences are to be drawn, it is essential that everyone be clear as to what information is being accepted to start with. In the case of questionable information, it is of course desirable to resolve the problem without using it. If this is impossible, the next best procedure is to accept the questionable information, clearly point out that you have done so, and proceed to give the solution.

The approach we have observed in solving the puzzle above can be of use in practical situations. The warmth and good fellowship with which many a social organization hopes to settle necessary business sooner or later gives way to the formulation of a set of by-laws. The well-known public admiration for bureaucracy notwithstanding, most of the grand goals of political oratory emerge as rules of a regulatory commission. It is tricky business to put into writing all the rules to be followed, even if by chance there is a will to do so. And all too often we discover only by sad experience the natural consequences of the rules we adopt.

Torn between a desire to describe realistic situations and a desire to use relatively simple situations, we consider two fictitious situations.

The Problem of Dormitory Governance

When State University changed some of its policies so that the Gertrude Smith Residence Hall for Women became the Smith Residence for People, the governing council of that dormitory decided to propose simultaneous changes in the way the committees of the council were set up. Several criticisms were made of the old system. It was noted that the old by-laws called for committees (such as the one to mete out punishment for violating the 10:00 P.M. weeknight curfew), that had not functioned for years. It was recalled that one fall, when apathy ran rampant, only one section of the dorm remembered to elect its representative, so that later in the year, when an issue arose, the dorm had a one-woman governing council. Some criticism was directed at two women who managed to dominate three key committees, and some was directed at two other women on the council who refused to serve on any committees. Finally, it was pointed out that one of the committees had a member who was not even an elected representative.

At the next meeting of the governing council, a member proposed that the new governing council be required to abide by the following rules.

Dormitory Governing Council Rules

Rule 1: Committees shall be composed of elected representatives.

Rule 2: The council shall not function unless at least two representatives have been elected.

Rule 3: Any two elected representatives must serve together on at least one and no more than one committee.

Rule 4: For each committee, there must be at least one representative not on that committee.

Rule 5: If A is a committee and x is a representative not on committee A, then x must serve on one and only one committee that has no members in common with committee A.

She defended her proposal in the following ways. In the first place, they were simple to check. Second, they were general enough for each year's governing council to set up the committees needed that year. Third, she pointed out that many of the situations criticized under the old arrangement couldn't happen with her system because they were precluded either explicitly or implicitly. She then proceeded to point out some of the features implicitly built into her proposal.

Consequences of the Rules

Consequence 1: Every representative must serve on at least two committees.

In the best manner of a good mathematics student, she argued as follows. Consider an arbitrary representative x. There is a second representative y (rule 2), and members x and y serve on a common committee, say A (rule 3). There must be a representative z not in A (rule 4), and since x and z have a common committee (rule 3), say B, we have x in both A and B.

Note that in the course of this argument she also showed the following.

Consequence 2: There must be at least three elected representatives.

She next reminded the group of the criticism that some committees were still part of the official structural even though no members had been assigned to serve on these committees for years. That, she claimed, could not happen if her suggested rules were enforced.

Consequence 3: There are no "empty" committees; that is, every committee has at least one member.

Again her argument was along the lines a mathematician might use, this time employing proof by contradiction. She supposed there was an empty committee, called E. Starting as in the argument for consequence 1, she pointed out that we have

$$x \text{ and } y \text{ in } A$$

$$x \text{ and } z \text{ in } B$$

This, however, violates rule 5, because x is not in the empty committee E, hence should belong to only one committee having no members in common with E. But x belongs to both A and B. This is a contradiction.

Not content with having proved that every committee has at least one member, she claimed that even more was true:

Consequence 4: Every committee must contain at least two members.

Again she supplied an argument to support her claim. You may wish to try supplying a convincing argument for the last assertion. (It will be worthwhile for you to try this not trivial problem, even if frustration drives you to peek ahead to Section 10.2 where an argument is given.) There are of course other questions about the rules that may come to the mind of the governing council. Several of them are suggested in the problems at the end of this section.

The Bus Companies

In the small, developing Imagin nation, the government agency of public transportation is charged with the regulation of bus companies operating between cities. This agency is not to regulate companies operating within a city; to fall under its jurisdiction, a company has to have a route between at least two cities. For the purposes of the government agency, a route is simply a group of cities served by a particular bus run; thus a company may list:

Route A: Pitstop, Center City, Posthole

Route B: Pitstop, Klondike

If a company claims to serve a certain two cities, it is required that these two cities appear on a common route, but that only one route list both cities. In an effort to force bus companies to provide service to smaller cities off the main highways, it is further required that, for any route listed by a company, it must serve a city not on that route. Finally, in an attempt to control the size of the companies the agency imposes the following restriction. If R is a route and x is a city served by a company, but not on route R, the company can have only one route through x that has no cities in common with R.

Regulations for Intercity Bus Companies

Regulation 1: A route is a collection of cities through which a bus passes on its run.

Regulation 2: A company must serve at least two cities.

Regulation 3: Any two cities served by a company must be on one and only one common route.

Regulation 4: Given any route, a company must serve at least one city not on that route.

Regulation 5: If R is a route and x is a city served by a company but not on R, there must be one and only one route through x having no cities in common with R.

The perceptive reader notes that, rhetoric aside, these regulations are a restatement of the rules of the dormitory governing council discussed above. There should be little trouble, once this observation has been made, in establishing the following operation guidelines.

Guideline 1: Every city served by a company must be on at least two routes.

Guideline 2: A company must serve at least three cities.

Guideline 3: Every route must include at least two cities.

In fact, every question asked about the dormitory regulations gives rise to a question about the bus regulations. And, more to the point, every answer to a question about one problem answers the corresponding question about the other. The economy thus achieved is a feature to which we return later.

Summary

Any collection of interrelated statements (clues to a puzzle, rules for a game, or regulations governing an organization) is likely to have unexpected implications. To discover them, we should be sure that we begin with a clear understanding of what information is given; and when the information is ambiguous, we should be clear as to how we have decided to interpret it. We then must discipline ourselves to use only the given information, nothing else, unless—stymied—we decide to use plausible idea, in which case we clearly identify our added assumptions. Finally, we should review information not used, asking whether it has been used unconsciously, whether it can be used to shorten the argument, or whether it really is extraneous.

A Good Argument

1. *Get a clear understanding of the given information.*
2. *Use nothing that is not given.*
3. *If you decide to make further assumptions, say so.*
4. *Review any given information that you didn't use.*

Self-Test

Choose the completion that seems best to you. If none of them seem quite correct to you, be prepared to tell why and to supply an ending of your own in class discussion.

1. In the problem of dormitory governance, we proved that every committee has at least one member.
   (a) This is actually obvious from rule 1.
   (b) In light of the fact that consequence 4 can be proved (stating that every committee must contain at least two members), it was a waste of time to prove that every committee has at least one member.
   (c) If we found a residence hall somewhere and noted that because of resignations, some committees had no members, then we could conclude that the rules proposed in our example were not the rules being used to govern that residence.
2. The logic used in solving puzzles
   (a) differs from the logic used in mathematics because mathematics deals with actual situations in the real world.
   (b) is no different than the logic used in mathematics.
   (c) is the key to finding the solution. If one can follow logic essential to the solution, then one can surely find the solution for himself.
3. In titling this section *The Consequences of Given Rules,* we meant to emphasize
   (a) that rules have consequences which, though perhaps not immediately evident, will be exactly the same for everyone who abides by those rules.
   (b) that everyone who studies the rules will discover the same set of consequences.
   (c) that while rules may look the same, different people can, in applying these rules, come to different and even contradictory conclusions.

Problem Set 10.1

1. Following the proofs used in discussing the dormitory governance problem, prove guidelines 1 and 2 of the regulations for intercity bus companies. Write out all the details, being sure that you understand each step.
2. As in Problem 1, write out the proof of guideline 3.
3. In the argument for consequence 3 of the dormitory governing rules, we began by assuming that there could be an empty committee. Is this not already in violation of Rule 1?
4. We have heard the following complaint about the proof of consequence 1. While x has been shown to be on two committees, y has been shown to be on only one committee. Yet the assertion says that *every* representative must serve on two committees. Have we really proved what we set out to prove? Why?

Problems 5–9 ask some further questions about the dormitory problem. They are of course questions about the bus regulations as well.

5. Prove consequence 4.

6. Although the stated requirements are that at least two representatives must be elected, it was easily shown (consequence 2) that at least three had to be elected if all rules were to be satisfied. Show that even more is true: If all the rules are to be satisfied, at least four representatives must be elected.

7. Show that there must be at least six committees if all the rules are to be satisfied.

8. Show that all the rules will be satisfied if four representatives are elected and they form six committees.

9. Interpret Problem 8 for the bus route problem. Draw a map indicating the four cities and the six routes.

10. Here is another problem about forming committees. Since it is a new problem based on rules quite different from those used in the dormitory governance example, you cannot use the results proved for that situation. A legislative group agrees that the following rules shall be enforced with regard to any committees set up.

(a) Any two committees must have at least one common member.
(b) Any two committees shall have at most one common member.
(c) Every member of the legislature must serve on at least two committees.
(d) No member of the legislature may serve on more than two committees.
(e) There shall be four committees.

How many members must there be in the legislature? How many members serve on each committee?

11. Four boys, Alan, Brian, Charles and Donald, and four girls, Eve, Fay, Gwen and Helen, are each in love with one of the others, and, sad to say, in no case in their love requited. Alan loves the girl who loves the boy who loves Eve. Fay is loved by the boy who is loved by the girl loved by Brian. Charles loves the girl who loves Donald. If Brian is not loved by Gwen, and the boy who is loved by Helen does not love Gwen, who loves Alan?

J. F. Hurley, Litton's *Problematical Recreations,* (New York: Van Nostrand, 1971), p. 66.

12. In our village we have a Mr. Carpenter, a Mr. Machinist, and a Mr. Smith. One is a carpenter, one a machinist, one a smith. None follows the vocation corresponding to his name. Each is assisted in his work by the son of one of the others. As with fathers, so with sons; none follows the trade that corresponds to his name. If Mr. Machinist is not a carpenter, what is the occupation of young Smith?

Oswald Jacoby, *Math for Pleasure,* (New York: McGraw-Hill, 1962), p. 91.

13. On the Island of Perfection there are four political parties—the Free Food, the Pay Later, the Perfect Parity, and the Greater Glory. Smith, Brown, and Jones were speculating about which of them would win the forthcoming election. Smith thought it would be either the Free Food Party or the Pay Later Party. Brown felt confident that it would certainly not be the Free Food Party. And Jones expressed the opinion that neither the Pay Later Party nor the Greater Glory Party stood a chance. It turned out that only one of them was right. Which party won the election?

E. R. Emmet, *Puzzles for Pleasure,* (New York: Emerson Books, 1972), p. 4.

14.* There are five houses, each of a different color. They are inhabited by five families who are of different nationalities, read different newspapers, keep different pets, and own different kinds of vehicles:
(a) The English live in the red house.
(b) The Spanish own the dog.
(c) The *Examiner* is read in the green house.
(d) The African reads the *Times.*
(e) The green house is immediately to the right of the brown house (as you face them).
(f) The owner of the antique car keeps goldfish.
(g) The people in the yellow house have a sports car.
(h) The *Daily News* is read in the middle house.
(i) The Norwegian lives in the first house on the left.
(j) The family with the station wagon lives in the house next to the family with the cat.
(k) The family with the sports car lives next door to the family that owns the horse.
(l) The Japanese drive a sedan.
(m) The Norwegian lives next door to the blue house.
(n) The family with the camper truck reads the *Herald.*
From the clues listed, determine who reads the *Tribune* and who owns the monkey.

For Research and Discussion

Here are the ballots of the six members of the governing board who have been asked to rate four alternatives a_1, a_2, a_3, a_4, under consideration.

Ballot 1	Ballot 2	Ballot 3	Ballot 4	Ballot 5	Ballot 6
a_1	a_3	a_3	a_4	a_1	a_3
a_4	a_2	a_4	a_3	a_4	a_4
a_2	a_1	a_1	a_1	a_3	a_1
a_3	a_4	a_2	a_2	a_2	a_2

How would you proceed if it was your job to tally the votes and announce as the group's decision a rank ordering of the four alternatives? Can you show that different reasonable systems of counting will have different outcomes? What rules should characterize "good" counting system?

10.2 Finite Geometries

Mr. Promo's Business Trip

Mr. Promo is planning a business trip that will take him from his home office in Chicago to visit some major customers on the West Coast. His supervisor asks him to prepare for her consideration several lists of customers that he might visit, subject to the following guidelines.

1. She would like him to visit three customers on this trip, meaning that he won't get to them all.
2. She would like to be able to find, for any two customers, one and only one of his lists that includes them both.
3. Any two lists she looks at should have at least one customer in common.

Can you tell from this information how many customers the company has on the West Coast?

It was seen in Section 10.1 that two lists of rules, the dormitory governing rules and the regulations for intercity bus companies, were, from our point of view, the same. That is, if the words "representative" and "committee" in the first list are replaced by "city" and "route," and if appropriate changes are made in grammar, we will get the second list. The consequences of the given rules are similarly related and, more important, the arguments establishing these consequences are exactly the same. That is to say, the arguments do not depend at all on whether we use the word "representative" or "city," or the word "committee" or "route." In fact, we could "abstract" the rules stated for the dormitories or the regulations for the buses by using nonsense words as follows.

Rule 1a: Lokes are collections of pokes.

Rule 2a: There are at least two pokes.

Rule 3a: Any two pokes must be members of one and only one common loke.

Rule 4a: For each loke L, there is a poke x not in L.

Rule 5a: If L is a loke and x is a poke not in L, then x is in one and only one loke M having no members in common with L.

We can now proceed to establish the consequences of these rules.

Consequence 1a: Every poke must be in at least two lokes.

Consider an arbitrary poke x. There is a second poke y (rule 2a), and the pokes x and y are in a common loke, say A (rule 3a). There must be a poke z not in A (rule 4a) and, since x and z are in a common loke (rule 2a), say B, we have x in both A and B.

Comparison shows that we have merely copied the proof of consequence 1 from Section 10.1. In the same way we can easily prove further results.

Consequence 2a: There must be at least three pokes.

Consequence 3a: Every loke contains at least one poke.

Since we did not prove consequence 4 in Section 10.1, let us do so now.

Consequence 4a: Every loke must contain at least two pokes.

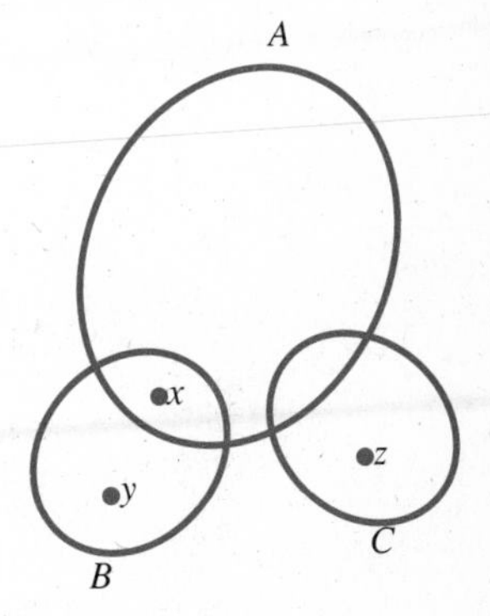

In the accepted way, we establish this by showing it to be true for an arbitrary loke that we call A. It follows from consequence 3a that A must contain some poke x, and we know from rule 4a that there is a poke y not in A. It follows in turn that there is a loke B containing x and y (from rule 3a), that there is a poke z not in B (rule 4a), and that there is a loke C that contains z and no points of B. Using dots to represent pokes, and circles to enclose pokes on the same lokes, the situation as we now have it can be pictured as in the margin. (Of course, z might be in A, but in that case we would be finished.)

Now if A and C have no common pokes, x, a poke not in C, will be a member of the two lokes A and B, neither of them having a poke in common with C. This violates rule 5a, so C and A must have some common poke, say w. This gives a second poke in A, establishing our assertion.

Except for the terminology employed, we have here an example of what is called a **finite geometry.** In finite geometries, the nonsense words are commonly called **points** and **lines;** the stated rules are called **axioms;** and the derived consequences are called **theorems.**

We worked our way into finite geometries slowly, because most students have a preconceived idea of what a point is and what a line is. They find it very difficult to let go of these ideas (thereby demonstrating, incidentally, that by 18 or 19, we are already quite set in our ways, resistant to new ideas). If we had stated consequence 4a in the form,

> Every line contains at least two points

instead of in the form,

> Every loke contains at least two pokes

most readers would not, in the beginning, see this as something needing proof. They would see it as an obvious statement about points and lines.

> In the beginning everything is self-evident, and it is hard to see whether one self-evident proposition follows from another or not. Obviousness is always the enemy of correctness. Hence, we must invent new and even difficult symbolism in which nothing is obvious.
>
> *Bertrand Russell*

> One must be able to say at all times, instead of points, straight lines, and planes—tables, chairs, and beer mugs.
>
> *David Hilbert*

Yet, in finite geometry, as in all of mathematics, the goal is to prove all consequences, showing them to be inevitable implications of the rules with which we start. It is no more acceptable to use preconceived ideas about points and lines than it would have been in the dormitory governance example to bring in personal opinions about the people who were elected representatives. The logic of any argument used to establish a theorem in finite geometry should not be affected if "point" and "line" are replaced with "poke" and "loke."

In the same way, the rules with which we begin are to be regarded as arbitrary, just as arbitrary as the bus regulations in the nation of Imagin. The first axiom, asserting that all lines are collections of points, should not be regarded as any more obvious than the assertion that all routes are collections of cities. Again, temptations to regard them as obvious can be dispelled by remembering that they can at any time be replaced by statements about pokes and lokes.

What we have said about points and lines being undefined terms and about axioms being arbitrary statements is true about all forms of geometry. Finite geometries are so named because, while employing the terminology and methods familiar from Euclidean geometry, their axioms can be satisfied by a finite collection of lines and points. Thus the axioms stated in terms of pokes and lokes at the beginning of this section are satisfied by a collection of four points and six lines, which can be represented pictorially. A series of dashes connects points that are to be thought of as lying on the same line, but it is to be noted that each line, represented by the dashes, contains just two points.

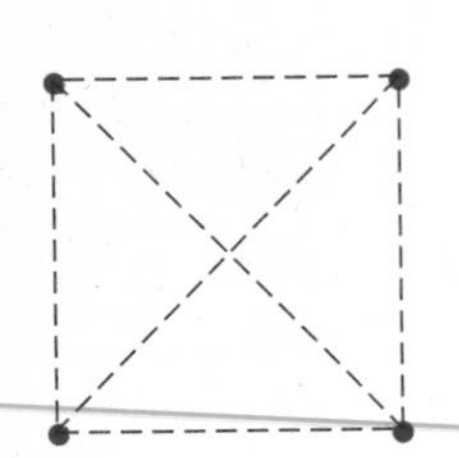

The problems at the end of this section give many examples of finite geometries, some of which have been studied extensively. The purpose of asking you to supply proofs is not to acquaint you with a lot of facts about finite geometry. Rather, writing proofs offers an unparalleled opportunity to practice saying precisely what we mean in a way that cannot be misunderstood. This is an opportunity, as the accompanying short story suggests, that should not be allowed to go to the dogs.

One's first attempt to prove a theorem often includes going up some blind alleys, including extraneous facts not necessary for the proof, etc. The proofs for which we ask, then, do not usually represent one's first effort. No author should be satisfied with his first draft; neither should a proof maker.

We are saying that a good proof should consist of attractive, well-reasoned grammatically correct paragraphs. Every step should be clear, concise, as well as precise.

Many mathematicians have claimed that the writing of such a proof is for them an esthetically rewarding experience. In this spirit we suggest several theorems on which you can work with reasonable hope of success. But to give it a fair try, a proof must be your own—your own argument written in your own style. You can then judge for yourself whether or not you derive any satisfaction from such a project.

Gauss always strove to give his investigations the form of finished works of art. He did not rest well until he had succeeded, and hence he never published a work until it had achieved the form he wanted. He used to say that when a fine building was finished, the scaffolding should no longer be visible.

Sartorius Von Watterhouser

You know that I write slowly. This is chiefly because I am never satisfied until I have said as much as possible in a few words, and writing briefly takes far more time than writing at length.

Carl Gauss

From the Minutes of a Borough Council Meeting:

Councillor Trafford took exception to the proposed notice at the entrance of South Park: "No dogs must be brought to this Park except on a lead." He pointed out that this order would not prevent an owner from releasing his pets, or pet, from a lead when once safely inside the Park.

The Chairman (Colonel Vine): What alternative wording would you propose, Councillor?

Councillor Trafford: "Dogs are not allowed in the Park without leads."

Councillor Hogg: Mr. Chairman, I object. The order should be addressed to the owners, not to the dogs.

Councillor Trafford: That is a nice point. Very well then: "Owners of dogs are not allowed in this Park unless they keep them on leads."

Councillor Hogg: Mr. Chairman, I object. Strictly speaking, this would prevent me as a dog-owner from leaving my dog in the backgarden at home and walking with Mrs. Hogg across the Park.

Councillor Trafford: Mr. Chairman, I suggest that our legalistic friend be asked to redraft the notice himself.

Councillor Hogg: Mr. Chairman, since Councillor Trafford finds it so difficult to improve on my original wording, I accept. "Nobody without his dog on a lead is allowed in this Park."

Councillor Trafford: Mr. Chairman, I object. Strictly speaking, this notice would prevent me, as a citizen, who owns no dog, from walking in the Park without first acquiring one.

Councillor Hogg (with some warmth): Very simply, then: "Dogs must be led in this Park."

Councillor Trafford: Mr. Chairman, I object: this reads as if it were a general injunction to the Borough to lead their dogs into the Park.

Councillor Hogg interposed a remark for which he was called to order; upon his withdrawing it, it was directed to be expunged from the Minutes.

The Chairman: Councillor Trafford, Councillor Hogg has had three tries; you have had only two. . . .

Councillor Trafford: "All dogs must be kept on leads in this Park."

The Chairman: I see Councillor Hogg rising quite rightly to raise another objection. May I anticipate him with another amendment: "All dogs in this Park must be kept on the lead."

This draft was put to the vote and carried unanimously, with two abstentions.

R. Graves and A. Hodge, *The Reader over Your Shoulder*,
(New York: Macmillan, 1961), pp. 149–150.

Mathematics must be beautiful . . . a mathematical proof should resemble a simple and clear cut constellation, not a scattered cluster in the Milky way.

G. H. Hardy

Summary

The strength of a logical argument does not depend on the meaning of the words used to name the objects of discussion. It does depend on the way the objects of discussion are related to one another. This dependence on interrelationships is sometimes easier to see if we can designate the objects of our discussion by names that do not carry any previous connotations; and because we work to free ourselves of the multiple, emotional connotations of normal language, mathematical argument affords us a unique opportunity to develop skills in clear communication.

Self-Test

Choose the completion that seems best to you. If none of them seem quite correct to you, be prepared to tell why and to supply an ending of your own in class discussion.

1. If we had used points and lines instead of representatives and committees in our discussion of dormitory governance
  (a) the proofs would have been easier; it would, for example, have been obvious that every line contains a point.
  (b) we would probably have gotten more confused, not realizing that we needed to prove statements that looked obvious.
  (c) we could at least have had the help of a picture because we know how to draw points and lines.

2. We have quoted Russell as saying "Obviousness is always the enemy of correctness."
  (a) This no doubt reflects his political views in which he scorned those who propose simplistic solutions to complex problems.
  (b) It is the spirit of this remark which led us to suggest that the theorems of finite geometry might be easier to prove if we refer to pokes and lokes instead of points lines.
  (c) By this he means that a deductive argument, if it is to be correct by a logician's standards, must be phrased in ways that will appear to be anything but obvious to the nonlogician.

3. How would you describe a finite geometry?
  (a) It is a study in which we concentrate our attention on just a finite subset of the points in the plane.
  (b) It might be misleading to use the term geometry. Perhaps it should be called the study of finite logical systems, since it is the study of a finite collection of elements related to each other by a few rules or axioms.
  (c) We temporarily blind ourselves to the fact that lines contain an infinite number of points, thereby enabling ourselves to focus on the finite number of points essential to a particular proof.

Problem Set 10.2

In Problems 1 through 3, we give the axioms for a finite geometry and then list several theorems in an order in which they can be proved. In attempting to prove a theorem, assume that all theorems listed previously in a problem have been proved. Thus inability to prove Theorem T4 in a list does not preclude trying to prove Theorem T5, etc. One problem, completed in the spirit we have described, (and illustrated with our proof of consequence 4 in this section) is a significant piece of work, worth more than hasty notes scribbled out as "solutions" to a number of problems.

1. ***Axioms:***

A1. There is at least one line.
A2. Each line contains at least two points.
A3. Any two points lie on exactly one common line.
A4. Given any line L, there exists a point not on L.
A5. Given any line L and any point x not on L, there are at least two lines through x having no point in common with L.

Theorems:

T1. There are at least 3 points.
T2. There are at least 3 lines.
T3. There are at least 5 points.
T4. There are at least 10 lines.

Draw a figure (similar to the one on page 328) that represents this geometry.

2. ***Axioms:***

A1. If x and y are any two points, there is at least one line containing both x and y.
A2. If x and y are any two points, there is at most one line containing both x and y.
A3. If L and M are any two lines, there is at least one point that lies on both L and M.
A4. There are exactly three points on each line.
A5. If L is any line, there is at least one point not on L.
A6. There exists at least one line.

Theorems:

T1. If L and M are any two lines, there is at most one point that lies on both L and M.
T2. Two lines have exactly one point in common.
T3. If x is any point, there is at least one line that does not contain x.
T4. Every point lies on at least three lines.
T5. There are precisely seven points.

Draw a figure that represents this geometry.

3. ***Axioms:***

A1. If x and y are distinct points, there exists at least one line containing x and y.
A2. If x and y are distinct points, there exists not more than 1 line containing x and y.
A3. Given a line L not containing a point x, there exists 1 and only 1 line M containing x and no points of L.
A4. Every line contains exactly 3 points.

A5. Given a line L, there is a point x not on L.
A6. There exists at least 1 line.

Theorems:

T1. There exist exactly 9 points.
T2. There exist exactly 12 lines.
T3. Corresponding to each line L, there are exactly 2 lines having no points in common with L.
T4. If M and N are lines neither of which have any points in common with a third line L, then M and N have no points in common.

Draw a figure that represents this geometry.

4. Let us say that in a finite geometry, two lines are parallel if they have no points in common.
(a) In the example used at the start of this section, substitute "line" for "loke," and "point" for "poke," and use the word "parallel" to state rule 5a.
(b) State axiom A5 in Problem 1 using the word "parallel."
(c) State theorem T2 in Problem 2 in terms of parallel lines.
(d) State axion A3 and theorems T3 and T4 in Problem 3 in terms of parallel lines.

5. In seeking to write things, even nonmathematical things, in a way that is clear, it is often most helpful to use the symbols and terminology of mathematicians.
(a) Consider the rules for the governing council stated in Section 10.1. Try to state rule 5 without resorting to the use of names like A for the committee and x for the representative.
(b) The faculty by-laws of Macalester College include the following provision (written, it must be admitted, by a mathematician).

In any election where a committee of n members is to be elected, the Advisory Council shall submit a slate of $2n$ nominees. Voters shall list n choices in order of preference, and in the counting of votes, a nominee listed first on a ballot shall be awarded n points, the one listed second shall receive $n - 1$ points, etc. The n nominees with the highest totals shall be declared winners, with ties settled by subsequent elections between nominees not previously elected.

Try to state this without using n, $n - 1$, etc.

For Research and Discussion

An entertaining and informative book [L. Lieber, *The Education of T. C. Mits,* (Norton, 1994)] introduces a finite geometry of 25 points in which the letters from A to Y are arranged in 25 blocks. Obtain a copy of this book and see how far you can develop this geometry; that is, see how many of the assertions in the book you can prove, and see if you can discover a few theorems of your own.

10.3 The Axiomatic Method

The Thog Problem

In the designs in the margin, there is a particular shape and a particular color, such that any of the designs that has one and only one of these features is called a Thog. The black diamond is a Thog. What can you say, if anything, about whether each of the remaining three is a Thog?

—from Wasou, P. C. "Self-Contradiction" in *Thinking: Readings in Cognitive Science*

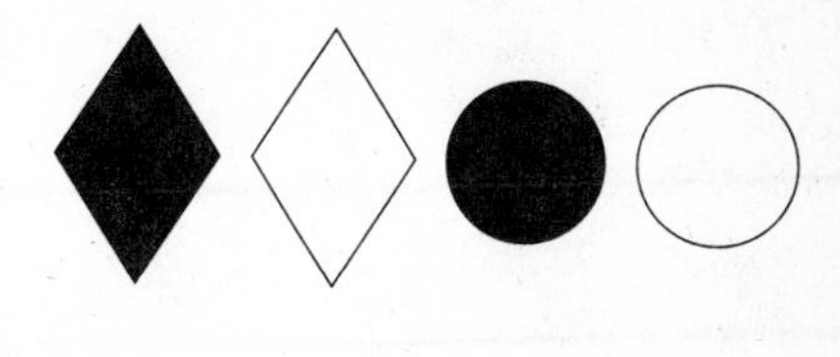

Undefined Terms

However clever we are, we don't learn a foreign language from a dictionary written in that language. A new term, whether in a foreign language or our own, must be explained to us in terms we already understand.

It's evident that the process of defining words in terms of other words must ultimately be circular. If we're to communicate with someone, we must give up the idea of defining each word or symbol. We have to depend on the other person sharing with us a common understanding of some familiar words. Stated more explicitly, some terms must be left undefined.

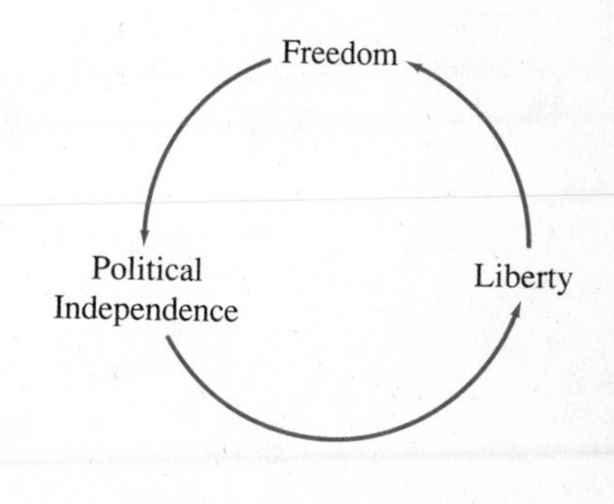

Recall that we felt it wise, when introducing finite geometry, to avoid using the words "point" and "line" because most people feel they know what a point is and what a line is. As we tried to show, however, such an intuitive notion is not necessary to prove the theorems. "Poke" and "loke" will do just as well. Perhaps "poke" and "loke" will do better, since there is no need to overcome an "It's obvious" feeling when asked to prove that every loke contains at least four pokes.

[Freedom is] the right to complain about the lack of it.

Will Rogers

Any logical discussion, mathematical or not, must ultimately rest on terms left undefined. The familiar cry, "Define your terms," if pushed too far, will lead to absurdity; and it is the one making the demand who is being absurd.

When I use a word, it means what I wish it to mean, neither more nor less.

Humpty Dumpty

Axioms

It is also clear from our work with finite geometry that nothing can be proved about points and lines (pokes and lokes) until some rules are stated which the points and lines are presumed to obey. When we discussed deductive reasoning, the only method of proof acceptable in mathematics, we emphasized that the method could not get off the ground unless there were previously accepted principles from which to work.

If we are to avoid circular reasoning, we must simply begin somewhere with statements we don't prove. These beginning statements are called **axioms.** We don't prove axioms any more than we try to prove the rules of chess. This is not because we are lazy; and it is not because they are obvious. We don't prove axioms because it can't be done. You can't prove everything.

We try to minimize the assumptions we make. In the finite geometry studied in Section 10.2, we assumed there were at least two points. We later proved that the axioms taken together implied that there were at least four points. We would have been studying the same geometry therefore if we had taken as our initial assumption the existence of at least four points. In stating axioms for a given system, however, the usual goal is to be able to derive all the theorems assuming as little as possible. However, the goal of reducing the assumptions to a bare minimum is sometimes set aside in order to make them as simple as possible to understand. For instance, the axioms given to a high school geometry class may include for clarity assertions that could be omitted and then derived as theorems. The effort to reduce and simplify the axioms is in the final analysis a matter of taste.

There is another consideration in setting up a system of axioms, which is decidedly not a matter of taste. We would like to be certain that it is not possible, by reasoning from the given axioms, to prove contradictory statements. If, in the axiom system discussed in Section 10.2 and in the previous paragraph, we had included an axiom stating that there were no more than three points, then (although we might not have immediately realized it), our axiom system would have been contradictory. The object is to avoid stating axioms that are contradictory.

Theorems

When the axioms have been stated, we can begin, in the manner of deductive reasoning, to deduce their implications. Guessing at these implications, the theorems of the subject, requires insight and imagination. A theorem is often named for the person who first suggests it, rather than for the one who first proves it.

Many mathematicians have reflected on the question of how one discovers a theorem. When asked how he made his discoveries. Newton said, "By always thinking unto them."

I believe that Newton could hold a problem in his mind for hours and days and weeks until it surrendered to him its secret. Then being a supreme mathematical technician he could dress it up, how you will, for purposes of exposition, but it was his intuition which was pre-eminently extraordinary—"so happy in his conjectures," said de Morgan, "as to seem to know more than he could possibly have any means of proving."

John Maynard Keynes

The mathematician Descartes is said to have done much of his work while resting in bed in the morning, a lifelong habit developed of necessity when he was a sickly child and apparently cultivated out of a desire for privacy when he became a mature thinker.

It takes keen insight to propose a theorem, and it often takes admirable ingenuity to offer a proof. But once a theorem is proved, anyone willing to take for granted the previously proved theorems, as well as the axioms, should be able to follow the proof step by step.

The Thog Problem

Fill in as much as you can.

Now, what about those Thogs? Are we handicapped by not knowing the precise definition of a Thog? Let's not dwell on what we don't know, thinking instead about what might be desirable features. The black diamond has one and only one of these features. Let's call an object a Hog if it has both desirable features, one of which is either to be black or to be diamond-shaped. Thus, a Hog is either a black circle or a white diamond.

Possibility 1 A Hog is a black circle. Then the black diamond and the white circle are both Thogs; and we might as well call the white diamond a Dog.

Possibility 2 A Hog is a white diamond. Then the black diamond and the white circle are Thogs, and the black circle becomes the Dog.

Either way, we see that besides the black diamond, the white circle is a Thog, and the other two are not Thogs.

Pure mathematics consists entirely of such assertions as that, if such and such a proposition is true of anything, then such and such another proposition is true of that thing. It is essential not to discuss whether the first proposition is really true, and not to mention what the anything is of which it is supposed to be true . . . . If our hypothesis is about anything and not about some one or more particular thing, then our deductions constitute mathematics. Thus mathematics may be defined as the subject in which we never know what we are talking about, nor whether what we are saying is true.

Bertrand Russell

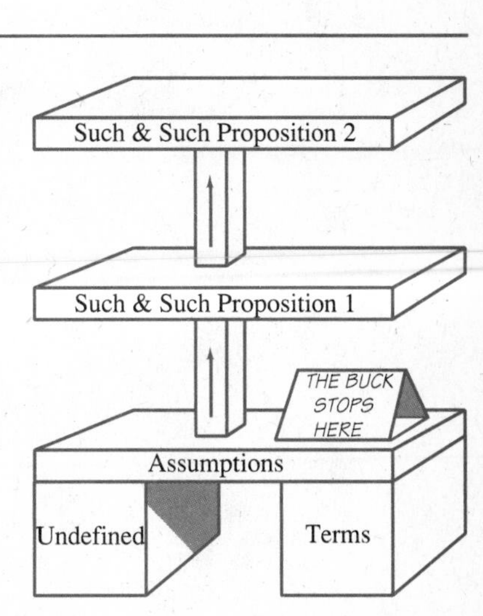

Summary

In any logically organized argument, there must be some terms that are not defined and some assumptions that are not proved. The trick is to be certain that one cannot deduce contradictory statements from the assumptions. Beyond this, it is desirable to keep the assumptions to a minimum. One form of mathematical insight is the ability to discover the implications (the theorems) that follow from the assumptions. These things, undefined terms, assumptions that relate them in a logically consistent way, and the implications of the assumptions, constitute what is called an axiomatic system.

Self-Test

Choose the completion that seems best to you. If none of them seem quite correct to you, be prepared to tell why and to supply an ending of your own in class discussion.

1. If we hope to organize any body of information into a systematic logical system.
   (a) all of our terms should be defined as carefully as possible, since failure to use terms in the same way is a root cause of much misunderstanding.
   (b) we may as well face from the beginning the impossibility of defining all of our terms.
   (c) then the arguments of this section underscore the difficulty we shall have in defining our terms, but do not excuse us from trying.

2. The critical question to ask about a theorem is:
   (a) Does it follow, without exception and by the accepted rules of deductive argument, from the axioms?
   (b) Is it true?
   (c) Who discovered it?

3. An axiom is a statement
   (a) so obvious that people are willing to accept it without proof.
   (b) which is simply assumed to be true.
   (c) which cannot be proved to be either true or false.

Problem Set 10.3

1. Illustrate the futility of trying to define everything by finding in your dictionary a series of definitions that take you in a circle. A simple way to find examples in most dictionaries is to look up a unit of foreign currency: krone, krona, guinea, etc.

2. Another problem with definitions, not discussed explicitly in the text, is that words often have a multiplicity of meanings. The problem can be illustrated:
   (a) Write down all the definitions you can think of for "fair."
   (b) Consider now the possible meanings of the sentence, "She is fair." Construct other examples, beginning with a familiar word having multiple meanings.

3. In the light of this section, how might you respond to the question, What is truth?

4. In any discussion between two people on a given subject, certain of the basic words used simply have to be accepted as being commonly understood. What might these terms be in a discussion of religion? Economics? Political ideology?

In Problems 5 through 8, we give a set of hypotheses. In each case give

(a) A conclusion that is valid in terms of the given hypotheses, but one that not everyone accepts as true.

(b) A conclusion, the truth of which may seem so apparent to some as to lead them to claim (falsely) that the conclusion is valid.

The idea is of course to provoke discussion. Many answers are possible.

5. *Hypotheses:* Useful education should prepare a student for a job. A liberal arts education doesn't prepare a student for a job. This course is intended only for students pursuing a liberal arts education.

6. *Hypotheses:* Poor entertainment is a waste of time.
Entertainment that encourages us to identify with unrealistic life situations is poor entertainment.
Television serials cause us to identify with life situations that are unrealistic.

7. *Hypotheses:* Those who wish to improve themselves must be people who want to learn.
People who want to learn go to school.
Homer goes to school.

8. *Hypotheses:* People who have a kind nature show it by being helpful to children.
People who are helpful to children are able to exercise firm discipline.
Children find that Mr. Jones is very helpful.

9. The following statements are probably statements that you would call true. For each statement, write down a set of hypotheses from which the desired truth follows as a valid conclusion.
(a) You should not commit murder.
(b) You should not belch out loud at the table.
(c) You should exercise moderately each day.
(d) A mother cat should care for her kittens.

10. Follow the instructions in Problem 9.
(a) You should be willing to help a blind person across a street.
(b) You should not steal another student's bicycle.
(c) Smoking cigarettes is bad for your health.
(d) To be a good citizen, you must vote.

For Research and Discussion

We holds these truths to be self-evident, that all men are created equal, that they are endowed by their Creator with certain unalienable Rights, that among these are Life, Liberty and the pursuit of Happiness.

The Declaration of Independence

In the beginning God created the heaven and the earth.

Genesis 1:1

Some commentators have found it a useful approach to analyze basic documents of a country, or a religious faith, or of any social structure in terms of the axiomatic method. What are the undefined terms? What are the underlying assumptions for which no rationale is even attempted? What consequences are seen by the writers to follow? See Becker, C. L. *The Declaration of Independence* (New York: A. A. Knopf, 1942) for such an analysis of the US Declaration of Independence. Can you think of other documents that seem to lend themselves to this approach?

10.4 Models

The Boat Model

Some boys required 3 hours to go 20 miles downstream, and 10 hours to return. Assuming that they rowed at the same rate all the time, what was their rate of rowing in still water, and what was the rate of the stream?

Hart, W. W., *Essentials of Algebra,* Heath, 1943, p. 122

We have described an axiomatic system as a collection of undefined terms, assumptions called axioms, and derived consequences called theorems. We have stressed that theorems are to be proved on the basis of logic and are not to depend on preconceived notions we may have had about the undefined terms. The entire system may seem therefore to be artificial, as certain in its conclusions as human reason itself, but unrelated to anything in our world of experience. We now wish to show how axiomatic systems relate to our understanding of the world, and to correct impressions, so far allowed to grow, about how axiomatic systems actually evolve.

Models from Axiomatic Systems

If an axiomatic system is to say anything about the world around us, it is clear that we must begin by assigning meanings to the undefined terms. Moreover, this has to be done in such a way as to make the axioms into sensible statements which appear to be true. If we succeed in this, so that we are willing to consider the axioms true, it follows that the theorems will also be true. The artificial system that said nothing about the world of experience is thereby transformed into a system we think is at least a partial description of the world as it really is. We have a model.

Let us be specific. In Section 10.2 we have a set of axioms defining a finite geometry. They become sensible statements if we substitute "representative" and "committee" for the undefined terms "poke" and "loke." We may or may not find a legislative body somewhere for which these sensible statements appear to be true. If we find such a body, we will have a ready-made model for that particular situation. And if we don't find such a body, our axiomatic system will be in no way invalidated. Perhaps some other words ("city" and "route") can make the axioms into statements that appear to be true about something in the world around us. It is the potential of a given structure to serve as a model for many different real-world situations that makes the axiomatic method a versatile and attractive scientific tool.

In scientific work, it is not enough to be able to solve one's problems. One must also turn these problems around and find out what problems one has solved. It is frequently the case that, in solving a problem, one has automatically given the answer to another, which one has not even considred in the same connection.

Norbert Wiener

Axiomatic Systems from Models

Euclid thought his axioms were self-evident statements about the world as it really is. It is a relatively recent development to think of axioms as assumptions and, even when they are recognized as assumptions, a practicing scientist is likely to think of them as assumptions that describe the world of experience. Even the most abstract-looking axiom systems are more often than not generalizations of very common mathematical structures, structures first developed as tools for handling real-world problems.

For these reasons one often finds a mathematician asking if a theorem is true. Technically, it is meaningless to ask if an axiom or a theorem is true. (What is the sense of asking whether or not it is true that every loke contains three pokes?) But in practice, the working mathematician has in mind some model of the system under consideration. The question, correctly put, is, "With respect to the model that I have in mind, would the following statement be true?"

It is this interplay between an abstract axiomatic system and attempts to make it serve as a model that has enriched both mathematics and science. The concept of a "pure" mathematician constructing axiom systems without regard for the world of experience and an "applied" mathematician or scientist trying to find a model that explains his or her observations is a convenient distinction for discussion. It does not convey the way science has progressed historically, or the way in which mathematics has developed. Except for isolated instances, mathematical sciences have been carried forward by people seeking to use the mathematics already known and to develop new mathematics as it is needed to attack problems encountered in the world.

Models are tentative. It is to be stressed that our identification of undefined terms with real objects in the world, being done in such a way that the axioms *appear* to be true statements, involves a value judgment. What seems to be true to one observer may not seem true to another. What seems true at one period in history may not seem true at another. Theorems, which were certain so long as we worked in the artificial axiomatic framework where validity was the only concern, lose their certainty when taken as statements supposed to be true about the real world.

Models may have to be revised or abandoned for any number of reasons: persistent discrepancies between the model and what can be observed, predictions based on the model that don't square with experience, and new observations—perhaps made possible with the development of more sophisticated instruments. For these reasons, acceptance of any model should be tentative.

At this point an enigma presents itself which in all ages has agitated inquiring minds. How can it be that mathematics, being after all a product of human thought which is independent of experience, is so admirably appropriate to the objects of reality? Is human reason, then, without experience, merely by taking thought, able to fathom the properties of real things?

In my opinion the answer to this question is, briefly, this: as far as the propositions of mathematics refer to reality, they are not certain; and as far as they are certain, they do not refer to reality. It seems to me that complete clarity as to this state of things became common property only through that trend in mathematics which is known by the name of "axiomatics."

A. Einstein

The classic example of our continuing effort to model what we observe is the attempt to understand, explain, predict, and make use of the apparent motions of the stars and planets as viewed from the earth. Here we see models proposed, accepted, adjusted to conform to new observations, and ultimately abandoned in favor of new ones. The traditional reluctance to abandon models has in this case been compounded by religious and philosophic arguments as well.

A Classic Example

The creation of a coherent system to explain and predict the apparent motions of the planets and stars as viewed from the earth is certainly a significant triumph of human intellect. That the problem had attracted attention from the most ancient times and that the theory is still undergoing modification in this century give an indication of the enormous time and energy that have gone into its study.

Early views of a fixed and flat earth covered by a spherical celestial dome were studied by the Greeks, who devised a real model in the fourth century B.C. which accounted at least approximately for the rough observations then available. The earth was viewed as fixed with a sphere containing the fixed stars rotating about it. The "seven wanderers" (the sun, moon, and five planets) moved in between. The Greeks' concern was to construct combinations of uniform circular motions centered in the earth by which the movements of the seven wanderers among the stars could be represented. Each body was moved by a set of interconnecting rotating spherical shells. This system was adopted by Aristotle, who introduced 55 shells to account for observed motions. This real model based on geometry was capable of reproducing the apparent motions, at least to a degree consistent with the accuracy of the contemporary observations. However, since it kept each planet a fixed distance from the earth, it could not account for the varying brightness of the planets as they moved.

This system was modified by Ptolemy, the last great astronomer at the famous observatory at Alexandria, in the second century A.D. In its simplest form the Ptolemaic system can be described as follows: Each planet moved in a small circle (epicycle) in the period of its actual motion through the sky, while simultaneously the center of this circle moved around the earth on a larger circle. The basic model was capable of repeated modification to account for new observations, and such modifications in fact took place. The result was that by the thirteenth century the model was extremely complicated, 40–60 epicycles for each planet, without commensurate effectiveness.

By the beginning of the sixteenth century there was widespread dissatisfaction with the Ptolemaic system. Difficulties resulting from more numerous and more refined observations forced repeated and increasingly elaborate revision of the epicycles on which the Ptolemaic system was based. As early as the third century B.C. certain Greek philosophers had proposed the idea of a moving earth, and as the difficulties with the Ptolemaic point of view increased, this alternative appeared more and more attractive. Thus in the first part of the sixteenth century the Polish astronomer Copernicus proposed a heliocentric (sun-centered) theory in which the earth, among the other planets, revolved about the sun. However, he retained the assumption of uniform circular motion—an assumption with a purely philosophical basis—and consequently he was forced to continue the use of epicycles to account for the variation in apparent velocity and brightness of the planets from the earth.

The next step, and a very significant one, was taken by Johannes Kepler. During the years 1576–1596 a Swedish astronomer, Tycho Brahe, had collected masses of observational data on the motion of the planets. Kepler inherited Brahe's records and undertook to modify Copernican theory to fit observations. He was particularly bothered by the orbit of Mars, whose large eccentricity made it very difficult to fit into circular orbit-epicycle theory. He was eventually led to make a very creative step, a complete break with the circular orbit hypothesis. He posed as a model for the motions of the planets the following three "laws":

1. The planets revolve around the sun in elliptical orbits with the sun at one focus (1609).
2. The radius vector from the sun to the planet sweeps out equal areas in equal times (1609).
3. The squares of the periods of revolution of any two planets are in the same ratio as the cubes of their mean distances to the sun (1619).

These laws are simply statements of observed facts. Nevertheless, they are perceptive and useful formulations of these observations. In addition to discovering these laws, Kepler also attempted to identify a physical mechanism for the motion of the planets. He hypothesized a sort of force emanating from the sun which influenced the planets. This model described very well the accumulated observations and set the stage for the next refinement, due to Isaac Newton.

All models developed up to the middle of the seventeenth century involved geometrical representations with minimal physical interpretation. The fundamental universal law of gravitation provides at once a physical interpretation and a concise and elegant mathematical model for the motion of the planets. Indeed, this law, when combined with the laws of motion, provides a description of the motion of all material particles. The law asserts that every material particle attracts every other material particle with a force which is directly proportional to the product of the masses and inversely proportional to the sqaure of the distance between them. In this framework the motion of a planet could be determined by first considering the system consisting only of the planet and the sun. The latter problem involves only two bodies and is easy to solve. The resulting predictions, the three laws of Kepler, are good first approximations since the sun is the dominant mass in the solar system and the planets are widely separated. However, the law of gravitation asserts that each planet is, in fact, subject to forces due to each of the other planets, and these forces result in perturbations in the predicted elliptical orbits. The mathematical laws proposed by Newton provide such an accurate mathematical model for planetary motion that they led to the discovery of new planets. One could examine the orbit of a specific planet and take into account the influence of all the other planets on this orbit. If discrepancies were observed between the predictions and observations, then one could infer that these discrepancies were due to another planet, and estimates could be obtained on its size and location. The planets Uranus, Neptune, and Pluto were actually discovered in this manner. However, even this remarkable model does not account for all the observations made of the planets. Early in this century small perturbations in the orbit of Mercury, unexplainable in Newtonian terms, provided some motivation for the development of the theory of relativity. The relativistic modification of Newtonian mechanics apparently accounts for these observations. Nevertheless, one should not view this model as ultimate, but rather as the best available at the present time.

The laws of Newton, viewed as a mathematical model, have provided an extremely effective tool to the physical sciences. The concepts of force, mass, velocity, etc., can be made quite precise and the model can be studied from a very abstract point of view. Although the social and life sciences do not yet have their equivalents of Newton's laws, the utility of mathematical models in the physical sciences gives hope that their use may contribute to the development of other sciences as well.

Daniel P. Maki and Maynard Thompson, *Mathematical Models and Applications,* (Englewood Cliffs, N.J.: Prentice-Hall, 1973) pp. 7–9. Reprinted by permission of Prentice-Hall, Inc.

Models in Science

We have made virtue of necessity. The features of the axiomatic method forced on us by logic turn out to be exactly the features desired for a scientific model.

Because some terms are left undefined, an abstract axiomatic structure is adaptable to the needs of scientists with widely varying interests. Points and lines may be variously interpreted as representatives and committees, cities and routes, dots and segments, stakes and chains, stars and light paths, etc.

A great step forward was taken in science when it became clear that the principles guiding work in a given area of science are not laws of the universe, but assumptions we are making. They are not self-evident truths; they are our present perceptions. This attitude toward basic principles is obviously in accord with the idea of writing down some axioms to begin with. It is instructive to note that this is precisely the way in which Newton began his epochmaking work, *Principia.* He refused, even when urged, to offer explanations as to why certain assumptions (for example, his dictum, "To every action there is an equal and opposite reaction") should be so. He patterned his work after Euclid, merely stating his principles as axioms and then showing their consequences.

We have emphasized that the axiomatic method is utterly dependent on human reason in deriving the implications of the axioms. Some philosophers of science have seen this as a limitation, arguing that the universe may be a chaotic place and that our method of studying it limits us to understanding only those things that happen to fall into patterns agreeable to our thinking processes. But others see faith in an orderly universe as the very backbone of science. They point out that one of the compelling reasons for abandoning the Ptolemaic model of astronomy was that, with its increasingly complex system of 40 to 60 epicycles for each planet, there developed a feeling that things just couldn't be that complicated. Poincaré adds to the argument for simplicity the example of someone who has taken a series of readings and

Without the belief that it is possible to grasp the reality with our theoretical constructions, without the belief in the inner harmony of our world, there could be no science.

L. Infeld and A. Einstein

plotted points (as in the figure). "Who," he asks, "would fail to draw through those points as smooth a curve as possible?" And what does this smooth curve represent except the experimenter's fundamental faith in an orderly universe?

The possibility of deducing implications merely by thinking is of course one of the principal attractions of the axiomatic method to a scientist. The discoveries of the planets Uranus, Neptune, and Pluto by analysis of Newton's model stand as remarkable examples of the power of the method. Not to be overlooked, however, are the hundreds of less celebrated accomplishments achieved in the same way. The marvels of electronics in communications owe much to the analysis of models; so do bridges, electric coffee pots, internal-combustion engines, vaccines, atomic power, wristwatches, etc. We seldom appreciate the power of a method that enables us to predict how things will turn out without always having to try them.

More laws are vain where less will serve.

Robert Hooke

The grand aim of all science is to cover the greatest possible number of empirical facts by logical deduction from the smallest possible number of hypotheses or axioms.

A. Einstein

Let us comment on the tendency in axiomatic thinking to want to minimize the number of axioms we use. While no logical difficutly is encountered if we state as axioms things that can be proved as theorems, this is not esthetically pleasing; and much effort has been directed toward reducing the number of axioms needed in various mathematical structures. Such efforts have practical benefits of course; when a structure is being considered as a possible model in science, the fewer axioms to be checked as to whether they seem to be true, the better. But beyond this, the effort to state the minimal number of axioms goes to the heart of what many see as the ultimate aim of sicence. To those who believe that science is possible because the universe is orderly, no goal less than discovery of the basic (and presumably simple) underlying causes of what we observe is worthy of the scientific effort.

Another Classic Model

We should not abandon our discussion of classic examples of mathematical modeling without referring to **The Boat Model** that opened this section. It is in its own way classic, some variation of it having appeared in algebra books studied by generations of students. The version used at the beginning of this section comes from the text from which the author studied in high school.

The problem invariably appears in the section on linear equations in two variables. It is intended that the student will let s = the speed of the stream, r = the rate of rowing in still water, and use the information given to write

$$3(r + s) = 20$$
$$10(r - s) = 20$$

Solution of this system gives $r = \frac{13}{3}$, $s = \frac{7}{3}$ miles per hour.

The old texts did not mention the modeling aspects of these problems, but that was opportunity lost. Indeed, the greater lesson of the problem is lost, for the method of actually solving the problem (deliberately suppressed here) is surely forgotten by vast numbers of students who might have remembered if it had been stressed that

(a) we begin with some data from the field that is, in all probability, not precise.
(b) we identify our symbols r and s with ideas that have meanings in the world of our experience.
(c) we make numerous simplifying assumptions (a boat can be rowed at a constant rate for hours, the rate of the current is the same everywhere in the river, the boat gets rowed in a straight line, etc.).

It has become quite fashionable to denigrate such problems in recent times for their artificiality, but this sort of "back of the envelope" figuring is used everyday by people estimating their arrival time as they cruise down a freeway and anticipate moving at a different rate once they hit the city, or as they estimate on the basis of time elapsed so far on a repetitious job how much longer it will take. The essence of modeling is to get an estimate of what might really happen, with accuracy that is suitable to the purposes intended.

And Finally

For illustrative purposes, we have confined our discussion of modeling to either artifically simple situations or to the classic examples. Most scientists are concerned with something in between. An engineer for a power utility wants to know before a multimillion dollar atomic generating plant is built on a certain river whether or not the plant can be operated without heating the river above levels specified by environmental control regulations. This task requires more than well-intentioned rhetoric about the value of rivers; it calls for a model that will predict the river temperature under a variety of conditions. The variables that may be involved are numerous (air temperature, river temperature upstream from the plant, humidity and wind conditions, cloud cover, accumulated ground heat, river flow in cubic units per second, electrical demands at various times of the day, etc.), and the way in which they enter into determining the river temperature is open to speculation. (How much attention, if any, must be paid to surface evaporation on a dry, windy day?) Anyone facing such a problem develops a quick appreciation for the art of making simplifying assumptions, a willingess to accept approximate solutions, and an admiration for those who have solved similar problems with enough generality so that their methods can be easily adapted to a new situation.

In the same way [as an artist] the researcher . . . depicts the world in scientific laws and concepts.

Siu

Modeling plays an increasingly important role in the social sciences. Pollsters and political analysts try to develop models that will enable them to predict public preferences, whether for toothpaste or for political candidates. Economists seek models for an economy that will enable them to predict the effect of certain monetary policies. Psychologists want models for human learning patterns (so for some reason they study rats); and on it goes.

The mere collection of facts is not science. An investigation takes on the characteristics of a scientific endeavor when enough is known so that certain hypotheses can be formulated, that is, when a model can be constructed and subjected to testing.

Summary

A mathematical model is obtained from an axiomatic system by assigning meanings to the undefined terms in such a way as to make the axioms appear to be true statements. Axiomatic systems do not grow in a vacuum; they commonly develop as attempts to model something observed in the world around us.

Self-Test

Choose the completion that seems best to you. If none of them seem quite correct to you, be prepared to tell why and to supply an ending of your own in class discussion.

1. Which of the following situations does not illustrate that people respond to a "belief in the inner harmony of our world"?
(a) The reasons for abandoning the Ptolemaic model of the motions of the planets and stars.
(b) The idea that we perceive order in the world because our thinking processes are orderly.
(c) The tendency to draw as smooth a curve as possible through plotted points obtained as readings in an experiment (and perhaps even to apply a "fudge factor" to the points that don't "line up").

2. Scientific investigation
(a) is ultimately based on assumptions.
(b) is the one area of human activity in which we can have a high degree of certainty.
(c) is so highly specialized that work in one area rarely relates to work in another area.

3. The distinction between the pure mathematician who works on axiomatic systems and the applied mathematician who uses the axiomatic systems to model natural phenomena
(a) was easier to see in historic times when day-to-day affairs were not so intertwined with technology.
(b) is more evident since our understanding of the axiomatic method has isolated it as a topic of study.
(c) has never been accurate, and serves only as a convenience for discussing the nature of scientific inquiry.

Problem Set 10.4

1. Using a simplified model, calculate the
   (a) Volume of your arm from your wrist to your elbow.
   (b) Volume of your head.
2. Using a simplified model, calculate the
   (a) Area of your arm (wrist to elbow).
   (b) Area of your head.
3. Estimate the number of apples in a bushel.
4. Estimate the number of marbles with a $\frac{1}{2}$ inch diameter that could be placed into a 1 gallon jug.
5. Jack Strop finds that it takes 45 seconds to fill a $2\frac{1}{2}$-gallon bucket with his garden hose. How long will it take him, using the same hose, to fill his new circular 30-foot-diameter plastic pool to a depth of 4 feet?
6. Estimate the cost of fuel needed to drive a medium size automobile from Chicago to Los Angeles.
7. A globe is of course a model of the earth. Use a globe to estimate the shortest distance (if you are flying) between the following cities.
   (a) San Francisco and Cape Town.
   (b) Calcutta and Los Angeles.
   (c) New York and Sydney.
   (d) New Orleans and Tokyo.
8. Again using a globe, estimate the distance, traveling by boat, between the cities listed in Problem 7.
9. A theorem of geometry tells us that, if the sides of a triangle have lengths a, b, and c related by $c^2 = a^2 + b^2$, the angle opposite the side of length c is a right angle. The fact is used in many models. How can "point" and "line" be interpreted in the following situations?
   (a) A carpenter is setting forms in which to pour concrete for a patio and wishes to verify that the planks are at right angles.
   (b) Several of the neighbors have gotten together to lay out a Little League baseball field. They seek a method for making sure the foul lines are at right angles.
10. We introduced our discussion of finite geometries by talking about rules for governing a dormitory, then by discussing bus company regulations, and finally by talking about points and lines. By assigning still other meanings to "point" and "line," other potential models can be obtained. Do so. (For example, would the axioms stated at the beginning of Section 10.2 appear to be sensible statements if "pokes" represented letters and "lokes" represented envelopes?)

SOME COMMON FORMULAS

A cube of side a has surface area S and volume V given by

$$S = 6a^2$$
$$V = a^3$$

A sphere of radius r has surface area S and volume V given by

$$S = 4\pi r^2$$
$$V = \frac{4}{3}\pi r^3$$

A cylinder of radius r and height h has surface area S (not counting the circular bases) and volume V given by

$$S = 2\pi rh$$
$$V = \pi r^2 h$$

$$1 \text{ bushel} = 2150 \text{ in.}^3$$
$$1 \text{ gallon} = 277 \text{ in.}^3$$

For Research and Discussion

Identify in a field of interest to you a model that was once accepted, but has since been set aside. This setting aside of one model (paradigm) for another has been likened to a political revolution in a book that is recommended reading for students in a wide variety of disciplines. See Kuhn, *The Structure of Scientific Revolutions,* 2nd ed., (The University of Chicago Press, 1970).

chapter 11

Geometries as Models

What are we to think of the question: Is Euclidean geometry true? It has no meaning. . . . One geometry cannot be more true than another; it can only be more convenient.

H. POINCARÉ

Beauty

Most people will readily agree that mathematics is useful, that it is important, or even that it represents a very high level of human achievement. But beautiful? Could anyone, other than perhaps an overzealous mathematician, describe the subject as beautiful? What is there that one would call beautiful?

Some would answer that as the language that enables us to apprehend the intricate workings of the world around us, mathematics is the source of the most fundamental beauty of all. Henry Poincare wrote:

> Without [mathematical analysis] most of the intimate analogies of things would have remained forever unknown to us; and we should have been forever ignorant of the internal harmony of the world . . . It is this harmony then which is the sole objective reality, the only truth we can attain; and when I add that the universal harmony of the world is the source of all beauty, it will be understood what price we should attach to the slow and difficult progress which little by little enables us to know it better.

Others find beauty in the arguments themselves, independent of whether the result has application or not. For them, there is beauty in a compelling argument, gracefully phrased to move purposefully through possible distractions to expose the nubbin of the matter. G. H. Hardy wrote:

> Mathematics must be beautiful . . . A mathematical proof should resemble a simple and clear cut constellation, not a scattered cluster in the Milky Way.

Sometimes it is the result itself that is called beautiful. Starting with the expression $(a + b)$, perform the repeated multiplications necessary to raise it to higher and higher powers:

$$(a + b) = 1a + 1b$$
$$(a + b)^2 = 1a^2 + 2ab + 1b^2$$
$$(a + b)^3 = 1a^3 + 3a^2b + 3ab^2 + 1b^3$$
$$(a + b)^4 = 1a^4 + 4a^3b + 6a^2b^2 + 4ab^3 + 1b^4$$
$$(a + b)^5 = 1a^5 + 5a^4b + 10a^3b^2 + 10a^2b^3 + 5ab^4 + 1b^5$$

Now focus attention on the coefficients of the expressions on the right side, arranging them in a symmetric pattern.

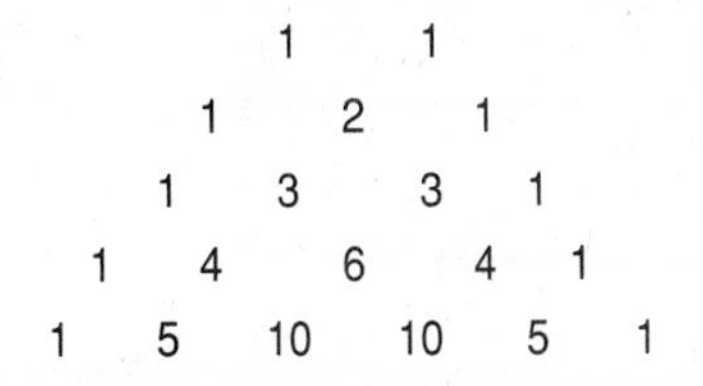

Suddenly, from what seemed a mindless repetitive exercise, there emerges an array of numbers that can be continued in an obvious way to a 6th line, a 7th line, etc., not by the laborious process of algebraic multiplications, but by the expedient observation that each number is the sum of the two numbers symmetrically placed on the line just above it.

There are other symmetries to notice. Examine the numbers lined up on any diagonal; look at the sum of the numbers on any horizontal line; keep looking. So rich in its properties as to have its own name (Pascal's Triangle), this set of numbers has been the subject of whole books. Quite apart from its usefulness in applied mathematics, it exhibits an intrinsic beauty of its own, the sort of thing W. F. White had in mind when he wrote:

> The beautiful has its place in Mathematics for here are triumphs of the creative imagination, beautiful theorems, proofs, and processes whose perfection of form has made them classic. He must be a 'practical' man who can see no poetry in mathematics.

The beauty of mathematics is not, of course, the beauty of a romantic symphony, nor the beauty of a rose that might be called gorgeous. But it is an achievement of human enterprise that Bertrand Russell compared to a beautiful sculpture.

> Mathematics possess not only truth, but supreme beauty—a beauty cold and austere, like that of a sculpture, without appeal to any of our weaker nature, sublimely pure, and capable of a stern perfection such as only the greatest art can show.

11.1 Euclid's Work

Equal Opportunity Triangles

Begin with any ΔABC. Draw as we have done in the figure at the left all the angle trisectors, and let D, E, and F be the intersections of adjacent trisectors. Then, starting with the same ΔABC, construct on each edge an equilateral triangle, having D, E, and F as centers. In each case, use the points D, E, and F as vertices of a triangle. What do these triangles have in common? Try the same constructions for a different initial ΔABC

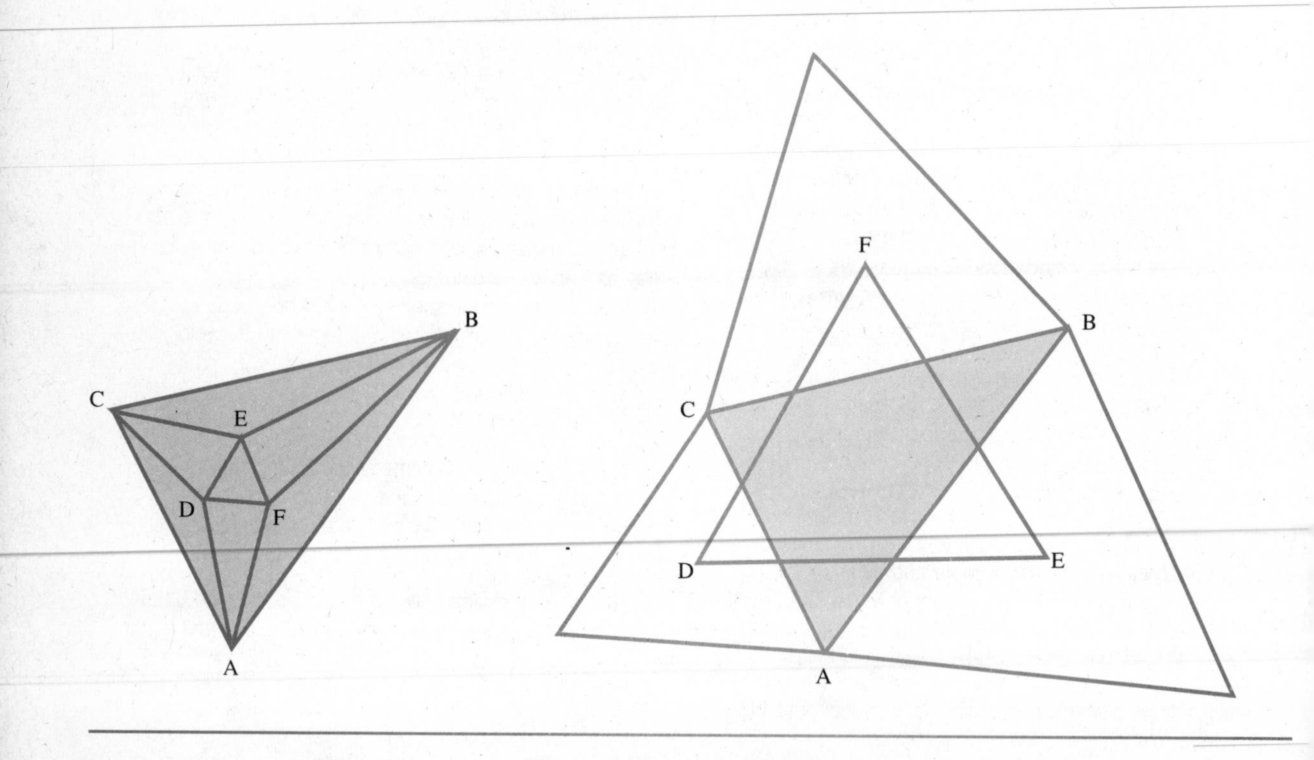

Except for the Bible, probably no book has had a greater influence on Western culture than Euclid's *Elements*. And, like the Bible, it is a book that modern readers are more likely to praise than to open. Our purpose here therefore is to open Euclid's book, at least to the extent of summarizing part of what it says.

The first surprise to most readers, if they were to actually look at Euclid's *Elements,* would be to find that geometry is not the only topic treated. Euclid was a collector rather than an originator. Surely he was a man of mathematical talent, but his great contribution was to organize previously known mathematics so that everything followed from a few assumptions. Euclid's *Elements* is really a collection of 13 books, only 5 of which are devoted to plane geometry. Other topics include proportions, number theory, and solid geometry.

Even the idea of beginning with assumptions did not originate with Euclid. Aristotle had discussed the nature of axioms somewhat earlier, and his influence can be seen in Euclid's work. For instance, Aristotle pointed out that one need not worry about the truth of the assumptions, but both he and Euclid surely felt that the things about which they wrote were the stuff of reality. If one did not need to worry about whether or not the assumptions were true, it was because the derived consequences could be checked to see if they agreed with reality.

Whatever he thought about their truth, Euclid did have in mind the goal of making as few assumptions as possible. This is clear because he went to the trouble of proving many theorems that seem no less self-evident than the statements he took as assumptions.

Euclid failed, however, to appreciate the necessity of leaving some terms undefined. Book I begins with 23 definitions, and his troubles are apparent from reading the first few:

A point is that which has no part.

A line is a breadthless length. The extremities of a line are points.

A straight line is a line that lies evenly with the points of itself.

Do these definitions really define? What is a "breadthless length"? What does it mean to "lie evenly"? And if these questions are answered, will the answers not use other terms needing definition, eventually leading us right back where we started?

A definition of a straight line that leads us in circles is indeed an unfortunate beginning for a book destined to serve for centuries as a model of logical thinking. This makes it easier for us, however, to agree with the modern viewpoint; some basic terms must be left undefined.

Euclid's Postulates (Axioms)

After listing a barrage of definitions, Euclid wrote down 10 grand assumptions. He called the first 5 "postulates" and the second 5 "common notions." We use the modern word "axiom" to describe them all. It is Euclid's choice of axioms that has stamped his name in history. Here they are, stated in their simple elegance.

Axiom 1 A straight line can be drawn from any point to any point.

Axiom 2 It is possible to extend a finite straight line indefinitely in a straight line.

Axiom 3 A circle can be drawn with any point as center and any radius.

Axiom 4 All right angles are equal.

Axiom 5 (Euclid's Famous Fifth, the Parallel Postulate) If a straight line intersects two straight lines so that the interior angles on the same side of it sum to less than two right angles, then the two lines, if extended indefinitely, will meet on that side on which the two angles are less than two right angles.

Axiom 6 Things that are equal to the same thing are also equal to one another.

Axiom 7 If equals be added to equals, the wholes are equals.

Axiom 8 If equals be subtracted from equals, the remainders are equal.

Axiom 9 Things that coincide with one another are equal to one another.

Axiom 10 The whole is greater than the part.

For 300 B.C., they are remarkably clear. But to help our readers, we suggest that Euclid meant Axiom 1 to say that a *unique* line (i.e., just one line) can be drawn between any two different points. Similarly Axiom 2 should say that any finite straight line has a *unique* extension. Finally Axiom 5 (the parallel postulate) became the subject of a controversy which continued for 2000 years, a controversy we discuss in Section 11.2. It was restated by John Playfair in a form most of us find more understandable.

Axiom 5 (after John Playfair) Given a line L and a point P not on L, there exists one and only one line M containing P and no points of L.

It is interesting to note that Euclid clearly distinguished between the definition of a figure and the proof from his axioms that such a figure can be constructed. This seemingly fine distinction led to some of the most famous problems in the history of mathematics, for it caused people to ask whether such and such a thing can be drawn using only a straightedge (to join any two points or to extend a segment indefinitely) and a compass (to draw a circle centered at any point, using any radius). Three of the most famous problems of this type are:

a. Trisect a given angle.
b. Construct a square with an area equal to the area of a given circle.
c. Construct a cube with a volume that is twice the volume of a given cube.

These problems attracted the efforts of many famous mathematicians, and many amateurs as well. Attempts to solve them have led to some very deep and useful mathematical ideas. They have also led to many wrong "solutions" and much frustration, for they are very hard. They are in fact worse than hard. We now know that they are all impossible. No one will ever solve any of them.

Euclid's Theorems

We list below several of Euclid's theorems. In stating these theorems, we assume the reader is familiar with definitions from elementary geometry (perpendicular and parallel lines, isosceles and equilateral triangles, etc.), and we have phrased them in ways that should sound familiar to the modern ear. Otherwise, the theorems and the order in which they are listed follow Euclid's *Elements*.

We have included proofs of several of the theorems for purposes of illustration. Most proofs are left as exercises for the reader.

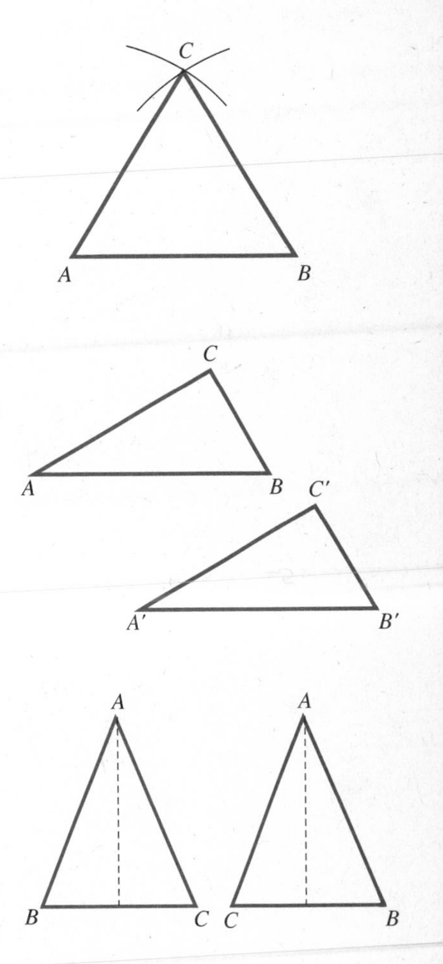

Theorem 1 On a given finite straight line, an equilateral triangle can be constructed.

PROOF Let the given line segment be AB. With A as center and AB as radius, construct a circle; construct a similar circle with B as center (Axiom 3). Let the circles intersect at C. Then ABC is the required triangle.

Theorem 4 (Side Angle Side) If two sides and the included angle of one triangle are equal to the corresponding parts of a second triangle, the triangles are congruent.

PROOF Let the two triangles be designated ΔABC and $\Delta A'B'C'$, the notation chosen so that $\angle A = \angle A', AB = A'B'$ and $AC = A'C'$. Imagine ΔABC moved so that the two equal sides and the equal included angles coincide. In particular, then B will coincide with B' and C with C'. Through these two points only one line can be drawn (this is where Euclid needs Axiom 1 in a form stating the uniqueness of the line). $BC = B'C'$ and the triangles coincide, as was to be proved.

Theorem 5 The base angles of an isosceles triangle are equal.

PROOF Let the given triangle be designated ΔABC, with $AB = AC$. The triangle can be "flipped" so that angle B lies where C was, and conversely. Then side AB corresponds to AC; and side AC corresponds to AB. Moreover, these corresponding sides and their included angles, being angle A in each case, are equal. Thus, by Theorem 4, the triangles ΔABC and ΔACB are congruent. It follows that the corresponding angles $\angle ABC$ and $\angle ACB$ are equal.

Theorem 6 If two angles of a triangle are equal, then the sides opposite these angles are also equal.

Theorem 8 (Side Side Side) If three sides of a triangle are equal to three corresponding sides of a second triangle, the triangles are congruent.

Theorem 9 Any angle can be bisected.

Theorem 10 Any given line segment can be bisected.

Theorem 11 If p is a point on a line l, then at least one line can be drawn through p perpendicular to l.

Theorem 12 If p is a point not on a line l, then at least one line can be drawn through p perpendicular to l.

Theorem 13 If two lines intersect, any two adjacent angles have a sum of 180° (two right angles).

Theorem 15 If two lines intersect, then the opposite angles (sometimes called vertical angles) are equal.

Theorem 16 If one side of a triangle is extended, then the exterior angle is greater than either of the opposite interior angles.

Theorem 17 The sum of any two angles of a triangle is less than 180° (two right angles).

Theorem 26 (Angle Angle Side) If two angles and a side of one triangle are equal to the corresponding parts of a second triangle, then the triangles are congruent.

Theorem 27 If a straight line falling on two other straight lines makes the alternate interior angles equal, then the two straight lines are parallel.

Theorem 29 If a straight line falls on two parallel lines, it makes the alternate interior angles equal to each other.

Theorem 30 Lines parallel to the same line are parallel to each other.

Theorem 32 The sum of the interior angles of a triangle is 180° (two right angles).

Theorem 47 (Pythagoras) In a right triangle, the square of the length of the hypotenuse is equal to the sum of the squares of the other two sides.

This last theorem is certainly one of the best known and most useful theorems in plane geometry. There are several independent proofs attributable to ancient writers, mathematicians ranging from the most prominent to the obscure, and even an American president.

A Proof in the Classical Style

Students in a plane geometry class are often required to use a format which emphasizes that each step must be justified by an axiom, a definition that has been given, or a theorem already proved. The method is illustrated by our proof that if it is given in the figure to the right that $AC = BC$ and $AD = BE$, then it follows that $CD = CE$.

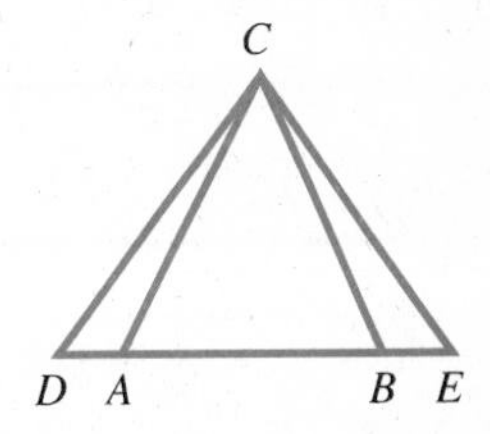

Statement	Reason
1. $AC = BC$ and $AD = BE$	1. Given
2. ΔABC is isosceles	2. Definition of isosceles
3. $\angle CAB + \angle CAD = 180°$ $\angle CBA + \angle CBE = 180°$	3. Theorem 13
4. $\angle CAB + \angle CAD = \angle CBA + \angle CBE$	4. Axiom 6
5. $\angle CAB = \angle CBA$	5. Theorem 5
6. $\angle CAD = \angle CBE$	6. Axiom 8
7. $\Delta CAD \cong \Delta CBE$	7. Theorem 4
8. $CD = CE$	8. Definition of congruent triangles

A Practical Result

We conclude this section with an example that deserves to be better known. Contrary to popular belief, it was well understood at the time of Columbus' voyage (1492) that the earth was spherical. Pythagoras (560–480 B.C.) taught that the earth was a sphere, and Aristotle (384–322 B.C.) argued not only for a spherical earth, but gave some (far too small) estimates of its size.

The first estimates we have that are accompanied by a description of their derivation are attributed to Eratosthenes (275–194 B.C.). His starting assumptions were

- The earth is spherical.
- Light from the sun travels in parallel rays.

The data reported (and cited by commentators as an early example of data "fudged" by later recorders) were as follows.

- Syene (the modern Aswan) is directly south of Alexandria (check a map; what do you think?)
- The distance from Syene to Alexandria is 5000 stadia (so we literally get a stadium rather than a ballpark estimate here) which is 500 miles.

- At the time of the summer solstice, the sun was directly overhead in Syene (which would mean Syene is on the Tropic of Cancer; check your map again) but appeared to be 1/50 of a great circle, that is about 360/50 degrees, off of the vertical in Alexandria (how did they measure this?).

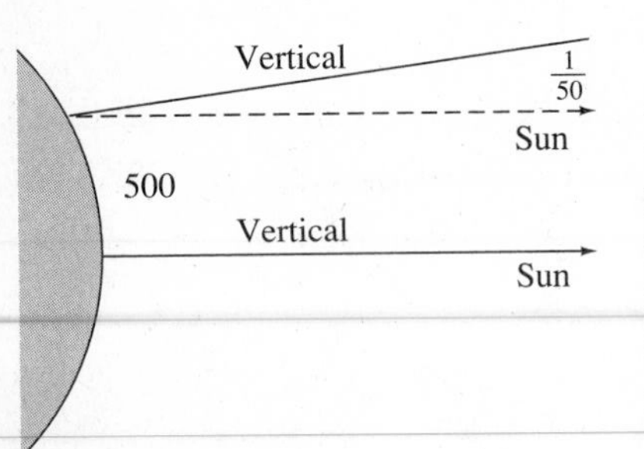

All of this is open to debate for the reasons we have been noting as we went along. That is, of course, a part of the modeling process. Assumptions about the world in which we live and measurements from the field are always open to speculation and improvement. But then comes the mathematical reasoning, based on the model, and this part is not questioned. Using the parallel lines provided by the sun's rays, the equality of vertical angles (Theorem 15), and the equality of alternate interior angles (Theorem 29), he would have arrived at the figure indicated.

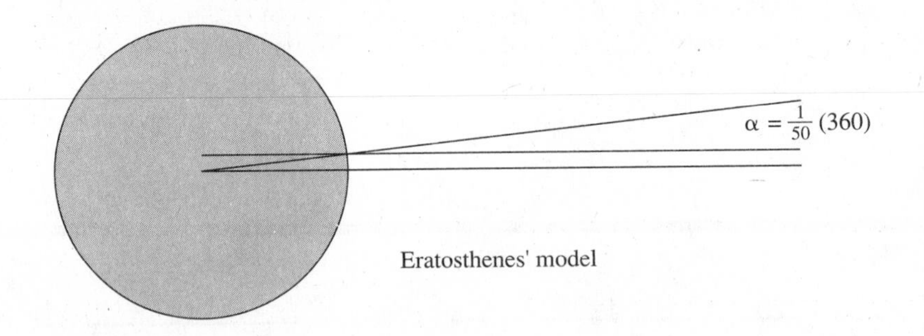

Eratosthenes' model

The angle at the center of the earth intercepts an arc of 500 miles. Working with proportions then, one writes

$$\frac{1}{50} = \frac{500}{\text{circumference}}$$

from which we get

$$\text{circumference} = 500(50) = 25{,}000 \text{ miles}$$

(Modern geographers believe the earth to be pumpkin-shaped but, taking it as a sphere for easy calculation, the accepted value for the circumference today is 24,890 miles.)

A Beautiful Result

We develop elsewhere the idea that those who create new mathematics explain their sense of satisfaction in terms not unlike those used by a poet, a composer, or a printer. But what of the things that the mathematician creates. Are they themselves the stuff of art? Can they awaken a sense of wonder, admiration, awe, in the eye of the beholder—especially the beholder who is not trained in mathematics?

The writer, unencumbered by any training or ability in the visual arts, well recalls coming upon *The Hay Wain* by John Constable in The National Gallery in London. Why does a scene in which there really is quite a lot of activity overwhelm me with such a sense of peace? Why is it that I am so impressed by the intricate detail in a picture that is more than six feet across? It was a mesmerizing moment that would not let go, remaining still a moment as my actual time in front of the picture stretched into minutes.

Is there anything so captivating to be found in mathematics? Edna St. Vincent Millay must have thought so when she wrote,

> "Euclid alone has looked on beauty bare."

What, one wonders, caught the fancy of the poet? It is like wondering what painting in The National Gallery would capture her fancy. I can only tell you which result I would take you, the reader, to see in the gallery of beautiful theorems in Euclidean geometry. I would point out Morley's Theorem.

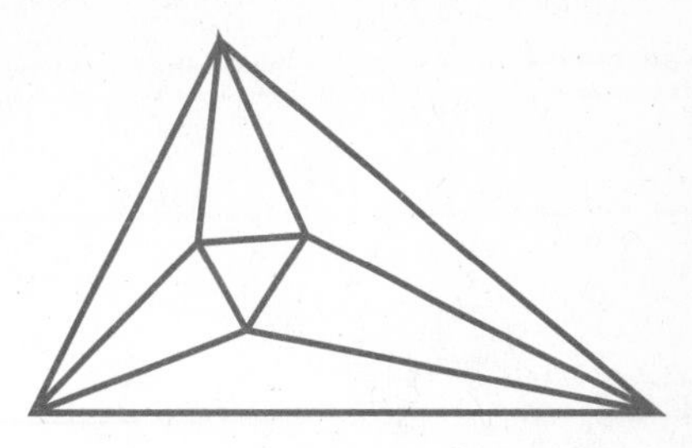

Morley's Theorem
Starting with an arbitrary triangle, draw all the angle trisectors. The intersections of adjacent trisectors form vertices of an equilateral triangle.

Summary

Most people in our society have heard of Euclid's *Elements,* but their misconceptions about it are many. You should now be able to give better answers to the questions below than might be expected from folks having only an ordinary education (i.e., not having read this book, or one like it).

What is the subject of Euclid's *Elements?*

Were most theorems of plane geometry known before the time of Euclid?

What is Euclid's chief contribution to the development of mathematics?

Does Euclid's work, as is sometimes alleged, continue to set the standard for logical exposition?

In Euclid's eyes, what was the point?

In addition to providing some insights into Euclid's work, this chapter also serves as a reminder of certain concepts such as congruent triangles, isosceles triangles, and right triangles, and facts associated with them.

Self-Test

Choose the completion that seems best to you. If none of them seem quite correct to you, be prepared to tell why and to supply an ending of your own in class discussion.

1. Euclid proved some statements that are no less obvious than others that he took for axioms.
(a) From this we deduce that Euclid understood that nothing is obvious.

(b) It is therefore clear that not everything which seems obvious to the modern mind seemed obvious in Euclid's time.
(c) Euclid does seem to have had in mind the goal of making as few assumptions as possible, and then proving everything else.

2. Euclid's first theorem asserted that on a given finite straight line, an equilateral triangle could be constructed.
(a) We now know that there is really nothing to prove here; this is simply a restatement of the definition of an equilateral triangle.
(b) This exemplifies the subtle but important distinction that Euclid made between defining a figure and proving that such a figure does exist.
(c) This theorem is technically wrong since two distinct equilateral triangles can be constructed, one on each side of the line.

3. Attempts by amateurs to trisect an arbitrary angle using only a compass and straightedge are routinely ignored by mathematicians
(a) because it is unlikely that an amateur will solve a problem that has defied mathematicians for centuries.
(b) because they know without reading them that they are wrong.
(c) because we now have highly accurate tools to do the job, rendering the restriction to compass and straightedge obsolete.

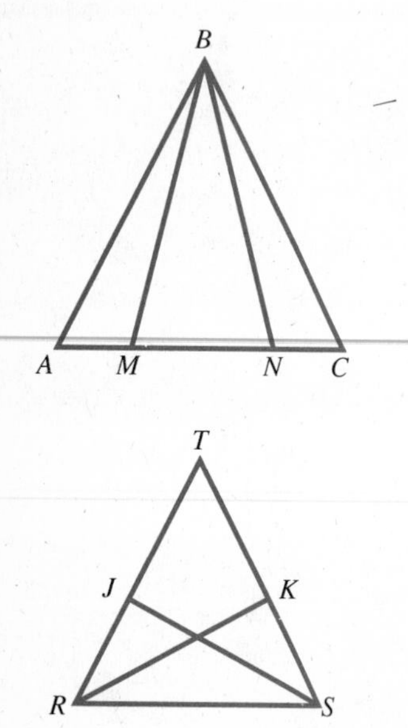

Problem Set 11.1

In the following problems, use any of Axioms 1 through 10 and any of the stated theorems to prove the result.

1. Two isosceles triangles are congruent if a leg and the angle opposite the base of one are equal to the corresponding parts of the other.

2. Two right triangles are congruent if a side and the hypotenuse of one equal the side and hypotenuse of the other.

3. If AM is the perpendicular bisector of BC, then ΔABC is isosceles.

4. The medians to the equal sides of an isosceles triangle are equal.

5. If ΔABC of the top figure in the margin is isosceles and $\angle ABM = \angle CBN$, then ΔBMN is isosceles.

6-17. *Prove Theorems 6 through 17 in this section. In proving any theorem on the list, you may assume previously listed theorems but not subsequent ones.*

18. In the bottom figure in the margin, $\angle TRS = \angle TSR$, and SJ and RK bisect angles S and R, respectively. Prove that $\angle TJS = \angle TKR$.

19. In the bottom figure in the margin, RK and SJ bisect the base angles of the isosceles triangle RST. Prove that $RJ = SK$.

For Research and Discussion

What do mathematicians believe to be the most beautiful results in their discipline? This question was posed in a survey conducted in 1988, and the results were reported in an article [Wells, D., "Are these the most beautiful?,"

The Mathematical Intelligencer, Vol. 12, No. 3, 1990, pp. 37–41] that also tried to identify the characteristics that make for beauty. If you consult the article, you will find a fair number of results known to you from reading this book, but (alas) you won't find Morley's triangle, nor will you find the theorem hinted at in the problem that heads this section, known in the literature as Napoleon's theorem. See if by consulting the article cited, conducting your own survey among the mathematics faculty at your school, or browsing through this or some other book, you can identify a theorem that strikes you as beautiful.

11.2 Non-Euclidean Geometry

A New Perspective

In projective geometry, parallel lines recede to an ideal point, also called the point at infinity Two triangles are said to be in perspective if their corresponding vertices each lie on a common line through the ideal point, as in the figure in the margin. Here is a chance to discover another beautiful theorem. What can you say about the corresponding sides of two triangles in perspective

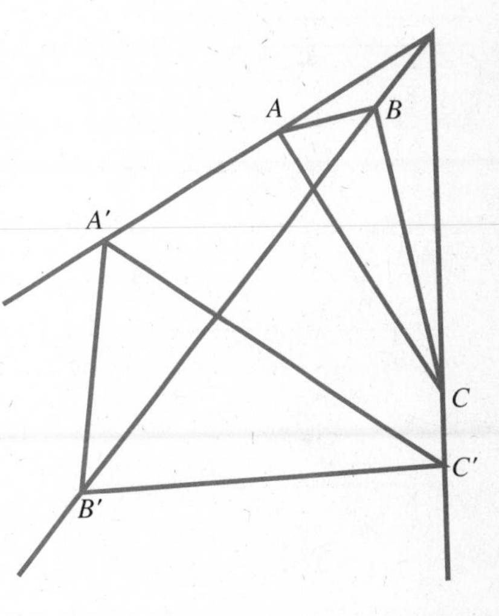

In many ways, the development of non-Euclidean geometry is one of the most significant of human achievements. Not only did it transform our understanding of geometry, but it affected our whole concept of science, and ultimately of reality.

Difficulties with the Parallel Postulate

From the very beginning, Euclid's fifth axiom was troublesome. The most casual reader of his *Elements* is certain to notice it just because it is much longer than the others; and if he goes the extra step of actually reading the axioms, he will see that it is much less intuitive than the other nine. Euclid himself leaves evidence of being dissatisfied with the parallel postulate, since he proceeds as far as he can without using it. Those who came after Euclid were also unhappy with this postulate, and their efforts to remedy the situation serve in some ways to underscore Euclid's genius in including it in the form that he did.

The mathematician Jean le Rond d'Alembert wrote in 1759 that the problem of the parallel postulate was "the scandel of the elements of geometry." In 1763 a mathematician named Klügel wrote a doctoral thesis exposing errors in 28 different "proofs of the parallel postulate."

There were two basic approaches to the parallel postulate from 300 B.C. to 1800 A.D. One idea was to deduce it as a theorem from the other nine axioms. The other was to replace it by something that would more directly appeal to intution. It was the latter effort that led Playfair to propose in 1795 the form mentioned in Section 11.1, and repeated here for convenience.

Playfair's Axiom Given a line L and a point P not on L, there exists one and only one line M containing P and no points of L.

One of the most systematic, and in a sense most fruitful, approaches to the problem was made by a Jesuit professor of mathematics named Girolamo Saccheri (1667–1733). His idea was to begin with all of Euclid's assumptions, except the troublesome Axiom 5, and to prove as many theorems as possible. This leads to what is called **absolute geometry.** The first 28 theorems of Euclid in which he avoided using Axiom 5 are therefore theorems of absolute geometry. Proceeding in this way, Saccheri eventually defined what is now called a **Saccheri Quadrilateral.** This is a quadrilateral $ABCD$ in which angle A and angle B are right angles and the lengths of sides AD and BC are equal; that is, $AD = BC$. The reader's immediate reaction is probably to call such a figure a rectangle. That, however, is a response that comes either from what seems apparent from a picture, or else from a theorem of Euclidean geometry. Neither reason is allowed here. The picture is, we recall, just one of many possible models of the axiom system; and to prove that the figure is a rectangle, one needs the very axiom that has been omitted in absolute geometry. Working with only the nine axioms of absolute geometry (which, as we have observed, allow us to use the first 28 theorems of Euclid), we prove the following.

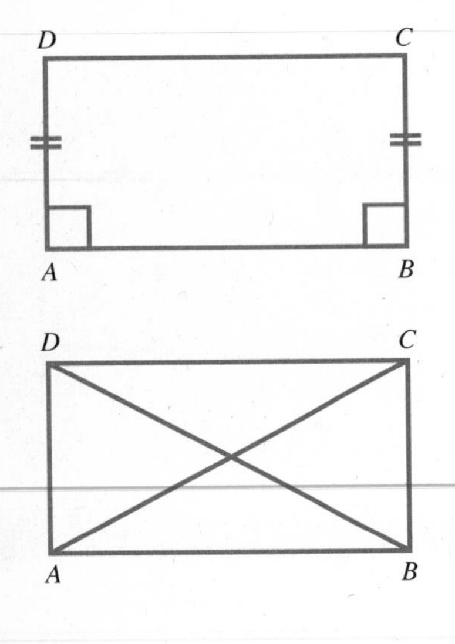

Theorem (Saccheri) If $ABCD$ is a Saccheri Quadrilateral, $\angle C = \angle D$.

PROOF. We can draw lines AC and BD (Axiom 1). Then ΔABC is congruent to ΔABD (Theorem 4), and so we have $\angle CAB = \angle DBA$, $\angle ACB = \angle BDA$, and $AC = BD$. Using the information that $\angle CAB = \angle DBA$ and the given fact that $\angle DAB = \angle CBA$, together with Axiom 8, we have $\angle DAC = \angle CBD$. Another appeal to Theorem 4 then gives ΔACD congruent to ΔBDC, hence $\angle ACD = \angle BDC$. Finally, since $\angle ACB = \angle BDA$ and $\angle ACD = \angle BDC$, we have from Axiom 7 that $\angle BCD = \angle ADC$; i.e., $\angle C = \angle D$.

With this theorem proved, Saccheri moved on to "prove" the troublesome Axiom 5 by considering three possibilities: (1) the equal angles C and D are right angles; (2) they are obtuse angles; or (3) they are acute angles. He then showed that case 1 can be true if and only if Euclid's Parallel Postulate holds. It remained to show that both case 2 and case 3 led to contradictions, and he was able to do this for case 2. Angles C and D could not be obtuse, for a contradiction would result.

He was one step from the long-sought proof. He only had to show that case 3 also led to a contradiction. But, try as he would, no contradiction could be obtained. He did get a series of theorems which seemed ridiculous. For instance, he proved that, if case 3 could be established, then

1. In the Saccheri Quadrilateral, $CD > AB$.
2. In any right triangle, the sum of the interior angles is less than 180°.

The Parallel Road to Destruction

Wolfgang Bolyai, father of John Bolyai who did pioneering work in non-Euclidean geometry, did nothing to encourage his son's efforts. Quite the contrary! He once wrote to his son as follows:

You must not attempt this approach to parallels. I know this way to its very end. I have traversed this bottomless night, which extinguished all light and joy of my life. I entreat you, leave the science of parallels alone. . . . I thought I would sacrifice myself for the sake of the truth. I was ready to become a martyr who would remove the flaw from geometry and return it purified to mankind. I accomplished monstrous, enormous labors; my creations are far better than those of others and yet Ihave not achieved complete satisfaction. . . . I turned back when I saw that no man can reach the bottom of the night. I turned back unconsoled, pitying myself and all mankind.

I admit that I expect little from the deviation of your lines. It seems to me that I have been in these regions; that I have traveled past all reefs of this infernal Dead Sea and have always come back with broken mast and torn sail. The ruin of my disposition and my fall date back to this time: I thoughtlessly risked my life and happiness—aut Caesar aut nihil.

One must ask, however, why these theorems seem ridiculous. The answer is disconcerting. They are ridiculous because they violate what seems apparent from a picture, or because they contradict theorems from Euclidean geometry. These are the very arguments we rejected when the Saccheri Quadrilateral first appeared to us to be a rectangle. If we would not allow them before, we should not allow them now. Saccheri was apparently tiring, however (he died the same year his book was published), because after several theorems of this type he concluded that case 3 was impossible, since it led to conclusions that were "repugnant to the nature of straight lines." He published his work in a book modestly titled, *Euclid Freed of Every Defect*.

The best of thinkers are on occasion guilty of those lapses in logic that allow them to prove what they believe.

Saccheri's conclusion was the only unfortunate part of his work. The fact is that Saccheri's theorem is a theorem of absolute geometry, but that the axioms of absolute geometry do not allow one to decide in favor of either

1. The equal angles C and D are right angles, or
3. The equal angles C and D are acute angles.

One of these can be chosen arbitrarily and added to the list of axioms. If case 1 is chosen, we are back to Euclidean geometry; if case 3 is chosen, we get a different geometry, but nevertheless a geometry that is self-consistent.

It had, as a matter of fact, occurred to several people in the late eighteenth century that it might be possible to obtain a logically consistent geometry even if one replaced Euclid's Parallel Postulate by something contradictory to it. Three names associated with this revolutionary idea are J. H. Lambert (1728–1777), C. F. Schweikart (1780–1859) , and F. A. Taurinus (1794–1874). But these three apparently never thought that such axiom tampering could be related to the physical world, which they still believed to be just the way Euclidean geomerty described it. It was the great mathematician Gauss who first grasped the full significance of what was blowing in the wind. He traced his ideas to a period when he was about 15 and wrote to a friend, "I am becoming more and more convinced that the [physical] necessity of our [Euclidean] geometry cannot be proved, at least not by nor for human reason." Gauss did not publish his thinking on this subject, but his ideas probably had some influence on the work of two younger men who did publish their ideas.

Lobatchevsky (1793–1856) and Bolyai (1802–1860)

Nikolai Lobatchevsky was a Russian who had for one of his teachers a good friend of Gauss. He wrote several papers about geometry before writing the full-scale treatment in 1840 that came to Gauss' attention and subsequently made Lobatchevsky famous. John Bolyai was the son of the Hungarian mathematician Wolfgang Bolyai. John began to fill in as lecturer to his father's university classes when he was 13, was an accomplished violinist, and once accepted the challenge to duel 13 fellow soldiers—the understanding being that he could have a short period of repose with his violin between each duel. R. Bonola records that he left all 13 lying in the square. Mightier than his sword was his pen, however, for in a 20-page appendix to one of his father's books, he immortalized himself with his "Essays on Elements of Mathematics for Studious Youths" (1832). It is significant to note that his father was also a friend of Gauss.

Like many a son, John Bolyai chose to ignore his father's advice, and his progress on "this infernal Dead Sea" soon had his father writing in an altogether different vein. "If you have really succeeded in the question, it is right that no time be lost in making it public for two reasons: first because ideas pass easily from one to another, who can anticipate its publication; and secondly, there is some truth in this, that many things have an epoch, in which they are found at the same time in several places, just as the violets appear on every side in the spring."

Since both Lobatchevsky and Bolyai were indirectly influenced by Gauss, it is perhaps not surprising that they had parallel ideas. Indeed, Gauss wrote (without much generosity, it is sometimes said) of John Bolyai's work that he was unable to praise it, for to do so would be to praise his own work. Bolyai and Lobatchevsky did develop the details of this new geometry independently, however. In fact, John Bolyai thought when he first read Lobatchevsky's work that it had been copied from him. The almost simultaneous appearance of their work is often cited as an example of the kind of thing that happens after the groundwork has been laid for a new idea.

The geometry of Lobatchevsky and Bolyai is obained by adding to the axioms of absolute geometry one more axiom.

Lobatchevskian Parallel Axiom Given a line l and a point p not on l, there exist at least two lines containing p and no points of l.

The "repugnant" theorems of Saccheri then appear (along with the theorems of absolute geometry) as theorems in this new geometry. For instance, in Lobatchevskian geometry we are able to prove the following.

Theorem The sum of the interior angles of a triangle is less than 180°.

We can almost sense that by now the reader is ready, logic or no logic, to side with Saccheri; these results violate good sense. Once again therefore we pause to remind the reader that "line" and "point" are undefined concepts, and that he or she is not to place on them the usual associations. Let us emphasize this by suggesting meanings that can be assigned to these words that will enable us to draw pictures in which the Lobatchevskian Parallel Axiom seems to be true. We restrict our attention to what, in familiar Euclidean terms, represents the inside of a circle (and not the boundary or rim). By "point" we mean a dot inside the circle. A "line" is a chord of the circle (minus its end points of course), and a "line segment" is the usual Euclidean "streak" joining any two (interior) points of the circle. It is easily seen that the nine axioms of absolute geometry seem to be true and, if we define parallel lines as lines having no point in common, it will be clear that through a point p not on a line l at least two lines (such as m_1 and m_2 in the figure) can be drawn parallel to l.

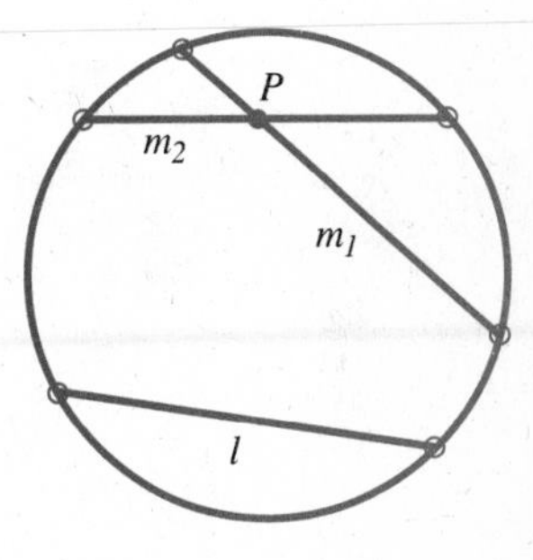

It now appears that there are an infinite number of lines through p that are parallel to l. This is so. In Lobatchevskian geometry one has the following.

Theorem Given a line l and a point p not on l, there exist an infinite number of lines through p parallel to l.

Surely this is a new geometry. We stress that it is as logical as Euclid's geometry.

Riemann (1826–1866)

Once it was understood that one could, without jumping the tracks of straight thinking, abandon Euclid's postulate on parallel lines, others were quick to branch off. Bernhard Riemann was a student of (guess) Gauss; he was to make major contributions to many areas of mathematics, but it was in geometry that he gave the lecture that was to qualify him to teach at Göttingen in 1854. His idea was to take a direction opposite that of Lobatchevsky in replacing the troublesome parallel postulate.

Riemannian Parallel Axiom There are no parallel lines.

Since the existence of parallel lines can be proved in absolute geometry, it is clear that Riemann could not make this assumption without abandoning more than Euclid's Parallel Postulate. This can be done in any of several ways. We outline only one. The idea is to restate Axiom 1 so that the line determined by two points need not be unique.

Axiom 1 (Riemann) Given any two points, there is at least one line containing them.

Riemann's statement has the virtue of saying exactly what he meant; he means to allow the possibility of more than one such line. With these changes, we again obtain a consistent geometry, and we again get theorems that differ in obvious ways from those of Euclidean geometry. In particular, having followed the work of Saccheri and his concern with the way in which interior angles add up, we note that we now have the "impossible" situation in which the two upper angles in the Saccheri Quadrilateral are obtuse angles.

Theorem The sum of the interior angles of any triangle exceeds 180°.

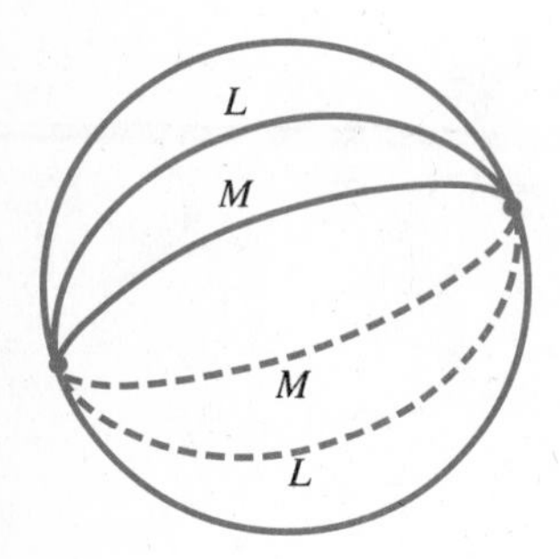

Once again we can assign meanings to the words "point" and "line" in such a way as to give us a picture in which the Riemannian axioms appear to be true. We think of the surface of a ball. Points are dots on his surface, and lines are great circles (the intersection formed by the ball and a plane cutting through the center). All the axioms appear to be true. In particular, there are no parallel lines (lines having no points in common); in fact, all lines intersect twice. Moreover, while any two points lie on at least one line (great circle), some pairs of points—those lying at opposite ends of a diameter—lie on an infinite number of distinct lines.

Summary

Two basic approaches were used in attacking the "scandal" of Euclid's Parallel Postulate:

Deduce it from the other nine.

Replace it by something more intuitive.

Following the first line of attack, Saccheri defined a quadrilateral later associated with his name. With only the axioms of absolute geometry, he showed that either the parallel postulate could be derived or "repugnant" results were derived. Two mathematicians, Lobatchevsky and Bolyai, both having connections with Gauss, showed that an axiom contradicting the parallel postulate led to a consistent geometry and enabled one to prove the "repugnant" theorems of Saccheri. Riemann later started with still another substitute for Euclid's Parallel Postulate and obtained another self-consistent geometry. We have stated the substitute axioms and described for each geometry a model that satisifies the axioms.

Self-Test

Choose the completion that seems best to you. If none of them seem quite correct to you, be prepared to tell why and to supply an ending of your own in class discussion.

1. Euclid's treatment of the fifth axiom
(a) is now known to be wrong.
(b) shows that he had difficulty in treating an idea if he could not visualize it.
(c) exhibits his genius, first because he recognized that this needed to be an axiom, and second because he stated it without making ambiguous or foolish statements about infinity.

2. The development of non-Euclidean geometries is often ranked as one of the great intellectual achievements of mankind. One reason for this is
(a) it explained why Euclidean geometry could not be used for navigation on a spherical earth.
(b) it dramatically demonstrated that the results we obtain depend upon the assumptions we make, and that replacing an assumption with its negation can result in a system no less logical than the original.
(c) it demonstrated that even so revered a work as Euclid's *Elements* might have logical errors.

3. Efforts to prove Euclid's fifth axiom from the other nine
(a) were doomed to failure because no such proof will ever be found.
(b) may be resumed if geometry ever becomes a "fashionable" topic again.
(c) resulted in numerous "proofs" that turned out to have errors in them.

Problem Set 11.2

1. What evidence is there that Euclid was dissatisfied with his fifth postulate, the Parallel Postulate?

2. What reasons can you give for thinking that Gauss was the person principally responsible for the development of non-Euclidean geometry?

3. Can you think of instances in other branches of science where essentially the same idea is "found at the same time in several places, just as the violets appear on every side in the spring"?

4. The finite geometries discussed in Section 10.2 are non-Euclidean geometries. Contrast the axioms of finite geometries with those of Euclidean geometry. How many are the same? Where do they differ? Do you suppose finite geometries appeared before or after the work of Bolyai and Lobatchevsky?

Problems 5 and 6 relate to Lobatchevskian geometry in which we assume all the results of absolute geometry plus the Lobatchevskian Parallel Axiom, which, it will be recalled, leads to the following.

Theorem The sum of the interior angles of a triangle is less than 180°.

5. Use the theorem quoted above to show that in Saccheri's Quadrilateral the equal angles C and D must be acute.

6. Are the following statements true or false in Lobatchevskian geometry? (Refer to the model provided in the text.)
   (a) If two distinct lines m and n are both parallel to a third line, lines m and n are parallel to each other.
   (b) Rectangles do not exist.
   (c) Parallel lines are everywhere equidistant.

7. Are the following statements true or false in Riemannian geometry? (Refer to the model provided in the text.)
   (a) Two distinct triangles can have the same three vertices.
   (b) Two points determine either one line or infinitely many lines.
   (c) A triangle can have two angles of 90°.

For Research and Discussion

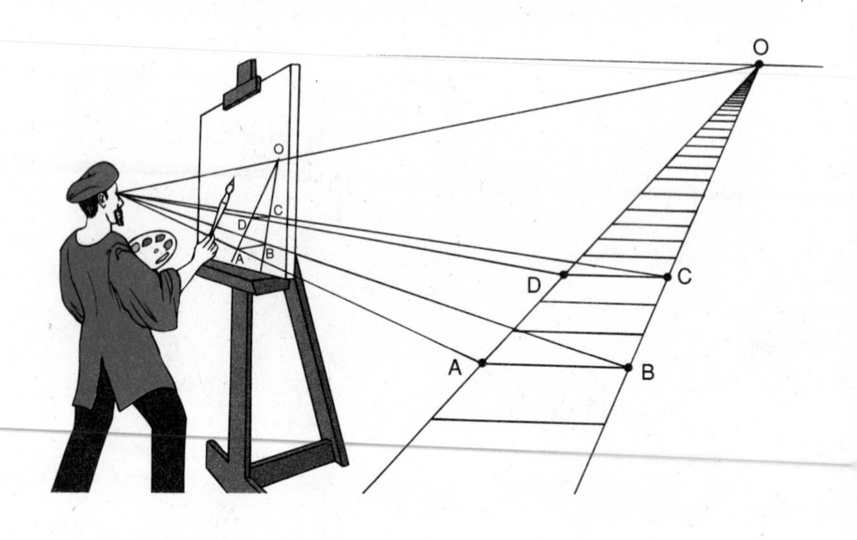

Mighty is geometry; joined with art, resistless.

Euripides

Much has been written about the role of perspective in Renaissance paintings. Find several examples where a famous painting is shown with the superimposed lines that guided the artist. Sources to begin with are:

Kline, M. *Mathematics for the Liberal Arts,* Addison-Wesley, 1967, Chapter 10.

Encyclopaedia Britannica, 15th ed., S.V. "projective geometry."

11.3 Lessons from Non-Euclidean Geometry

Working from an Arbitrary Starting Point

In the false bottom of a chest which had belonged to a pirate, the following instructions were found for finding treasure buried on a certain island. "Start from the gallows and walk to the large white rock, counting your paces. At the rock, turn left through a right angle and walk the same number of paces. Mark the spot with your knife. Return to the gallows. Count your paces to the black rock, turn right through a right angle, and again walk the same distance. The treasure is midway between you and your knife."

When searchers got to the island, they found the rocks, but no trace of the gallows. What advice could you give them that would help to locate the treasure? (This is an "old chestnut" familiar to the readers of mathematical puzzle books.)

We can almost hear the reader saying, "Well, your point is made. One does not have to assume that, through a point outside a line l, one and only one parallel to l can be drawn. One can assume with Lobatchevsky that millions of such lines are possible, or with Riemann that no such lines are possible. Since none of these assumptions lead to contradictions, any of them can be used to provide drill work for people who like logical drills. Very cute! Now let's get back to the plane facts of Euclid which correspond to the world the way it really is."

Such a response would be understandable; we may indeed appear to have belabored our point. At the same time, we as authors can be excused if we take some satisfaction in such a response, for it suggests that our readers have acquired an understanding of geometry that goes quite beyond that of most people. Yet we aim for more. After all, Lambert, Schweikart, and Taurinus apparently realized about 1800 that games could be played with the axioms in such a way that they would still yield logically consistent results. Our goal is to understand Gauss' observation that "the physical necessity of Euclidean geometry cannot be proved."

Any attempt to verify that a certain geometry is the so-called true geometry must identify the terms of geometry (points, lines, etc.) with objects in our world of experience. Then the following difficulties are encountered:

1. The truth of even the simplest axiom is difficult to ascertain.
2. The crucial axiom about parallel lines, depending as it does on extending lines indefinitely, reaches beyond the limits of our finite experience.
3. Attempts to check theoretical results (theorems) against facts are inevitably limited by our inability to measure things without error.
4. Ascertaining facts in the real world often involves assumptions about the real world which, if wrong, may explain why we have a lack of agreement between theoretical predictions and observations.

We illustrate each of these difficulties. With respect to the first, consider the common attempt to identify "points" with dots, and "lines" with streaks on a piece of paper. Look at two dots with a magnifying glass. Is it clear that only one line can be drawn through them? Is there any difference between passing through a point and touching a point?

Two tracks, laid for a railroad car, never touch of course. But they follow the curvature of the earth, so they aren't straight lines (not even in the sense of Riemann's geometry as interpreted on the surface of a ball). Now imagine a plane that extends into space. Lay two tracks on it for a railroad car; that is, lay them so they never touch. Are these lines straight? How do you know? Can you—can anyone—follow such lines indefinitely far?

We have the following contrasting theorems in three geometries.

Euclidean: The interior angles of a triangle sum to 180°.

Lobatchevskian: The interior angles of a triangle sum to less than 180°.

Riemannian: The interior angles of a triangle sum to more than 180°.

Neither of the last two theorems tells us what deviation from 180° is to be expected. If you experimented in Exercise 12 of Section 9.1 with summing the interior angles of a triangle, you are well aware of the difficulty. Suppose we measure and get 179°59′58″. Is Lobatchevskian geometry thereby established as the true geometry? An experiment of this kind was once attributed to Gauss who, the story went, stationed observers on three mountain peaks and had them measure the angles of the triangle so formed. The figure 179°59′58″ given was the one attributed to them. Though it now seems certain that the story was not true, it illustrates the difficulty. Is an observed deviation due to observational error, or is the geometry of Lobatchevsky the real geometry? When telescope making became more refined, a similar experiment was performed using stars as three points of a triangle. This experiment, following the theoretical work of Einstein, suggested that a form of Riemannian geometry may be the true geometry.

Finally, we note that any of the measurements that try to use large triangles, whether in the clouds or in the stars, rely on the assumption that light travels in a straight line. What if it doesn't? What if it bends? That is, what if light coming from B follows the dotted line to A? In measuring the angle from due east to B, the observer then looks in the direction AB' to see B. A grievous error has been made if the observer thinks that EAB has been measured.

It should now appear as something of an understatement to say that it is difficult to determine what is the "true" geometry. The better view seems to be that we should think of any geometry as a model, and that we should at any given time use the model that gives results most in accord with observations.

Learning from Non-Euclidean Geometry

The introduction of non-Euclidean geometry laid to rest once and for all the notion that Euclid's postulate about parallel lines could be proved. And it did more than this. It forced on us the idea that axioms are not self-evident truths; and consequently it resulted in a reevaluation of the relation between scientific theories and reality. We have come to see any theory as a model which we use as a tentative explanation of things that can be observed.

If Euclid could be wrong, so might anyone else.

Besides altering the way in which we understand the world around us, non-Euclidean geometries have made us think about the very nature of human reason. We have more clearly distinguished between validity and truth. It has become more obvious that our conclusions depend on the initial assumptions, and that these are more arbitrary than was once thought, even if the goal is to obtain results useful in understanding what we observe around us.

The conclusions of right thinking are valid, but they may be wrong.

> I attach special importance to the view of geometry I have just set forth, because without it I should have been unable to formulate the theory of relativity.
>
> A. Einstein

Finally, non-Euclidean geometry provided the foundation for a concept of space and time that was to become very important for Einstein's work. This is the one result least likely to be understood (we, for example, have not even hinted at just how Riemannian geometry is related to the theory of relativity). Yet, it is probably this "practical result" that would most persuasively justify non-Euclidean geometry in the eyes of most people (if they knew there were such a thing as non-Euclidean geometry). That which produces visible power unfortunately obscures the power of ideas that affect our whole understanding of the world around us.

Summary

We have listed four difficulties that stand in the way of deciding which geometry is the "true" geometry. These same problems occur whenever we attempt to verify a scientific theory and, when this lesson was driven home by the study of geometry, it changed our entire understanding of science. Gone is the expectation of finding a law that describes once and for all the way the world really is. With new force, it has become clear that "now we see through a glass darkly," and that any theory should be regarded as a tentative model.

Self-Test

Choose the completion that seems best to you. If none of them seem quite correct to you, be prepared to tell why and to supply an ending of your own in class discussion.

1. Experiments to accurately measure the sum of the interior angles of a triangle are of interest because
   (a) it is known that the sum is 180°, so the experiments afford an opportunity to check the accuracy of the measuring instruments.
   (b) a reliable result might settle the argument as to whether "real" geometry is Euclidean, Lobatchevskian, or Riemannian.
   (c) the foundations of Euclidean geometry would be destroyed if the sum could be shown to be different from 180°.

2. Non-Euclidean geometry
   (a) has no practical applications.
   (b) has stymied progress in geometry because it cast doubts on the foundations of the subject.
   (c) has helped us to understand that our assumptions are not so self-evident as we once thought.

Problem Set 11.3

1. Discuss the nature of scientific laws such as those below. How are they verified? What possible questions exist about their applicability? Within what limits can they be used?
   (a) The force exerted by a spring is proportional to the distance it has been displaced (Hooke's Law).

(b) To every action there is an equal and opposite reaction (Newton's Third Law).
(c) An attack of appendicitis is accompanied by a rise in the white corpuscle count.

2. Think of some simple statements of fact about which you have recently had difficulty in coming to agreement with someone. For example:
(a) This board will be strong enough.
(b) There's plenty of gas to get to the next town.
(c) The radiator won't freeze tonight.

Short of subjecting onself to the inconvenience if not the danger of being wrong, how are such disputes settled?

3. Suppose that, as we move further from the center of the earth, the size of all material objects decreases proportionally. Could we detect this? How? Would this cause a finite universe to look infinite?

4. In a novel called *Flatland,* (New York: Dover Publications 1952) Edwin A. Abbott describes a world in which everything happens in a plane and the characters have no perception of a third dimension. The two-dimensional characters can be perceived from our vantage point outside the plane as circles (the educated class—well rounded), triangles, etc., but they fail to see any of this. Can you think of any way they could learn to distinguish between characters of different shapes?

5. Try to recall and state in your own words the four difficulties that we have said confront anyone who wishes to decide on the "true" geometry. Can you restate these without referring to geometry so that they are applicable to attempts to verify any theory? Do these difficulties apply, for example, to one who wishes to prove that there is no God?

For Research and Discussion

The problem for Research and Discussion that closed Section 10.1 pictured the ballots of six people trying to rank four options (or candidates). You were asked to devise different reasonable ways to count the votes that would give different results, and to think of properties that should characterize a good system. One set of rules that has been proposed is as follows:

1. The rules should, for any set of submitted ballots, be able to provide a rank ordering of candidates.

2. If every voter prefers x over y, then the final rank ordering should show x above y.

3. If in a second election with the same candidates, all voters keep a certain subset of candidates in the same relative order, then the results of the two elections should show this subset of candidates in the same relative order.

Kenneth Arrow showed that the consequences of these modest rules are very surprising. Look up the result; discuss its implications. See if you can follow the proof.

Luce, R. D. and Raiffa, H., *Games and Decisions,* (New York John Wiley, 1957), Chapter 14.
Malkevitch, J. and Meyer, W., *Graphs, Models, and Finite Mathematics,* (Englewood Cliffs, N.J.: Prentice-Hall, 1974), Chapter 10.

11.4 Lessons from Euclidean Geometry

What's Wrong Here?

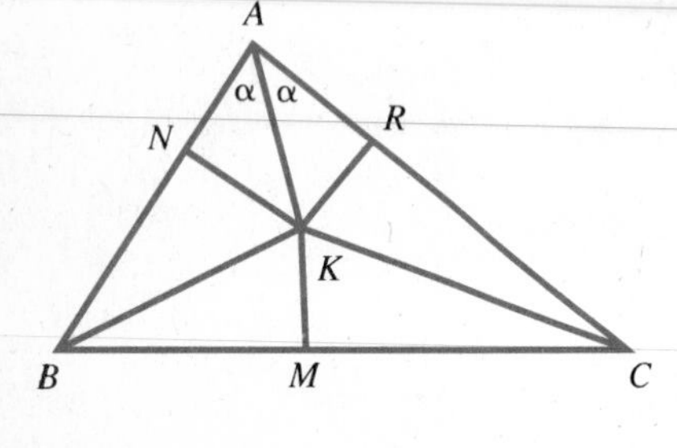

See if you can find an error in the proof of what appears to be a ridiculous assertion.

Theorem Every triangle is isosceles.
PROOF. Beginning with an arbitrary triangle ΔABC, we draw the bisector of the angle at A (Theorem 9) and the perpendicular bisector of BC (Theorems 10 and 11), designating their intersection by K. Next draw BK and CK (Axiom 6), and from K drop perpendiculars KN and KR to sides AB and AC, respectively (Theorem 12). Then ΔBMK and ΔCMK are congruent (SAS), so $BK = CK$. Also ΔAKN and ΔAKR are congruent (AAS), so $NK = RK$. Thus ΔBKN is congruent to ΔCKR (Problem 2, Section 10.1), giving us $BN = CR$. Using once again the fact that ΔAKN is congruent to ΔAKR, we have $NA = RA$. Thus $BA = CA$ (Axiom 7).

Let no one ignorant of geometry enter this door.
Plato

Many claims have been made about the value of studying Euclidean geometry. In this section we highlight what we believe are the main contributions, apart from the actual facts of plane geometry, that it makes to a general education.

Euclid's Weaknesses

Judged by present standards and understanding, there are defects in Euclid's *Elements*. It is our purpose to identify some of these, not to criticize Euclid (as some have) but to set in bold relief things we think have been learned.

We should not try to define every term.

We have already pointed out that Euclid never realized that some terms must be left undefined. His attempts stand in fact as a classic example of the futility of trying to define everything. This realization came rather late in intellectual history, however, and it is not surprising that Euclid stumbled over the idea. Now we do realize it, however, so there is no reason for an educated person, whatever his or her field, to think that every term should be defined. The political scientist can choose, without apology, to use a few basic terms (freedom, democracy, etc.) without defining them, as if everyone understood them. The theologian need not blush over his or her inability to define God. For some terms, there can be no definition except associations that develop through the ways in which the term is used (the assumptions that are made about it).

An Exaggeration

It has been customary when Euclid, considered as a textbook, is attacked for his verbosity or his obscurity or his pedantry, to defend him on the ground that his logical excellence is transcendent, and affords invaluable training to the youthful powers of reasoning. This claim, however, vanishes on a close inspection. His definitions do not always define, his axioms are not always indemonstrable, his demonstrations require many axioms of which he is unconscious. A valid proof retains its demonstrative force when no figure is drawn, but very many of Euclid's earlier proofs fail before this test. . . . The value of his work as a masterpiece of logic has been very grossly exaggerated.

Bertrand Russell (1902)

It is also likely that Euclid believed his assumptions to be true statements about the world as it really is. One appreciates the nature of his starting assumptions only after seeing that it is possible to start with different assumptions, that the different assumptions lead to a different but completely consistent system, and that the new system may also explain in a useful way things we observe in the world about us. Once this is seen, the lesson ought not to be lost on an educated person. The Declaration of Independence of the United States reasons from principles the writers took to be self-evident. The economic theory one accepts ultimately rests on a view of humans and how they can or will behave. The Christian reads in the Bible that "he that cometh to God must believe that He is," and similar assumptions undergird any system of religious thought. Any effort to develop a theory or explanation of anything ultimately rests on assumptions that cannot be proved.

Any logical argument ultimately rests on assumptions.

Euclid's biggest problem, however, involved not his attitude toward the axioms and postulates he stated, but the many assumptions he made without realizing he had done so. Consider the proof given above that all triangles are isosceles. Did you find an error? Sooner or later, you will come to the conclusion that the picture is drawn incorrectly. And it is true that, if you start with the segments joining the dots A, B, and C shown in the problem opening this section, and if you construct with a straightedge and compass the bisector of the angle at A and the perpendicular bisector of segment BC, you will find that they intersect outside the triangle.

David Hilbert
(1862-1943)

You may, indeed you should, object that proofs are not supposed to depend on pictures. Euclid thought his proofs did not depend on the pictures. Yet the only thing we seem to find wrong with the proof is that the picture appears to be incorrect.

Actually, the argument above is not easily fixed up by drawing a better picture; indeed, with nothing more than the axioms stated by Euclid to guide you, it is not possible to decide whether a particular picture is right or wrong. To resolve the difficulty, nothing less will do than to add another axiom to Euclid's list.

Euclid did not of course try to prove this theorem. He just naturally (and unconsciously) made such further assumptions as he needed to exclude ridiculous theorems.

Thus is illustrated the biggest flaw in Euclid's work. Time and again he unconsciously used assumptions he had not stated. Our purpose is not to identify all his hidden assumptions. That is no easy job. The celebrated mathematician David Hilbert undertook this project, publishing his first effort in 1899; in 1930 he was still at it, that being the year in which he brought out his seventh revision of the project. When he had finished, the undefined terms had increased to 8 in number. He had grouped axioms according to the topic they dealt with, listing a total of 20 in five different groupings. It suffices to say that Euclidean geometry has turned out to be more complicated than Euclid expected.

One must take great care if all the assumptions being made are to be identified.

Again there is a warning to the person who would be well-educated. It takes a great deal of effort to identify all the assumptions we are making.

Euclid's Strengths

The endurance of his work is more elegant testimony to Euclid's strength than anything we can say. Nevertheless, we wish to comment on a few matters in particular.

Euclid understood that the goal is to make the minimal possible number of assumptions.

Euclid's first theorem states that, on a finite straight line, an equilateral triangle can be constructed. If one thinks in terms of the "dots" and "streaks" usually used to represent points and lines, this assertion is hardly less obvious than Euclid's Axiom 3 which assures us that a circle can be drawn with any point as center and with any radius. Yet Euclid stated one assertion as a postulate and the other as a theorem. Clearly, Euclid was not trying to prove only things not obvious. He did have in mind the idea of assuming as little as possible and then deriving the rest.

Euclid understood that a postulate about parallel lines was necessary, potentially troublesome, and possible to state unambiguously.

The fifth postulate, Euclid's Parallel Postulate, has the mark of genius on it. In the first place, Euclid saw that such an assumption was necessary. As we have seen, generations of mathematicians have doubted this, and some have even offered (incorrect) proofs that this assumption is unnecessary. Euclid was right! Second, while aware that he needed such a postulate, he also recognized that its character was somehow different, and he avoided using it as long as possible. Finally, in his statement of the axiom, he carefully avoided mentioning infinity or otherwise introducing much of the vagueness (and sometimes outright nonsense) which even to this day appears in the writing of mystics who wish to relate the fate of parallel lines to immortality, etc.

Finally, let us recognize that, while modern critics can identify flaws in his work, Euclid wrote a book singularly influential in the area of science. Centuries later, Newton modeled his *Principia* after the style of Euclid's *Elements,* and it is still the basis of geometry courses taught in modern high schools. Moreover, the work of correcting such flaws as Euclid left has itself led to a deeper, more useful understanding of the nature of science.

Euclid understood what a valid proof is, and he wrote a book as enduring as human reason itself. The work he left has inspired some of the greatest of humankind's intellectual achievements.

Summary

The criticisms made of Euclid's work merely illustrate what we have been saying throughout this part of the book. He did not seem to realize that some terms must be left undefined, and he probably felt that his axioms described the world the way it really was. His most serious error, however, was in failing to identify all his assumptions. We illustrated this by showing that, if we use only the assumptions he listed (instead of relying, as he did, on others besides), there is no easy way to exclude ridiculous theorems. However, we have called attention in marginal notes to some profound ideas that Euclid handled with a skill that fully justifies his reputation.

Hats Off to This Man

Morris Kline (*Mathematics for Liberal Arts,* Addison-Wesley, Reading Mass. 1967, p. 51) attributes the following story to Charles W. Eliot, former President of Harvard University.

President Eliot entered a crowded restaurant and handed his hat to the attendant in the check room. He got no hat check, and so when he returned sometime later to reclaim his hat, he was quite amazed to see the attendant at once pick his hat out of scores of hats on the racks. "How did you know that was my hat?" asked President Eliot. "I didn't," replied the attendant. "Why, then, did you hand it to me?" was Eliot's second question. "Because," said the attendant, "you handed it to me, sir."

Self-Test

Choose the completion that seems best to you. If none of them seem quite correct to you, be prepared to tell why and to supply an ending of your own in class discussion.

1. This section contains a proof, based on Euclid's axioms, that every triangle is isosceles. This proof is included
(a) to show that even Euclid was capable of some pretty silly mistakes.
(b) to show that disclaimers notwithstanding, Euclid's work did depend on the pictures.
(c) to emphasize that Euclid failed to identify all the assumptions that he made.

2. Which of the following statements is not true of Euclid?
(a) He made a futile effort to define all of his terms.
(b) He thought his axioms were true statements about the world which all reasonable people would take to be self-evident.
(c) He saw no reason to prove a statement that appeared to be self-evident.
(d) He unconsciously assumed many things that were not stated in his listed axioms.

3. Which of the following is the greatest liability of Euclid's *Elements?*
(a) The awe inspiring authority of his work inhibited explorations of alternatives, thus unwittingly holding back progress in the study of geometry.
(b) It might have had far greater effect if the results could have been presented in a more readable style that played down the role of rigorous deduction.
(c) Based on what he wrote, there is no way to exclude from the list of possible theorems some that are perfectly ridiculous.

Problem Set 11.4

In Problems 1 through 4, list the implicit assumptions being made.

1. My son got a poor grade in mathematics, so I called his teacher. The teacher feels that the kid works hard and that he is probably doing about as well as possible in view of his low aptitude scores in mathematics.

2. My daughter got a poor grade in band. I guess that's to be expected. No one in our family has much ability in music.

3. I would have been embarrassed about not writing home for so long were it not for the fact that my folks are terrible correspondents themselves.

4. He should be ashamed of himself. He took advantage of Jim's ignorance when he sold him that lemon of a car.

Problems 5–8 are continuations of the correspondingly numbered problems in Problem Set 10.3. In each case we have given an invalid conclusion which may appear to some to be true. What minimal addition or change in the hypotheses will make the conclusion valid?

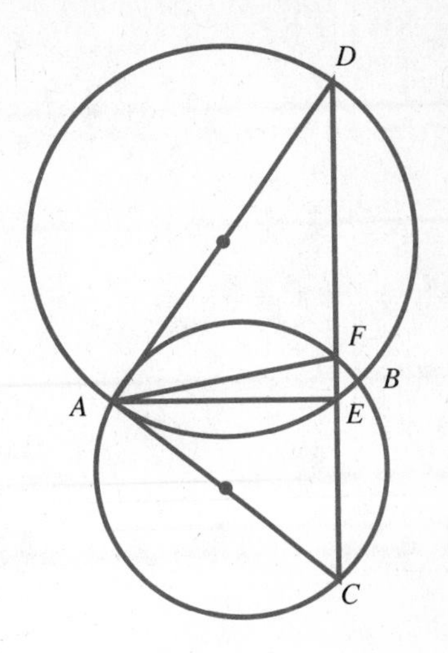

5. This course can be useful in preparing for certain jobs.
6. Some television serials are useful because they help us escape from the problems of real life.
7. Homer wants to learn.
8. Mr. Jones has a kind nature.
9. It can be proved in Euclidean geometry that, if a triangle is inscribed in a semicircle (understanding that two vertices lie at the end points of the diameter), the triangle is a right triangle. Using this theorem, we prove the following.

   Theorem. There exists a triangle having two right angles.
   PROOF. Draw two circles meeting at points A and B. Draw diameters AC and AD of the respective circles and designate by E and F the points where the line CD intersects the two circles. Now ΔAFC is a right triangle, as is ΔAED (because both are inscribed in semicircles). Then ΔAEF is the required triangle having two right angles.

   What is wrong with this proof?

For Research and Discussion

We wrote in this section "there is a warning to the person who would be well-educated. It takes a good deal of effort to identify all the assumptions we are making." There are a good many people who profit from counseling that has as its aim the identification of things they are unconsciously assuming. Try to identify assumptions you make that undergird your world view, that is your religious beliefs, your beliefs about social organization, your own sense of values, etc.

part IV
Abstracting from the Familiar

We point to three bulbs in a ceiling fixture and tell the toddler, "Three." We say the same thing about the cat's kittens, some envelopes on the table, and the eggs boiling on the stove. From what is common about the bulbs, the kittens, the envelopes, and the eggs, the child is expected to perceive the meaning of "three." The concept of number is an abstraction.

Back in Chapter 10, we introduced finite geometries by referring first to the governance of a dormitory and then to the regulation of a bus company. From what was common to the rules and regulations, we abstracted a set of statements about pokes and lokes, later changed to points and lines. Emphasizing that the logic of our arguments was not to depend on preconceived notions about points and lines, we derived consequences that can apply to any number of specific situations, depending on what real objects are identified with the terms "point" and "line." Our reasoning was abstract, a matter of the logical relationships between the terms, and not related to the meaning of the terms.

Abstraction is a central feature of mathematics. We ask what essential properties are shared by different mathematical systems. At the same time we look for basic properties that make two systems distinct. Finally, when the essential properties that characterize a system have been identified, one looks for the inescapable implications of these properties.

In this, the fourth part of the book, we examine the basic properties of the number systems with which we are familiar, and then those of some less familiar systems. Against the background of these examples, we state the axioms for two important algebraic structures called rings and fields (no connection with circus rings or farmer's fields). In proving several theorems that must hold in these abstract settings, we obtain insights into why the familiar rules of arithmetic must be as they are. We also see that algebra, no less than geometry, is an axiomatic system in which all the results rest on a sparse set of assumptions we call axioms.

The interplay between generality and individuality, deduction and construction, logic and imagination—this is the profound essence of live mathematics. Any one or another of these aspects of mathematics can be at the center of a given achievement. In a far reaching development all of them will be involved. Generally speaking, such a development will start from the "concrete" ground, then discard ballast by abstraction and rise to the lofty layers of thin air where navigation and observation are easy; after this flight comes the crucial test of landing and reaching specific goals in the newly surveyed low plains of individual "reality." In brief, the flight into abstract generality must start from and return again to the concrete and specific.

RICHARD COURANT

Fraulein Noether was the most significant creative mathematical genius thus far produced since the higher education of women began. In the realm of algebra, in which the most gifted mathematicians have been busy for centuries, she discovered methods which have proved of enormous importance in the development of the present-day younger generation of mathematicians. Pure mathematics is, in its way, the poetry of logical ideas. One seeks the most general ideas of operation which will bring together in simple, logical and unified form the largest possible circle of formal relationships. In this effort toward logical beauty spiritual formulae are discovered necessary for the deeper penetration into the laws of nature.

Born in a Jewish family distinguished for the love of learning, Emmy Noether, who, in spite of the efforts of the great Göttingen mathematician, Hilbert, never reached the academic standing due her in her own country, none the less surrounded herself with a group of students and investigators at Göttingen, who have already become distinguished as teachers and investigators. Her unselfish, significant work over a period of many years was rewarded by new rulers of Germany with a dismissal, which cost her the means of maintaining her simple life and the opportunity to carry on her mathematical studies. Farsighted friends of science in this country were fortunately able to make such arrangements at Bryn Mawr College and at Princeton that she found in America up to the day of her death not only colleagues who esteemed her friendship but grateful pupils whose enthusiasm made her last years the happiest and perhaps the most fruitful of her entire career.

A. EINSTEIN

Emmy Noether
(1882-1935)

chapter 12

Number Systems

Numbers are an indispensable tool of civilization, serving to whip its activities into some sort of order. . . . The complexity of a civilization is mirrored in the complexity of its numbers.

PHILIP J. DAVIS

What's New?

Every once in a while a colleague from another department lets slip a comment that betrays his or her opinion that teaching mathematics must be a snap in the sense that one need not be concerned with keeping up with the field. After all, whatever escaped Euclid's attention was picked up and finished off by Newton. Once you have learned the stuff, that's about it, isn't it?

Computers? Well, yes, of course they are relatively new and murder to keep up with, but everybody's got that problem. Anyhow, they are mainly used for wordprocessing, creating graphics, and (now that you mention it) for calculating. But since you have made the point that calculating is not what mathematics is all about, and since computers are the domain of the computer scientists, why should rapid changes in computing be germane to a discussion of keeping up with new mathematics?

The fact is that computer science as a discipline sits firmly on a foundation of mathematics that includes very old topics like binary and modular arithmetic, and Boolean algebra, to mention a few topics that are newly spruced up to meet the demands of a new technology. The same demands have stimulated the flowering of whole areas of mathematics that barely existed not so long ago: graph theory, finite state machines, and data structures for example.

The availability of computers has, moreover, made it possible to approach old problems with new tools. New tools include finite elements, wavelets, numerical methods in the solution of differential equations, and much more. Classic problems (can any map be colored with any four colors if no countries with a common border are to be the same color?) have been solved using the computer to check out possibilities far too numerous to have ever been done by hand.

More, much more, could be said about the way the computer is affecting the development of mathematics, but to do so would run the risk of suggesting that all new mathematics is related to the development of the computer, and that's not true at all. Medical technology has raised new problems in such diverse areas as fluid dynamics, dispersion (as with the assimilation of a drug into the blood stream), and pattern recognition. Modern manufacturing processes have given rise to great advances in statistical sampling to enhance quality control, the mathematics that underlies robotics and control theory, and operations research. Communications today also make use of operations research as well as error correcting codes and queueing theory.

Linear programming was developed as a discipline during World War II. Its applications now reach from agriculture to the management of traffic in a busy harbor, from inventory control to the operation of a petroleum cracking plant. Game Theory, another discipline that traces its roots to the period of World War II, is still being developed as a tool for economic and political analysis. And a person browsing in a bookstore might find among the art books one that reproduces beautiful examples of work done with fractals, having no idea of the connections with the geometrical ideas of Mandelbrot or with the study of dynamical systems.

Research mathematicians reading this short essay addressed to the proposition that mathematics is a growing field, infused almost everywhere by recent discoveries, will be disappointed at what has not been mentioned, since we have omitted most of what they consider the most significant advances of the last decade. But then a little criticism of such an effort is to be expected. It's no snap trying to keep up with advances being made in the field.

12.1 The Counting Numbers

Prime Spaces

A prime number is an integer having no factor other than itself and 1. Between the prime numbers of 23 and 29 we have five consecutive integers 24, 25, 26, 27, 28 that are nonprime. Another run of five consecutive nonprimes is given by 32, 33, 34, 35, 36. Can you find six consecutive nonprimes? How about seven? Can you, for any integer n, find n consecutive nonprime integers?

We have pointed out in our opening comments that a number like three is an abstraction. The collection of all such numbers (one, two, . . . , nine, ten, . . .) called the **counting numbers,** is the first example of an abstract mathematical system that we encounter in life. And though we meet it early, it seems to hold a fascination for people of all ages. Some, to be sure, are more fascinated than others. The most zealous come to be known as number theorists, or students of number theory.

The questions asked in number theory often appear to have no possible application and to be deceptively easy, and the outsider may legitimately wonder why they are asked at all. On the other hand, no branch of mathematics has had more appeal for outsiders—perhaps just because humans are intrigued by questions that are easy to understand, and turn out to be surprisingly difficult to answer. With the hope that you too may find some intriguing questions, we begin our discussion of abstract mathematical systems with the system of counting numbers.

Around and Around We Go

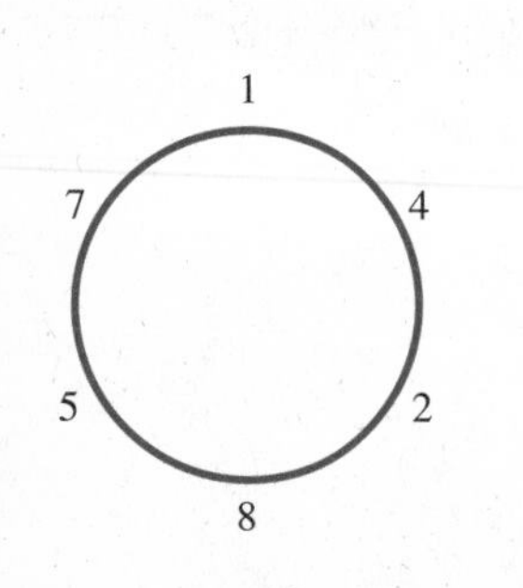

Beginning at 1, and reading clockwise, we have 142857.

$$\begin{array}{r} 142857 \\ \times 2 \\ \hline 285714 \end{array}$$

Beginning at 2, and reading clockwise, we have 285714

Similarly, each of the following products can be read around the circle.

$$\begin{array}{r} 142857 \\ \times 3 \\ \hline 428571 \end{array} \quad \begin{array}{r} 142857 \\ \times 4 \\ \hline 571428 \end{array} \quad \begin{array}{r} 142857 \\ \times 5 \\ \hline 714285 \end{array} \quad \begin{array}{r} 142857 \\ \times 6 \\ \hline 857142 \end{array}$$

Divisors and Multiples

We start with a review of some elementary notions. If three counting numbers are related by

$$ab = c$$

c is a **multiple** of a, and a is a **divisor** or **factor** of c. Similarly of course c is a multiple of b, and b is a divisor of c.

Given a set of two or more counting numbers, they may or may not have a common divisor greater than 1. The largest counting number that is a divisor of all of them (even if that largest number is 1) is called the **greatest common divisor** (g.c.d.) of the set. For example, the g.c.d. of

$$A = \{8, 12, 20\}$$

is 4, while the g.c.d. of

$$B = \{12, 20, 35\}$$

is 1.

Take any three-digit number, say 241 for example. Write it down again to obtain a six-digit number, such as 241,241. This six-digit number is always divisible by 7,11, and 13. Why?

Every finite set of numbers has a common multiple. Thus a common multiple of set A is $8 \cdot 12 \cdot 20 = 1920$, and a common multiple of set B is $12 \cdot 20 \cdot 35 = 8400$. Once a multiple has been found, there is no trouble finding a bigger one; just double or triple or whatever. The challenge lies in the other direction. What is the **least common multiple** (l.c.m.)? That is, what is the smallest number that is a multiple of each number in the set? For set A, the answer is 120; try to find the answer for B.

The l.c.m. is first encountered when one is learning to add fractions. In that setting, one needs the l.c.m. of the denominators, more commonly called the least common denominator.

The problem of finding the g.c.d. and the l.c.m. for a set becomes increasingly difficult as the numbers get larger, or the set contains more numbers, or both. Try, for example, to find the g.c.d. and the l.c.m. of the set

$$C = \{42, 63, 273, 364\}$$

To proceed most efficiently with such problems, it is best to express each member of the given set as a product of the primary building blocks of multiplication, the prime numbers. A **prime number** is a counting number greater than 1 that is divisible only by itself and by 1. The first few primes are

$$2, 3, 5, 7, 11, 13, 17, 19, 23, 29, \ldots$$

A counting number greater than 1 is either prime or it isn't. If it isn't, it has divisors which in turn are either prime or not prime. Building on this fact, we see that any number can be written as a product of primes. For example, the members of set C can be written

$$42 = 6 \cdot 7 = 2 \cdot 3 \cdot 7$$
$$63 = 3 \cdot 21 = 3 \cdot 3 \cdot 7$$
$$273 = 3 \cdot 91 = 3 \cdot 7 \cdot 13$$
$$364 = 2 \cdot 182 = 2 \cdot 2 \cdot 91 = 2 \cdot 2 \cdot 7 \cdot 13$$

The last paragraph is not likely to startle anyone and, to be truthful, neither is this one. This however, is in a sense a great tribute to the way in which arithmetic has been developed and taught, for we are dealing here with nothing less than the fundamental idea of arithmetic—and, behold, it seems obvious! After pointing out that a nonprime can be expressed as a product of primes, it remains only to say that, except for the order in which these primes are written down, this expression is unique; that is, there is only one such expression for the number.

The Fundamental Theorem of Arithmetic Every nonprime counting number (greater than 1) can be written as a product of primes and, except for the order in which the factors are written, this expression is unique.

The meaning of the statement about uniqueness can be grasped by contrasting this theorem with a statement about sums. Every nonprime counting number can be written as a sum of primes, but in this case the expression is not unique. For example,

$$\begin{aligned} 27 &= 23 + 2 + 2 \\ &= 19 + 5 + 3 \\ &= 7 + 7 + 7 + 3 + 3 \end{aligned}$$

Once the members of a set of counting numbers have each been expressed as a product of primes (or simply left alone in the case of numbers that are prime to start with), the g.c.d. is easily determined. If no prime appears as a factor of each number in the set (as in set B below), the g.c.d. is 1. If there is a prime p that occurs in every number, p is a factor of the g.c.d. If there is only one such p (as in set C below), p is the g.c.d. If there is another prime q that occurs in every number (and we allow the possiblity that $q = p$, as in set A below where every number has 2 as a factor twice), the g.c.d. has factors of p and q. The g.c.d. is the product of all primes that occur in every number, with a prime repeated k times in the g.c.d. if it occurs k times as a factor of each number in the given set.

For the sets A, B, and C, we have

$$A = \{8, 12, 20\} = \{2{\cdot}2{\cdot}2,\ 2{\cdot}2{\cdot}3,\ 2{\cdot}2{\cdot}5\}$$
$$\text{g.c.d.} = 2 \cdot 2 = 4$$
$$B = \{12, 20, 35\} = \{2{\cdot}2{\cdot}3,\ 2{\cdot}2{\cdot}5,\ 5{\cdot}7\}$$
$$\text{g.c.d.} = 1$$
$$C = \{42, 63, 273, 364\}$$
$$= \{2{\cdot}3{\cdot}7,\ 3{\cdot}3{\cdot}7,\ 3{\cdot}7{\cdot}13,\ 2{\cdot}2{\cdot}7{\cdot}13\}$$
$$\text{g.c.d.} = 7$$

The prime factorization makes it easy to find the l.c.m. of a given set too. Begin by writing down the product of all distinct primes that appear in any of the given numbers. For set C, this means that we begin with the product of 2, 3, 7, and 13. Now consider the first prime p written in the product. Does it occur more often than once as a factor of some number in the given set? If it occurs twice (or three times or k times) as a factor of some number in the given set, it must occur twice (or three times or k times) in the l.c.m. Similarly consider each of the distinct primes.

Having written down 2, 3, 7, and 13 for the set C, we next observe that 2 occurs twice as a factor in one of the given numbers; so does 3. Hence, for the set C, we have an l.c.m. of $2 \cdot 2 \cdot 3 \cdot 3 \cdot 7 \cdot 13$. We use the same method to write down the l.c.m. for the sets A and B:

$$\text{l.c.m. for } A = 2 \cdot 2 \cdot 2 \cdot 3 \cdot 5 = 120$$
$$\text{l.c.m. for } B = 2 \cdot 2 \cdot 3 \cdot 5 \cdot 7 = 420$$
$$\text{l.c.m. for } C = 2 \cdot 2 \cdot 3 \cdot 3 \cdot 7 \cdot 13 = 3276$$

Questions About Primes

No even number greater than 2 is prime; no multiple of 3 greater than 3 is prime. Each time we find a prime, all larger multiples of it are nonprime. It seems that it is very hard for a large number to be prime. Could it be that there is only a finite number of primes? No! Almost 2300 years ago, Euclid found an ingenious way of showing that there are infinitely many primes. His argument has come to serve as a classic example of proof by contradiction; it is outlined in Problem 20.

Consecutive odd primes like 11 and 13, 29 and 31, etc. are called twin primes. Are there infinitely many twin primes? Euclid didn't know. Neither does anyone else.

Twin primes are separated in the counting sequence by a single nonprime. In **Prime Spaces** we pointed out that 23 and 29 are separated by five consecutive nonprimes, and asked if you could find six consecutive nonprimes; or seven?

One approach to this problem is simply to examine the integers, looking for gaps between primes. This sort of examination will ultimately bring you to the observation that between the primes 89 and 97, there are seven consecutive nonprimes. This answers the question, but the idea of extending this method to answer the question for any n tires even the imagination. Another approach would be welcomed.

Reflect on your solution! Is your method the best possible?

Suppose we could construct six consecutive nonprimes, and do it in a way that could clearly be carried out for any n. That holds more promise. Consider then the following list of six integers (in which we make use of the factorial notation $7! = 7 \cdot 6 \cdot 5 \cdot 4 \cdot 3 \cdot 2 \cdot 1$)

$$n_1 = 7! + 2 = 5042 = 2 \cdot 2521$$
$$n_2 = 7! + 3 = 5043 = 3 \cdot 1681$$
$$n_3 = 7! + 4 = 5044 = 4 \cdot 1261$$
$$n_4 = 7! + 5 = 5045 = 5 \cdot 1009$$
$$n_5 = 7! + 6 = 5046 = 6 \cdot 841$$
$$n_6 = 7! + 7 = 5047 = 7 \cdot 721$$

The first number, being the sum of two integers that each have a factor of 2, obviously has a factor of 2. The next integer similarly must have a factor of 3, etc. Best of all, this construction can clearly be modified to construct n consecutive nonprime numbers. Our question is answered in the affirmative.

The numbers 1, 4, 7, 10, . . . are generated by the formula $t_n = 3n - 2, n = 1, 2, 3, \ldots$. Is there a similar formula that generates all the primes? Attempts have been made and can be summarized concisely; they have simply failed.

Well, if we can't find a formula that generates all the primes, can we find one that generates nothing but primes? Some attempts to answer this question have not simply failed; they have failed in notable ways.

To illustrate what we are after, consider

$$p_n = n^2 - n + 41$$

Table 12-1

n	P_n
1	41
2	43
3	47
4	53
5	61
6	71
7	83
⋮	⋮
40	1601
41	1681

The first few values are shown at the right. It looks promising. In fact, the formula works for $n = 1, 2, \ldots, 40$. But alas, $p(41) = 1681 = (41)(41)$. See **How Many Are Enough** in Section 9.1.

There is a more notable failure. A lawyer named Pierre de Fermat (1601–1665), probably the most famous amateur mathematician who ever lived, once suggested that the formula

$$f_n = 2^{2^n} + 1$$

always produces primes. He had some evidence, for

$$f_1 = 5 \qquad f_2 = 17 \qquad f_3 = 257 \qquad f_4 = 65{,}537$$

are all primes. Besides this he had enormous prestige. But he was wrong. In 1732, Leonhard Euler, the gifted Swiss mathematician, discovered without the aid of a calculator, mind you, that

$$f_5 = 4{,}294{,}967{,}297 = (641)(6{,}700{,}417)$$

For different reasons, mathematicians have studied the formula

$$g_n = 2^n - 1$$

It is conjectured that, while it doesn't always generate primes, at least it generates infinitely many. Modern electronic computers have been used to test the primeness of g_n for unbelievably large values of n, and the search goes on. There is no more certain way to make your text outdated than to report the largest prime number currently known.

Pierre de Fermat (1601–1665)
Pierre de Fermat spent his entire working career in the service of the state. He never achieved great heights in his profession, but in his avocation of mathematics, it was quite another story. The simple elegance of his work in number theory at first obscures the depth of his thinking. In analysis he anticipated some of the ideas that Newton was to use so successfully, and he is generally acknowledged to have been one of the greatest mathematicians of the seventeenth century.

Summary

We have reviewed the definitions of several terms and perhaps introduced a few new ones. You should now be able to write down a short, clear sentence that defines each of the following:

Counting numbers
Multiple
Divisor (or factor)
Greatest common divisor (g.c.d.)
Least common multiple (l.c.m.)
Prime

The most important theorem mentioned in this section, indeed one of the most important in elementary mathematics, is the Fundamental Theorem of Arithmetic. You should be able to state it and explain what it means.

We have also learned some things about prime numbers: There is an infinite number of primes; no simple formula is known that will generate all the primes, nor is a simple formula known that will generate only primes, though there have been some notable conjectures along these lines.

Problem Set 12.1

1. Find the g.c.d. and l.c.m. for each of the following.
(a) {10, 12, 16} (b) {9, 15, 18} (c) {8, 9, 25}
2. Find the g.c.d. and l.c.m. for the following sets.
(a) {4, 5, 6} (b) {6, 7, 8} (c) {8, 10, 16}
3. Write the prime factorization for each of the following numbers.
(a) 40 (b) 52 (c) 78 (d) 60 (e) 70
(f) 126 (g) 252
4. Write the prime factorization for each of the following.
(a) 36 (b) 63 (c) 84 (d) 40
(e) 140 (f) 175
5. For each of the following sets, find the g.c.d. and the l.c.m. (Use the information in Problem 3).
(a) {40, 52, 78} (b) {60, 70, 126, 252}
6. For each of the following sets, find the g.c.d. and the l.c.m. (Use the information in Problem 4).
(a) {36, 63, 84} (b) {40, 140, 175}
7. The notion of least common denominator is useful in adding fractions. Perform the following operations and reduce your answer to the simplest form.
(a) $\frac{1}{4} + \frac{1}{6} + \frac{3}{8}$ (b) $\frac{3}{4} + \frac{9}{16} - \frac{5}{24}$
(c) $\frac{17}{30} + \frac{9}{14}$ (d) $\frac{7}{24} + \frac{11}{60} - \frac{7}{90}$
8. Follow the directions in Problem 7.
(a) $\frac{1}{5} + \frac{3}{10} + \frac{8}{15}$ (b) $\frac{1}{6} + \frac{3}{8} + \frac{11}{12}$
(c) $\frac{5}{90} + \frac{38}{600}$ (d) $\frac{8}{9} - \frac{3}{65} + \frac{13}{79}$
9. Here are some helpful divisibility rules.
   1. A number is divisible by 2 if and only if its last digit is even.
   2. A number is divisible by 3 if and only if the sum of its digits is divisible by 3; for example, 14,562 is divisible by 3 because $1 + 4 + 5 + 6 + 2 = 18$ is divisible by 3.
   3. A number is divisible by 5 if and only if its last digit is 0 or 5.

   Use these facts to help you find the prime factorization of
(a) 10,800 (b) 52,650 (c) 38,775
10. Follow directions in Problem 9 for the following numbers.
(a) 3572 (b) 11,775 (c) 29,920
11. Find the smallest number with eight different divisors.
12. Find all prime number years between 1950 and 2000.
13. What is the first value of $n > 1$ for which $g_n = 2^n - 1$ is not prime?
14. What is the first value of n for which $k_n = n^2 + n + 17$ fails to be prime?
15. Suppose we want to tile a floor measuring 12 by 15 feet. Only square tiles are available in sizes 4, 5, 8, or 9 inches on a side. If we wish to use only whole tiles, all of the same size, what size should we order?

16.* The prime factorization of a number N can be written

$$N = p_1^{a_1} p_2^{a_2} \cdots p_n^{a_n}$$

Determine the p's and a's for $N = 30$, 48, and 180. Let d_N be the number of divisors of N. Find d_N for $N = 30$, 48, and 180. Note that, in these three cases,

$$d_N = (a_1 + 1)(a_2 + 1) \cdots (a_n + 1)$$

Try to prove that this is true in general.

17. The distinct integers a, b, and c each factor into a product of three (not necessarily distinct) primes. Their greatest common divisor is 7; their least common multiple is 252. What is their sum?

18.* The Sieve of Eratosthenes is a method for finding all the primes less then a given number n. The procedure is as follows. Write down in order all the counting numbers from 2 through n. Circle 2 and cross out all other multiples of 2 in the list. Then circle the next prime, 3, and cross out all other multiples of 3 in the list. Continue this process, moving each time to the next prime (the next number in order not already crossed out). The process is stopped when a number greater than $\sqrt{n}$ has been circled. The primes are those numbers not crossed out (whether circled or not). Use this method to determine the prime numbers less than 200.

19.* We discussed attempts to find an elementary formula generating all the primes (simple failure) and atempts to find a formula generating nothing but primes (some notable failures). There is a formula due to P. G. L. Dirichlet (1805–1859) known at least to generate an infinite number of primes. It says that, if a and b are counting numbers with no common divisor greater than 1, then $D_n = an + b$ generates infinitely many primes. Show that, under the same conditions, D_n also generates an infinite number of nonprime numbers. Note that $F_n = 2n - 1$ also generates an infinite number of primes. Why do you suppose, then, that anyone should think Dirichlet's Theorem is worth attention?

20.* Here, in outline form, is Euclid's argument showing that the number of primes is infinite. Suppose it is not. Then all the primes can be listed: 2, 3, 5, 7, 11, . . . , n, where n is the largest prime. Form the number

$$k = (2 \cdot 3 \cdot 5 \cdot 7 \cdot 11 \cdots n) + 1$$

Observe that 2 is not a factor of k, since if it were, that is, if $k = 2r$ for some integer r, we would have

$$2r - 2 \cdot 3 \cdot 5 \cdot 7 \cdot 11 \cdots n = 1$$
$$2(r - 3 \cdot 5 \cdot 7 \cdot 11 \cdots n) = 1$$

This is impossible. (Why?) Similarly, 3 is not a factor of k. Neither is any other of the listed primes. Thus there is some prime greater than n, perhaps k itself.

21.* Choose some set $a_1, a_2 \ldots, a_{101}$ of 101 numbers from the set $k = \{1, 2, \ldots, 199, 200\}$. Then for some i and j, a_i divides a_j. Hint: write each of the 101 selected numbers as a power of 2 (perhaps the 0 power) times an odd integer. That is, write $a_i = 2^{P_i} q_i$. Each of the 101 odd integers q_i is chosen from the set $\{1, 3, 5, \ldots, 199\}$ of 100 odd integers. Use the pigeonhole principle.

For Research and Discussion

In discussing the Fundamental Theorem of Arithmetic, we pointed out that it would be easy to express any number as a sum of primes. Consider this question: Can any even integer greater than 2 be expressed as a sum of two primes? After you have thought about it, go to a library, find a book on number theory, and look up Goldbach's conjecture.

12.2 Extending the Number System

Getting to the Root of the Thing

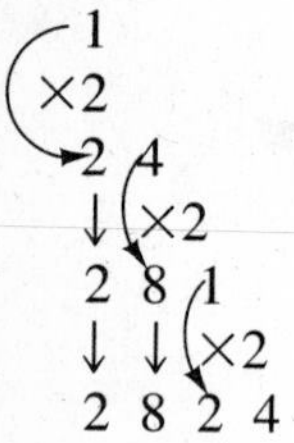

```
 1                     1. 4  1  4
×2                   √2.00 00 00
 2 4                  1
 ↓ ×2                 1 00
 2 8 1                  96
 ↓ ↓ ×2                  400
 2 8 2 4                 281
                         11900
                         11296
                           604
```

The algorithm on the left can be used to calculate square roots. Can you see how it works? Try to calculate the next two digits. Then square your answer to see how close it is to 2.

If carried out far enough, will you obtain a repeating decimal?

One way to introduce the solving of equations to youngsters is to ask what number, when added to 3, gives 7. Then repeat the question while writing

$$3 + ? = 7$$

Later still, since mathematicians believe that x is superior to other letters that might be employed, write

$$3 + x = 7$$

It is not unreasonable to think that from among the **counting numbers**

$$1, 2, 3, 4, \ldots,$$

youngsters in the early grades, perhaps aided by their fingers, will be able to select the correct answer.

Unless they have been introduced to the concept of negative numbers, however, it would be unreasonable to expect an answer to the question, what number added to 3 gives 2?

$$3 + x = 2$$

The youngsters we have in mind can be forgiven if they say that it can't be done. The solution of this problem requires that students have more at their fingertips, namely, the **integers**

$$\ldots, -4, -3, -2, -1, 0, 1, 2, 3, 4, \ldots$$

The integers, required to expand the class of equations that can be solved, are said to be an extension of the system of counting numbers. The integers, in the same way, must be extended to a larger system if we are to solve

$$3x = 2$$

For this purpose, we need the **rational numbers** (quotients of integers) including such numbers as

$$\ldots, -\frac{3}{2}, -\frac{4}{3}, -1, -\frac{2}{3}, -\frac{1}{2}, -\frac{1}{3}, 0, \frac{1}{3}, \frac{1}{2}, \frac{2}{3}, 1, \frac{4}{3}, \frac{3}{2}, \ldots$$

A rational number does not have a closest neighbor since between any two of them, there is always another one; for example, between $\frac{3}{80}$ and $\frac{4}{80}$ we have

$$\frac{\frac{3}{80} + \frac{4}{80}}{2} = \frac{3 + 4}{2(80)} = \frac{7}{160}$$

The system is said to be dense; if every rational number were represented on the number line by a red dot, the result would appear to us as a solid red line.

As plentiful as the rational numbers seem to be, their inadequacies were a matter of concern even to ancient mathematicians. They knew, from the Theorem of Phythagoras, that the lengths y and z of the legs of a right triangle are related to the length x of the hypotenuse by $x^2 = y^2 + z^2$. This meant that if $y = z = 1$, then the length of the hypotenuse had to satisfy

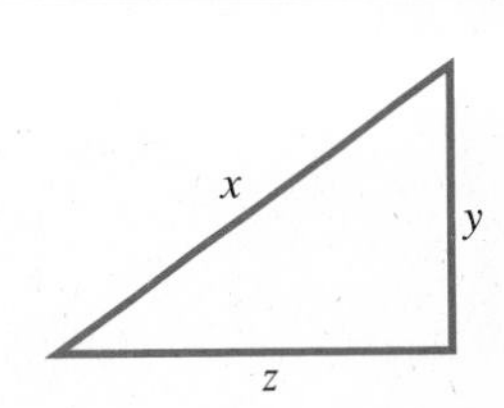

$$x^2 = 2$$

The problem was, however, that no rational number, when squared, equaled 2. Our proof of this fact, a classic example of proof by contradiction, is a variation of a proof given by Euclid.

We begin by assuming that there is a rational number r which, when squared, gives 2. This number r is by definition the quotient of two positive integers m and n. That is, we suppose

$$r = \frac{m}{n} \quad \text{and} \quad \frac{m^2}{n^2} = 2$$

Multiplying both sides of the latter by n^2 gives

$$m^2 = 2n^2$$

Now by the Fundamental Theorem of Arithmetic (Section 12.1), m and n can each be written as a product of a unique set of prime numbers. Thus $m = p_1 \cdot p_2 \cdots p_j$, and $n = q_1 \cdot q_2 \cdots q_k$, where the p's and q's are prime numbers (not necessarily all different). When we substitute these expressions in $m^2 = 2n^2$, we get

$$p_1 \cdot p_1 \cdot p_2 \cdot p_2 \cdots p_j \cdot p_j = 2 \cdot q_1 \cdot q_1 \cdot q_2 \cdot q_2 \cdots q_k \cdot q_k$$

Next we count the number of times the prime number 2 appears on each side of the equation. On the left where each prime factor of m appears twice, the prime number 2 must occur an even number of times (possibly zero times); on the right, there is an odd number of 2's. This is impossible. Our original assumption must have been wrong. The number r cannot be rational.

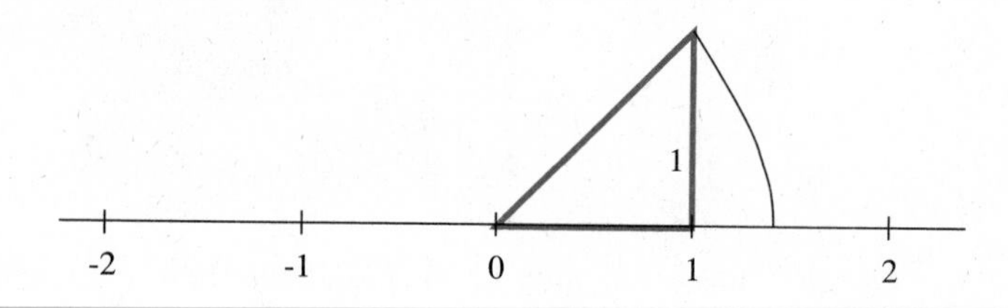

Here then, is the problem that troubled the ancient mathematicians. Return to the line on which all the rational numbers have been colored red. Using 0 as one end point, lay off the hypotenuse of the right triangle with equal sides of length 1. The end of the hypotenuse will not coincide with a red dot. If one only knows about rational numbers (which is what the ancients had to work with), what number measures the length of the hypotenuse? Such line segments came to be known as incommensurables.

Clearly we again need a larger system of numbers if we are to be able (conceptually) to label all of the points on the number line. The collection of numbers that is large enough to assign a unique number to each point on the line is called the set of **real numbers,** and the labeled line is called the real line. It must include all the rationals, since they certainly label points. But it must also include $\sqrt{2}$ (the number which when squared gives 2), π (pi, the circumference of a circle of diameter 1), and many more numbers.

The real numbers are conveniently represented using decimal notation. Recall that the familiar division algorithm enables us to represent any rational number by a decimal:

```
     0.77272
22)17.000000
   15 4
    1 60
    1 54
        60
        44
       160
       154
         60
         44
         16
```

$$\frac{17}{22} = 0.772\overline{72}$$

The bar indicates that the pattern of 72 repeats indefinitely. We say that $\frac{17}{22}$ is represented by a **repeating decimal.**

As a matter of fact, any rational number can be represented by a repeating decimal. The repeating pattern may be very simple, as in

$$\frac{1}{4} = 0.25000\ldots$$

$$\frac{1}{3} = 0.3333\ldots$$

or it may be more complicated, as in the case of

$$\frac{2}{7} = 0.2857\ldots$$

where the repeating sequence is not yet evident. We are certain that it will repeat, however, because, if we examine the remainders in the division algorithm (shown in boldface below),

```
   0.2857
7)2.0000
  1 4
    60
    56
     40
     35
      50
      49
       1
```

it is clear that there are only seven possibilities: 0, 1, 2, . . . , 6. Thus, after no more than seven steps, we will find a remainder already obtained—at which point the process will begin to repeat. Actually, as you will discover (Problem 3), the repetition occurs in this example after six steps. A similar argument for any rational number a/b shows that its decimal representation has to repeat.

As indicated in Problems 7 through 10, the converse is also true. Any decimal that ultimately falls into a repeating sequence of digits represents a rational number.

The nonrational real numbers (called the **irrational numbers**) are, therefore, the numbers that are represented by nonrepeating decimals. Our opening problem, **Getting to the Root of the Thing,** indicates in a rough (some would say lumpy) manner how to find the decimal representation for $\sqrt{2}$. We claim, though we do not intend to prove, that this procedure will produce a decimal—a nonrepeating decimal—whose square is 2 (Why must the decimal representation be nonrepeating?).

The real numbers, defined above as the collection of numbers required to (conceptually) label every point on a line, may also be described as the collection of numbers that can be represented using decimal notation, either as repeating or nonrepeating decimals. We ask our readers not to worry too much about several logical points. How exactly is an unending decimal to be understood? How are such things to be added or multiplied? These are hard questions which we must dismiss with a simple statement of assurance ("proof" by authority) that they do have perfectly logical answers.

We return to the recurring question of whether we now have enough numbers to solve any equation. One is not long in discovering that the real numbers do not suffice to solve $x^2 = -1$. For this, the so-called **imaginary number** i has been invented. It has the property that $i^2 = -1$, and it is used in combination with real numbers a and b to form the system of **complex numbers,** numbers of the form $a + bi$. Such numbers are added and multiplied according to the rules

$$(a + bi) + (c + di) = (a + c) + (b + d)i$$
$$(a + bi)(c + di) = (ac - bd) + (ad + bc)i$$

The second product is remembered by writing

$$\begin{aligned}(a + bi)(c + di) &= a(c + di) + bi(c + di) \\ &= ac + adi + bci + bdi^2 \\ &= ac + (ad + bc)i + bd(-1) \\ &= (ac - bd) + (ad + bc)i\end{aligned}$$

You may assume by now that this is an unending process; invent more numbers, find an equation that still can't be solved, invent more numbers, etc. The remarkable fact about algebra is, however, that the process stops once the complex numbers have been invented.

Fundamental Theorem of Algebra Any equation of the form

$$a_x x^n + \cdots + a_1 x + a_0 = 0$$

has a solution in the system of complex numbers.

Summary

We have progressively extended the collection of numbers available to solve equations. Each new system is said to be an extension of the previous system because the new system always includes the previous one as a subset.

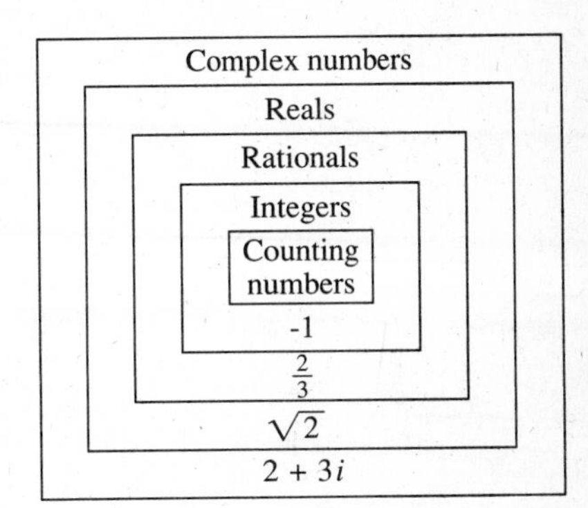

Equation	*System Needed for Solution*
$3 + x = 7$	Counting numbers: 1, 2, 3, 4, . . .
$3 + x = 2$	Integers: . . . , −4, −3, −2, −1, 0, 1, 2, 3, 4, . . .
$3x = 2$	Rationals: any number expressible as a quotient a/b of two integers, $b \neq 0$
$x^2 = 2$	Reals: any number expressible using decimal notation
$x^2 = -1$	Complex numbers: Numbers of the form $a + bi$ where a and b are real, $i^2 = -1$

The rational numbers are the real numbers that have a repeating decimal representation. The real numbers represented by nonrepeating decimals are called irrational. Numbers of the form bi where $b \neq 0$ are called imaginary, and numbers of the form $a + bi$ are called complex. According to the fundamental theorem of algebra, the table above is complete. No equation of the form $a_n x^n + \cdots + a_1 x + a_0 = 0$ will require for its solution a larger system than the complex numbers.

Problem Set 12.2

1. Perform the indicated operations.
(a) 0.375 + 12.24 (b) 12.24 − 0.375
(c) (0.375)(0.15) (d) 0.375/0.15

2. Perform the indicated operations.
(a) 56.13 + 1.245 (b) 1.245 − 56.13
(c) (0.037)(6.42) (d) 0.3144/0.24

3. Continue the division process for $\frac{2}{7}$, which was started in the text, until the repeating pattern is evident. Then write $\frac{2}{7}$ as a decimal using the bar notation.

4. Find the decimal representation of $\frac{2}{13}$.

5. Write each of the following as a repeating decimal using the bar notation.
(a) $\frac{3}{8}$ (b) $\frac{19}{8}$ (c) $\frac{5}{11}$ (d) $\frac{47}{12}$

6. Write as repeating decimals using the bar notation.
(a) $\frac{3}{16}$ (b) $\frac{17}{16}$ (c) $\frac{5}{9}$ (d) $\frac{31}{9}$

7. Every repeating decimal represents a rational number. To show that this is true for $0.\overline{36}$, let $x = 0.\overline{36} = 0.363636\ldots$. Then $100x = 36.363636\ldots$. Substract x from $100x$ and simplify as follows.

$$\begin{aligned} 100x &= 36.363636\ldots \\ x &= 0.363636\ldots \\ \hline 99x &= 36 \\ x &= \frac{36}{99} = \frac{9 \cdot 4}{9 \cdot 11} = \frac{4}{11} \end{aligned}$$

We multiplied x by 100 because x was a decimal that repeated in a two-digit group. If the decimal had repeated in a three-digit group, we would have multiplied by 1000. Try this technique to obtain the rational number corresponding to $0.\overline{147}$.

8. Follow the pattern indicated in Problem 7 to find the rational number represented by $0.153\overline{153}$.

9. Find the rational numbers represented by the given repeating decimals.
(a) $0.575757\ldots$ Hint: Multiply by 100.
(b) $0.2575757\ldots$ Hint: Multiply by 10 and then use what you learned in part (a).

10. Use the hints in Problem 9 to find rational numbers represented by the following.
(a) $0.696969\ldots$ (b) $0.5696969\ldots$

11. Consider the infinite decimal $0.121121112\ldots$.
(a) Following the pattern that you see, what are the next 10 digits?
(b) Does this infinite decimal represent a rational or irrational number?

12. Write down the first few digits (enough to establish a pattern) of an infinite decimal that you are sure represents an irrational number.

13. The sum of two rational numbers is always rational. Is the sum of two irrational numbers always irrational?

14. The product of two rational numbers is always rational. Is the product of two irrational numbers necessarily irrational?

15. Show that $\sqrt{2} + \frac{3}{4}$ is irrational. Hint: Let $\sqrt{2} + \frac{3}{4} = r$ and suppose r is rational. Look for a contradiction.

16. Show that the sum of an irrational number and a rational number is necessarily irrational. See the hint in Problem 15.

17. Show that the product $\frac{3}{4} . \sqrt{2}$ is irrational.

18. Show that the product of a nonzero rational number and an irrational number is always irrational.

19. Which of the following are rational and which are irrational? (See Problems 13 through 18.)
(a) $\sqrt{2} + 1$ (b) $3\sqrt{2}$
(c) $\sqrt{2}(\sqrt{2} + 1)$ (d) 0.12
(e) $0.\overline{12}$ (f) $0.123456789101112\ldots$
(g) $\sqrt{2}(0.\overline{25})$ (h) $(0.\overline{25})(0.\overline{34})$

20. What is the smallest positive integer?

21. Write a positive rational number smaller than 0.0000001. What is the smallest positive rational number?

22. $(0.0000001)\sqrt{2}$ is irrational. Write a smaller positive irrational number. What is the smallest positive irrational number?

23. Show that between any two different rational numbers there is another rational number. (Actually, there are infinitely many.)

24. Show that between any two rational numbers there is an irrational number.

25. It is known that π is irrational. What does this mean about its decimal expansion?

26. Here is the beginning of the decimal expansion of π:

$$\pi = 3.14159\ldots$$

Is $\pi - \frac{22}{7}$ positive or negative?

27.* Use the Pythagorean Theorem to show that $\sqrt{5}$ measures a length.

28.* Use the algorithm displayed at the beginning of this section to find the first four digits in the decimal expansion of $\sqrt{5}$.

29.* Verify by substitution that $1 + i$ is a solution to $x^2 - 2x + 2 = 0$.

30.* Verify by substitution that $1 - i$ is a solution to $x^2 - 2x + 2 = 0$.

For Research and Discussion

High school students learn that

$$\text{if } ax + b = 0, \text{ then } x = -\frac{b}{a}$$

$$\text{if } ax^2 + bx + c = 0, \text{ then } x = \frac{-b \pm \sqrt{b^2 - 4ac}}{2a}$$

The next logical step is to find a formula for the general third degree equation

$$\text{if } ax^3 + bx^2 + cx + d = 0, \text{ then ? ? ?}$$

The search for a formula for a solution to the third degree equation gave rise to one of the more shameful episodes in the history of mathematics. See what you can learn about it from books on algebra or the theory of equations. Key names involved are Cartan and Tartaglia. Good references:

Sondheimer, E. and Rogerson, A, *Numbers and Infinity* (Cambridge University Press, 1981).

Dobbs, D. and Hanks, R, *A Modern Course on the Theory of Equations,* 2nd ed., (Washington, New Jersey: Polygonal Pub., 1992.)

12.3 Modular Number Systems

Casting Out Nines

To check the addition on the left above, first add the digits for each number. For example, $8 + 3 + 5 + 6 = 22$. Then cast out as many 9's as you can. Two 9's can be squeezed out of 22, leaving 4. Do this for each number. The check comes in observing that, when all the 9's are squeezed out of $2 + 2 + 4 + 2 + 1$ and $4 + 1 + 6$, you get 2 in both cases. What is behind this check?

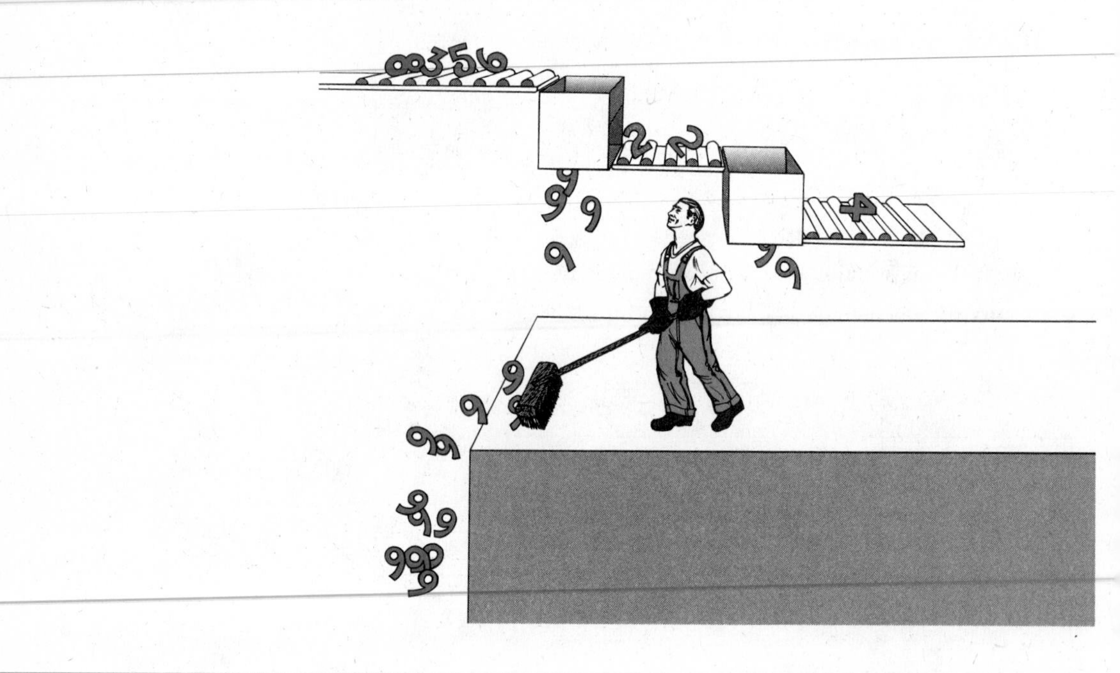

Suppose that a doctor gives you a pill at 11:00 A.M. and tells you to take one every 4 hours. What time should you take the next pill?

Suppose you are canning pickles, following a recipe that calls for soaking cucumbers in brine for 30 hours. If you put them into brine at 10:00 A.M., at what time the next day should you take them out?

These are common problems we all face, and somehow we solve them without writing anything down. If we did, we might be a bit surprised.

$$11 + 4 = 3 \quad \text{(for pills)}$$
$$10 + 30 = 4 \quad \text{(for pickles)}$$

What is going on? Actually the procedure is quite simple. One adds just as usual but then casts away as many 12's as possible. Here in a familiar setting are the ingredients of a strange new mathematical system; we'll call it **clock arithmetic.** It merits investigation.

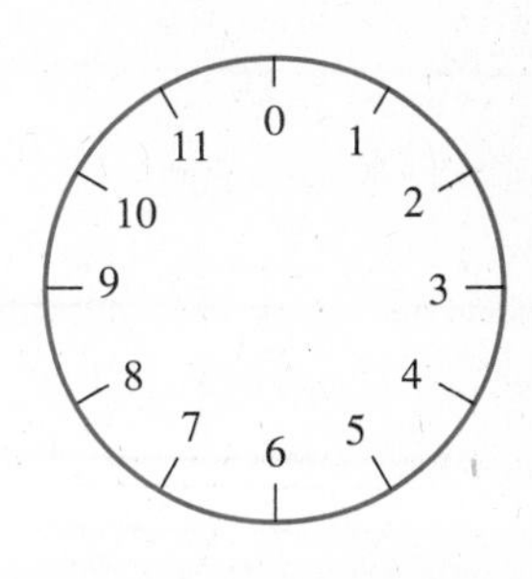

Right away we notice the special role of 12. It acts like 0; that is, you can add 12 to any number on the clock and you are right back where you started. For this reason, we'll rub out 12 on our clock and replace it with 0. But having done this, we observe that 12 is not the only number that behaves like 0; so do −12, 24, 36, 48, . . . , and a host of others. Similarly, −9, 3, 15, 27, etc., all act like 3 in clock arithmetic.

Table 12-2

+	*0*	*1*	*2*	*3*	*4*	*5*	*6*	*7*	*8*	*9*	*10*	*11*
0	0	1	2	3	4	5	6	7	8	9	10	11
1	1	2	3	4	5	6	7	8	9	10	11	0
2	2	3	4	5	6	7	8	9	10	11	0	1
3	3	4	5	6	7	8	9	10	11	0	1	2
4	4	5	6	7	8	9	10	11	0	1	2	3
5	5	6	7	8	9	10	11	0	1	2	3	4
6	6	7	8	9	10	11	0	1	2	3	4	5
7	7	8	9	10	11	0	1	2	3	4	5	6
8	8	9	10	11	0	1	2	3	4	5	6	7
9	9	10	11	0	1	2	3	4	5	6	7	8
10	10	11	0	1	2	3	4	5	6	7	8	9
11	11	0	1	2	3	4	5	6	7	8	9	10

Table 12-3

×	*0*	*1*	*2*	*3*	*4*	*5*	*6*	*7*	*8*	*9*	*10*	*11*
0	0	0	0	0	0	0	0	0	0	0	0	0
1	0	1	2	3	4	5	6	7	8	9	10	11
2	0	2	4	6	8	10	0	2	4	6	8	10
3	0	3	6	9	0	3	6	9	0	3	6	9
4	0	4	8	0	4	8	0	4	8	0	4	8
5	0	5	10	3	8	1	6	11	4	9	2	7
6	0	6	0	6	0	6	0	6	0	6	0	6
7	0	7	2	9	4	11	6	1	8	3	10	5
8	0	8	4	0	8	4	0	8	4	0	8	4
9	0	9	6	3	0	9	6	3	0	9	6	3
10	0	10	8	6	4	2	0	10	8	6	4	2
11	0	11	10	9	8	7	6	5	4	3	2	1

The situation reminds us of one that we faced with fractions. Recall that $\frac{3}{4}, \frac{6}{8}, \frac{9}{12}, \ldots$ all represent the same number. In any calculation, we can use any one of these representatives as well as another. Actually we tend to favour $\frac{3}{4}$; it's the reduced form of the fraction. Similarly, in clock arithmetic, we may think of $-9, 3, 15, 27, \ldots$ as representing the same number, with 3 being the reduced form or, as we prefer to say, the **principal representative.**

To put it slightly differently, we are using the clock to group the integers into 12 classes, each with its principal representative (shown in boldface below).

$$\begin{array}{l} \ldots, -36, -24, -12, \mathbf{0}, 12, 24, 36, \ldots \\ \ldots, -35, -23, -11, \mathbf{1}, 13, 25, 37, \ldots \\ \ldots, -34, -22, -10, \mathbf{2}, 14, 26, 38, \ldots \\ \qquad \vdots \\ \ldots, -25, -13, -1, \mathbf{11}, 23, 35, 47, \ldots \end{array}$$

The members of any one class differ by multiples of 12.

Using the principal representatives, we can construct addition and multiplication tables for clock arithmetic. We simply add and multiply as usual and then cast away enough 12's to get back into the principal set.

Now we can amuse ourselves with all kinds of questions. For example, can we solve equations in this new mathematical system? The answer is, sometimes yes; but the reader should be prepared for some surprises. At the moment, it is best to search for solutions by trial and error. Try substituting each of the numbers from 0 to 11 in the equations

a. $x + 11 = 3$
b. $5x = 8$
c. $9x + 4 = 10$
d. $4x + 1 = 6$

If you were careful, here is what you found out. The number 4 is the only solution to equations a and b; the numbers 2, 6, and 10 all satisfy equation c; and equation d doesn't have any solution at all.

Arithmetic Modulo m

Arithmetic based on the clock is amusing and has practical uses (recall the pills and the pickles). However, its real importance lies more in what it suggests than in what we have seen so far. First, there is nothing sacred about the number 12. Dividing the clock into 12 units was an arbitrary decision, extending back into ancient time. There is nothing wrong with 8, 10, or 50-hour clocks. And thinking about other possibilities will provide us with a tool for solving a class of problems called Diophantine equations (Section 12.4). Second, the various systems provide very simple illustrations of the abstract mathematical systems we intend to study in Chapter 14.

Before we plunge in all the way, consider one more familiar example. Suppose that, for a year that starts on Tuesday, we make a mammoth calendar numbering the days from 1 to 365. Which numbers

Table 12-4

S	*M*	*T*	*W*	*T*	*F*	*S*
		1	2	3	4	5
6	7	8	9	10	11	12
13	14	15	16	17	18	19
20	21	22	23	24	25	26
27	28	29	30	31	32	33
34	35	36	37	38	39	40
⋮	⋮	⋮	⋮	⋮	⋮	⋮
363	364	365				

represent the same day of the week, say Sunday? Clearly it is the numbers 6, 13, 20, 27, . . . , numbers that differ by a multiple of 7. How do we know that the day numbered 365 is a Tuesday? Because 365 differs from 1 by a multiple of 7, that is,

$$365 - 1 = 7 \cdot 52$$

The dictionary says that to modulate is to regulate or tone down in accord with some rule. To modulate by 7 is to tone down by a multiple of 7. Here then is the precise definition toward which we have been leading. We say that **a is equal to b modulo 7** if $a - b$ is a multiple of 7; that is,

$$a \equiv b \bmod 7 \qquad \text{if } a - b = 7k$$

for some integer k. Note the triple bars ($\equiv$) which we use consistently to distinguish this new kind of equality from the ordinary one ($=$). Thus

Pat signs a 60-day note on Friday. On what day of the week will it come due?

$$27 \equiv 6 \bmod 7 \qquad [\text{since} \quad 27 - 6 = 7(3)]$$
$$19 \equiv 5 \bmod 7 \qquad [\text{since} \quad 19 - 5 = 7(2)]$$
$$-8 \equiv 13 \bmod 7 \qquad [\text{since} \quad -8 - 13 = 7(-3)]$$

This new type of equality shares all the important properties of ordinary equality. For example,

1. $a \equiv a$.
2. If $a \equiv b$, then $b \equiv a$.
3. If $a \equiv b$ and $b \equiv c$, then $a \equiv c$.
4. If $a \equiv b$ and $c \equiv d$, then $a + c \equiv b + d$.
5. If $a \equiv b$ and $c \equiv d$, then $a \cdot c \equiv b \cdot d$.

Property 4 is the familiar statement, "Equals added to equals give equals." To illustrate, $27 \equiv 6 \bmod 7$ and $19 \equiv 5 \bmod 7$; sure enough $27 + 19 \equiv 6 + 5 \bmod 7$ (since $46 - 11 = 7 \cdot 5$).

Now what is done for 7 or 12 can be done for any modulus m. We write

$$a \equiv b \bmod m \qquad \text{if } a - b = m \cdot k$$

for some integer k, and properties 1 through 5 always hold (see Problems 14 through 17).

Arithmetic Modulo 9

$$\begin{array}{r} 928 \\ 9\overline{)8356} \\ \underline{81} \\ 25 \\ \underline{18} \\ 76 \\ \underline{72} \\ \textcircled{4} \end{array}$$

It is time to respond to the question posed in **Casting Out Nines.** What is behind the arithmetic checking process?

Let's begin by doing some reductions modulo 9. Take the number 8356, for example. Modulo 9, it is equal to one of the numbers from the principal set {0, 1, 2, 3, 4, 5, 6, 7, 8}. For such a large number one can remove multiples of 9 by long division. Just divide by 9 and find the remainder (see margin). Thus

$$8356 \equiv 4 \bmod 9$$

Reflect: Can the solution be made shorter?

As the reader can check, this method works for any modulus whatever.

Long division is okay, but for arithmetic modulo 9 there is a better way. All our readers will recall that 8356 is just a compact way of writing

$$8(1000) + 3(100) + 5(10) + 6$$

But $10 \equiv 1 \bmod 9$, $100 = 10^2 \equiv 1^2 = 1 \bmod 9$, $1000 = 10^3 \equiv 1^3 = 1 \bmod 9$, etc. By the properties above, especially properties 4 and 5,

$$8(1000) + 3(100) + 5(10) + 6 \equiv 8 + 3 + 5 + 6 = 22$$

In fact, this reasoning shows that *any number is equal to the sum of its digits modulo 9.* This is particularly useful for large numbers, but of course it works for small numbers as well. Thus

$$22 \equiv 2 + 2 = 4$$

But then most of us can cast 9's out of 22 directly.

Now we see why the check in our cartoon worked. Here it is again:

$$\begin{array}{r} 8356 \equiv 22 \equiv 4 \\ 6148 \equiv 19 \equiv 1 \\ \underline{7917 \equiv 24 \equiv 6} \\ 22{,}421 \equiv 11 \equiv \textcircled{2} \end{array}$$

It is really just that old friend, "Equals added to equals must give equals," interpreted for the new kind of equality. If we don't get equality in this process, we've made a mistake.

Property 5 means that casting out 9's can also be used to check multiplication. Here is an example:

> *Remember that we are multiplying, so we multiply the 8 by the 1.*

$$
\begin{array}{r}
287 \equiv 17 \equiv 8 \\
\underline{37} \equiv 10 \equiv \underline{1} \\
2009 \\
\underline{861} \\
10619 \equiv 17 \equiv \textcircled{8}
\end{array}
$$

Summary

Two integers a and b are said to be equal modulo the positive interger m if and only if there is an integer k such that

$$a - b = mk$$

Equality modulo m has all the properties normally associated with equality (equals can be added to or multiplied by equals, things equal to the same thing equal each other, etc.) The numbers 0, 1, 2, . . . , $m - 1$ are called principal representatives of numbers in the system modulo m and, in a given system, the reader should have no trouble representing a sum or a product by a principal representative; thus

$$17 + 12 \equiv 10 \bmod 19$$
$$17 \cdot 12 \equiv 14 \bmod 19$$

Finally, the reader should master the techniques of checking ordinary computations by making use of arithmetic modulo 9.

Problem Set 12.3

1. Make addition and multiplication tables for a 7-hour clock. Now solve the following equations in this arithmetic.

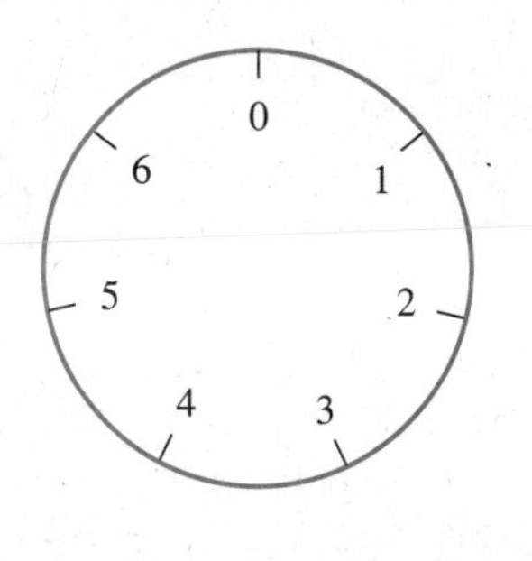

   (a) $x + 4 = 3$ (b) $2x = 3$
   (c) $2x + 3 = 4$ (d) $3x + 2 = 1$

2. Make addition and multiplication tables for a 6-hour clock and solve the equations in Problem 1.

3. What day of the week will it be 93 days after Wednesday? 193 days after Wednesday? Hint: Use a 7-unit clock and think of Wednesday as the fourth day of the week.

4. What time will an ordinary clock show 93 hours after it shows 4? 193 hours after it shows 4?

5. Reduce each of the following, giving the answer from the principal set. Hint: To reduce 77 mod 8, divide 77 by 8, obtaining a remainder of 5. Then $77 \equiv 5 \bmod 8$.

(a) 98 mod 5 (b) 981 mod 5
(c) $(98 + 981)$ mod 5 (d) $(98 \cdot 981)$ mod 5
(e) 492 mod 7 (f) 9811 mod 7
(g) $(492 + 9811)$ mod 7 (h) $(492 \cdot 9811)$ mod 7

6. Follow the directions in Problem 5.

(a) 47 mod 14 (b) 891 mod 14
(c) $(47 + 491)$ mod 14 (d) $(47 \cdot 491)$ mod 14
(e) 75 mod 8 (f) 750 mod 8
(g) $(75 + 750)$ mod 8 (h) $(75 \cdot 750)$ mod 8

7. Find all solutions in the appropriate principal set to the following equations.

(a) $x + 9 \equiv 2 \bmod 12$ (b) $3x \equiv 9 \bmod 12$
(c) $5x + 1 \equiv 4 \bmod 12$ (d) $7x + 2 \equiv 6 \bmod 12$

8. Solve each of the equations in Problem 7 with modulo 12 replaced by modulo 11.

9. Use the easy method to reduce each of the following.

(a) 3451 mod 9 (b) 623,852 mod 9
(c) $(4562 + 7321 + 9876)$ mod 9
(d) ((62,381)(92,734)) mod 9

10. Reduce each of the following.

(a) 25,763 mod 9 (b) 742,316 mod 9
(c) $(2576 + 4321)$ mod 9 (d) ((52,345)(8743)) mod 9

11. Some of the following calculations may be incorrect. Use the method of casting out 9's to identify them.

(a) $\begin{array}{r} 3417 \\ 2985 \\ 6321 \\ \underline{4173} \\ 16886 \end{array}$ (b) $\begin{array}{r} 9625 \\ 7163 \\ 3582 \\ \underline{1473} \\ 21843 \end{array}$ (c) $\begin{array}{r} 371 \\ \underline{816} \\ 2226 \\ 371 \\ \underline{2958} \\ 301736 \end{array}$ (d) $\begin{array}{r} 433 \\ \underline{721} \\ 433 \\ 866 \\ \underline{3031} \\ 312293 \end{array}$

12. Follow the instructions in Problem 11.

(a) $\begin{array}{r} 8614 \\ 2375 \\ 1627 \\ \underline{3748} \\ 16364 \end{array}$ (b) $\begin{array}{r} 5312 \\ 4871 \\ 2638 \\ \underline{1549} \\ 14270 \end{array}$ (c) $\begin{array}{r} 4916 \\ \underline{27} \\ 34312 \\ \underline{9832} \\ 132632 \end{array}$ (d) $\begin{array}{r} 9162 \\ \underline{38} \\ 73296 \\ \underline{27386} \\ 347156 \end{array}$

13. The casting out 9's check does not detect certain calculation errors. Can you identify them?

14. Show that if $a \equiv b \bmod 12$ and $b \equiv c \bmod 12$, then $a \equiv c \bmod 12$. Hint: By hypothesis, $a - b = 12k$ and $b - c = 12j$ for some integers k and j. Thus $a - c = a - b + b - c = 12k + 12j$. Now what?

15. Show that, if $a \equiv b \bmod m$ and $b \equiv c \bmod m$, then $a \equiv c \bmod m$. Hint: See Problem 14.

16. Show that, if $a \equiv b \bmod 12$ and $c \equiv d \bmod 12$, then $a + c \equiv b + d \bmod 12$.

17. Show that, if $a \equiv b \bmod 7$ and $c \equiv d \bmod 7$, then $a \cdot c \equiv b \cdot d \bmod 7$.

18. In ordinary arithmetic, $-a$ is the number that when added to a gives 0. For example, $-2 + 2 = 0$ and $-5 + 5 = 0$. We call $-a$ the **additive inverse** of a. Consider now a 7-hour clock with numbers 0, 1, 2, 3, 4, 5, 6. Note that 5 is the additive inverse of 2, since $5 + 2 = 0$. Find the additive inverse of
(a) 1 (b) 3 (c) 4 (d) 5

19. In ordinary arithmetic, $1/a$ is the number that when multiplied by a gives 1. For example, $\frac{1}{4} \cdot 4 = 1$ and $\frac{1}{9} \cdot 9 = 1$. We call $1/a$ the **multiplicative inverse** of a. Consider a 7-hour clock with numbers 0, 1, 2, 3, 4, 5, 6. In this arithmetic, 3 is the multiplicative inverse of 5, since $3 \cdot 5 = 1$. Find the multiplicative inverse of
(a) 2 (b) 3 (c) 4 (d) 6

20. On a 12-hour clock, try to find the multiplicative inverse of each of the numbers 0 through 11. You will discover that several of them do not have a multiplicative inverse.

21.* Since casting out 9's doesn't detect errors caused by inverting digits (see Problem 13), a method called casting out 11's has been suggested. Can you devise this system? Hint: $10 \equiv -1 \bmod 11$.

For Research and Discussion

For various choices of $a < 7$ compute $a^7 \bmod 7$. For various choices of $a < 5$ compute $a^5 \bmod 5$. Try other experiments that these computations suggest. Make a conjecture. Try to prove it. Look up Fermat's Theorem in a book on number theory. Explore the same computations using a nonprime modulus. Then look up Euler's Theorem.

12.4 Equations with Integer Answers

A Regular Cut-Up

Hugo Hardback, the head librarian, reported to the board that in furnishing the new library, he had spent \$3409 for tables and chairs. With the temerity that sometimes seizes people in positions of power, one board member asked how close they had come to the planned seating capacity. Hardback testily pointed out that, since the board had previously approved the purchase of tables costing \$288 each and chairs costing \$19 each, it should not be necessary to ask how many chairs had been purchased.

115 chairs @ \$119 = \$2185
$4\frac{1}{4}$ tables @ \$288 = \$1224
\$3409

If we let s represent the number of chairs and t represent the number of tables bought by the librarian described in the problem above,

$$19s + 288t = 3409$$

There are plenty of solutions to this equation. For instance, if we choose $s = 115$, a little computation gives $288t = 1224$, or

$$t = \frac{1224}{288} = \frac{17}{4} = 4\frac{1}{4}$$

Since t represents the number of tables, however, this answer leaves something to be desired; more literally, it leaves something not desired, namely $\frac{1}{4}$ table.

Sometimes a modest amount of thought about a problem makes it clear that only integer answers are acceptable. Such problems are called **Diophantine problems.** It will perhaps not surprise the reader to learn that modular arithmetic is often useful in solving such problems, and it is to a brief introduction of this whole idea that we now turn.

Suppose that in the equation

$$19s + 288t = 3409$$

we reduce everything modulo 19. Since $19 \equiv 0$, $288 \equiv 3$, and $3409 \equiv 8 \bmod 19$, this leads us to

$$3t \equiv 8 \bmod 19$$

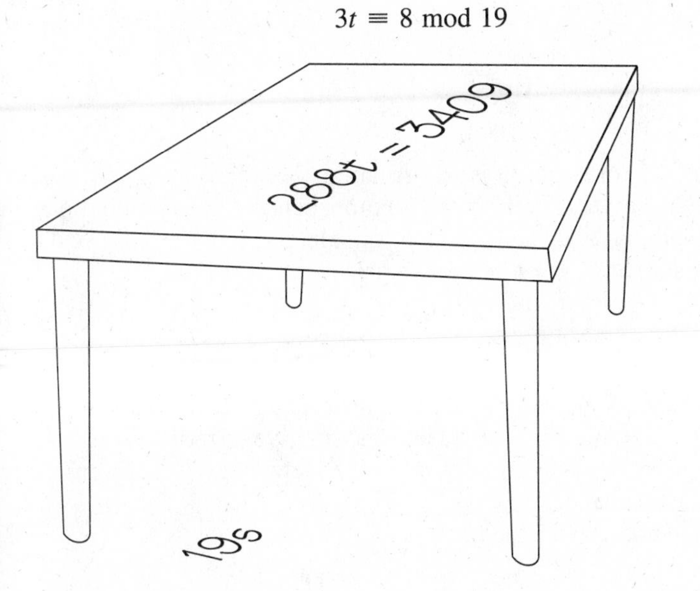

Now we saw in Section 12.3 that not all such equations have a solution and that some have several. Happily, we can settle the matter in a

Experiment, Guess, Demonstrate

reasonably small number of trials (in this case, no more than 19). In due time, after discovering that 0, 1, . . . , 8 do not work, we come to try $t = 9$. We get

$$3 \cdot 9 = 27 \equiv 8 \bmod 19$$

Thus $t \equiv 9 \bmod 19$; that is,

$$t = \ldots, -10, 9, 28, 47, \ldots$$

Diophantus

Diophantus was a Greek mathematician who lived about 250 A.D. While not all of his manuscripts have come down to us, we have enough of his work to know that he was interested in algebraic problems that have integral answers. Problems of this kind are for this reason called Diophantine problems. Of his personal life, we have only a description in the form of a problem. It seems that Diophantus spent $\frac{1}{6}$ of his life as a boy, grew a beard after $\frac{1}{2}$ more, and married after yet another $\frac{1}{7}$. A son was born 5 years later, but died in the prime of life. Four years after his son died. Diophantus died, having lived twice as long as his son had lived. Can you determine the number of years he lived?

should all work. It appears that there are many solutions to our original problem. Let us try $t = 9$. Then

$$\begin{aligned} 19s + 2592 &= 3409 \\ 19s &= 817 \\ s &= 43 \end{aligned}$$

Certainly $t = 9$ and $s = 43$ is one solution.

Since negative answers make no more sense than fractional ones in our problem, we only need to look for values of s that correspond to positive choices of t. For $t = 28$, we get

$$\begin{aligned} 19s + 8064 &= 3409 \\ s &= -245 \end{aligned}$$

Since larger values of t will give us a number larger than 8064, it is clear that all other possible positive values of t will give negative values of s. The only possible conclusion is that the new library addition has 9 tables and 43 chairs.

Consider again the equation

$$19s + 288t = 3409$$

We proceeded by reducing everything in sight modulo 19. Given an equation, it would of course be correct to reduce both sides using any modulus we please. The incentive to use 19 is clear; in this way we made one of the unknowns drop out of the equation. The same thing could have been accomplished using 288. This would have given

$$19s \equiv 241 \bmod 288$$

The only drawback is that, since our only method of solution is trial and error, we would be faced with 288 trials. Of course (since we know the answer—having worked it out above), if we proceed systematically, trying $s = 1, 2, \ldots,$ we will be rewarded on the forty-third trial. This is scant comfort to anyone not being paid by the hour, and it explains why we used 19 as the modulus.

If you seek integer solutions to

$$rx + sy = t$$

where r, s, and t are given positive integers, select the smaller of r and s and then reduce the equation modulo this number. You will be left with an equation of the form

$$az \equiv b \bmod m$$

We know that such equations may have no solution, a unique solution, or several solutions in the set $\{0, 1, 2, \ldots, m - 1\}$. If there are solutions, they can be found by trial and error.

Example with Two Large Coefficients

If we attempt to solve a problem such as

$$87x + 281y = 4983$$

the procedure outlined above results in

$$20y \equiv 24 \bmod 87$$

The number of possibilities to be tried is still too large to be taken seriously, so we appeal to the definition of equality modulo 87 and note that there must be an integer k such that

$$20y = 24 + 87k$$

If we apply the same techniques to this equation, reducing everything modulo 20 (the smallest coefficient), we get

$$0 \equiv 4 + 7k \bmod 20$$

Since $-4 \equiv 16 \bmod 20$, we have

$$7k \equiv -4 \equiv 16 \bmod 20$$

We can now find the solution (if there is one) by trial and error or, if 20 trials still seem more than we wish to try, the same procedure can be repeated again. Either way, we very shortly obtain

$$k \equiv 8 \bmod 20$$

Substituting the value $k = 8$ in $20y = 24 + 87k$ gives

$$20y = 24 + 87(8) = 720$$

or

$$y = 36$$

Thus the solution to $20y \equiv 24 \bmod 87$ is $y \equiv 36 \bmod 87$. Finally, substitution of $y = 36$ into the original problem gives

$$\begin{aligned} 87x + 281(36) &= 4983 \\ 87x &= -5133 \\ x &= -59 \end{aligned}$$

From $y \equiv 36 \bmod 87$, we can obtain other solutions as well. A partial listing gives

y	...	-51	36	123	...
x	...	222	-59	-340	...

It is noted that, as was to be expected, the values obtained for x are equal modulo 281. If the problem giving rise to this equation were such that only positive integers would be acceptable answers, we would report that there are no answers, since in the pairing of answers indicated in the table, either x or y is always negative.

Summary

Given an equation of the form

$$ax + by = c$$

in which a, b, and c are integers and in which we are interested only in finding integer values for x and y, we can use modular arithmetic to find solutions if they exist. We simply reduce the equation modulo a or modulo b, whichever is the most convenient (which usually means using the one closest to zero). The resulting modular equation in one unknown can then be investigated for possible solutions.

Problem Set 12.4

1. Find all integer solutions to the following Diophantine equations.
(a) $2x + 5y = 2$
(b) $15x + 16y = 17$
(c) $37x + 25y = 8$
(d) $38x - 14y = 43$
(e) $117x + 86y = 157$

2. Find all integer solutions.
(a) $x - y = 7$ (b) $13x + 14y = 15$
(c) $41x - 18y = 11$ (d) $26x + 54y = 119$
(e) $87x - 137y = 908$

3. Find all integer solutions that satisfy both the equations

$$3x - y + t = 0$$
$$2x - 3y + 5t = 17$$

Hint: Begin by eliminating t.

4. Find the integer solutions of the pair of equations

$$-2x + 5y - 3t = -3$$
$$3x - 2y + t = -2$$

Hint: Begin by eliminating t.

5. Homer and Horatio put up a sign offering to wash and clean the interior of a car for $4, or to wash and wax it for $25. Homer takes all the $4 jobs, and Horatio the $25 jobs. They agree that each will take the money for his own customers, but they use a common box in which to keep their money. After a weekend of hard work they find that they have $143, but they can't remember who gets what. Can you help them?

6. When the first edition of this book was being written, the Mathematics Association of America sold books in the Studies in Mathematics Series to members for $5 per book. At the registration desk at a regional meeting, books in this series were on sale. At the same table, members paid for their registration and lunch ticket ($8). The person collecting the money reported that she had collected $866, only then to be told that the funds were to be kept separate, the $8 fees going to the local arrangements committee and the $5 payments going to the regional treasurer. She had seven checks for $5, 74 checks for $8, and four checks for $13. Could you have helped her figure out how much money goes to the local committee? What if it can be ascertained from the box in which the books were brought to the meeting that no more than 20 books were sold?

7. The company that sells the tables and chairs that Hugo Hardback's library bought reports receipts of $20,603 for a given month. If the company expects orders of about six chairs to each table, what guess would you give as to the number of tables and chairs sold that month?

8. In another month (see Problem 7), receipts were $19,445. How many each of tables and chairs were sold that month?

9.* In a short story entitled "Coconuts," author Ben Williams told of two businessmen bidding for the same contract. One, in order to distract the other, who had a penchant for mathematical puzzles, sent him the following puzzle just as the deadline for bids was approaching.

> Five men and a monkey were shipwrecked on a desert island, and they spent the first day gathering coconuts for food. They piled them all up together and then went to sleep for the night.

> But when they were all asleep one man woke up, and he thought there might be a row about dividing the coconuts in the morning, so he decided to take his share. So he divided the coconuts into five piles. He had one coconut left over, and he gave that to the monkey, and he hid his pile and put the rest all back together.
>
> By and by the next man woke up and did the same thing. And he had one left over, and he gave it to the monkey. And all five of the men did the same thing, one after the other; each one taking a fifth of the coconuts in the pile when he woke up, and each one having one left over for the monkey. And in the morning they divided what coconuts were left, and they came out in five equal shares. Of course each one must have known there were coconuts missing; but each one was as guilty as the other, so they didn't say anything. How many coconuts were there in the beginning?

The story, along with this now famous puzzle, appeared in the October 9, 1926, issue of *The Saturday Evening Post*. It is reported that some 2000 letters arrived in the week after the story appeared asking for the solution. It is not reported how many of these requests were thrown to a monkey, but editor George Lorimer did send the following wire to author Williams: "FOR THE LOVE OF MIKE, HOW MANY COCONUTS? HELL POPPING AROUND HERE."

For Research and Discussion

We pose the following question. Can you discover a rule or rules that will enable us to tell, from looking at $a \neq 0$, b, and m, whether or not there are solutions to $ax \equiv b \bmod m$? To begin with, look at some examples. You already have a multiplication table for arithmetic modulo 12. Make similar tables for modulo 5, 6, and 9. In many years of posing this problem for classes to work on, we have heard the following suggested rules. Try to decide which ones are true, and which are false.

(a) If m is odd, then $ax \equiv b \bmod m$ will have a solution no matter how $a \neq 0$ and b are chosen.

(b) There will always be a solution if a is odd.

(c) There will always be a solution if a is odd and b is even.

(d) There will always be a solution if m is prime.

(e) We can be certain of a solution only if m is prime.

Some of the above are false. Some are true. None give the entire story. Make a conjecture of your own. Discuss it with others.

chapter 13

The System of Matrices

It is true that Fourier has the opinion that the principal object of mathematics is the public utility and the explanation of natural phenomena; but a scientist like him ought to know that the unique object of science is the honor of the human spirit and on this basis a question of the theory of numbers is worth as much as a question about the planetary system.

C. G. J. JACOBI

What's This Good For?

You pose several questions, each one interesting in itself, but seemingly unrelated to each other. You know, however, that they are related in unexpected ways, and as you carefully develop the material that draws the necessary ideas together, perhaps over a period of several class hours, you look forward to the moment when the curtain will be pulled aside to reveal the interconnections, when it will be seen that the questions can be made to tumble like a row of carefully placed dominoes.

At last the great moment arrives. You lay bare the culmination of your masterfully crafted chain of great ideas, and a hand goes up. "What," asks the inquiring young mind, "is this good for?" You wonder if Rossini had to justify the *William Tell Overture* by explaining that it would one day serve to introduce The Lone Ranger, or if Winslow Homer ever pointed out that his paintings could be used to cover up unsightly cracks in a plaster wall.

Some well-known mathematicians have gloried in the fact that their work had no practical uses. G. H. Hardy, recognized in his lifetime as one of the world's leading mathematicians, wrote in his *Mathematician's Apology,*

> I have never done anything useful. No discovery of mine has made, or is likely to make, directly or indirectly, for good or ill, the least difference to the amenity of the world. . . . I have added something to knowledge, and helped others to add more; and that these somethings have a value which differs in degree only, and not in kind, from that of creations of the great mathematicians, or of any of the other artists, great or small, who have left some kind of memorial behind them.

Paul Halmos attempted to explain his motivations for doing mathematics in an essay titled "Mathematics As a Creative Art." Many mathematicans would like to explain what mathematics is good for by echoing Jacobi. When charged with not applying his considerable talent to practical problems, he defended himself with the observation that the goal is not to serve public utility, but to honor the human spirit.

The majority of mathematicians, however, probably work with the hope that what they are doing will, in some way, turn out to be useful. George Birkhoff, a person who gave strong leadership to the development of mathematical research in the United States, wrote,

> A simple abstraction without present application is [not] to be regarded as without value. All abstractions are significant if they possess beauty; and the experience of the race shows that such abstractions are almost certain sooner or later to prove useful.

The history of science suggests that one cannot separate mathematics as an art form from mathematics as a tool that will find application in science. For those who take the extreme view of Hardy, this may be a dismaying fact, but the truth is that even the useless number theory in which Hardy took such delight has found application in cryptography, in error-correcting codes, and in other areas Hardy never anticipated. His appraisal of Einstein's work provides an example that ought to warn anyone who ventures an opinion as to whether or not a particular idea will have practical application. Quoting again from his *Apology* written in 1940,

> No one has yet discovered any warlike purpose to be served by . . . relativity, and it seems very unlikely that anyone will do so for many years.

Mathematics does have its applications. It has more applications than the average mathematician knows about, certainly more than the mathematician can explain. This does not excuse the classroom teacher from knowing some of the applications of the discipline, nor from making the effort to know more. Neither does the lack of an immediate application excuse the young student from trying to understand something that is in itself the culmination of an intense creative effort.

I am tempted, when asked about the usefulness of something I have presented in the mathematics classroom, to use a response I once heard attributed to Dr. Van Allen when he was asked what was the practical use of knowing about the Van Allen radiation belt that envelops our planet: It has for many years provided me with a very nice living.

13.1 Matrices

Tuning In

Except for Sunday evenings when we listen to a favorite program on a religious radio station, the radio in our kitchen is always tuned to either KSJN or WCAL (referred to henceforth as K or W), and once we choose a station in the morning, we tend to leave the dial set for the day. If the dial is set to K in the morning, the probability that it will still be there when we shut the radio off is $\frac{3}{4}$, but if it is set to W in the morning, the probability of its still being there in the evening is just $\frac{1}{2}$. If we are equally likely to start Monday morning with either K or W, what is the probability that the dial will be on K when we shut the radio off on (a) Monday night? (b) Tuesday night?

The home do-it-yourselfer occasionally discovers among collected "junk" something that can be fashioned into just the tool that is needed. Scientists sometimes have the same luck. While working to develop a certain theory, they suddenly discover that the very mathematical tools they need have been available for a long time.

Matrices, the subject of this chapter, have been "discovered" several times. Much of what we know about the subject today was being used in the early 1800s by people using the tools of calculus and differential equations to study celestial mechanics. Then in 1858, Arthur Cayley wrote a paper often cited as the place where matrices were invented. Nevertheless, it apparently came as something of a surprise to Werner Heisenberg when he realized in 1925 that matrices were just what he needed to develop his idea of quantum mechanics. And it is safe to say that all of the early workers would be surprised if they could now see the uses to which matrices are put in modern physics, economics, business, statistics, and systems research.

Here, then, is an example that nicely portrays what we wish to say about the role of abstraction in mathematics. First arising in a specific application, later developed as an object of mathematical curiosity, a subject ultimately becomes an indispensable tool in applied mathematics. The abstraction lifts the subject from the context in which it first arises, identifies the crucial ideas, and develops their consequences, sometimes with and sometimes without reference to particular applications. Ultimate applications may not even be anticipated as the subject is developed.

Some Definitions

A matrix is a rectangular array (box) of numbers arranged in rows and columns:

$$A\begin{bmatrix} -1 & 3 & 0 \\ 4 & -2 & 5 \end{bmatrix} \qquad B = \begin{bmatrix} 2 & 0 \\ -3 & 1 \\ 0 & 4 \end{bmatrix}$$

Matrix A is said to be 2×3 (read "2 by 3"), meaning it has two rows and three columns; matrix B is 3×2. Two matrices are said to be equal if and only if they have the same dimensions (same number of rows and same number of columns) and if their corresponding entries are equal.

There are applications in which the matrices are very large, but we for the most part confine ourselves to 2×2 matrices.

You as a reader now have a choice that roughly corresponds to two approaches to learning mathematics. You may try **Tuning In,** the subsection below that approaches matrix multiplication via an application intended to motivate an otherwise mysterious definition. Or, you may skip over to the subsection, "Multiplication of Matrices," that describes the mechanics of matrix multiplication, taking the purist position that a mathematical structure is worthy of study whether it has applications or not.

In the first approach, you run the risk of getting bogged down in details of the problem that obscure what is for us the main point, how to multiply two matrices together. In the second approach, you must resist the temptation to be critical of a definition that when first encountered seems so arbitrary and capricious.

Perhaps that best strategy is to read **Tuning In** quickly, remembering as you go that motivation is the main point, and that what you really need to know will be explained in the next subsection.

Tuning In

The behavior described in tuning our radio can be summarized as follows.

End of the Day

$$\text{Start the day}\quad \begin{matrix} \\ \text{K} \\ \text{W} \end{matrix}\begin{matrix} \text{K} & \text{W} \\ \left[\begin{matrix} \frac{3}{4} & \frac{1}{4} \\ \frac{1}{2} & \frac{1}{2} \end{matrix}\right] \end{matrix}$$

The matrix that, depending on the station where we begin, gives the probability of where we shall end, is called the **transition matrix.** Let us indicate the equally likely choices of either K or W on Monday morning by writing

$$[\text{K} \quad \text{W}] = \begin{bmatrix} \frac{1}{2} & \frac{1}{2} \end{bmatrix}$$

A tree diagram helps us see how to use the transition matrix to determine the probability of being tuned to K Monday night.

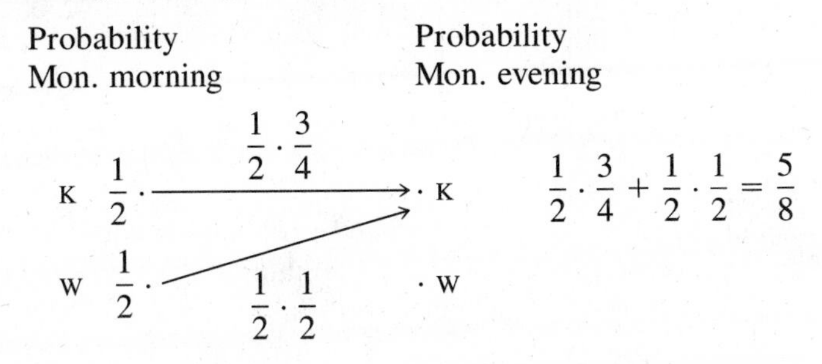

If we had focused on being tuned to W on Monday night, our diagram would have been

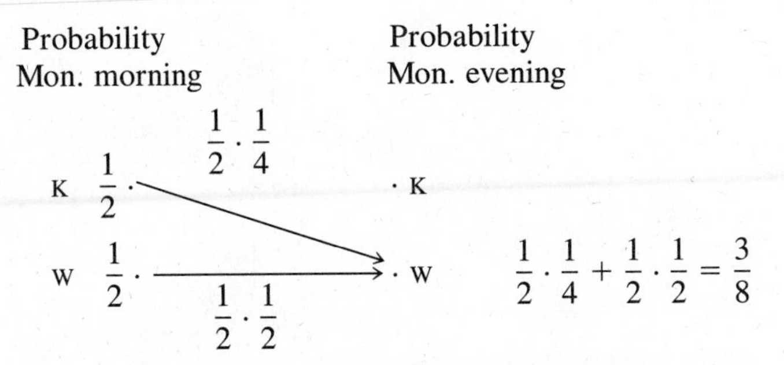

Since the probability of being on one station or the other is 1, it is confirming to observe that $\frac{5}{8} + \frac{3}{8} = 1$. The computations just discussed can be summarized this way.

$$\begin{bmatrix} \frac{1}{2} & \frac{1}{2} \end{bmatrix} \begin{bmatrix} \frac{3}{4} & \frac{1}{4} \\ \frac{1}{2} & \frac{1}{2} \end{bmatrix} = \begin{bmatrix} \frac{1}{2} \cdot \frac{3}{4} + \frac{1}{2} \cdot \frac{1}{2} & \frac{1}{2} \cdot \frac{1}{4} + \frac{1}{2} \cdot \frac{1}{2} \end{bmatrix}$$

$$= \begin{bmatrix} \frac{5}{8} & \frac{3}{8} \end{bmatrix}$$

Notice the multiplication of the 1×2 matrix by the 2×2 matrix. The elements in the first row (yes, the only row) of the first matrix are multiplied by the corresponding elements in the first column of the second matrix, and added to give the first entry in the product matrix. Similarly describe for yourself how the second entry of the product matrix was obtained.

Since the ending probabilities of Monday night are the beginning probabilities of Tuesday morning, the same computational scheme now gives us

$$\begin{bmatrix} \frac{5}{8} & \frac{3}{8} \end{bmatrix} \begin{bmatrix} \frac{3}{4} & \frac{1}{4} \\ \frac{1}{2} & \frac{1}{2} \end{bmatrix} = \begin{bmatrix} \frac{5}{8} \cdot \frac{3}{4} + \frac{3}{8} \cdot \frac{1}{2} & \frac{5}{8} \cdot \frac{1}{4} + \frac{3}{8} \cdot \frac{1}{2} \end{bmatrix}$$

$$= \begin{bmatrix} \frac{21}{32} & \frac{11}{32} \end{bmatrix}$$

Again we are pleased to note that the probability of being on one station or the other is $\frac{21}{32} + \frac{11}{32} = 1$. Our real interest, however, is to see how we might get one transition matrix to take us from Monday morning to Tuesday night. If in

$$\begin{bmatrix} \frac{5}{8} & \frac{3}{8} \end{bmatrix} \begin{bmatrix} \frac{3}{4} & \frac{1}{4} \\ \frac{1}{2} & \frac{1}{2} \end{bmatrix} = \begin{bmatrix} \frac{21}{32} & \frac{11}{32} \end{bmatrix}$$

we replace $\begin{bmatrix} \frac{5}{8} & \frac{3}{8} \end{bmatrix}$ by $\begin{bmatrix} \frac{1}{2} & \frac{1}{2} \end{bmatrix} \begin{bmatrix} \frac{3}{4} & \frac{1}{4} \\ \frac{1}{2} & \frac{1}{2} \end{bmatrix} = \begin{bmatrix} \frac{5}{8} & \frac{3}{8} \end{bmatrix}$, we get

$$\begin{bmatrix} \frac{1}{2} & \frac{1}{2} \end{bmatrix} \begin{bmatrix} \frac{3}{4} & \frac{1}{4} \\ \frac{1}{2} & \frac{1}{2} \end{bmatrix} \begin{bmatrix} \frac{3}{4} & \frac{1}{4} \\ \frac{1}{2} & \frac{1}{2} \end{bmatrix} = \begin{bmatrix} \frac{21}{32} & \frac{11}{32} \end{bmatrix},$$

How now shall we get a single 2 × 2 matrix that multiplies $[\frac{1}{2} \ \ \frac{1}{2}]$ to give $[\frac{21}{32} \ \ \frac{11}{32}]$? Follow the same pattern of using rows from the first 2 × 2 matrix, columns from the second. Thus, the entries are

$$\begin{bmatrix} \frac{3}{4} \cdot \frac{3}{4} + \frac{1}{4} \cdot \frac{1}{2} & \frac{3}{4} \cdot \frac{1}{4} + \frac{1}{4} \cdot \frac{1}{2} \\ \frac{1}{2} \cdot \frac{3}{4} + \frac{1}{2} \cdot \frac{1}{2} & \frac{1}{2} \cdot \frac{1}{4} + \frac{1}{2} \cdot \frac{1}{2} \end{bmatrix} = \begin{bmatrix} \frac{11}{16} & \frac{5}{16} \\ \frac{10}{16} & \frac{6}{16} \end{bmatrix}.$$

It only remains to verify that

$$\begin{bmatrix} \frac{1}{2} & \frac{1}{2} \end{bmatrix} \begin{bmatrix} \frac{11}{16} & \frac{5}{16} \\ \frac{10}{16} & \frac{6}{16} \end{bmatrix} = \begin{bmatrix} \frac{21}{32} & \frac{11}{32} \end{bmatrix},$$

and multiplication according to the rules we have been using confirms that it is so. We are ready to formally define the multiplication of matrices.

Multiplication of Matrices

The product of two 2 × 2 matrices is again a 2 × 2 matrix, obtained according to the following rule.

$$\begin{bmatrix} a & b \\ c & d \end{bmatrix} \cdot \begin{bmatrix} A & B \\ C & D \end{bmatrix} = \begin{bmatrix} aA + bC & aB + bD \\ cA + dC & cB + dD \end{bmatrix}$$

The definition is difficult to remember without a few pointers. In this case, one's fingers will do. To obtain the entry in the first row and first column of the product, place the left index finger on a and the right one on A. Slide the left finger along the first (top) row and the right finger down the first (left) column. The fingers in turn point to aA and bC. Their sum is the desired entry.

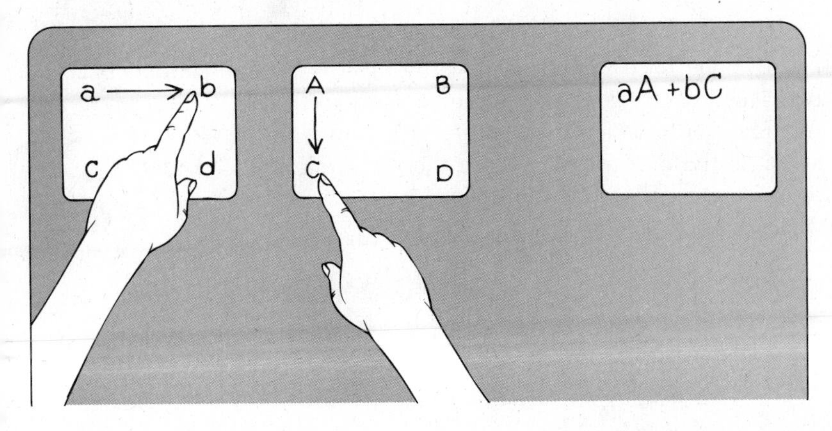

Similarly, to get the entry in the second row and first column of the product, place the left index finger on c, preparing to slide it along the second row. Place the right index finger on A, preparing to slide it down the first column simultaneously. This time the fingers pick out the products cA and dC.

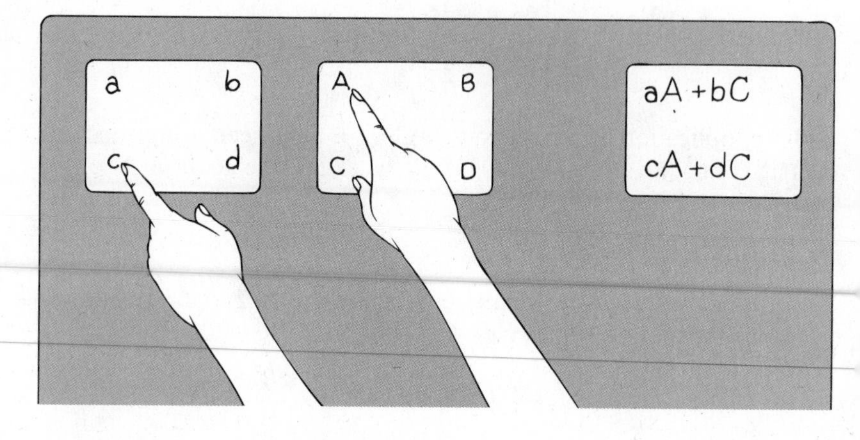

Use your fingers as well as your head to follow this numerical example:

$$\begin{bmatrix} 2 & 3 \\ 1 & 4 \end{bmatrix} \begin{bmatrix} -3 & -1 \\ 0 & 5 \end{bmatrix} = \begin{bmatrix} 2(-3) + 3(0) & 2(-1) + 3(5) \\ 1(-3) + 4(0) & 1(-1) + 4(5) \end{bmatrix} = \begin{bmatrix} -6 & 13 \\ -3 & 19 \end{bmatrix}$$

Though our definition covers only the multiplication of 2×2 matrices, the "finger rule" points the way to the multiplication of any two matrics in which the number of columns in the left matrix is the same as the number of rows in the right. Thus 2×3 matrix A and 3×2 matrix B can be multiplied:

$$AB = \begin{bmatrix} -1 & 3 & 0 \\ 4 & -2 & 5 \end{bmatrix} \begin{bmatrix} 2 & 0 \\ -3 & 1 \\ 0 & 4 \end{bmatrix}$$

$$= \begin{bmatrix} (-1)(2) + 3(-3) + 0(0) & (-1)(0) + 3(1) + 0(4) \\ 4(2) + (-2)(-3) + 5(0) & 4(0) + (-2)(1) + 5(4) \end{bmatrix}$$

$$= \begin{bmatrix} -11 & 3 \\ 14 & 18 \end{bmatrix}$$

Addition of Matrices

Contrary to the rule for multiplication (actually, it is the rule for multiplication that is contrary), the definition for the addition of matrices is straightforward. Any two matrices with the same dimensions are added by simply adding corresponding terms. For the 2×2 matrices in which we are primarily interested,

$$\begin{bmatrix} a & b \\ c & d \end{bmatrix} + \begin{bmatrix} A & B \\ C & D \end{bmatrix} = \begin{bmatrix} a + A & b + B \\ c + C & d + D \end{bmatrix}$$

Summary

In this section we have introduced the concept of a matrix, explained what we mean by the dimensions of a matrix, and stated the conditions under which we call two matrices equal. Definitions for addition and multiplication have been given, the first being straightforward and the second at first seeming peculiar. It should be pointed out that matrices, together with the definitions of addition and multiplication, acquaint us with another mathematical system.

This new mathematical system has some strange properties, especially with respect to multiplication. These have been left for you to discover, and it is therefore strongly recommended that you do at least all the odd-numbered problems in order to see what is to be discovered.

Problem Set 13.1

$$A = \begin{bmatrix} 1 & 1 \\ 1 & 2 \end{bmatrix} \quad B = \begin{bmatrix} 5 & 7 \\ 2 & 3 \end{bmatrix} \quad C = \begin{bmatrix} 3 & -1 \\ -5 & 2 \end{bmatrix} \quad D = \begin{bmatrix} 5 & -3 \\ -3 & 2 \end{bmatrix} \quad I = \begin{bmatrix} 1 & 0 \\ 0 & 1 \end{bmatrix}$$

1. Find AB and BA. Find BC and CB. What have you learned?

2. Find AC and CA. Find CD and DC. What have you learned?

3. In Problem 1, you found AB and BC. Call the answers M and N, respectively. Find MC and AN.

4. In Problem 2 you found AC and CD. Call the answers R and S, respectively. Find RD and AS.

5. In Problem 3, you found $MC = (AB)C$ and $AN = A(BC)$. In the same way find $(BC)D$ and $B(CD)$.

6. In Problem 4, you found $RD = (AC)D$ and $AS = A(CD)$. In the same way find $(CD)B$ and $C(DB)$.

7. Find $B + C$, and then $A(B + C)$. Does this equal $AB + AC$?

8. Find $C + D$, and then $(C + D)B$. Does this equal $CB + DB$?

9. Find AI, IA, DI, and ID.

10. Find BI, IB, CI, and IC.

11. (a) Let $E = \begin{bmatrix} 2 & -1 \\ -1 & 1 \end{bmatrix}$. Find AE and EA.

(b) Let $F = \begin{bmatrix} 2 & 1 \\ 5 & 3 \end{bmatrix}$. Find CF and FC.

12. (a) Let $G = \begin{bmatrix} 3 & -7 \\ -2 & 5 \end{bmatrix}$. Find BG and GB.

(b) Let $H = \begin{bmatrix} 2 & 3 \\ 3 & 5 \end{bmatrix}$. Find DH and HD.

13. Given a matrix K, the matrix L is called the inverse of K with respect to multiplication if $KL = I$. Try to find an inverse with respect to multiplication for each of the following.

(a) $\begin{bmatrix} 7 & 4 \\ 5 & 3 \end{bmatrix}$ (b) $\begin{bmatrix} 8 & 5 \\ 5 & 3 \end{bmatrix}$ (c) $\begin{bmatrix} 4 & 7 \\ 2 & 4 \end{bmatrix}$ (d) $\begin{bmatrix} 4 & 2 \\ 6 & 3 \end{bmatrix}$

14. As in Problem 11, try to find an inverse with respect to multiplication for each of the following.

(a) $\begin{bmatrix} 5 & 7 \\ 2 & 3 \end{bmatrix}$ (b) $\begin{bmatrix} 3 & 5 \\ 2 & 3 \end{bmatrix}$ (c) $\begin{bmatrix} 4 & 3 \\ 2 & 2 \end{bmatrix}$ (d) $\begin{bmatrix} 6 & 9 \\ 2 & 3 \end{bmatrix}$

15. Compute the following products.

(a) $\begin{bmatrix} 3 & 1 \\ 0 & 2 \\ 1 & 4 \end{bmatrix} \begin{bmatrix} 1 & 2 & 1 \\ 3 & 1 & 0 \end{bmatrix}$ (b) $\begin{bmatrix} 3 & 1 & 0 \\ 2 & 0 & 2 \end{bmatrix} \begin{bmatrix} 1 & -1 & 2 \\ 2 & 1 & 0 \\ 0 & 1 & -1 \end{bmatrix}$

16. Compute the following products.

(a) $\begin{bmatrix} 1 & 2 \\ 0 & 1 \\ 2 & 2 \\ 1 & 0 \end{bmatrix} \begin{bmatrix} 3 & 1 & 0 \\ 1 & 2 & 4 \end{bmatrix}$ (b) $\begin{bmatrix} 1 & 3 & -1 \\ 0 & 1 & 2 \end{bmatrix} \begin{bmatrix} 1 & 2 \\ 4 & 1 \\ -1 & 0 \end{bmatrix}$

17. What must be true of the dimensions k, l, m, and n of two matrices

$$\overset{l}{k\begin{bmatrix} & \end{bmatrix}} \quad \overset{n}{m\begin{bmatrix} & \end{bmatrix}}$$

if they are to be multiplied? What will be the dimensions of the product?

For Research and Discussion

A system (such as a radio dial) that begins in one of n states (such as being tuned to one of n stations) and moves according to known probabilities to one of the other n states is, subject to satisfaction of certain conditions, called a **Markov Process.** Look up a careful description of such a process, and make a list of applications that can be modeled as Markov Processes. What properties always characterize a transition matrix?

13.2 Properties of Matrix Multiplication

Such Things Ought Not To Be

The number 0 has the distinguishing feature that adding it to any number x does not affect x; $x + 0 = x$. The 2×2 matrix $O = \begin{bmatrix} 0 & 0 \\ 0 & 0 \end{bmatrix}$ is the zero matrix because adding it to any matrix A does not affect A.

$$A + O = \begin{bmatrix} a_1 & a_2 \\ a_3 & a_4 \end{bmatrix} + \begin{bmatrix} 0 & 0 \\ 0 & 0 \end{bmatrix} = \begin{bmatrix} a_1 & a_2 \\ a_3 & a_4 \end{bmatrix} = A$$

But odd things can happen. Find examples to show that

(1) we can find matrices $A \neq O$, $B \neq O$, such that $AB = O$.
(2) we can find a matrix $A \neq O$ such that $A^2 = O$.
(3) we may not conclude from $AB = AC$ that $B = C$.

Perhaps it is appropriate to remind our readers again of the spirit in which we are proceeding, for some may wonder why anyone cares about the properties of matrix multiplication. Research of the kind we are trying to describe in this chapter often seeks the answer to a question, even when there is no apparent "payoff" for finding it. We are reminded of the time that we took a group of students on a tour of the Argonne National Laboratories. A resident scientist explained the elaborate apparatus he was constructing to determine the speed with which a particle was traveling. When he finished his lecture, a student asked the obvious question, "Why would anyone want to know?"

Slowly withdrawing the pipe from his mouth and using it to punctuate his remarks, he gave his answer. "I don't know why anyone would want to know, but someday, somebody, for some reason, might ask. And if they do" (here the pointer swung down with a flourish that almost left our scientist friend impaled on the stub end of his pipe) "I'm going to know the answer."

It is this spirit of curiosity, this willingness to pause and examine the unusual, that often carries science forward. And it is on this same curiosity that we now rely in urging the reader to further explore some of the peculiarities of matrix multiplication as they emerged from Problem Set 13.1.

Matrix multiplication is not commutative.

One of the things quickly learned from a few examples is that there are matrices A and B for which $AB \neq BA$. That is, matrix multiplication is not **commutative.** We all know that it makes a difference as to whether we take off our slippers and step into the bath water, or vice versa, but it comes as a shock to most people to find that there are useful mathematical systems in which $AB \neq BA$.

Such unconventional behaviour surely arouses suspicions about other laws long taken for granted. For example, is the following always true?

$$(AB)C = A(BC)$$

Matrix multiplication is associative.

This is called the **associative** property, and it does hold for matrix multiplication. Though examples don't prove anything (a point we have stressed repeatedly), they do suggest things. In the case of Problems 3 through 6 in Problem Set 13.1, they suggest something that is true. Since the proof involves cumbersome notation, we ask the reader to take our word for it, even though such advice is contrary to our warnings about proof by authority.

The matrix

$$I = \begin{bmatrix} 1 & 0 \\ 0 & 1 \end{bmatrix}$$

plays an important role in matrix multiplication:

$$\begin{bmatrix} 1 & 0 \\ 0 & 1 \end{bmatrix}\begin{bmatrix} a & b \\ c & d \end{bmatrix} = \begin{bmatrix} a & b \\ c & d \end{bmatrix} = \begin{bmatrix} a & b \\ c & d \end{bmatrix}\begin{bmatrix} 1 & 0 \\ 0 & 1 \end{bmatrix}$$

$I = \begin{bmatrix} 1 & 0 \\ 0 & 1 \end{bmatrix}$

is the multiplicative identity.

It commutes with any matrix, and it acts the way the number 1 acts in ordinary multiplication. It is called the identity with respect to multiplication, or simply the **multiplicative identity.**

Given a matrix K, the matrix L is called the inverse of K with respect to multiplication (or the **multiplicative inverse**) if $LK = I$. Thus, since

$$\begin{bmatrix} 2 & -5 \\ -3 & 8 \end{bmatrix}\begin{bmatrix} 8 & 5 \\ 3 & 2 \end{bmatrix} = \begin{bmatrix} 1 & 0 \\ 0 & 1 \end{bmatrix}$$

we say that

$$\begin{bmatrix} 2 & -5 \\ -3 & 8 \end{bmatrix}$$

is the inverse of

$$\begin{bmatrix} 8 & 5 \\ 3 & 2 \end{bmatrix}$$

After observing a few matrices and their inverses, most students are ready to make a guess:

$$\text{The inverse of } \begin{bmatrix} a & b \\ c & d \end{bmatrix} \text{ may be } \begin{bmatrix} d & -b \\ -c & a \end{bmatrix}.$$

We have tried in the past to encourage guessing as a part of mathematics. We have also tried to encourage the checking of guesses against numerous examples. The guess above suggests that

$$\text{The inverse of } \begin{bmatrix} 5 & 6 \\ 2 & 3 \end{bmatrix} \text{ may be } \begin{bmatrix} 3 & -6 \\ -2 & 5 \end{bmatrix}.$$

It may be, but alas it isn't. The product gives

$$\begin{bmatrix} 3 & -6 \\ -2 & 5 \end{bmatrix}\begin{bmatrix} 5 & 6 \\ 2 & 3 \end{bmatrix} = \begin{bmatrix} 3 & 0 \\ 0 & 3 \end{bmatrix}$$

This has the merit of 0's in the right places, but it suffers the defect of having 3's where we wanted 1's. Some insight is gained by checking the guess in its general form, rather than for a particular matrix. That is, try multiplying

$$\begin{bmatrix} d & -b \\ -c & a \end{bmatrix}\begin{bmatrix} a & b \\ c & d \end{bmatrix} = \begin{bmatrix} ad - bc & 0 \\ 0 & ad - bc \end{bmatrix}$$

Evidently the guess works only so long as $ad - bc = 1$. This explains why it worked for some examples but not others in Problems 11 through 14 in Problem Set 13.1.

For the given matrix

$$K = \begin{bmatrix} a & b \\ c & d \end{bmatrix}$$

let us set $ad - bc = D$. The number D is called the **determinant** of matrix K. After some experimenting (yes, one can conduct experiments in mathematics), it usually occurs to the venturesome soul (the one who tries to find out instead of waiting to be told) to try what we shall call

$$K^{-1} = \begin{bmatrix} d/D & -b/D \\ -c/D & a/D \end{bmatrix}$$

as a possible inverse matrix. This in fact works as long as $D \neq 0$. When $D = 0$, there is no multiplicative inverse for K.

Since matrix multiplication is not commutative, it comes as something of a surprise that K^{-1} works on either side; that is, $K^{-1}K = KK^{-1} = I$.

A word about notation is in order. Students who recall their algebra well enough may remember that 3^{-1} means $\frac{1}{3}$. It does not follow that

K^{-1} means 1 over K. When K is a matrix, this would result in a very odd-looking creature. One should think of K^{-1} as the matrix that multiplies K to give I. It does not hurt to think of 3^{-1} as the number that multiplies 3 to give 1.

Let us state formally what we have learned about the multiplicative inverse of a matrix

$$K = \begin{bmatrix} a & b \\ c & d \end{bmatrix}$$

We can find the multiplicative inverse if $D \neq 0$.

Theorem Set $D = ad - bc$. If $D = 0$, no inverse exists. If $D \neq 0$, the inverse is

$$K^{-1} = \begin{bmatrix} d/D & -b/D \\ -c/D & a/D \end{bmatrix}$$

Once the right guess is made, the proof is easy. Simply multiply $K^{-1} \cdot K$ and $K \cdot K^{-1}$. The products will both be I, every time.

Finally we mention the distributive law which holds for matrices (as suggested by Problems 7 and 8 in Problem Set 13.1).

$$A(B + C) = AB + BC$$

Matrices satisfy the distributive law.

The only caution to be exercised in using the distributive property is to be sure to pay attention to the order of multiplication. Thus

$$(B + C)A = BA + CA$$

but

$$(B + C)A \neq AB + AC$$

Summary

We have stressed the following properties for 2×2 matrices:

1. Multiplication is associative; that is, for every A, B, and C, $A(BC) = (AB)C$.
2. Multiplication is not commutative; there exist matrices A and B for which $AB \neq BA$.
3. There is a multiplicative identity, which we denote by I, satisfying $AI = IA = A$.
4. If a matrix A has a nonzero determinant, there is a multiplicative inverse A^{-1} satisfying

$$A \cdot A^{-1} = A^{-1} \cdot A = I$$

5. The distributive laws hold:

$$A \cdot (B + C) = AB + AC$$
$$(B + C) \cdot A = BA + CA$$

The reader should, by choosing arbitrary 2×2 matrices, be able to illustrate any of these properties.

While we have not stressed the same properties for addition, it is easy to verify that matrix addition is associative and commutative, that there is an additive identity, and that every matrix has an additive inverse (see Problems 17 and 18).

Problem Set 13.2

$$A = \begin{bmatrix} 1 & -2 \\ 2 & -3 \end{bmatrix} \quad B = \begin{bmatrix} -1 & 0 \\ 3 & 1 \end{bmatrix} \quad C = \begin{bmatrix} 2 & -1 \\ 1 & 4 \end{bmatrix} \quad D = \begin{bmatrix} 1 & 4 \\ 2 & 0 \end{bmatrix}$$

1. Use matrices A, B, and C to illustrate the associative law for multiplication.

2. Use matrices B, C, and D to illustrate the associative law for multiplication.

3. Use matrices A and B to show that multiplication is not commutative.

4. Use matrices C and D to show that multiplication is not commutative.

5. Use matrices A, B, and C to illustrate the distributive law $A(B + C) = AB + AC$.

6. Use matrices, B, C, and D to illustrate the distributive law $B(C + D) = BC + BD$.

7. Find the multiplicative inverse of
(a) A (b) C (c) AC
Is the inverse of AC equal to $A^{-1}C^{-1}$?

8. Find the multiplicative inverse of
(a) B (b) D (c) BD
Is the inverse of BD equal to $B^{-1}D^{-1}$?

9. Find $C^{-1}A^{-1}$. Find the inverse of CA. Compare your answers with those for Problem 7.

10. Find $D^{-1}B^{-1}$. Find the inverse of DB. Compare your answers with those for Problem 8.

11. Let det A denote the determinant of A. Find det A, det B, and det AB.

12. Using the notation of Problem 11, find det C, det D, and det CD.

13. Use the distributive law to show that $(A + B)^2 = A^2 + AB + BA + B^2$. Illustrate this with matrices A, B, C, and D.

14. Use matrices C and D to show that $(C + D)^2 \neq C^2 + 2CD + D^2$.

15. Illustrate that $A(BC) = (AB)C$ using the 3×3 matrices

$$A = \begin{bmatrix} 1 & 0 & -1 \\ 2 & 1 & 1 \\ 3 & 0 & -2 \end{bmatrix} \quad B = \begin{bmatrix} 2 & 1 & 1 \\ 1 & 3 & 2 \\ -1 & 0 & 0 \end{bmatrix} \quad C = \begin{bmatrix} 0 & 2 & -2 \\ -1 & 1 & 0 \\ 2 & 0 & 3 \end{bmatrix}$$

16. Illustrate that $(BA)C = B(AC)$ using the 3×3 matrices in Problem 15.

17. What is the additive identity in the system of 2×2 matrices?

18. For a given 2×2 matrix

$$\begin{bmatrix} a & b \\ c & d \end{bmatrix}$$

what is the additive inverse?

19. Does $\det (A + B) = \det A + \det B$?

Something to Discover

$$M = \begin{bmatrix} a & b \\ c & d \end{bmatrix}$$

The determinant of M is defined to be the number $ad - bc$.

$$A = \begin{bmatrix} 3 & 5 \\ 2 & 4 \end{bmatrix} \qquad B = \begin{bmatrix} 5 & 2 \\ 6 & 3 \end{bmatrix}$$

For instance, the determinant of A is 2, and the determinant of B is 3. Find the product AB. What is its determinant?

For Research and Discussion

The study of determinants historically preceded the study of matrices, and much is known about them. We have already hinted at the multiplicative property which says that if $AB = C$, then $(\det A)(\det B) = \det C$.

1. Can you prove the multiplicative property?
2. Refer to **Such Things Ought Not To Be** Explain why many of the matrices used in constructing such peculiar examples must be matrices for which $\det A = 0$.
3. Let $A(a_1, a_2)$ and $B(b_1, b_2)$ be two points in the plane. How does the area of ΔOAB compare with the determinant of the matrix $\begin{bmatrix} a_1 & a_2 \\ b_1 & b_2 \end{bmatrix}$?

Consider special cases; in this case, that would mean locating A and B to get triangles for which it is easy for you to find their area.

What other properties of 2×2 determinants can you discover?

13.3 Some Applications

A 3 × 3 Box of Help

Check the following multiplication of two 3×3 matrices.

$$\begin{bmatrix} 1 & 2 & -1 \\ 2 & 1 & -1 \\ -1 & 3 & -1 \end{bmatrix} \begin{bmatrix} 2 & -1 & -1 \\ 3 & -2 & -1 \\ 7 & -5 & -3 \end{bmatrix} = \begin{bmatrix} 1 & 0 & 0 \\ 0 & 1 & 0 \\ 0 & 0 & 1 \end{bmatrix}$$

Does this help you to solve the system of equations

$$\begin{aligned} 2x - y - z &= 2 \\ 3x - 2y - z &= 5 \\ 7x - 5y - 3z &= -3 \end{aligned}$$

Our presentation has been somewhat in the spirit of the historical development of matrices. Much was known about their properties before there were many applications, but the applications have been both numerous and important. This tendency for applications to follow theoretical developments is a feature that, for some people, justifies abstract research. To others, such justification is unnecessary, a kind of insult to the integrity of intellectual creativeness—like defending a symphony on the grounds that workers in a factory produce faster when it is played as background music.

Useless

I have never done anything "useful" . . . I have helped to train other mathematicians, but mathematicians of the same kind as myself, and their work has been, so far at any rate as I have helped them to it, as useless as my own. Judged by all practical standards, the value of my mathematical life is nil. . . . Time may change all this. No one foresaw the applications of matrices and groups and other purely mathematical theories to modern physics, and it may be that some "highbrow" applied mathematics will become useful in as unexpected a way; but the evidence so far points to the conclusion that in one subject as in another, it is what is commonplace and dull that counts for private life.

G. H. Hardy, A Mathematician's Apology, (Cambridge University Press, 1967)

G. H. Hardy was generally acknowledged during his lifetime (1877–1947) to be one of the world's leading mathematicians. A student of number theory, he delighted in the esoteric quality of his work.

Having lectured a bit in the last two sections on the subject of being willing to explore matrices in the almost total absence of any practical motivation, we now tip our hand by including at least a few applications in this closing section. Even at the risk of putting ourselves at odds with so eminent a mathematician as G. H. Hardy, we admit to feeling that the most interesting mathematics does have applications. It doesn't seem right, then, to leave the topic of matrices without pointing the way to at least some of the practical uses to which they can be put.

Solving Systems of Equations

In Section 2.6, while trying to solve a little puzzle problem about grazing oxen attributed to Isaac Newton, we were confronted with the necessity of solving the set of equations

$$10s + 40g = 144$$
$$10s + 90g = 189$$

We are now able to rewrite this system of equations as a single matrix equation:

$$\begin{bmatrix} 10 & 40 \\ 10 & 90 \end{bmatrix} \begin{bmatrix} s \\ g \end{bmatrix} = \begin{bmatrix} 144 \\ 189 \end{bmatrix}$$

How do we solve an equation of the form $AZ = B$? The first answer to the question is usually, "Divide by A." The problem, however, if A, Z, and B are matrices, is that we don't have a definition for matrix division.

We do have a definition for multiplication. Can we solve $AZ = B$ by multiplication? Of course we can if we know the multiplicative inverse of A. And as it happens, we do know how to find A^{-1} for a 2×2 matrix if $\det A \neq 0$. This suggests the pattern. Multiply both sides by A^{-1}:

$$\begin{aligned} A^{-1}(AZ) &= A^{-1}B \\ (A^{-1}A)Z &= A^{-1}B \\ IZ &= A^{-1}B \\ Z &= A^{-1}B \end{aligned}$$

Note that we multiplied both sides of the equation on the left by A^{-1}. We did not write

$$AZA^{-1} = BA^{-1}$$

This would of course be correct, for it obeys the golden rule of algebra. It is correct, but it is not helpful, because we can't put A^{-1} next to A since A and Z don't commute.

What you do to one side, you must do to the other.

Neither did we write

$$A^{-1}AZ = BA^{-1}$$

This wouldn't even be correct, for we have not obeyed the golden rule. We multiplied the left side by A^{-1} on the left, and the right side by A^{-1} on the right.

We are now ready to solve our matrix equation

$$\begin{bmatrix} 10 & 40 \\ 10 & 90 \end{bmatrix} \begin{bmatrix} s \\ g \end{bmatrix} = \begin{bmatrix} 144 \\ 189 \end{bmatrix}$$

The determinant of the coefficient matrix is 500, so the inverse is

$$\begin{bmatrix} \dfrac{9}{50} & -\dfrac{4}{50} \\ -\dfrac{1}{50} & \dfrac{1}{50} \end{bmatrix}$$

and multiplication of both sides of our equation on the left gives

$$\begin{bmatrix} \dfrac{9}{50} & -\dfrac{4}{50} \\ -\dfrac{1}{50} & \dfrac{1}{50} \end{bmatrix} \begin{bmatrix} 10 & 40 \\ 10 & 90 \end{bmatrix} \begin{bmatrix} s \\ g \end{bmatrix} = \begin{bmatrix} \dfrac{9}{50} & -\dfrac{4}{50} \\ -\dfrac{1}{50} & \dfrac{1}{50} \end{bmatrix} \begin{bmatrix} 144 \\ 189 \end{bmatrix}$$

or

$$\begin{bmatrix} s \\ g \end{bmatrix} = \begin{bmatrix} \dfrac{9}{50} & -\dfrac{4}{50} \\ -\dfrac{1}{50} & \dfrac{1}{50} \end{bmatrix} \begin{bmatrix} 144 \\ 189 \end{bmatrix} = \begin{bmatrix} \dfrac{54}{5} \\ \dfrac{9}{10} \end{bmatrix}$$

Again, as in Section 2.6, we get $s = \frac{54}{5}$ and $g = \frac{9}{10}$.

One More Example

Consider the simultaneous equations

$$x + y = 21$$
$$2x + 4y = 70$$

We are now able to rewrite these as a single matrix equation:

$$\begin{bmatrix} 1 & 1 \\ 2 & 4 \end{bmatrix} \begin{bmatrix} x \\ y \end{bmatrix} = \begin{bmatrix} 21 \\ 70 \end{bmatrix}$$

The inverse of

$$\begin{bmatrix} 1 & 1 \\ 2 & 4 \end{bmatrix} \text{ is } \begin{bmatrix} \frac{4}{2} & -\frac{1}{2} \\ -\frac{2}{2} & \frac{1}{2} \end{bmatrix}$$

Multiplying both sides on the left by this inverse, we have

$$\begin{bmatrix} \frac{4}{2} & -\frac{1}{2} \\ -\frac{2}{2} & \frac{1}{2} \end{bmatrix} \begin{bmatrix} 1 & 1 \\ 2 & 4 \end{bmatrix} \begin{bmatrix} x \\ y \end{bmatrix} = \begin{bmatrix} \frac{4}{2} & -\frac{1}{2} \\ -\frac{2}{2} & \frac{1}{2} \end{bmatrix} \begin{bmatrix} 21 \\ 70 \end{bmatrix}$$

$$\begin{bmatrix} 1 & 0 \\ 0 & 1 \end{bmatrix} \begin{bmatrix} x \\ y \end{bmatrix} = \begin{bmatrix} \frac{84 - 70}{2} \\ \frac{-42 + 70}{2} \end{bmatrix}$$

$$\begin{bmatrix} x \\ y \end{bmatrix} = \begin{bmatrix} 7 \\ 14 \end{bmatrix}$$

Again, as in Section 2.6, we get $x = 7$ and $y = 14$.

The method just described works with a system of three equations in three unknowns, and more generally for n equations in n unknowns. Our problem of course is that we only know how to find inverses of 2×2 matrices. Since our purpose here is just to show why people become interested in studying matrices, we pursue this no further.

Summary

A system of n equations in n unknowns can be written as a matrix equation

$$AZ = B$$

If we know the multiplicative inverse A^{-1} of matrix A (being careful to multiply both sides on the left, since matrix multiplication is not commutative), we can solve the system:

$$A^{-1}AZ = A^{-1}B$$
$$Z = A^{-1}B$$

This is of practical use to us only in the case in which $n = 2$, since this is the only case for which we have learned how to find the multiplicative inverse A^{-1}. (You can find out about larger n's in books on matrix theory.)

There are numerous other applications in which the product of two matrices naturally displays desired information. These may involve multiplying nonsquare matrices, as some of the problems illustrate.

Problem Set 13.3

1. Solve the following sets of equations by matrix methods.

(a) $3x - 5y = 19$, $4x - 7y = 26$ (b) $3x + 4y = 4$, $x - 2y = -7$ (c) $3x + 4y = 5$, $2x - 5y = -12$

2. Solve the following sets of equations by matrix methods.

(a) $5x + 7y = -1$, $2x + 3y = 0$ (b) $4x - y = 8$, $6x + 3y = 3$ (c) $2x + 3y = 8$, $5x - 2y = 1$

3. Use matrix methods to solve the system

(a)
$$\begin{aligned} x - 2y + 3z &= 3 \\ x - y + 4z &= -2 \\ x + 6z &= 5 \end{aligned}$$

(b)
$$\begin{aligned} x - 2y + 3z &= 4 \\ x - y + 4z &= -3 \\ x + 6z &= 1 \end{aligned}$$

Hint: Multiply

$$\begin{bmatrix} -6 & 12 & -5 \\ -2 & 3 & -1 \\ 1 & -2 & 1 \end{bmatrix} \begin{bmatrix} 1 & -2 & 3 \\ 1 & -1 & 4 \\ 1 & 0 & 6 \end{bmatrix}$$

4. Use matrix methods to solve the systems

(a)
$$\begin{aligned} 2x + 9y + 13z &= 4 \\ 3x + 5y + 7z &= -2 \\ -x + 2y + 3z &= 5 \end{aligned}$$

(b)
$$\begin{aligned} 2x + 9y + 13z &= -1 \\ 3x + 5y + 7z &= 7 \\ -x + 2y + 3z &= 3 \end{aligned}$$

Hint: Multiply

$$\begin{bmatrix} 1 & -1 & -2 \\ -16 & 19 & 25 \\ 11 & -13 & -17 \end{bmatrix} \begin{bmatrix} 2 & 9 & 13 \\ 3 & 5 & 7 \\ -1 & 2 & 3 \end{bmatrix}$$

5. The Kindle Company, which sold the tables and chairs to Librarian Hardback (Section 12.4), has like most companies raised prices periodically. The prices for various years are indicated below. Find the cost of Hardback's orders (9 tables and 43 chairs) for each of the years indicated.

	Table	*Chair*
1970	195	12
1975	239	16
1980	288	19
1985	325	23
1990	349	26

6. The Kindle Company (see Problem 5) has warehouses in Posthole (P), Center City (C), and Klondike (K). The number of tables and chairs commonly stocked by each warehouse is indicated in the table. Find the value of the inventory in each warehouse for the years indicated in Problem 5.

	P	C	K
Tables	10	65	15
Chairs	75	350	100

7. Steady Eddy's Pizzeria offers four choices of pizza. The Cheese Special requires 1 unit of dough, 3 units of cheese, and 1 unit of tomato sauce. The Sausage requires 1 unit of dough, 1 unit of cheese, 1 unit of sausage, and 1 unit of tomato sauce. The Super Sausage calls for 1 unit of dough, 1 unit of cheese, $\frac{3}{2}$ units of sausage, 1 unit of mushrooms, and 1 unit of tomato sauce. The Large Special calls for 2 units of dough, $\frac{5}{2}$ units of cheese, 2 units of sausage, 2 units of mushrooms, and $\frac{5}{2}$ units of tomato sauce. Depending on whether Eddy buys supplies in small or large amounts he pays, per unit, for dough \$0.35 or \$0.28, for cheese \$0.35 or \$0.28, for sausage \$0.79 or \$0.63, for mushrooms \$0.18 or \$0.15, and for tomato sauce \$0.12 or \$0.09. Display these figures in two matrices in such a way that their product shows the cost of each choice of pizza for each of the two purchasing options.

For Research and Discussion

It is one thing for an individual to revel in a choice to work hard on things that have no evident practical purpose, but might it be another thing if that person is supported with public tax money? If the answer seems simple to you, you are lacking information on one side or the other.

On the side of support, look further into Hardy's admission that matrices have proved useful in applications. Note particularly the use made of matrices in George Dantzig's development of linear programming. What event motivated Dantzig's work, and when was it developed relative to Hardy's comments?

Note also that when Einstein did his work, an accurate measure of the speed of light proved most helpful. What motivated the physicist Albert Michelson to have the answer ready when Einstein needed it?

On the side of questioning such support, look into the array of research projects supported in your own state university. Do the same for projects supported by the National Science Foundation. Read in particular the debates among scientists about NSF support of "big science."

How will you, as an informed citizen, react to the use of your tax money to support basic research?

References:

Lieber, L., *The Education of T. C. Mits,* (Norton, 1942), Chapter 5.
Jaffe, B., *Michelson and the Speed of Light* (N.Y.: Doubleday, 1960).

chapter 14

Algebraic Structures

To carry out his role of abstractor, the mathematician must continually pose such questions as "What is the common aspect of diverse situations?" or "What is the heart of the matter?" He must always ask himself, "What makes such and such a thing tick?" Once he has discovered the answer to these questions and has extracted the crucial simple parts, he can examine these parts in isolation. He blinds himself, temporarily, to the whole picture, which may be confusing.

PHILIP DAVIS AND WILLIAM CHINN

Something Lasting

One of the most satisfied workers interviewed by Studs Terkel for his book *Working* was a stone mason. He drew his satisfaction from the fact that his work was about as lasting as anything a human being can do. "I can't imagine a job where you go home and maybe go by a year later and you don't know what you've done. My work, I can see what I did the first day I started. All my work is set right there in the open and I can look at it as I go by. It's something I can see for the rest of my life. . . . It's always there. Immortality as far as we're concerned."

Such satisfactions are available to the mathematician. Schoolchildren today still learn Euclid's geometry. A theorem proved is proved forever. Ideas are more lasting than the work of stonemasons and sculptors.

> Masterpieces of sculpture once shattered are difficult to restore or even remember. The greater ideas of mathematics survive and are carried along in the continual flow, permanent additions immune to the accident of fashion.
>
> E. T. Bell

In this respect mathematics differs from the natural sciences that it serves. Physicists once thought about light as being transmitted by ether, now defined in the dictionary as "a hypothetical invisible substance postulated in older theory . . ." Similarly, chemists once thought about fire as a process that produced phlogiston, another substance defined in the dictionary as "an imaginary element formerly believed to be given off by anything burning." Changing paradigms in the natural sciences have the effect of changing everything, including the vocabulary as well as the problems that people work on.

Barry Mazur commented on this aspect of mathematics shortly after Andrew Wiles announced in 1993 that he had proved Fermat's Last Theorem. He wrote,

> But as for Fermat's Last Theorem, here we have a very precise and very technical assertion scrawled in a margin in 1637. Three and a half centuries later, this technical assertion, unmodified in any way, is finally established.
>
> I wonder whether there could be any comparably "stable" technical assertion, for example in Physics or in Chemistry, announced in the middle of the seventeenth century—and established (in its own terms) at the end of ours. The very vocabulary of those sciences has changed so much in the interim: the year 1637 is roughly 140 years before Priestley, and Lavoisier grappled with the discovery, and invention of Oxygen.
>
> The dogged stability of the technical statement of Fermat's Last Theorem, waiting patiently for three and a half centuries for its affirmation, deserves some mention.

Whether or not Wiles' proof will stand the intense scrutiny it is getting as this is written is irrelevant to our point. Mathematicians are still grappling with the problem exactly as it was posed over 350 years ago.

Not only does the mathematician have the opportunity to create something lasting, but he or she also enjoys the sense of being part of a long chain of contributors, all aware that they do not replace, but rather build upon the work of their predecessors.

> The majority of ideas we deal with were conceived by others, often centuries ago.
>
> D. Mach

> If I have seen a little further than others, it is because I stood on the shoulders of giants.
>
> Isaac Newton

14.1 Basic Concepts of Algebra

May We Substitute?

1. Suppose Willie Hitit and Claude Candue have equal batting averages going into a game in which they both get 1 hit in 4 official times at bat. Will they necessarily have the same batting average after the game?
2. Will we get into any logical difficulties if we agree to call two matrices A and B equal if and only if $\det A = \det B$?
3. Amy and Beth, while comparing their stamp collections, decide to trade, each taking one of the other's stamps. Since no money was involved, they called it an equal trade. If it was equal, why make it?
4. *Homer wonders if Amy is the woman he will marry.* Amy is, in fact, the woman he will marry. May we, therefore, substitute "Amy" in the italicized sentence for, "the woman he will marry?" Only, it seems, if we can make sense of, "Homer wonders if Amy is Amy."

What do we mean when we say that two things are equal? When may we substitute equals for equals?

Matrix multiplication, we have seen, exhibits some properties that surprise us. We saw, for example, that for two matrices A and B it is generally the case that $AB \neq BA$. From examples such as

$$\begin{bmatrix} 2 & -1 \\ -4 & 2 \end{bmatrix}\begin{bmatrix} 1 & -1 \\ -1 & 3 \end{bmatrix} = \begin{bmatrix} 2 & -1 \\ -4 & 2 \end{bmatrix}\begin{bmatrix} 2 & -4 \\ 1 & -3 \end{bmatrix}$$

we learned that $AB = AC$ does not imply $B = C$. we have also seen situations such as

$$\begin{bmatrix} 2 & -1 \\ -4 & 2 \end{bmatrix}\begin{bmatrix} 3 & 1 \\ 6 & 2 \end{bmatrix} = \begin{bmatrix} 0 & 0 \\ 0 & 0 \end{bmatrix}$$

in which $AB = 0$ but $A \neq 0$ and $B \neq 0$.

These surprising properties are not unique to the system of matrices. In the system of integers modulo 12, we saw the nonsense that would result from canceling the 3's in the expression $3 \cdot 5 \equiv 3 \cdot 9$. And divisors of zero became commonplace as we got accustomed to such products as $3 \cdot 4 \equiv 6 \cdot 2 \equiv 9 \cdot 8 \equiv 0$.

We could introduce other algebraic systems exhibiting other surprising properties, but more examples are unlikely to change our general expectations. We continue to be surprised by anything that violates what our familiarity with arithmetic and high school algebra has conditioned us to expect. And this is wholesome. Our purpose here is not to confuse you about things you feel you understand, but to identify the basic principles that underlie both arithmetic and common algebra.

We have drawn our examples from the algebra of matrices and from modular arithmetic. These systems, as well as the number systems reviewed in Chapter 12 and common high school algebra, illustrate the concept of an algebraic structure. Like any system of deductive thought, algebraic structures begin with undefined terms and axioms. As was the case with the various geometric structures we studied (finite geometries and Euclidean and non-Euclidean geometries), most algebraic structures share a common terminology and some common axioms. And as was the case with geometries, algebraic structures differ from one another by virtue of certain variations in the axioms. Our purpose in this section is to introduce some of the terms and assumptions common to a variety of algebraic structures.

The **elements** of algebraic structures, that is, the objects about which we state axioms, are of course left undefined. They are designated by single letters, and one of the things you must keep in mind is that an element designated by b can be a matrix, a member of the set of integers modulo m, or something you've never heard of. Therefore you must be careful not to perform any operation on b (like assuming that $1/b$ makes sense) not specifically allowed by an axiom.

Binary Operations

Not knowing what kind of elements we are talking about, it is hard to say how they are to be combined (added, multiplied, intersected, etc.), but it is generally the case that there is at least one way to combine two elements to get another. Without specifying just how this is to be done, i.e., leaving it undefined, we refer to this combining of two terms as a **binary operation.** Suppose, for example, that the elements with which we are working are the rational numbers. We can create a binary operation $*$ by defining

$$a * b = \frac{a + b}{2}$$

$4 * 6 = 5$

$7 * 2 = \frac{9}{2}$

You have not seen such a definition before and, to be truthful, once you have put aside this book. you'll probably not see it again. In the spirit of abstract mathematics, however, this should not prevent us from investigating the properties of $*$ (Problems 4 through 5). It does what a binary operation is supposed to do; given two elements of a set, it tells us how to combine them to get a third element.

It is possible that a binary operation between two elements of a set may produce an element not in the set. Thus, if we start with the set of integers and use the operation $*$ defined above, we may obtain an element no longer in the set ($7 * 2$ is not an integer). Again, beginning with the set of integers and using the operation $\div$ of division, we are quickly taken outside our original system. Sometimes, however, we stay inside the system. The binary operation of ordinary addiition, $+$, applied to any two integers always gives another integer. We say the integers are closed with respect to addition. A set S is **closed** with respect to a binary operation $\circledast$ if, for any two elements x and y in S, the element $x \circledast y$ is again in S.

An analogy can be drawn between the notion of a closed set and the biblical notion in which two creatures coming together (a binary operation) produce a creature of their own kind. Mutations then correspond to nonclosed sets, in which a binary relationship produces an element not in the set. Since two odd integers, when they multiply, bring forth another odd integer, we say the odd integers are closed with respect to multiplication. This same set of odd integers is not closed, however, with respect to addition. Do you see why?

Equality

Another idea common to many algebraic structures, indeed to most mathematical structures (geometry, logic, etc.) is the notion of equality. Again the specific meaning of the term depends on the context. A youngster in the sixth grade understands

$$\frac{2}{3} = \frac{10}{15}$$

but she would think it quite peculiar if she saw $15 = 3$ written on the blackboard. Yet in arithmetic modulo 12, this is correct (though we have preferred to write $15 \equiv 3$). It means one thing to say $A = B$ if A and B are matrices; it means something else if A and B represent lengths of sides of triangles.

Whatever the specific meaning attached to the relationship (which we here designate by $\mathscr{E}$), there are four properties (axioms) that mathematicians always expect of an **equivalence relation.**

Determinative Given two elements r and s, we must have a clear rule which enables us to determine whether or not $r \mathscr{E} s$.

Reflexive For all elements r, it must be that $r \mathscr{E} r$.

Symmetric Whenever $r \mathscr{E} s$, it must be that $s \mathscr{E} r$.

Transitive If $r \mathscr{E} s$ and $s \mathscr{E} t$, then $r \mathscr{E} t$.

When we begin with the set of all triangles, the idea of congruence is an equivalence relation. In this situation, we usually write $r \cong s$. If we begin with the same set of triangles, the notion of similarity will also pass the tests (satisfy the axioms) listed above. When this is the idea of equivalence that concerns us, we commonly write $r \sim s$.

Well-Defined Binary Operations

There is one more idea, common to many algebraic structures, that should be mentioned while we are discussing binary operations and notions of equality. We illustrate it by using a method for "adding" fractions which, while wrong by commonly accepted standards, is nevertheless popular. Consider the rule

$$\frac{a}{b} \text{ “+” } \frac{c}{d} = \frac{a + c}{b + d}$$

Having said already that this rule is wrong by commonly accepted standards, let us point out that there are applications in which this is exactly the correct rule. Suppose, for example, that about midseason the mighty Casey has been at bat 261 times and has hit safely 88 times. His average is $88/261 = .337$. Supppose, further, that in the remainder of the season he goes to the bat 232 times and gets 65 hits. Then his average is found by writing

$$\frac{88}{261} \text{ “+” } \frac{65}{232} = \frac{88 + 65}{261 + 232} = \frac{153}{493} = .310$$

This application of "addition" notwithstanding, it is still the case that most of the time we do not accept this definition as a reasonable way to add fractions, and there are several good reasons for this. One of them is that there are few situations in life in which this kind of "addition" gives results that correspond to our expectations. We wish, however, to stress another good reason for rejecting this definition. We noted above that

$$\frac{2}{3} = \frac{10}{15}$$

Now consider the consequences of "adding" $\frac{2}{5}$ to both sides. On the left we get

$$\frac{2}{3} \text{ “+” } \frac{2}{5} = \frac{4}{8} = \frac{1}{2}$$

while on the right we get

$$\frac{10}{15} \text{ “+” } \frac{2}{5} = \frac{12}{20} = \frac{3}{5}$$

Surely if we "add" the same element to equals, the result should be equal, and on these grounds (along with others, as we said before), this kind of addition is disappointing.

Thus is illustrated our last idea, namely, that a binary operation $\circledast$ between elements of a set in which there is an equivalence relation $\mathscr{E}$, should be **well-defined** with respect to $\mathscr{E}$. That is, if

$$r \mathscr{E} s$$

we definitely want it to be true that, for any element t,

$$(r \circledast t) \mathscr{E} (s \circledast t)$$

and

$$(t \circledast r) \mathscr{E} (t \circledast s)$$

Note the care with which we have stated our definition. We have *not* said that, if $r \mathscr{E} s$, then $(r \circledast t) \mathscr{E} (t \circledast s)$. What is the difference?

Our study of geometry conditions us to expect that theorems can be proved from axioms; that is, the axioms say more than is immediately evident. This is as true in algebra as elsewhere. Note that our definition of a well-defined operation requires only that we be able to operate (multiply, add, or whatever) on equals with the same element. However, this implies the commonly accepted proposition that we can operate on equals with equals (e.g., add equals to equals and multiply equals by equals). We conclude with a demonstration of this fact, our first proof of an abstract algebraic theorem.

Theorem If $r \mathscr{E} s$, $t \mathscr{E} u$, and $\circledast$ is well-defined with respect to $\mathscr{E}$, then $(r \circledast t) \mathscr{E} (s \circledast u)$.

PROOF From the fact that $r \mathscr{E} s$,

$$(r \circledast t) \mathscr{E} (s \circledast t)$$

and from the fact that $t \mathscr{E} u$,

$$(s \circledast t) \mathscr{E} (s \circledast u)$$

Now from the transitive property of $\mathscr{E}$, we have

$$(r \circledast t) \mathscr{E} (s \circledast u)$$

Summary

Algebraic structures consist of elements, one or more methods of combining these elements by a binary operation, and a notion of equivalence between elements. A set may or may not be closed with respect to a binary operation. Different sets of elements call for different concepts of equivalence, and there may be more than one concept of equivalence for the same set of elements. Any concept that is to be described as an equivalence relation, however, must satisfy four conditions: determinative, reflexive, symmetric, and transitive. Finally, we have noted that, given an equivalence relation $\mathscr{E}$, we always require that any binary operation that is to be used shall be well-defined, a requirement that underlies the common assertion that equals can be multiplied by (or added to) equals.

Problem Set 14.1

1. Consider the binary operation of $\div$ (division) defined on the set of rational numbers.
(a) Is $\div$ commutative?
(b) Is $\div$ associative?

2. Consider the binary operation of $\times$ (multiplication) defined on the integers modulo 12.
(a) Is $\times$ commutative?
(b) Is $\times$ associative?

3. Consider the operation $\blacksquare$ defined on the set of rationals by $a \blacksquare b = a + b + ab$ (so, for example, $2 \blacksquare 3 = 11$).
(a) $(4 \blacksquare 7) \blacksquare -2 =$
(b) $4 \blacksquare (7 \blacksquare -2) =$
(c) Is $\blacksquare$ associative?
(d) Is $\blacksquare$ commutative?

4. In this section we defined $*$ on the set of rationals by $a * b = \frac{1}{2}(a + b)$.
(a) $(4 * 7) * -2 =$
(b) $4 * (7 * -2) =$
(c) Is $*$ associative?
(d) Is $*$ commutative?

5. The operation "+" defined by (a/b) "+" $(c/d) = (a + c)/(b + d)$ was seen to be not well-defined since, while

$$\frac{2}{3} = \frac{10}{15}$$

$$\frac{2}{3} \text{ "+" } \frac{2}{5} \neq \frac{10}{15} \text{ "+" } \frac{2}{5}$$

What about $*$ defined in Problem 4 above? Does $\frac{2}{3} * \frac{2}{5}$ equal $\frac{10}{15} * \frac{2}{5}$?

6. What about $\blacksquare$ as defined in Problem 3? Does $\frac{2}{3} \blacksquare \frac{2}{5}$ equal $\frac{10}{15} \blacksquare \frac{2}{6}$?

7. Is the set of positive integers closed under the following operations?
(a) Ordinary addition, $+$.
(b) Ordinary multiplication, $\times$.
(c) The operation $*$ defined in Problem 4.
(d) The operation $\blacksquare$ defined in Problem 3.

8. Is the set of integers that are multiples of 3 closed under the operations listed in Problem 7?

9. Consider the set of all cars parked in a certain lot, Which of the following relationships are determinative?
(a) Has the same number of cylinders as.
(b) Is the same color as.
(c) Uses the same size tires as.
(d) Is as pretty as.

10. Consider the set of all children in a certain classroom. Which of the following relationships are determinative?
(a) Has the same color hair as.
(b) Is as talented as.
(c) Is as tall as.
(d) Is the same sex as.

11. Consider the set of all U.S. citizens. Decide whether the listed relationships are reflexive; symmetric; transitive.
(a) Enjoys the same hobby as.
(b) Is married to.
(c) Is a first cousin of.
(d) Is as tall as.
(e) Lives within a mile of.
(f) Has ridden in the same car pool as.

12. Consider the set of all counting numbers. Decide whether the listed relationships are reflexive; symmetric; transitive.
(a) Is a divisor of.
(b) Is a multiple of.
(c) Has as many distinct prime factors as.
(d) Is a factor of the same number as.
(e) Has the same number of distinct prime factors as.

13. Beginning with the set of all cities shown on a map of the United States, let us agree that city X is separated from city Y if the distance between them is at least 100 miles. Is the relation "separated from" reflexive? Symmetric? Transitive?

14. Again beginning with the set of all cities on a map, let us say that city X is a neighbor of city Y if cities X and Y are within 100 miles of each other. Is the relation, "is a neighbor of" reflexive? Symmetric? Transitive?

15. Can you think of a relationship between the cities on a map of the United States that is reflexive and transitive but not symmetric?

16. Can you think of a relationship between the children in an elementary school classroom that is symmetric and transitive but not reflexive?

17.* For a fixed integer k, let $\equiv$ designate the relationship of equality modulo k. Show that $\equiv$ is an equivalence relation.

18.* Consider the set of all points (a, b) in the plane. We will say that $(a, b) \approx (c, d)$ if and only if $a + d = b + c$. For example, $(3, 7) \approx (11, 15)$. Show that $\approx$ is an equivalence relation.

19.* Referring to Problem 15, show that multiplication is well-defined modulo k; that is, show that, if $a \equiv b$, then $ac \equiv bc$.

20.* Referring to Problem 16, define the multiplication of two points as follows.

$$(a, b) \cdot (e, f) = (ae + bf, af + be)$$

Show that multiplication is well-defined with respect to $\approx$.

For Research and Discussion

The subject we have been calling abstract algebra is still listed in many college catalogs as "Modern Algebra." Because it has, as mathematical topics go, a recent history, tracing its introduction into the United States curriculum gives one a good idea of how the US has become a world leader in mathematics in the last fifty to sixty years. Try to find answers to the following questions.

1. Virtually all American students over a ten- to fifteen-year period studied the subject from a widely used text, *A Survey of Modern Algebra* Who were the authors? Where did they get their ideas?

2. How did it happen that Emmy Noether, the person generally credited with founding the subject, came to America?

3. What other mathematicians played key roles during the rise of American mathematics. Where and by whom were they trained?

References

More Mathematical People, Albers, Alexanderson, and Reid, editors, Harcourt Brace Jovanovich, 1990.

A Century of Mathematics in America, P. Duran, editor, *American Mathematical Society,* Vol 1, 1988.

14.2 Mathematical Rings

Old Idea, New Surroundings

In the system of integers modulo 30, if 17 is added to 13, we get 0. If 9 is added to 21, we get 0. Consider now the products.

$$\begin{array}{r} 17 \\ \underline{9} \\ 153 \equiv 3 \end{array} \qquad \begin{array}{r} 13 \\ \underline{21} \\ 13 \\ \underline{26} \\ 273 \equiv 3 \end{array}$$

What familiar property of high school algebra is here illustrated in a less familiar system?

An algebraic structure consists of a set of elements with a notion of equality and at least one binary operation, all being subject to certain axioms. But what should we take as axioms? We are guided to an answer by examining the common properties of the number systems we studied in Chapters 12 and 13. This process of recognizing and isolating the structural features common to a variety of situations is the essence of what we call *abstraction.*

The abstract theory of rings is entirely a product of the twentieth century. David Hilbert (1862–1943) coined the term "ring"; but it was Emmy Noether (1882–1935) who gave the theory an axiomatic basis and developed it to full flower. She is rightly called the founder of modern abstract algebra.

Recall five mathematical systems studied in those earlier chapters: the integers, the rational numbers, the real numbers, the modular systems, the system of 2×2 matrices. In each of them, we have a notion of equality and two binary operations called addition and multiplication. What basic properties do all these systems have in common?

First, we note that both operations are **associative,** that is,

$$a + (b + c) = (a + b) + c$$
$$a(bc) = (ab)c$$

The situation is quite different when we look at commutativity. For while the **commutative** law for addition

$$a + b = b + a$$

always holds, the corresponding law for multiplication may fail (recall that $ab \neq ba$ for matrices).

In each case there is an **additive identity,** that is, an element that acts like zero:

$$a + \mathbf{0} = \mathbf{0} + a = a$$

But we quickly point out that this zero may be different in different contexts. This is why we have used a boldfaced zero. It reminds us that we have to think of it abstractly. For in the integers **0** means the ordinary 0, whereas in the matrix system

$$\mathbf{0} = \begin{bmatrix} 0 & 0 \\ 0 & 0 \end{bmatrix}$$

Similarly, we always have a multiplicative identity **1** which acts like 1 and therefore satisfies

$$a\mathbf{1} = \mathbf{1}a = a$$

Here **1** is the ordinary number 1 when we are considering the integers. But when we are working with matrices, it is

$$\mathbf{1} = \begin{bmatrix} 1 & 0 \\ 0 & 1 \end{bmatrix}$$

When we come to inverses, we must be especially careful. The **additive inverse** of a is denoted by $\overline{a}$ (reminding us of negative a); it satisfies

$$a + \overline{a} = \overline{a} + a = \mathbf{0}$$

For the integers, $\overline{a}$ is simply $-a$, while if

$$a = \begin{bmatrix} -2 & -4 \\ 1 & 2 \end{bmatrix} \quad \text{then} \quad \overline{a} = \begin{bmatrix} 2 & 4 \\ -1 & -2 \end{bmatrix}$$

However, in the integers modulo 12, if $a \equiv 3$, then $\overline{a} \equiv 9$. We have to be careful, but in each of our five systems we always have $\overline{a}$.

When we consider the corresponding notion for multiplication, we have real trouble. Borrowing the notation we used for matrices, we want for each $a \neq \mathbf{0}$ an element a^{-1} called the **multiplicative inverse** satisfying

$$aa^{-1} = a^{-1}a = \mathbf{1}$$

We may not get it. In the integers modulo 12, 4 has no multiplicative inverse; neither does

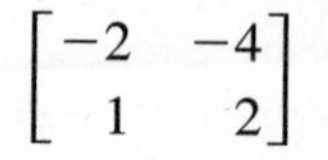

$$\begin{bmatrix} -2 & -4 \\ 1 & 2 \end{bmatrix}$$

in the matrix system, since its determinant is 0. However, when m/n is a rational number, the reciprocal n/m is its multiplicative inverse.

Finally, we mention the two **distributive** laws which hold in all five systems:

$$a(b + c) = ab + ac$$
$$(a + b)c = ac + bc$$

The formulas enclosed in the boxes are shared by all our systems. These are the properties that we isolate and make the axioms for an abstract structure that we call a ring.

Definition of a Ring

A mathematical **ring** is a set of elements with a notion of equality (=) that is closed with respect to two well-defined binary operations called addition, ⊕, and multplication, ⊗. These operations are subject to the following axioms.

Axioms for a Ring		
Name	***Law for Addition***	***Law for Multiplication***
Associative law	$a \oplus (b \oplus c) = (a \oplus b) \oplus c$	$a \otimes (b \otimes c) = (a \otimes b) \otimes c$
Commutative law	$a \oplus b = b \oplus a$	
Existence of identity	There is an element $\mathbf{0}$ satisfying $a \oplus \mathbf{0} = \mathbf{0} \oplus a = a$ for all a	There is an element $\mathbf{1}$ satisfying $a \otimes \mathbf{1} = \mathbf{1} \otimes a = a$ for all a
Existence of inverses	For each a, there is an element $\overline{a}$ satisfying $a \oplus \overline{a} = \overline{a} \oplus a = \mathbf{0}$	
Distributive laws	$a \otimes (b \oplus c) = (a \otimes b) \oplus (a \otimes c)$ $(a \oplus b) \otimes c = (a \otimes c) \oplus (b \otimes c)$	

There are two conspicuous gaps in the chart, corresponding to two properties which are missing in some of our examples, the commutative law of multiplication and the existence of multiplicative inverses. Filling in these two gaps is the main job in Section 14.3.

Some Theorems That Are True in Any Ring

As in geometry, so in algebra, axioms imply other rules—called theorems—which can be proved. Our first theorem deals with addition and justifies what is often called "canceling common terms."

Theorem 1 If $r \oplus t = s \oplus t$, then $r = s$.

Since one of our goals is to emphasize how algebra, like geometry, rests on axioms, we stress this similarity by writing the proof in the style sometimes used in elementary geometry (Section 11.1)

Statement	***Reason***
1. $r \oplus t = s \oplus t$	1. Given
2. $(r \oplus t) \oplus \overline{t} = (s \oplus t) \oplus \overline{t}$	2. Existence of additive inverse $\overline{t}$; add equals to equals
3. $r \oplus (t \oplus \overline{t}) = s \oplus (t \oplus \overline{t})$	3. Addition is associative
4. $r \oplus \mathbf{0} = s \oplus \mathbf{0}$	4. $\overline{t}$ is the additive inverse of t
5. $r = s$	5. $\mathbf{0}$ is the additive identity

We all know that in ordinary arithmetic zero times anything is zero. This fact holds in an arbitrary ring.

Theorem 2 For any r, it is true that $r \otimes \mathbf{0} = \mathbf{0}$ and $\mathbf{0} \otimes \mathrm{r} = \mathbf{0}$.

Statement	*Reason*
1. For any s, $\mathbf{0} \oplus s = s$	1. $\mathbf{0}$ is the additive identity
2. $\mathbf{0} \oplus (r \otimes \mathbf{0}) = r \otimes \mathbf{0}$	2. Since statement 1 is true for any s, it is true for $s = r \otimes \mathbf{0}$
3. $\mathbf{0} \oplus \mathbf{0} = \mathbf{0}$	3. Statement 1 is true for $s = \mathbf{0}$
4. $r \otimes (\mathbf{0} \oplus \mathbf{0}) = r \otimes \mathbf{0}$	4. Multiply equals by equals
5. $\mathbf{0} \oplus (r \otimes \mathbf{0}) = r \otimes (\mathbf{0} \oplus \mathbf{0})$	5. Transitive law for equality (things equal to the same thing, in this case $r \otimes \mathbf{0}$, are equal to each other)
6. $\mathbf{0} \oplus (r \otimes \mathbf{0}) = (r \otimes \mathbf{0}) \oplus (r \otimes \mathbf{0})$	6. Use the distributive law on the right side of statement 5
7. $\mathbf{0} = r \otimes \mathbf{0}$	7. Use Theorem 1 with $t = r \otimes \mathbf{0}$

This proves only half of the theorem, namely, $r \otimes \mathbf{0} = \mathbf{0}$. It is tempting to say that the other half, $\mathbf{0} \otimes r = \mathbf{0}$, follows from the commutative law, but alas we don't have the commutative law for multiplication in a ring. There is a better way (longer but correct). The demonstration above can be modified (a small change will do) so that it takes care of the second result. See if you can find the secret (Problem 13).

Now we are ready to give a logical demonstration of a fact that often puzzles high school students, namely that $(-a)(-b) = ab$. That may seem quite obvious to you for the integers. But is the corresponding fact for matrices obvious? Suppose

$$a = \begin{bmatrix} 2 & 1 \\ -3 & 4 \end{bmatrix} \qquad b = \begin{bmatrix} 1 & -1 \\ -4 & 3 \end{bmatrix}$$

Then

$$ab = \begin{bmatrix} -2 & 1 \\ -19 & 15 \end{bmatrix}$$

The additive inverses of a and b are

$$\bar{a} = \begin{bmatrix} -2 & -1 \\ 3 & -4 \end{bmatrix} \qquad \bar{b} = \begin{bmatrix} -1 & 1 \\ 4 & -3 \end{bmatrix}$$

You should compute $\overline{a}\overline{b}$ to see that you really do get the same answer as ab. In any case, we are now going to give a proof, an incontestable demonstration, that this holds in any ring whatever. This time we write the proof in a format more common in algebra.

Theorem 3 $\overline{a} \otimes \overline{b} = a \otimes b$

PROOF. First note that

$$\begin{aligned}(\overline{a} \otimes \overline{b}) \oplus (\overline{a} \otimes b) &= \overline{a} \otimes (\overline{b} \otimes b) \qquad \text{(distributive law)}\\ &= \overline{a} \oplus \mathbf{0} \qquad \text{(additive inverse)}\\ &= \mathbf{0} \qquad \text{(Theorem 2)}\end{aligned}$$

Also, and for exactly the same reasons,

$$\begin{aligned}(a \otimes b) \oplus (\overline{a} \otimes b) &= (a \oplus \overline{a}) \otimes b\\ &= \mathbf{0} \otimes b\\ &= \mathbf{0}\end{aligned}$$

Thus

$$(\overline{a} \otimes \overline{b}) \oplus (\overline{a} \otimes b) = (a \otimes b) \oplus (\overline{a} \otimes b)$$

since both are equal to the same thing, namely, **0**. Finally, from Theorem 1,

$$\overline{a} \otimes \overline{b} = a \otimes b$$

Incidentally, the problem that opens this section is an illustration of Theorem 3. See if you can understand why.

Our list of theorems could go on indefinitely, but perhaps three are enough to indicate the flavor of modern abstract algebra. Readers who want to know more about rings can check any book on modern algebra or abstract algebra. Those who do, however, must be warned that while we have for simplicity followed a well known text [G. Birkhoff and S. MacLane, *A Survey of Modern Algebra,* 3rd ed., (Macmillan, 1965), p. 346], most authors do not require that a ring must have a multiplicative identity.

Summary

When we examine the five mathematical systems (integers, rationals, reals, modular numbers, matrices) for common basic properties, we are led to the axioms, which appear in the table on page 447. These axioms allow us to make most (but not all) of the manipulations that are familiar from arithmetic and elementary algebra. We must only make sure that we avoid anything that depends on the commutative law of multiplication and the existence of multiplicative inverses.

+	*0*	*1*	*2*	*3*	*4*
0	0	1	2	3	4
1	1	2	3	4	0
2	2	3	4	0	1
3	3	4	0	1	2
4	4	0	1	2	3

×	*0*	*1*	*2*	*3*	*4*
0	0	0	0	0	0
1	0	1	2	3	4
2	0	2	4	1	3
3	0	3	1	4	2
4	0	4	3	2	1

Problem Set 14.2

1. The addition and multiplication tables for the integers modulo 5 are shown in the margin. If $x = 2$, $y = 3$, find
(a) $\overline{x}$ (b) $\overline{y}$ (c) $\overline{x}\,\overline{y}$ (d) xy

2. Do the calculations in Problem 1 if $x = 3$ and $y = 4$. What theorem is illustrated by these calculations?

3. In the integers modulo 5, find, if possible,
(a) 0^{-1} (b) 1^{-1} (c) 2^{-1} (d) 3^{-1}
(e) 4^{-1}

4. In the integers modulo 3, find each of the following.
(a) $\overline{1}$ (b) 1^{-1} (c) $\overline{2}$ (d) 2^{-1}

5. In the integers modulo 12, find, if possible,
(a) $\overline{3}$ (b) 3^{-1} (c) $\overline{5}$
(d) 5^{-1} (e) $\overline{11}$ (f) 11^{-1}

6. Do Problem 5 in the integers modulo 15.

7. In the ring of 2×2 matrices, let

$$x = \begin{bmatrix} 1 & 2 \\ 1 & 3 \end{bmatrix} \qquad y = \begin{bmatrix} 5 & 5 \\ 1 & 1 \end{bmatrix}$$

Find, if possible,
(a) $\overline{x}$ (b) x^{-1} (c) $\overline{y}$ (d) y^{-1}

8. Do Problem 7 with

$$x = \begin{bmatrix} 1 & -1 \\ 1 & -3 \end{bmatrix} \qquad y = \begin{bmatrix} 2 & -2 \\ 1 & 4 \end{bmatrix}$$

9. The counting numbers are not a ring. Which ring axioms fail?

10. Which integers have multiplicative inverses that are integers?

11. In elementary algebra, we learn that

$$(a + b)(a - b) = a^2 - b^2$$

Is this valid in any ring? Why or why not?

12. Why does

$$(a + b)^2 = a^2 + 2ab + b^2$$

fail to be valid in an arbitrary ring?

13. We proved only half of Theorem 2. Complete the proof by showing that $\mathbf{0} \otimes r = \mathbf{0}$. Hint: Start with $s \oplus \mathbf{0} = s$ and then let $s = \mathbf{0} \otimes r$. Mimic the earlier proof.

14. Show that $\overline{a \oplus b} = \overline{a} \oplus \overline{b}$. Hint: Show first that $(\overline{a} \oplus \overline{b}) \oplus (a \oplus b) = (\overline{a \oplus b}) \oplus (a \oplus b)$. Then use Theorem 1.

15. Show that $a \otimes \bar{b} = \overline{a \otimes b}$ by justifying each of the following steps.

$$\begin{aligned}
(\overline{a \otimes b}) \oplus (a \otimes b) &= \mathbf{0} \\
(a \otimes \bar{b}) \oplus (a \otimes b) &= a \otimes (\bar{b} \oplus b) \\
&= a \otimes \mathbf{0} \\
&= \mathbf{0} \\
(a \otimes \bar{b}) \oplus (a \otimes b) &= (\overline{a \otimes b}) \oplus (a \otimes b) \\
a \otimes \bar{b} &= \overline{a \otimes b}
\end{aligned}$$

16. Show that $\bar{a} \otimes b = \overline{a \otimes b}$. Hint: See Problem 15.

17. Show that $\bar{\bar{a}} = a$.

18.* Consider the collection of all subsets of the ordinary Euclidean plane. Let $\oplus$ be $\cup$ and $\otimes$ be $\cap$. Which of the ring axioms are satisfied?

For Research and Discussion

Examine a number of elementary algebra books. What sort of explanations are offered for the rule that $(-1)(-1) = 1$? Which ones appeal to you? How, if you were a teacher of students first encountering this idea, would you explain it?

14.3 Mathematical Fields

Distinctive Products

Think about arithmetic modulo m. For some integer $a < m$, form the $m - 1$ products

$$\begin{gathered}
a \cdot 1 \\
a \cdot 2 \\
\vdots \\
a \cdot (m - 1)
\end{gathered}$$

expressing each product with an integer between 0 and m^{-1}. How many distinct numbers will appear as products?

	Mod 12	*Mod 13*
$3 \cdot 1$	3	3
$3 \cdot 2$	6	6
$3 \cdot 3$	9	9
$3 \cdot 4$	0	12
$3 \cdot 5$	3	2
$3 \cdot 6$	6	5
$3 \cdot 7$	9	8
$3 \cdot 8$	0	11
$3 \cdot 9$	3	1
$3 \cdot 10$	6	4
$3 \cdot 11$	9	7
$3 \cdot 12$		10

Four distinct numbers appear as products modulo 12.

Twelve distinct numbers appear as products modulo 13.

Teachers who have graded too many papers develop somewhat sarcastic references to certain errors that crop up from one student generation to the next. One well-known abomination is the universal law of cancellation. It says that if the same number appears several times in an expression, it can be cancelled. Examples of this law, used with unfortunate success, are

$$\frac{16}{64} = \frac{1}{4} \qquad \text{(cancel the 6's)}$$

$$\frac{9^3 + 4^3}{9^3 + 5^3} = \frac{9 + 4}{9 + 5} \qquad \text{(cancel the 3's)}$$

The unfortunate thing about these examples is that a totally unjustified "cancellation" yields a correct result.

There are, of course, places where a cancellation can be justified. We learned about one in Section 14.2. If $a + b = a + c$, then $b = c$. The corresponding cancelation law for multiplication may not hold. For we learned that, in the matrix system, $a \cdot b = a \cdot c$ does not imply $b = c$. However, if a, b, and c are ordinary numbers, even this multiplicative cancelation is okay. This leads us to a study of systems in which the cancelation of common factors is valid. It has something to do with the two gaps we left in the axiom table on page 447.

Definition of a Field

A mathematical **field** is a set of elements with a notion of equality (=), which is closed with respect to two well-defined binary operations called addition ($\oplus$) and multiplication ($\otimes$). These operations obey the following axioms.

Axioms for a Field		
Name	*Law for Addition*	*Law for Multiplication*
Associative law	$a \oplus (b \oplus c) = (a \oplus b) \oplus c$	$a \otimes (b \otimes c) = (a \otimes b) \otimes c$
Commutative law	$a \oplus b = b \oplus a$	$a \otimes b = b \otimes a$
Existence of identity	There is an element **0** satisfying $a \oplus \mathbf{0} = \mathbf{0} \oplus a = a$ for all a	There is an element **1** satisfying $a \otimes \mathbf{1} = \mathbf{1} \otimes a = a$ for all a
Existence of inverses	For each a, there is an element $\overline{a}$ satisfying $a \oplus \overline{a} = \overline{a} \oplus a = \mathbf{0}$	For each $a \neq \mathbf{0}$, there is an element a^{-1} satisfying $a \otimes a^{-1} = a^{-1} \otimes a = \mathbf{1}$
Distributive laws	$a \otimes (b \oplus c) = (a \otimes b) \oplus (a \otimes c)$ $(a \oplus b) \otimes c = (a \otimes c) \oplus (b \otimes c)$	

Everything is as it was for a ring, except that we now require the commutative law for multiplication and the existence of multiplicative inverses.

Have we met any mathematical systems that satisfy all these axioms? We certainly have. The rational numbers, the real numbers, and the complex numbers are fields. So are the integers modulo 7. However, the integers modulo 12 are not a field (some numbers, such as 4 for instance, have no multiplicative inverse). The system of matrices is not a field for two reasons; the commutative law fails, and multiplicative inverses don't always exist.

Some Theorems That Are True in Any Field

Here's the theorem about cancelation of factors:

Theorem 1 If $a \otimes b = a \otimes c$ and $a \neq \mathbf{0}$, then $b = c$.
PROOF Since $a \neq 0$, it has a multiplicative inverse a^{-1}. Multiply by a^{-1} on the left of each side in the given statement:

$$a^{-1} \otimes (a \otimes b) = a^{-1} \otimes (a \otimes c)$$

Next use the associative law for multiplication:

$$(a^{-1} \otimes a) \otimes b = (a^{-1} \otimes a) \otimes c$$

Since $a^{-1} \otimes a = \mathbf{1}$, we get

$$\mathbf{1} \otimes b = \mathbf{1} \otimes c$$

or

$$b = c$$

Our second result is sometimes referred to as "no divisors of 0."

Theorem 2 If $a \otimes b = \mathbf{0}$, then either $a = \mathbf{0}$ or $b = \mathbf{0}$.
PROOF If both a and b are $\mathbf{0}$, there is nothing to prove. If $a \neq \mathbf{0}$, it has an inverse a^{-1}. Multiply on the left in $a \otimes b = \mathbf{0}$ by a^{-1} to obtain

$$a^{-1} \otimes (a \otimes b) = a^{-1} \otimes \mathbf{0}$$
$$(a^{-1} \otimes a) \otimes b = a^{-1} \otimes \mathbf{0}$$
$$\mathbf{1} \otimes b = a^{-1} \otimes \mathbf{0}$$
$$b = \mathbf{0}$$

A completely similar argument works to show that, if $b \neq \mathbf{0}$, then $a = \mathbf{0}$.

Recall from your study of the rational numbers that $1/(1/b) = b$. Here is the corresponding fact, true in any field:

Theorem 3 If $b \neq \mathbf{0}$, then $(b^{-1})^{-1} = b$.
PROOF To prove this, start with the fact that

$$b^{-1} \otimes (b^{-1})^{-1} = \mathbf{1}$$

Then multiply on the left by b and make the obvious simplifications.

An Example

We have claimed that the integers modulo 7 form a field. Perhaps a little more explanation should be given. Symbolize this system by {0, 1, 2, 3, 4, 5, 6}; we called these the principal representatives in Section 12.3. It is easy to make addition and multiplication tables for this system. They are shown below. Now we can verify all the axioms.

+	*0*	*1*	*2*	*3*	*4*	*5*	*6*
0	0	1	2	3	4	5	6
1	1	2	3	4	5	6	0
2	2	3	4	5	6	0	1
3	3	4	5	6	0	1	2
4	4	5	6	0	1	2	3
5	5	6	0	1	2	3	4
6	6	0	1	2	3	4	5

×	*0*	*1*	*2*	*3*	*4*	*5*	*6*
0	0	0	0	0	0	0	0
1	0	1	2	3	4	5	6
2	0	2	4	6	1	3	5
3	0	3	6	2	5	1	4
4	0	4	1	5	2	6	3
5	0	5	3	1	6	4	2
6	0	6	5	4	3	2	1

The question of inverses is always crucial. We suggest that you check on the following facts.

$$
\begin{array}{ll}
\overline{0} = 0 & 1^{-1} = 1 \\
\overline{1} = 6 & 2^{-1} = 4 \\
\overline{2} = 5 & 3^{-1} = 5 \\
\overline{3} = 4 & 4^{-1} = 2 \\
\overline{4} = 3 & 5^{-1} = 3 \\
\overline{5} = 2 & 6^{-1} = 6 \\
\overline{6} = 1 &
\end{array}
$$

We have observed that the integers modulo 12 are not a field, since in this system some elements don't have multiplicative inverses. This suggests an interesting and hard question. For which m's are the integers modulo m a field? There is no reason to think the answer is simple. But it is. It is a question we encourage you to explore for Research and Discussion at the and of this section.

Summary

It has been said that, while memorization without understanding is useless, understanding without memorization is hopeless. A student who is to succeed in algebra must memorize the axioms. In the case of a field, this is easy because of the symmetry of the axioms with respect to the two operations. Just remember that we have associativity, commutativity, identities, and inverses for both operations, and that the distributive laws hold.

Finally, we suggest that terms such as "cross-multiply," "cancel," "transpose," etc., should not be used unless one is certain that, if pressed to do so, he or she can describe exactly what has been done in terms of the axioms.

Problem Set 14.3

1. Find additive inverses for all elements and multiplicative inverses for all nonzero elements of {0, 1, 2, 3, 4}, the integers modulo 5. Then convince yourself that this system forms a field.
2. Follow the directions in Problem 1 for the integers modulo 11.
3. In the integers modulo 13, find
   (a) $\overline{4}$ (b) 4^{-1} (c) $\overline{5}$ (d) 5^{-1}
4. Convince yourself that the integers modulo 13 form a field.
5. Theorem 1 says that, if $a \otimes b = a \otimes c$ and $a \neq \mathbf{0}$, then $b = c$. This means that we can cancel on the left. Show that we can also cancel on the right; that is, if $b \otimes a = c \otimes a$ and $a \neq \mathbf{0}$, then $b = c$.
6. Show that, if $a \otimes b \otimes c = \mathbf{0}$ in a field, at least one of a, b, and c is $\mathbf{0}$.
7. Show that the familiar fact from elementary algebra

$$(a + b)^2 = a^2 + 2ab + b^2$$

   is true in any field. Begin by writing it in the symbols we have been using in an abstract field.
8. Rewrite

$$(a + b)(a - b) = a^2 - b^2$$

   in the symbols of an abstract field and then show that it is always correct.

In Problems 9 through 14, you are to think of sets as being the elements of an algebraic structure with the operations of union ∪ *and intersection* ∩. *Think of* ∪ *as* ⊕ *and* ∩ *as* ⊗. *Let A, B, and C be the sets diagramed below.*

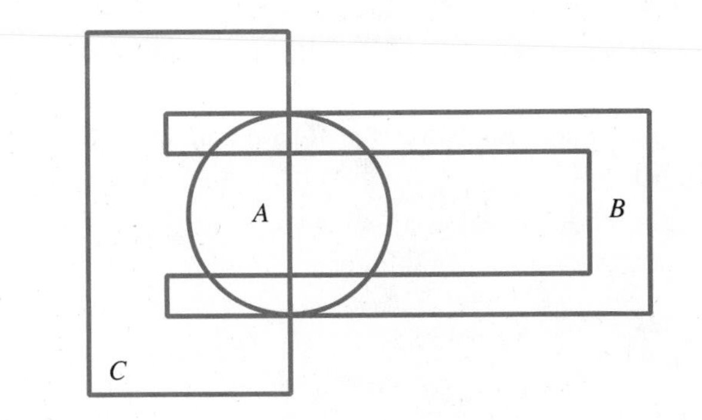

9. Make two copies of the set diagram above. On one shade $(A \cup B) \cup C$, and on the other $A \cup (B \cup C)$. What law is illustrated?

10. Make two more copies of the set diagram. On one shade $(A \cap B) \cap C$, and on the other $A \cap (B \cap C)$. What law is illustrated?

11. Again make two copies of the set diagram. On the shade $A \cap (B \cup C)$, and on the other $(A \cap B) \cup (A \cap C)$. What law is illustrated?

12. Follow the directions in Problem 11, this time shading $(B \cup C) \cap A$ and $(B \cap A) \cup (C \cap A)$.

13. If we consider the subsets of points on this page as the elements of an algebraic structure, what is the
(a) Multiplicative identity?
(b) Additive identity?

14. Refer to Problem 13. Does every subset have a multiplicative inverse? An additive inverse?

In Problems 15 through 20, we refer to the set C of all 2 × 2 matrices of the form

$$\begin{bmatrix} a & -b \\ b & a \end{bmatrix}$$

where a and b are real numbers. For example, C contains

$$\begin{bmatrix} 2 & -3 \\ 3 & 2 \end{bmatrix}, \quad \begin{bmatrix} 1 & 4 \\ -4 & 1 \end{bmatrix}, \quad \begin{bmatrix} -5 & \frac{1}{2} \\ -\frac{1}{2} & -5 \end{bmatrix}, \quad \begin{bmatrix} 1 & 0 \\ 0 & 1 \end{bmatrix}$$

15. Show that C is closed with respect to multiplication. Hint: Show that the product

$$\begin{bmatrix} a & -b \\ b & a \end{bmatrix}\begin{bmatrix} c & -d \\ d & c \end{bmatrix}$$

is again a matrix of a form belonging to C.

16. Show that C is closed with respect to addition. Notice the hint given in Problem 15.

17. In general, matrix multiplication is not commutative. However, show that, for matrices in C, it is.

18. Show that all nonzero matrices in C have multiplicative inverses.

19. Is C a field?

20. Show that $x^2 + 1 = 0$ has a solution in C.

For Research and Discussion

You either know by now or can easily establish that the integers mod 3, 5, 7 and 11 all form fields. From this you might conjecture that any set of integers modulo a prime number forms a field. You can prove this in either of two ways, both depending on the Euclidean Algorithm which you can find in any book on number theory. It says that, if d is the g.c.d. of the positive integers a and b, there are integers r and s such that

$$d = ra + sb$$

(the algorithm also provides a method of finding r and s). In particular, if the g.c.d. is 1,

$$1 = ra + sb$$

PROOF 1 Use the Euclidean algorithm to show that if a prime p divides ab, then either p divides a or p divides b. Use this fact to solve *Multiples of 3*. The same idea will show that for any $a < p$, $a \neq 0$, the sequence $a \cdot 1, \ldots, a(p-1)$ will consist of the distinct integers (mod p) $1, 2, \ldots, p-1$ in some order. In particular, since 1 is in the list, some product $a \cdot k = 1$; i.e. k is a multiplicative inverse of a.

PROOF 2 Return again to the Euclidean algorithm Let b be any nonzero element of the integers modulo p, where p is prime. Then there are integers r and s such that $rb + sp = 1$ or equivalently $rb - 1 = -sp$. Finish the argument. Finally, show that, if m is not prime, the integers modulo m are not a field.

14.4 Solving Equations

Only $7^4 = 2401$ Possibilities!

Suppose the numbers entered in a matrix are in the system of integers modulo 7. Then, for example,

$$\begin{bmatrix} 2 & 6 \\ 4 & 3 \end{bmatrix}\begin{bmatrix} 1 & 5 \\ 3 & 2 \end{bmatrix} = \begin{bmatrix} 6 & 1 \\ 6 & 5 \end{bmatrix}$$

Find four positive integers (no negatives and no fractions), a, b, c, and d, so that multiplication modulo 7 gives

$$\begin{bmatrix} a & b \\ c & d \end{bmatrix}\begin{bmatrix} 1 & 5 \\ 3 & 2 \end{bmatrix} = \begin{bmatrix} 1 & 0 \\ 0 & 1 \end{bmatrix}$$

Consider the following questions.

1. Can you find a rational number x such that $3x = 7$?
2. Can you find a matrix

$$x = \begin{bmatrix} x_1 & x_2 \\ x_3 & x_4 \end{bmatrix}$$

such that

$$\begin{bmatrix} 2 & -1 \\ 1 & 0 \end{bmatrix}\begin{bmatrix} x_1 & x_2 \\ x_3 & x_4 \end{bmatrix} = \begin{bmatrix} 3 & 1 \\ 2 & -2 \end{bmatrix}$$

3. Can you find a number x in the set $\{0, 1, \ldots, 11\}$ such that $4x \equiv 7 \bmod 12$?
4. Can you find a number x in the set $\{0, 1, \ldots, 10\}$ such that $4x \equiv 7 \bmod 11$?

Each problem really involves two related questions. The first is, Does a solution exist? (Since in some situations, like question 3, no solution exists.) The second is, If a solution exists, how to we find it?

In the setting of an abstract algebraic structure, each of the problems posed deals with an equation of the form $a \otimes x = b$. If the algebraic structure is a field, the two related questions can both be answered with dispatch. If $a \neq \mathbf{0}$, a solution exists, and we can give specific instructions for finding it. Simply multiply both sides by a^{-1}:

$$a^{-1} \otimes (a \otimes x) = a^{-1} \otimes b$$

Making obvious use of the axioms, we quickly see that

$$x = a^{-1} \otimes b$$

The key of course is the existence of a multiplicative inverse. In a field we are guaranteed that the inverse exists. We saw that the integers modulo 12 did not form a field, because not all elements had a multiplicative inverse, and this is precisely why we must answer no to question 3. There is no multiplicative inverse of 4 mod 12.

However, having observed that the integers modulo 11 form a field, we know that 4 does have a multiplicative inverse in this system. In fact $3(4) \equiv 1 \bmod 11$. Therefore we answer question 4 by writing

$$3(4x) \equiv 3(7) \bmod 11$$
$$x \equiv 10 \bmod 11$$

The reader may feel that we have only shifted the problem to another, equally difficult question. Instead of asking for a solution to $a \otimes x = b$, we are asking for the multiplicative inverse of a. If, as in arithmetic modulo 11, we are reduced to trial and error to find a^{-1}, why not just solve the given problem by trial and error?

Several answers can be given. In the first place, we often have available methods for finding inverses when they exist. An algorithm is available to help us in modular arithmetic if we are willing to invest some time in mastering the Euclidean algorithm. (See the Research and Discussion project at the end of Section 14.3; also problems 11 and 12 at the end of this section.) We know how to find the inverse of a 2×2 matrix when it exists.

Second, when we work in a field, we can always answer at least the first question: Does an answer exist? Yes. The trial-and-error method is not quite so discouraging when we are assured that some trial will not be an error.

Third, the procedure described focuses our attention on the right question. Even when working in a system that is not a field, such as the set of all 2×2 matrices, we know how to proceed. In question 2, whether or not we can find the desired matrix depends on whether or not we can find an inverse of

$$A = \begin{bmatrix} 2 & -1 \\ 1 & 0 \end{bmatrix}$$

We can. In fact,

$$A^{-1} = \begin{bmatrix} 0 & 1 \\ -1 & 2 \end{bmatrix}$$

Therefore the equation can be solved, and we know how to do it.

$$\begin{bmatrix} 0 & 1 \\ -1 & 2 \end{bmatrix}\begin{bmatrix} 2 & -1 \\ 1 & 0 \end{bmatrix}\begin{bmatrix} x_1 & x_2 \\ x_3 & x_4 \end{bmatrix} = \begin{bmatrix} 0 & 1 \\ -1 & 2 \end{bmatrix}\begin{bmatrix} 3 & 1 \\ 2 & -2 \end{bmatrix}$$

$$\begin{bmatrix} x_1 & x_2 \\ x_3 & x_4 \end{bmatrix} = \begin{bmatrix} 2 & -2 \\ 1 & -5 \end{bmatrix}$$

It must be noted that, in the case of a matrix equation $AX = B$, it is essential to multiply by A^{-1} on the left: $A^{-1}AX = A^{-1}B$. Multiplying on the right does not work. Why?

Finally, we suggest that, even in the most familiar setting, the rational numbers, the best instruction for solving

$$\frac{15}{4}x = \frac{9}{14}$$

is, "Multiply by the inverse of $\frac{15}{4}$" (for, while most people remember that, to divide, one inverts and multiplies, nowhere near so fine a percentage remembers which fraction to invert). This leads directly to

$$x = \frac{9}{14} \cdot \frac{4}{15} = \frac{3 \cdot 2}{7 \cdot 5} = \frac{6}{35}$$

Summary

The key to solving an equation of the form

$$a \otimes x = b$$

is to multiply both sides by a^{-1}. The trick, then, is to know whether an inverse exists, and how to find it when it does. If we know that our elements are members of a field, we can be certain that the inverse exists, hence the equation can be solved. Methods for actually finding the inverse have to be developed individually for various fields.

Problem Set 14.4

1. A beginning student in algebra, wishing to solve $x + 3 = -5$, decides to subtract 3 from both sides. He writes

$$\begin{array}{r} x + 3 = -5 \\ -3 \quad\; -3 \\ \hline \end{array}$$

On the left side he gets x, but on the right side he is momentarily confused. Then he recalls that, to subtract, you change the sign and add. He concludes that $x = -2$. What's wrong?

2. Solve for x the following equations from elementary algebra.

(a) $5x = 2(x - 4)$ (b) $-3(x - 2) = 4$

(c) $\frac{2}{3}x - \frac{1}{4} = \frac{11}{4}$ (d) $ax + b = bx + a$

3. Solve for x the following equations from elementary algebra.

(a) $2x - 3 = 5x + 2$ (b) $3 - 2(4 - x) = 0$

(c) $\frac{3}{4}x + \frac{7}{12} = \frac{1}{4}$ (d) $b - ax = a - bx$

4. The following calculations show all the steps in solving two equations in the field of integers modulo 5. Justify each step.

$$\begin{aligned} 2x &= 3 \\ 3(2x) &= 3 \cdot 3 \\ (3 \cdot 2)x &= 3 \cdot 3 \\ 1x &= 4 \\ x &= 4 \end{aligned} \qquad\qquad \begin{aligned} x + 2 &= 1 \\ (x + 2) + 3 &= 1 + 3 \\ x + (2 + 3) &= 1 + 3 \\ x + 0 &= 4 \\ x &= 4 \end{aligned}$$

5. Solve the equation $2x + 1 = 6$ in the field of integers modulo 7. Justify every step (see Problem 4).

6. Solve the equation $3x + 5 = 2$ in the field of integers modulo 11.

7. Solve the matrix equations

(a) $\begin{bmatrix} 3 & 1 \\ 0 & 2 \end{bmatrix}\begin{bmatrix} x_1 \\ x_2 \end{bmatrix} = \begin{bmatrix} 1 \\ 4 \end{bmatrix}$ (b) $\begin{bmatrix} 3 & 1 \\ 0 & 2 \end{bmatrix}\begin{bmatrix} x_1 & x_2 \\ x_3 & x_4 \end{bmatrix} = \begin{bmatrix} 2 & 4 \\ 3 & 2 \end{bmatrix}$

8. Solve the matrix equations

(a) $\begin{bmatrix} 6 & 2 \\ 2 & 1 \end{bmatrix} \begin{bmatrix} x_1 \\ x_2 \end{bmatrix} = \begin{bmatrix} 3 \\ 4 \end{bmatrix}$ (b) $\begin{bmatrix} 6 & 2 \\ 2 & 1 \end{bmatrix} \begin{bmatrix} x_1 & x_2 \\ x_3 & x_4 \end{bmatrix} = \begin{bmatrix} -2 & -3 \\ 5 & 1 \end{bmatrix}$

9. We can describe the process for finding the inverse of a 2×2 matrix in language that allows the entries to come from any field. To find the multiplication inverse of

$$\begin{bmatrix} a & b \\ c & d \end{bmatrix}$$

(a) Find $D = ad - bc$.
(b) Replace b and c by their additive inverses.
(c) Interchange a and d.
(d) Multiply each entry by D^{-1}, the multiplicative inverse of D.
Use this procedure to find the inverse of

$$\begin{bmatrix} 2 & 1 \\ 6 & 4 \end{bmatrix}$$

in the field of integers modulo 11. Then solve

$$\begin{bmatrix} 2 & 1 \\ 6 & 4 \end{bmatrix} \begin{bmatrix} x_1 & x_2 \\ x_3 & x_4 \end{bmatrix} = \begin{bmatrix} 9 & 3 \\ 6 & 2 \end{bmatrix}$$

in this system.

10. Using the method described in Problem 9, solve the problem that introduced this section.

11. The g.c.d. of 15 and 26 is 1. According to the Euclidean Algorithm (refer to Research and Discussion at the end of Section 14.3) there are integers r and s satisfying

$$1 = r(26) + s(15)$$

Check that $r = -4$ and $s = 7$ work. Thus

$$7(15) - 1 = 4(26)$$

which means that

$$7(15) \equiv 1 \bmod 26$$

(a) What is the multiplicative inverse of 15 in the integers modulo 26?
(b) Solve $15x \equiv 12 \bmod 26$.

12. Check that

$$1 = 9(37) - 4(83)$$

Use what you learned in Problem 11 to
(a) Find the multiplicative inverse of 37 mod 83.
(b) Solve $37x \equiv 16 \bmod 83$.

For Research and Discussion

The problem **Only 7^4 = 2401 Possibilities** can be solved by following the outline given in Problem 9. Moreover, this outline is one that, when seen, seems to be nothing more than one perfectly logical step after another; there are no surprises. Yet our experience is that very few students think their way to this solution, so few in fact that we are inclined to urge anyone who does it in this "natural" way to give serious consideration to majoring in mathematics.

The question that puzzles us is this. Why do so few get it? Is there really such a thing as a mathematical way of thinking? Can you, by talking to classmates, identify what went wrong for those who did not get it? Finally, for those of you who have come this far with us, and did solve **Only 7^4 = 2401 Possibilities** we have a question. Have you considered majoring in mathematics?

Answers

Problem Set 1.1 (Page 13)

1. 12. **3.** 1 hour and 20 minutes is 80 minutes. **5.** Give the fifth apple and the basket to the fifth girl. **7.** Never; the ship rises with the tide. **9.** Three. **11.** 30,000. **13.** Call the slices: A, B, and C. Put A and B in the pan. After 30 seconds turn A over, take B out, and put C in. After 1 minute, remove A, put B back in the pan on its unbrowned side, and turn C over.

15. **17.**

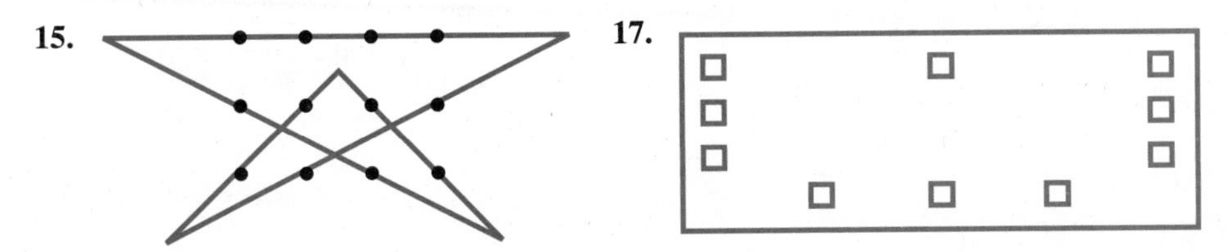

19. The North Pole is one solution; there are infinitely many others. Consider a circle of circumference 1 mile around the South Pole. Start from any point 1 mile north of this circle. There are still other solutions. Find them. **21.** Open rings 1, 2, and 3. Use ring 1 to link ring 6 to ring 7. Then use ring 2 to link ring 9 to ring 10. Finally, use ring 3 to link ring 12 to ring 13. **23.** Cut links 4 and 11; this gives subchains of length 1, 1, 3, 6, 12. **25.** "Unfolded," the cube is

Problem Set 1.2 (Page 22)

1. 91 feet. **3.** 11, counting the 2 it meets in the bus depots. **5.** July 14. **7.** 35. **9.** Send 2 boys across, send 1 boy back, send 1 soldier across, and send a second boy back. Now repeat this process until all 10 soldiers are across. **11.** Split the contents of the beaker between the two test tubes. Return the contents of one (i.e., 24 milliliters) to the beaker. Now split the contents of the remaining test tube between the two test tubes and pour one of them (i.e., 12 milliliters) into the beaker. Finally, split the contents of the remaining test tube between them and pour one of them (6 milliliters) into the beaker. **13.** 10. **15.** $\sqrt{108}$ miles. **17.** 65. **19.** Alphabetically by name. **21.** 43 **23.** 31.8 **25.** Dick shook 3 hands. **27.** 13,501. **29.** Three weighings suffice. A complete description would take too much space. Here are some hints. Number the coins 1 to 12. Weigh coins 1 through 4 against 5 through 8. If the pans balance, the bad coin comes from 9 through 12. It can be determined in two more weighings. If the pans do not balance in the first step, coins 9 through 12 are good. Suppose the side with coins 5 through 8 goes up. Then weigh 1, 2, 3, and 8 against 4, 9, 10, and 11. You finish. **31.** Take 1 coin from pile 1, 2 coins from pile 2, 3 coins from pile 3, etc. or a total of 55 coins, and put them on the scale. If all are good, they will weigh 308 grams. They will weigh less than this by the number of grams equal to the number of bad coins. This is also the number of the bad pile. Thus one weighing suffices.

Problem Set 1.3 *(Page 29)*

5. (r, s) may be (14, 4), (8, 6), or (6, 12). **7.** The "greedy algorithm" in which at any stage you color the largest possible square results in coloring only 41, whereas 42 is possible. **9.** $4\frac{4}{15}$ miles.

Problem Set 2.1 *(Page 37)*

1. Adding 6 feet to the circumference adds about 1 foot to the radius. You could slip through. **3.** Running half of the time is more running than running half of the distance. Homer will get there first. **5.** Suppose there are m people at the college. Put a label on each person corresponding to the number of friends he or she has. This label will be a number between 1 and $m - 1$. There are m labels, but at most $m - 1$ values. At least two must agree. **7.** $\frac{1}{1\cdot 2} = \frac{1}{2}$, $\frac{1}{1\cdot 2} + \frac{1}{2\cdot 3} = \frac{2}{3}, \ldots, \frac{1}{1\cdot 2} + \frac{1}{2\cdot 3} + \cdots + \frac{1}{99\cdot 100} = \frac{99}{100}$. **9.** It's fairly easy to find 33 sets of triples; the hard part is to show that there are no more. **11.** $2^7 = 128$.

Problem Set 2.2 *(Page 41)*

1. 3 girls and 4 boys in the family.

3.

8	1	6
3	5	7
4	9	2

5. Move left and right coins in top row directly down to third row. Then move bottom coin above top row. **7.** $(1 \times 2)(3 + 4)(5) + (6 + 7 + 8 + 9) = 100$. **9.** 25. **11.** Allison is the waitress, Betty is the bus driver, Claire is the clerk. **13.** A beat B, 2 to 0; A beat C, 5 to 1; B and C tied, 2 to 2.

15.

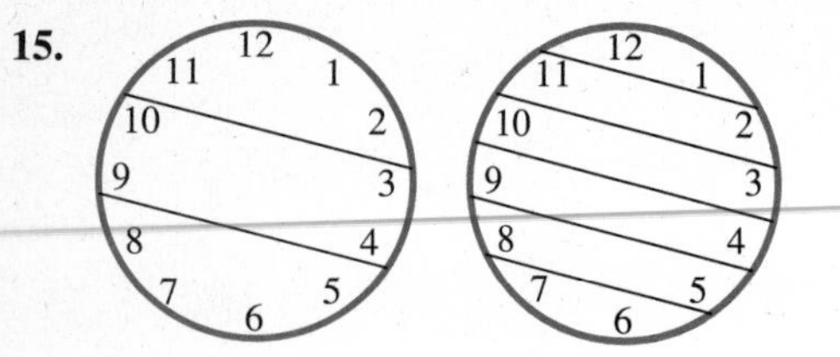

17. I am a male nurse. **19.** $(10t + u)tu = 100u + 10u + u$. Deduce $t < 4$ and use trial and error to obtain 37.

Problem Set 2.3 *(Page 47)*

1. 600. **3.** \$24.19. **5.** A = {1, 4, 6, 7} B = {2, 3, 5, 8} or conversely. **7.** 50¢.

9. $\begin{array}{r} 286 \\ \underline{286} \\ 527 \end{array}$ $\begin{array}{r} 286 \\ \underline{3210} \\ 3496 \end{array}$ or $\begin{array}{r} 236 \\ \underline{236} \\ 472 \end{array}$ $\begin{array}{r} 236 \\ \underline{9280} \\ 9516 \end{array}$ **11.** $\begin{array}{r} 927 \\ \underline{63} \\ 58401 \end{array}$

Problem Set 2.4 *(Page 52)*

1. 88. **3.** 24, 25. **5.** Lay the ruler across the board at an angle so the ends of the ruler coincide with the edges of the board. Mark at 2-inch intervals. **7.** 100 miles. **9.** Horace and Homer meet in 4 hours. Therefore Trot runs 120 miles. **11.** He drove a post in the ground directly across the river from the house and equally far from it. Then he sighted from the post to the barn. Where his line of sight intersected the river is the desired point. **13.** No. The central cube has six sides. It requires six cuts to slice its sides. **15.** A domino covers a black and white square. The removed squares are of the same color. Therefore the mutilated chessboard cannot be covered with dominoes. **17.** I noted my clock's time when I left and when I returned, so I knew how long I was gone. From this I subtracted the time I was at my friend's house and divided by 2. This gave me the time required to walk home, which I added to the time showing on my friend's clock when I left his home. **19.** Here is one argument. Suppose the string lies along the real line so that each point of it has a coordinate x between 0 and b. After stretching, each point has a new coordinate, $y = f(x)$. Graph $y = f(x)$ in the coordinate plane. The graph must cross the line $y = x$. The x coordinate of the crossing point is also the x coordinate of an unmoved point.

Problem Set 2.5 *(Page 62)*

1. **(a)** $\frac{1}{2}(x + x^2)$ **(b)** $3 + 2x$ **(c)** $x + (x + 2) + (x + 4)$ or $3x + 6$ **(d)** $4x$ **(e)** $x(2x - 3)$ **(f)** $x^2(\frac{1}{2}x + 3)$ **(g)** $5x + 30x + 3x - 4$. **3.** **(a)** $x, x + 1, x + 2$ **(b)** $x + (x + 1) + (x + 2) = 636$ **(c)** The three numbers: 211, 212, 213. **5.** **(a)** $s, s + 1, s + 2$ **(b)** $s + (s + 1) + (s + 2) = s + 63$ **(c)** The three numbers: 30, 31, 32. **7.** **(a)** length $= 2w + 4$ **(b)** $2w + 2(2w + 4) = 200$ **(c)** Dimensions: 32 feet by 68 feet. **9.** **(a)** $3x$ **(b)** $x + 15, 3x + 15, 3x + 15 = 2(x + 15)$ **(c)** 15 **11.** **(b)** $(10 - x)(0.08)$ **(c)** $(10 - x)(0.08) = 10(.05)$ **(d)** 3.75 gallons. **13.** 20 gallons. **15.** 9 feet. **17.** **(a)** Time downstream $= 6 - t$ **(b)** $3t = 7(6 - t)$ **(c)** 4.2 hours after starting, or at 10:12 P.M. **19.** 6.96 hours after starting, or at about 6:57 P.M. **21.** $23.25. **23.** 84.

Problem Set 2.6 *(Page 70)*

1. **(a)** $x + y = 17$ **(b)** $\frac{2}{3}x - y = 8$ **(c)** $x = 15, y = 2$. **3.** **(a)** Value of nickels is $5n$; Value of quarters is $25q$; $5n + 25q = 600$ **(b)** $n = 3q$ **(c)** $n = 45; q = 15$. **5.** **(a)** $x + y = 10$ **(b)** $264x + 384y = 3180$ **(c)** 4.5 pounds of Aromatic; 5.5 pounds of Caffineo. **7.** **(a)** width $= w + 8$; length $= l - 10$ **(b)** $l - 10 = w + 8$ **(c)** $lw = (w + 8)(l - 10)$ **(d)** $l = 50; w = 32$. **9.** **(a)** Rate downstream is $x + y$; Rate upstream is $x - y$ **(b)** $16 = x + y; 16 = 2(x - y)$ **(c)** $x = 12, y = 4$. **11.** 50 dresses, 28 coats. **13.** 9 children each getting $9000. **15.** 151.

Problem Set 3.1 *(Page 81)*

5. $\frac{3}{20}$. **7.** 1.

Problem Set 3.2 *(Page 87)*

7. 5.

Problem Set 4.1 *(Page 99)*

1. **(a)** 15, 18 **(b)** 9, 6 **(c)** $\frac{1}{16}, \frac{1}{32}$. **3.** **(a)** 20, 32 **(b)** $\frac{1}{18}, \frac{1}{40}$ **(c)** $\frac{1}{8}, \frac{1}{64}$ **(d)** 3, −9. **5.** **(a)** $a_n = 3n$ **(b)** $b_n = 21 - 3(n - 1) = 24 - 3n$ **(c)** $c_n = (\frac{1}{2})^{n-1}$. **7.** **(a)** 81 **(b)** 14 **(c)** $\frac{1}{4}$ **(d)** 5. **9.** **(a)** $a_n = a_{n-1} + 3$ **(b)** $b_n = b_{n-1} - 3$ **(c)** $c_n = c_{n-1}/2$. **11.** **(a)** 486, 1458 **(b)** 7, 1 **(c)** $\frac{6}{7}, \frac{7}{8}$ **(d)** 1/216, 1/343 **(e)** $\frac{3}{2}, \frac{3}{16}$ **(f)** 16, 26 **(g)** 48, 88 **(h)** 19, 23 **(i)** 28, 36. **13.** **(a)** 126 **(b)** 33 **(c)** 0 **(d)** 70 **(e)** 15.75 **(f)** 0. **15.** $\frac{1}{7} = 0.142857142857\ldots$, so $a_1 = 1, a_2 = 4, a_3 = 2, \ldots, a_{53} = 5$. $\frac{5}{13} = 0.384615384615\ldots$, so $a_1 = 3, a_2 = 8, a_3 = 4, \ldots, a_{53} = 1$. **17.** $a_n = n^2$.

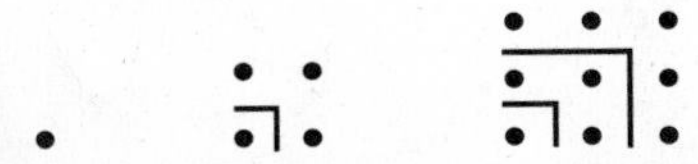

19. $a_n = 3 + 2^{n-1}$. **21.** 1225; 41,616; 1,413,721; 48,024,900.

Problem Set 4.2 (Page 106)

1. **(a)** 9, 11 **(b)** 17, 20 **(c)** 84, 80. **3.** **(a)** $d = 2$, $a_{40} = 1 + (39)2 = 79$ **(b)** $d = 3$, $b_{40} = 122$ **(c)** $d = -4$, $c_{40} = -56$. **5.** **(a)** $A_{40} = \frac{40}{2}(1 + 79) = 1600$ **(b)** $B_{40} = 2540$ **(c)** $C_{40} = 880$.
7. **(a)** 2550 **(b)** 2500 **(c)** 1683. **9.** $16 + 9(32) = 304$ feet, $16 + 48 + 80 + \cdots + 304 = 1600$ feet.
11. $\frac{21}{2}(40 + 60) = 1050$ centimeters. **13.** $7 + 14 + \cdots + 350 = \frac{50}{2}(7 + 350) = 8925$ cents $= \$89.25$.
15. 19,900. **17.** 5244. **19.** $1^3 + 2^3 + 3^3 + \cdots + n^3 = n^2(n + 1)^2/4$. **21.** $h_n = 1 + 3n(n - 1)$.
23. $97(6) = 582$.

Problem Set 4.3 (Page 113)

1. **(a)** 512, 2048 **(b)** $\frac{1}{3}, \frac{1}{9}$ **(c)** 3, −3. **3.** $a_{40} = 2(4)^{39}$, $b_{40} = 27(\frac{1}{3})^{39}$, $c_{40} = 3(-1)^{39}$.
5. **(a)** $A_5 = \dfrac{2(1 - 4^5)}{1 - 4} = \dfrac{2(1 - 1024)}{-3} = 682$ **(b)** 121/3 **(c)** 3. **7.** 1,024,000. **9.** $1 + 7 + 7^2 + 7^3 + 7^4 = 2801$. **11.** 2^{40} square inches $\approx 10^{12}$ square inches $\approx 10^{10}$ square feet ≈ 300 square miles. **13.** $a_{40} = 2.48328$, $A_{40} = 65.9737$. **15.** \$194,530.50 **17.** $(\frac{2}{3})^5 = 32/243$, 12 strokes. **19.** **(a)** $\frac{1}{3}$ **(b)** $\frac{7}{9}$ **(c)** 1. **21.** \$4 billion.

Problem Set 4.4 (Page 120)

1. 1500. **3.** 4000; 8000; 1,024,000. **5.** Neither. **7.** This is eight half-lives; $200(\frac{1}{2})^8 \approx 0.8$ gram. **9.** 200.
11. About 2085. **13.** 19,664,010.
15. **(a)** Town A: 3000, 4000, 5000, 6000; Town B: 1814, 2443, 3291, 4432. **(b)** 140 years.

Problem Set 4.5 (Page 126)

1. **(a)** 2.69158803 **(b)** 5.47356576 **(c)** 1.99900463 **(d)** 466.095714. **3.** **(a)** \$684.85 **(b)** \$724.46.
5. \$270.48. **7.** **(a)** between 6 and 7 years **(b)** 70 months **(c)** a little under 70 months.
9. **(a)** \$225.37 **(b)** 12.68% **11.** **(a)** 453.29 **(b)** 530.22. **13.** **(a)** \$689.98 **(b)** \$697.10 **(c)** \$697.77
15. \$1064.43. **17.** \$1341.21. **19.** \$774.34. **21.** **(a)** Most of the time Homer will be using much less than \$20, since he is steadily paying off his debt. **(b)** 18% compounded monthly. **23.** 15% compounded monthly.

Problem Set 4.6 (Page 133)

1. 2584, 4181, 6765. **3.** 7, 12, 20, 33, 54, 88. **5.** 1, 3, 8, 21, f_{2k}. **7.** 89. **9.** The two sides of a family tree do not overlap. For example, sisters and brothers never marry. **11.** $3 \cdot 5$, $5 \cdot 8$, $f_n \cdot f_{n+1}$.
13. **(a)** 1, 3, 4, 7, 11, 18, 29, 47, 76, 123 **(b)** 7, 11, 18; 21, 55, 144; 21, 55, 144; $g_n = f_{n-1} + f_{n+1}$; $f_{2n} = f_n \cdot g_n$.
15. **(a)** 19, 29 **(b)** 8, 22 **(c)** 16, 27. **17.** $\frac{1}{1}, \frac{2}{1}, \frac{3}{2}, \frac{5}{3}, \frac{8}{5}, \frac{13}{8}, \frac{21}{13}, \frac{34}{27}$; $k_n = \dfrac{f_{n+1}}{f_n}$.

Problem Set 5.1 (Page 142)

1. $3 \cdot 10 \cdot 4 = 120$. **3.** $4 \cdot 5 = 20$. **5.** $(6 + 8)4 = 56$. **7.** $5 \cdot 5 = 25$, $5 \cdot 4 = 20$.
9. $4 + 4 \cdot 3 + 4 \cdot 3 \cdot 2 + 4 \cdot 3 \cdot 2 \cdot 1 = 64$. **11.** $3 \cdot 9 \cdot 5 \cdot 4 \cdot 3 \cdot 2 \cdot 5 \cdot 4 \cdot 3$. **13.** $(10)^7$, $9(10)^6$.
15. $9 \cdot 10 \cdot 10$, $9 \cdot 9 \cdot 8$. **17.** $6 \cdot 10 \cdot 3 = 180$, $1 \cdot 7 \cdot 3 + 5 \cdot 10 \cdot 3 = 171$. **19.** **(a)** $7 \cdot 6 \cdot 5 \cdot 4 \cdot 3 = 2520$ **(b)** $8 \cdot 7 \cdot 6 \cdot 5 = 1680$ **(c)** 35 **(d)** 756. **21.** **(a)** n **(b)** $(n)(n)(n - 1) = n^3 - n^2$. **23.** **(a)** 1680 **(b)** 18,150 **(c)** $3^9 = 19{,}683$. **25.** $n!$. **27.** $(n/2)^n$. **29.** 19. **31.** 248. **33.** **(a)** 150 **(b)** 243.

Problem Set 5.2 (Page 149)

1. **(a)** $4! = 24$ **(b)** $5! = 120$ **(c)** $7!$ **3.** $8 \cdot 7 \cdot 6 \cdot 5, 8 \cdot 7 \cdot 6 \cdot 5 \cdot 4$. **5.** **(a)** $5!/2! = 60$ **(b)** $6!/2! = 360$ **(c)** $7!/(2!\,2!)$ **(d)** $10!/3!$. **7.** ${}_6P_6 + {}_6P_5 + {}_6P_4 + {}_6P_3 + {}_6P_2 + {}_6P_1 = 1956$. **9.** **(a)** $6!/3! = 120$ **(b)** 4^6. **11.** $10 \cdot 9 \cdot 8 = 720$. **13.** $2^{10} = 1024$. **15.** $9!/(4!\,3!\,2!)$. **17.** 70. **19.** $26^{10} + 26^9 + 26^8 + \cdots + 26^2 + 26 \approx 1.4681 \times 10^{14}$.

Problem Set 5.3 (Page 154)

1. **(a)** 120 **(b)** 126 **(c)** 455 **(d)** 4950 **(e)** 56 **(f)** $8 \cdot 7 \cdot 6 \cdot 5 \cdot 4 = 6720$. **3.** ${}_{50}C_3 = 19600$. **5.** **(a)** ${}_5C_3 = 10$ **(b)** ${}_5C_4 = 5$ **(c)** $10 + 5 + 1 = 16$. **7.** ${}_{13}C_5$, ${}_{12}C_4$. **9.** **(a)** ${}_9C_3 \cdot {}_6C_2 = 84 \cdot 15 = 1260$ **(b)** ${}_6C_3 \cdot {}_9C_2 = 20 \cdot 36 = 720$ **(c)** ${}_9C_5 + {}_6C_5 = 126 + 6 = 132$ **(d)** ${}_9C_3 \cdot {}_6C_2 + {}_9C_4 \cdot {}_6C_1 + {}_9C_5 = 2142$. **11.** ${}_{12}C_4 \cdot {}_8C_4$. **13.** 20,736. **15.** 32. **17.** 86. **19.** $b_n = a_n + a_{n-9} + a_{n-19} + a_{n-29} + \cdots$.

Problem Set 5.4 (Page 162)

1. **(a)** 10 **(b)** 84 **(c)** 210 **(d)** 495. **3.** There are ${}_5C_3 = 10$ such words. **5.** **(a)** ${}_9C_3 = 84$ **(b)** ${}_{12}C_3 = 220$ **(c)** ${}_{13}C_7$. **7.** **(a)** 55 **(b)** 56. **11.** Fibonacci Sequence. **13.** **(a)** $a^4 + 4a^3b + 6a^2b^2 + 4ab^3 + b^4$ **(b)** $c^5 - 5c^4d + 10c^3d^2 - 10c^2d^3 + 5cd^4 - d^5$ **(c)** $u^6 + 6u^5(2v) + 15u^4(2v)^2 + 20u^3(2v)^3 + 15u^2(2v)^4 + 6u(2v)^5 + (2v)^6$ or $u^6 + 12u^5v + 60u^4v^2 + 160u^3v^3 + 240u^2v^4 + 192uv^5 + 32v^6$.

Problem Set 5.5 (Page 168)

3. 6188 **5.** 330 **7.** **(a)** 5^{18} **(b)** 7315 **(c)** 96, 621, 525.

Problem Set 6.1 (Page 177)

1. **(a)** $\frac{1}{6}$ **(b)** $\frac{1}{2}$ **(c)** $\frac{1}{3}$ **(d)** $\frac{1}{2}$ **(e)** $\frac{1}{2}$. **3.** **(a)** $\frac{1}{8}$ **(b)** $\frac{3}{8}$ **(c)** $\frac{1}{2}$. **5.** **(a)** $\frac{1}{6}$ **(b)** $\frac{1}{6}$ **(c)** $\frac{20}{36} = \frac{5}{9}$. **7.** **(a)** 16 **(b)** $\frac{2}{16}$ **(c)** $\frac{13}{16}$. **9.** **(a)** States vary in population; the equal likelihood assumption fails. **(b)** A person may both drink and smoke; the disjointness assumption fails. **(c)** The two numbers should add to one; they don't. **(d)** Football frequently results in ties. **11.** **(a)** $\frac{16}{50}$ **(b)** $\frac{100}{240}$ **(c)** $\frac{120}{300}$ **(d)** $\frac{32}{300}$. **13.** $\frac{1}{4!} = \frac{1}{24}$. **15.** **(a)** $\frac{1}{2}$ **(b)** $\frac{1}{4}$ **(c)** $\frac{1}{13}$. **17.** **(a)** ${}_{26}C_3/{}_{52}C_3$ **(b)** ${}_{13}C_3/{}_{52}C_3$ **(c)** ${}_4C_1 \cdot {}_{48}C_2/{}_{52}C_3$ **(d)** ${}_4C_3/{}_{52}C_3$. **19.** **(a)** $\frac{1}{4} + \frac{1}{12} + \frac{1}{12} = \frac{5}{12}$ **(b)** $\frac{1}{3} + \frac{1}{4} + \frac{1}{6} + \frac{1}{12} = \frac{10}{12}$ **(c)** $\frac{11}{12}$. **23.** $4(12/{}_{52}C_5)$. **25.** .0004952. **27.** ${}_8C_5/{}_{52}C_5 \approx .0000215$. **29.** ${}_{34}C_5/{}_{52}C_5 \approx .10706$.

Problem Set 6.2 (Page 186)

1. 1/216. **3.** **(a)** $\frac{1}{4} \cdot \frac{1}{2} = \frac{1}{8}$ **(b)** $\frac{3}{4} \cdot \frac{1}{2} = \frac{3}{8}$ **(c)** $\frac{1}{4} \cdot \frac{1}{2} = \frac{1}{8}$ **(d)** $\frac{1}{4} \cdot \frac{5}{6} = \frac{5}{24}$ **(e)** $\frac{1}{2} \cdot \frac{2}{3} + \frac{1}{2} \cdot \frac{1}{3} = \frac{1}{2}$. **5.** **(a)** $\frac{5}{12} \cdot \frac{5}{12} = \frac{25}{144}$ **(b)** $\frac{5}{12} \cdot \frac{4}{11} = \frac{5}{33}$. **7.** **(a)** 0.9 **(b)** 0.5 **(c)** 0.8. **11.** $\frac{9}{10} \cdot \frac{9}{10} \cdot \frac{9}{10}$. **13.** $\frac{1}{2} \cdot \frac{4}{100} + \frac{1}{2} \cdot \frac{1}{100} = \frac{1}{40}$. **15.** $\frac{6}{10} \cdot \frac{5}{9} \cdot \frac{4}{8} = \frac{1}{6}$. **17.** **(a)** $\frac{2}{3} \cdot \frac{2}{3} \cdot \frac{2}{3} \cdot \frac{2}{3} = \frac{16}{81}$ **(b)** $\frac{1}{3} \cdot \frac{1}{3} \cdot \frac{1}{3} \cdot \frac{1}{3} = \frac{1}{81}$ **(c)** $\frac{17}{81}$ **(d)** $4 \cdot \frac{1}{3} \cdot \frac{2}{3} \cdot \frac{2}{3} \cdot \frac{2}{3} \cdot \frac{2}{3} = \frac{64}{243}$ **(e)** $\frac{64}{243} + \frac{8}{243} = \frac{8}{27}$. **19.** $\frac{6}{15} = 40\%$.

Problem Set 6.3 (Page 194)

1. **(a)** 6 **(b)** 15 **(c)** 20. **3.** **(a)** $(\frac{1}{2})^6 = \frac{1}{64}$ **(b)** ${}_6C_1(\frac{1}{2})^6 = \frac{6}{64}$ **(c)** ${}_6C_2(\frac{1}{2})^6 = \frac{15}{64}$ **(d)** ${}_6C_3(\frac{1}{2})^6 = \frac{20}{64}$ **(e)** $1 - (\frac{1}{64} + \frac{6}{64} + \frac{15}{64} + \frac{20}{64}) = \frac{22}{64}$. **5.** **(a)** ${}_{12}C_4(\frac{2}{3})^4(\frac{1}{3})^8$ **(b)** ${}_{12}C_6(\frac{2}{3})^6(\frac{1}{3})^6$. **7.** **(a)** .0162 **(b)** .2522. **9.** ${}_{10}C_0(\frac{1}{4})^0(\frac{3}{4})^{10} + {}_{10}C_1(\frac{1}{4})^1(\frac{3}{4})^9 + {}_{10}C_2(\frac{1}{4})^2(\frac{3}{4})^8 + {}_{10}C_3(\frac{1}{4})^3(\frac{3}{4})^7 = .0563 + .1877 + .2816 + .2503 = .7759$.

11. ${}_{10}C_3(.35)^3(.65)^7 + {}_{10}C_4(.35)^4(.65)^6 + \cdots + {}_{10}C_{10}(.35)^{10}(.65)^0 = .7383$. **13.** **(a)** $6(\frac{1}{6})^5$ **(b)** $6 \cdot {}_5C_4(\frac{1}{6})^4(\frac{5}{6})^1$ **(c)** $6 \cdot {}_5C_3(\frac{1}{6})^3(\frac{5}{6})^2$ **(d)** $6 \cdot 5 \cdot {}_5C_2(\frac{1}{6})^5$ **(e)** $2 \cdot 5! \cdot (\frac{1}{6})^5$. **15.** $(\frac{5}{6})^{20} + {}_{20}C_1(\frac{5}{6})^{19}(\frac{1}{6}) + {}_{20}C_2(\frac{5}{6})^{18}(\frac{1}{6})^2 + {}_{20}C_3(\frac{5}{6})^{17}(\frac{1}{6})^3 \approx .5665$. **17.** $1 - [(.9876)^{100} + {}_{100}C_1(.9876)^{99}(.0124) + {}_{100}C_2(.9876)^{98}(.0124)^2] \approx .1282$.

Problem Set 6.4 (Page 201)

1. **(b)** 9 **(c)** $\frac{9}{24}$ **(d)** $\frac{9}{24}$. **3.** $1 - Q_{20} = 1 - .41 = .59$. **5.** $\frac{99}{100} \cdot \frac{98}{100} \cdot \frac{97}{100} \cdots \frac{77}{100}$. **7.** Paul is right. Peter's four cases are not equally likely. **11.** **(a)** $1 - (.001)(.001) = .999999$ **(b)** $1 - [(.001)^4 + 4(.001)^3(.999)] = .999999996$. **13.** $\frac{1}{2} \cdot 1 + \frac{1}{4} \cdot 2 + \frac{1}{8} \cdot 4 + \frac{1}{16} \cdot 8 + \cdots$, which is infinite.

Problem Set 7.1 (Page 212)

1.

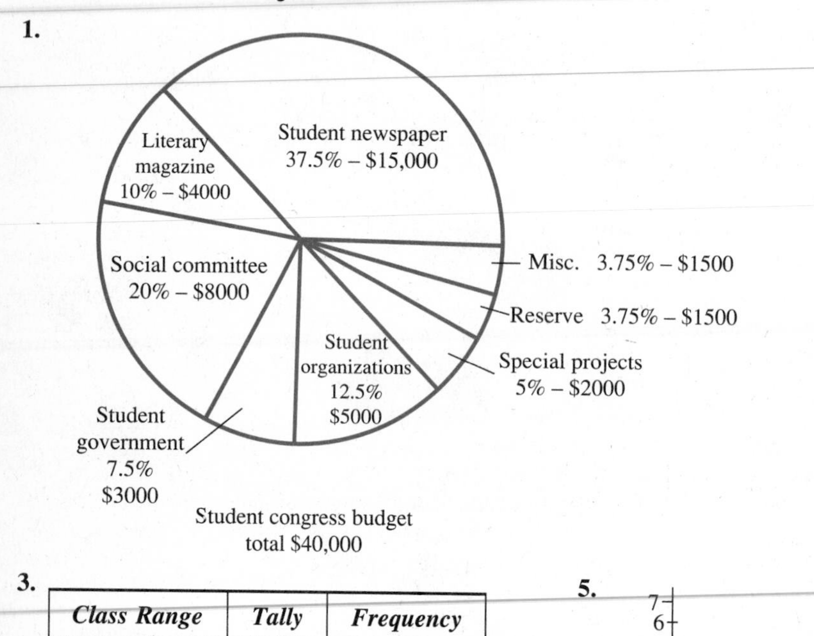

3.

Class Range	*Tally*	*Frequency*
24–29	II	2
30–35	III	3
36–41	I	1
42–47	IIII	4
48–53	~~IIII~~ II	7
54–59	~~IIII~~ I	6
60–65	III	3
66–71	II	2
72–77	II	2
78–83	III	3
84–89	II	2
90–95	I	1

5.

Frequency

Class intervals: 24-29, 30-35, 36-41, 42-47, 48-53, 54-59, 60-65, 66-71, 72-77, 78-83, 84-89, 90-95

7. Probably scores on a physics exam. Exam scores tend to cluster in the middle and thin out at the ends. It seems unlikely that the age of welfare recipients would cluster around 50. **9.** Make a trend line and a comparative bar graph. **11.** It will be four times as large. Doubling the dimensions of a cube will make the volume eight times what it was. **13. (a)** Uncertain; one needs to know if there are more male drivers and if they drive more miles. **(b)** Invalid. Ohio is much bigger and has more drivers. **(c)** Invalid. The foreign manufacturer probably began exporting to the United States recently. **(d)** Possibly invalid. It may have been an exceptionally hard test.

Problem Set 7.2 (Page 219)

1. (a) Mean 6.56, median 7, mode 8 **(b)** mean 16.22, median 16, mode 16.

3.

Word Length	*Tally*	*Frequency*
13	I	1
12		0
11	I	1
10	I	1
9	II	2
8	I	1
7	I	1
6	II	2
5	II	2
4	卌 卌 卌 I	16
3	卌 卌 II	12
2	卌 卌 II	12
1	II	2

5. 142.2 pounds. **7.** Mean $22,000; median $12,000. **9.** Not necessarily. The temperature may fluctuate wildly and still have a mean of 25°C. **11.** 3.31. **13.** 5.47 feet from the end. **15.** 150 mph. It can't be done. **17.** All students get the same score. **19.** 392.3. **21.** 2.602.

Problem Set 7.3 (Page 228)

1. Mean 7, standard deviation 1.90.

3.

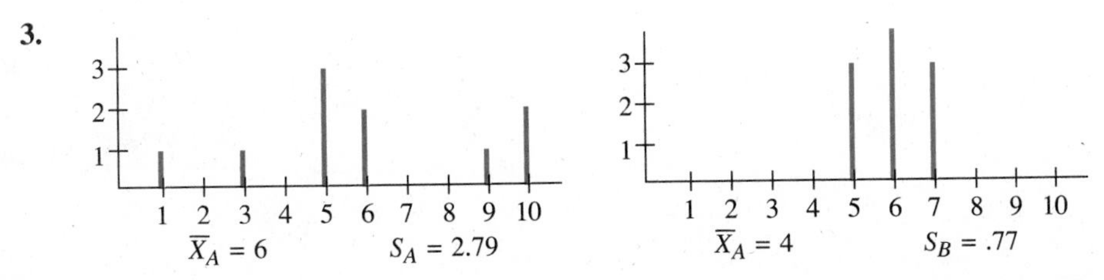

5. −2.38, −0.63, 0.75, 1.88 **7.** German −0.36, Math −1.30+ she did better in German. **9.** Mean 12, standard deviation 1.90; you add the number to the mean, but the standard deviation is unchanged. **11.** 4%, 16%.

13.

Word Length	*Tally*	*Frequency*
11	|	1
10		0
9	||	2
8	||	2
7	|||	3
6	||||	4
5	~~||||~~ |	6
4	~~||||~~ |||	8
3	~~||||~~ ~~||||~~ |||	13
2	~~||||~~ ~~||||~~	10
1	|||	3

Mean 4.12. Standard deviation 2.22

15. Zero. If the standard deviation is 0, $(x_1 - \bar{x})^2 + (x_2 - \bar{x})^2 + \cdots + (x_n - \bar{x})^2 = 0$. This implies that $x_i = \bar{x}$ for each i. **17.** $\bar{x} = 12.5$, $s_x = 1.3$. **19.** $\bar{z} = 12.8$, $s_z = 2.5$.

Problem Set 7.4 (Page 235)

1. (a) 72 **(b)** 80 **(c)** 55 **(d)** $\frac{25}{12}$. **3. (a)** $\sum_{i=1}^{10} i$ **(b)** $\sum_{i=1}^{10} i^2$ **(c)** $\sum_{i=1}^{37} y_i$ **(d)** $\sum_{i=1}^{n} (y_i - 3)^2$. **5. (a)** 4.31 **(b)** 1293 **(c)** $3 \sum_{i=1}^{100} x_i + \sum_{i=1}^{100} 1 = 3(431) + 100 = 1393$. **7.** $\bar{x} = 2$, $s = 6$. **9.** $\bar{x} = 7$, $s = 1.90$.

Problem Set 7.5 (Page 241)

1.

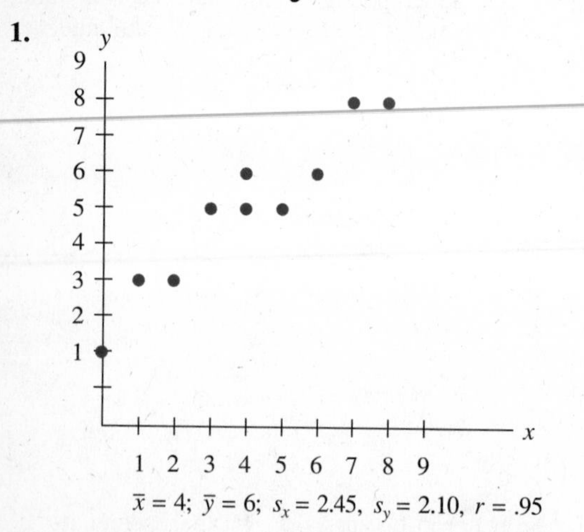

$\bar{x} = 4$; $\bar{y} = 6$; $s_x = 2.45$, $s_y = 2.10$, $r = .95$

3. (a) High negative correlation; y decreases as x increases. **(b)** Low negative correlation; y decreases as x increases. **(c)** Very low (and probably insignificant) correlation. **(d)** Quite high positive correlation; y increases as x increases. **5.** Author's guesses are: **(a)** high negative **(b)** high positive **(c)** low positive **(d)** high negative **(e)** low positive **(f)** low negative **(g)** low positive **(h)** high positive **(i)** low positive. **11.** $r_{xy} \approx -0.237$. This slight negative correlation is probably not significant. **13.** $r_{xy} \approx 0.976$. This is a very high positive correlation and indicates a strong linear relationship. Of course, the data lie on the graph of $y = \sqrt{x}$ which is definitely not a straight line. However, if you graph $y = \sqrt{x}$, you will note a linear tendency.

Problem Set 8.1 (Page 249)

1. (a) No **(b)** no **(c)** yes **(d)** yes **(e)** yes. **3. (a), (b), (c),** and **(e)** have Hamiltonian circuits. **5. (a)** Yes **(b)** no. **7. (a)** No **(b)** yes **(c)** yes. **9.** A network has an Euler path if and only if it has either zero or two odd vertices. **11. (a)** yes **(b)** yes **(c)** yes. **15.** 5.

Problem Set 8.2 (Page 255)

1. $d_1 = 27$, $d_2 = 0$, $d_3 = 6$, $d_4 = 3$, $d_5 = 3$, $d_6 = 1$; $1 \cdot 6 + 2 \cdot 3 + 3 \cdot 3 + 4 \cdot 1 = 27 - 2$. **3.** $V = 40$, $E = V - 1 = 39$.

5.

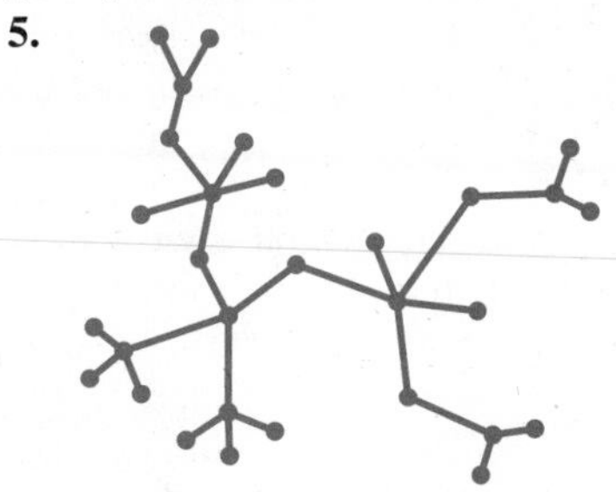

7. 50.

9.

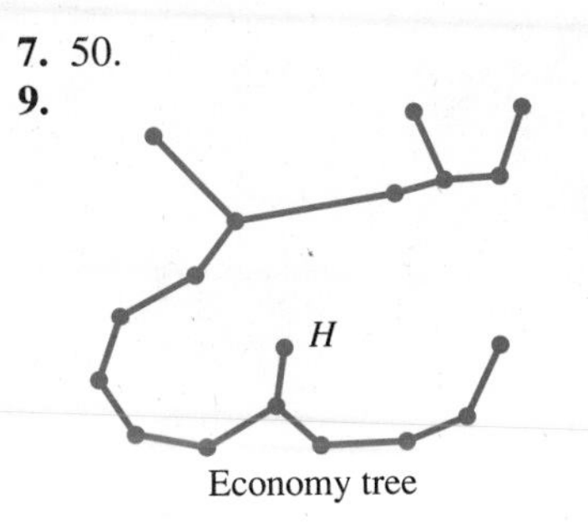

Economy tree

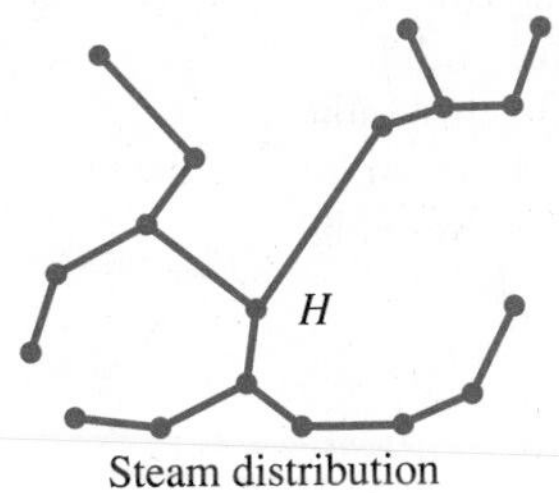

Steam distribution
(not only possibility)

Problem Set 8.3 (Page 263)

1. $F = 14$, $V = 15$, $E = 27$. **3.** $F = 6$, $V = 5$, $E = 9$. The same number of faces do not meet at each vertex.
5. $E = 9$.

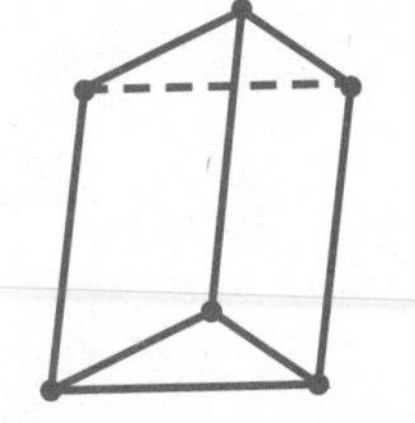

7. Suppose it did, that is, suppose $V = E$. Use Euler's Formula to arrive at the conclusion $F = 2$, which is impossible. **9.** Tetrahedron, no; cube, no; octahedron, yes; dodecahedron, no; icosahedron, no. **11.** $F = 8$, $V = 12$, $E = 18$; yes. **13.** Assume the polyhedron is flexible. Push out the dents. This doesn't change F, V, E.

Problem Set 8.5 (Page 278)

1. (a) Three colors **(b)** two colors **(c)** three colors **(d)** two colors. **3.** Four colors. **7.** 77. **9. (a)** 4 **(b)** 3 **(c)** 2 **(d)** 4 **(e)** 3.

Problem Set 9.1 (Page 288)

1. Obviously, there is no single set of answers. We have in mind such things as the following: **(a)** There was an earthquake last week in ______. Certain insulating materials are fire-resistant. Company X makes the best calculators. The use of toothpaste reduces cavities. **(b)** A certain road will break up every spring. If it rains, be it ever so gentle, traffic jams will build up at the bridge. The Wonder Department Store will have a January sale. **(c)** Two aspirin will relieve a headache. A certain spot in your yard is not a good place to grow tomatoes. Brand X ballpoint pens perform better than brand Y ballpoint pens. **5.** The numbers through $k = 27$ are classified as: *Odd type:* 2, 3, 5, 7, 8, 11, 12, 13, 17, 18, 19, 20, 23, 27; *Even type:* 4, 6, 9, 10, 14, 15, 16, 21, 22, 24, 25, 26. The odds lead the evens, 14 to 12. **7. (a)** The numbers on the left are the binomial coefficients; hence $(x + y)^3 = 1x^3 + 3x^2y + 3xy^2 + 1y^3$. Set $x = y = 1$. The natural and correct conjecture comes from the expansion of $(x + y)^n$. **(b)** The natural conjecture is shattered by circling primes in the next line. **9. (a)** The intersections of AB with ab, AC with ac and BC with bc will be in a straight line. This is Desargues' Two-Triangle Theorem. **(b)** The intersections of the three pairs of external tangents are colinear.

Problem Set 9.2 (Page 296)

1. (a) Not valid. **(b)** Valid. **(c)** Not valid. **3. (a)** Not valid. **(b)** Not valid. **(c)** Not valid. **(d)** Valid.

5. **(a)** Valid. **(b)** Not valid. **(c)** Not valid. **(d)** Not valid.

7. **(a)** Not valid. **(b)** Not valid. **(c)** Not valid.

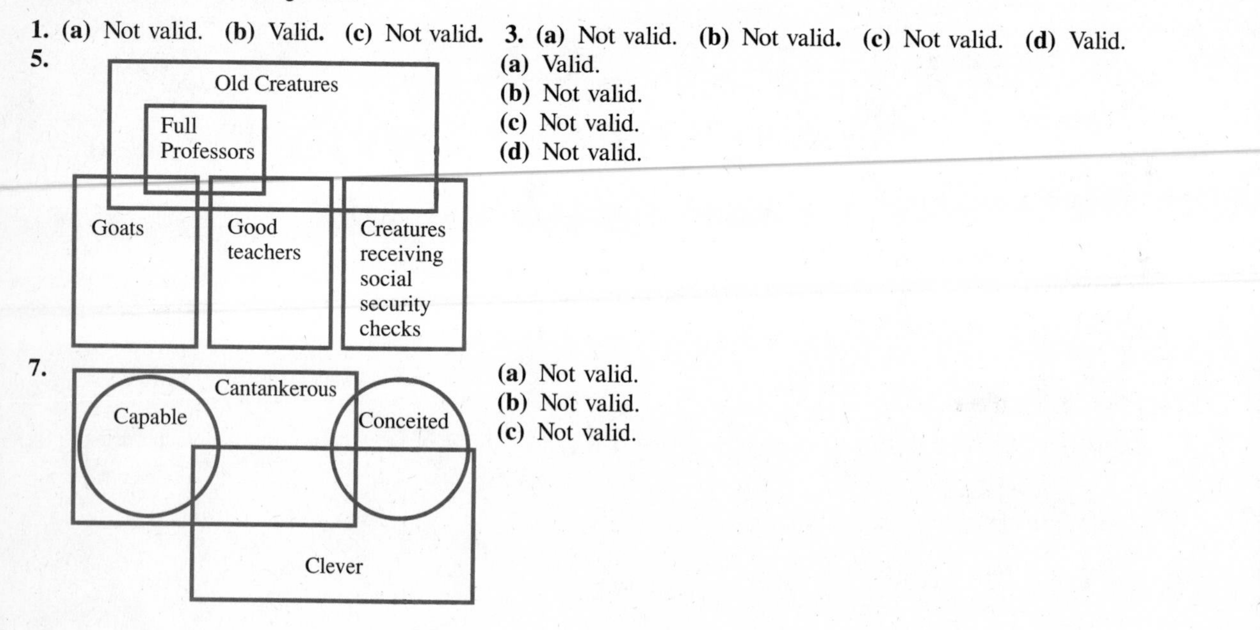

9.

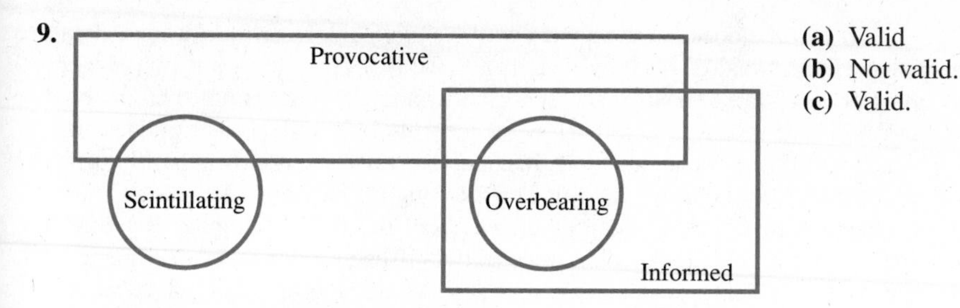

(a) Valid
(b) Not valid.
(c) Valid.

11. **(a)** This is not valid. (We need the added assumption that Hugo follows the best tactic.) **(b)** Horace is wrong. (He assumes the converse of the proposition.) **(c)** Valid. **13.** **(a)** Not valid. **(b)** Not valid. **(c)** Not valid. **(d)** Valid.

15. **(a)** *Hypotheses:* My home state collects taxes on income made in the state.
I conduct my business in another state.
Conclusion: I won't have to pay income tax in my state.

(b) *Hypotheses:* Dogs can't pull a sled that is too heavy.
If you take all that gear, the sled will be too heavy.
Conclusion: The dogs won't be able to pull the sled.

(c) *Hypotheses:* A sensitive person does not continually refuse offers of help.
Homer accepts any offer of help.
Conclusion: Homer is a sensitive person.

17.

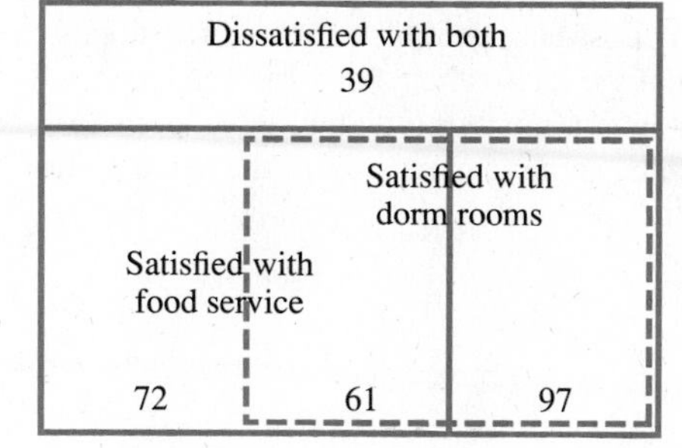

Total Students	
Satisfied with food service only	11
Satisfied with dorm rooms only	36
Satisfied with both	61
Satisfied with neither	39
	147

19. **(a)** *Hypotheses:* Moon rocks are made of blue cheese.
No one has touched blue cheese from the moon.
Conclusion: No one has touched a rock from the moon.

(b) *Hypotheses:* Moon rocks are made of green cheese.
No one has eaten blue cheese from the moon.
Conclusion: No one has eaten moon rocks.

(c) If the hypotheses are true and the argument valid, the conclusion must be true.

(d) *Hypotheses:* Moon rocks can only be obtained by going to the moon.
We have moon rocks.
Conclusion: We have been to the moon.

Problem Set 9.3 (Page 303)

1. **(a)** If the true place for a just man is in prison, then a government imprisons any person unjustly. **(b)** If a man is alone, then he is thinking or working. **(c)** If there are many things a man can afford to leave alone, then he is truly rich. **(d)** If a man hears a different drummer, then he does not keep pace with his companions. **(e)** If a man sweats easier than I do, then he must earn his living by the sweat of his brow. **3. (a)** If the true place for a just man is not in prison, then a government does not imprison any man unjustly. **(b)** If a man is alone, then he is not thinking or working. **(c)** If there is nothing that one can afford to leave alone, then he is not truly rich. **(d)** If a man does not hear a different drummer, then he keeps pace with his companions. **(e)** If a man does not sweat easier than I do, then he will not have to earn his living by the sweat of his brow. **5. (a)** The meaning of the word "depression" has shifted. **(b)** If *A* implies *B*, but *A* doesn't happen, no conclusion can be drawn about *B*. **(c)** If *A* implies *B* and *B* is true, it is not necessarily true that *A* is true. **7.** At a combined selling price of two-fifths of a dollar for one gizmo, the one who had been selling them for one-third of a dollar apiece gained $30(\frac{2}{5} - \frac{1}{3}) = \$2$, so his share of the sales is his expected earnings of \$10 plus \$2. The other exhibitor, however, lost $30(\frac{1}{2} - \frac{2}{5}) = \$3$, so his share is \$15 minus \$3. Nothing is missing. **9.** The farmer, in willing $\frac{1}{2} + \frac{1}{3} + \frac{1}{9} = \frac{17}{18}$ of his possessions, didn't provide for giving away all that he had, so the additional contributed horse was left over. **11. (a)** Both balls arrive simultaneously. Imagine a 15-pound weight falling. Will it fall faster (or slower) if it has heen sawed into two parts? **(b)** Try it. **13.** If it is true that every rule has an exception, so must the rule in part (a); hence some rule must have no exception. A similar difficulty is encountered with parts (b) and (c).

Problem Set 9.4 (Page 313)

1. $(2n)^2 = 4n^2 = 2(2n^2)$. **3.** $(3n + 1)^2 = 9n^2 + 6n + 1 = 3(3n^2 + 2n) + 1$. $(3n + 2)^2 = 9n^2 + 12n + 4 = 3(3n^2 + 4n + 1) + 1$. **5.** $2n(2n + 1) = 2(2n^2 + n)$. **7.** Let the odd number be $2n + 1$. Its square is $4n^2 + 4n + 1$. If n is even, $n = 2k$. $4n^2 + 4n + 1 = 4(2k^2) + 4(2k) + 1 = 8[2k^2 + k] + 1$ If n is odd, $n = 2k + 1$ and $4n^2 + 4n + 1 = 4[4k^2 + 4k + 1] + 4[2k + 1] + 1 = 8[2k^2 + 3k + 1] + 1$ **9.** Suppose an arbitrary angle has been divided into six equal subangles, using only a compass and straightedge. Then, taken in adjacent pairs, they would be trisectors, violating the fact that one cannot so construct angle trisectors. **11.** Suppose n is not odd, so that $n = 2k$. Then $n^2 = 2(2k^2)$ is not odd, contradicting what we know about n^2. **13.** Imagine each maple tagged with a slip showing the number of leaves. Let n be the maximum number on a tag. By hypotheses, there are more than n trees, hence more than n tags. By the pigeonhole principle, some tags must have the same number written on them. **15.** Yes. Again the argument is by the pigeonhole principle.

Problem Set 10.1 (Page 323)

11. Gwen loves Alan. **13.** The Greater Glory Party.

Problem Set 10.2 (Page 330)

1. **3.**

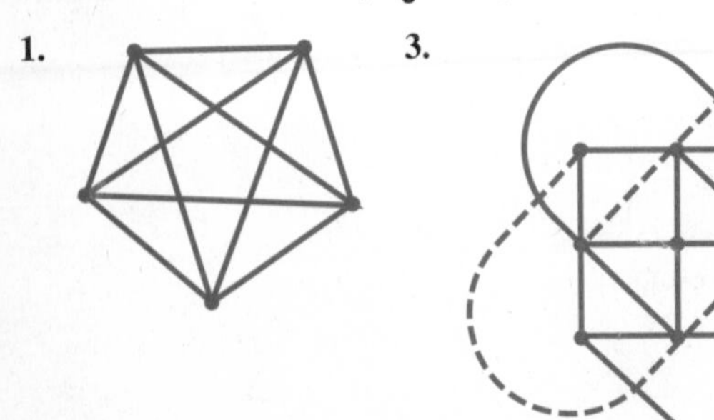

Problem Set 10.3 (Page 336)

5. (a) This course is not useful education. **(b)** This course can be useful in preparing for certain jobs. **7. (a)** Those who wish to improve themselves go to school. **(b)** Homer wishes to improve himself.

Problem Set 10.4 (Page 347)

1. (a) My arm from wrist to elbow measures 28 centimeters, and it appears to have a radius of about 4 centimeters. With a cylinder as a model, the volume appears to be $\pi 4^2 \cdot 28 = \frac{22}{7}(4^2)(28) = 22(4^3) = 1408$ cubic centimeters. **(b)** With a sphere as a model and approximating the radius of my head to be $r = 12$ centimeters, the volume appears to be $\frac{4}{3}\pi \cdot 12^3 = 7235$ cubic centimeters.

Problem Set 11.4 (Page 376)

1. There should be a correlation between aptitude and grade; a person with low aptitude should expect a low grade. **3.** Since my folks are poor correspondents, then I am excused for not writing to them. **5.** Change the second hypothesis to: A liberal arts education should also prepare students for a job. **7.** Homer wishes to improve himself. **9.** Draw the picture carefully; we encountered the same problem in our proof that all triangles are isoceles.

Problem Set 12.1 (Page 388)

1. (a) g.c.d. = 2, l.c.m. = 240 **(b)** g.c.d. = 3, l.c.m. = 90 **(c)** g.c.d. = 1, l.c.m. = 1800. **3. (a)** $2 \cdot 2 \cdot 2 \cdot 5$ **(b)** $2 \cdot 2 \cdot 13$ **(c)** $2 \cdot 3 \cdot 13$ **(d)** $2 \cdot 2 \cdot 3 \cdot 5$ **(e)** $2 \cdot 5 \cdot 7$ **(f)** $2 \cdot 3 \cdot 3 \cdot 7$ **(g)** $2 \cdot 2 \cdot 3 \cdot 3 \cdot 7$. **5. (a)** g.c.d. = 2, l.c.m. = 1560 **(b)** g.c.d. = 2, l.c.m. = 1260. **7. (a)** $\frac{19}{24}$ **(b)** $\frac{53}{48}$ **(c)** $\frac{127}{105}$ **(d)** $\frac{143}{360}$. **9. (a)** $2^4 \cdot 3^3 \cdot 5^2$ **(b)** $2 \cdot 3^4 \cdot 5^2 \cdot 13$ **(c)** $5^2 \cdot 3 \cdot 11 \cdot 47$. **11.** 24. **13.** $n = 4$. **15.** 4 inches or 9 inches. **17.** 133.

Problem Set 12.2 (Page 395)

1. (a) 12.615 **(b)** 11.865 **(c)** 0.05625 **(d)** 2.5. **3.** $0.\overline{285714}$. **5. (a)** $0.375\overline{0}$ **(b)** $2.375\overline{0}$ **(c)** $0.\overline{45}$ **(d)** $3.91\overline{6}$. **7.** $\frac{49}{133}$. **9. (a)** $\frac{19}{33}$ **(b)** $\frac{17}{66}$. **11. (a)** 1 2 11 2 111 2 1111 2 11111 2 . . . . **(b)** This is an irrational number. **13.** Yes. **15.** Note that $r - \frac{3}{4}$ is rational **17.** Let $r = \frac{3}{4}\sqrt{2}$ and note that if r is rational so is $\frac{4}{3}r$. **19. (a)** irrational **(b)** irrational **(c)** irrational **(d)** rational **(e)** rational **(f)** irrational **(g)** irrational **(h)** rational. **21.** 0.0000001/2 will do. There is no smallest rational. **23.** If r and s are rational, so is their average $(r+s)/2$. **25.** The decimal expansion of π is nonrepeating.

Problem Set 12.3 (Page 403)

1. (a) 6 **(b)** 5 **(c)** 4 **(d)** 2. **3.** Friday, Sunday. **5. (a)** 3 **(b)** 1 **(c)** 4 **(d)** 3 **(e)** 2 **(f)** 4 **(g)** 6 **(h)** 1. **7. (a)** 5 **(b)** 3, 7, 11 **(c)** 3 **(d)** 4. **9. (a)** 4 **(b)** 8 **(c)** 6 **(d)** 5. **11. (a)** 16,896 **(b)** okay **(c)** 302736 **(d)** 312193. **13.** See Problem 21. **15.** If $a = b + km$ and $b = c + rm$, then $a = c + (k + r)m$. **17.** $a = b + 7k$ and $c = d + 7m$, so $ac = bd + 7[bm + kd + 7mk]$. **19. (a)** 4 **(b)** 5 **(c)** 2 **(d)** 6.

Problem Set 12.4 (Page 410)

1. **(a)** $x = 1 \bmod 5$, $y = 0 \bmod 2$

x	$\cdots$	-4	1	6	11	$\cdots$
y	$\cdots$	2	0	-2	-4	$\cdots$

(b) $x = -1 \bmod 16$, $y = 2 \bmod 15$

x	$\cdots$	-17	-1	15	$\cdots$
y	$\cdots$	17	2	-13	$\cdots$

(c) $x = 9 \bmod 25$, $y = -13 \bmod 37$

x	$\cdots$	-41	-16	9	$\cdots$
y	$\cdots$	61	24	-13	$\cdots$

(d) No integer solutions exist.

(e) $x = 55 \bmod 86$, $y = -73 \bmod 117$

x	$\cdots$	-31	55	141	$\cdots$
y	$\cdots$	44	-73	-190	$\cdots$

3.

x	$\cdots$	-1	1	3	$\cdots$
y	$\cdots$	2	15	28	$\cdots$
t	$\cdots$	5	12	19	$\cdots$

5. Homer gets $4(17) = \$68$; Horatio gets $25(3) = \$75$. **7.** 47 tables, 311 chairs. **9.** Coconuts $= 3121 \bmod 15{,}625$; *i.e.* 18,746; 34,371; etc.

Problem Set 13.1 (Page 421)

1. $AB = \begin{bmatrix} 7 & 10 \\ 9 & 13 \end{bmatrix}$ $BA = \begin{bmatrix} 12 & 19 \\ 5 & 8 \end{bmatrix}$; $BC = \begin{bmatrix} -20 & 9 \\ -9 & 4 \end{bmatrix}$ $CB = \begin{bmatrix} 13 & 18 \\ -21 & -29 \end{bmatrix}$.

3. $MC = \begin{bmatrix} -29 & 13 \\ -38 & 17 \end{bmatrix} = AN.$

5. $(BC)D = \begin{bmatrix} -20 & 9 \\ -9 & 4 \end{bmatrix}\begin{bmatrix} 5 & -3 \\ -3 & 2 \end{bmatrix} = \begin{bmatrix} -127 & 78 \\ -57 & 35 \end{bmatrix} = \begin{bmatrix} 5 & 7 \\ 2 & 3 \end{bmatrix}\begin{bmatrix} 18 & -11 \\ -31 & 19 \end{bmatrix} = B(CD).$

7. $A(B+C) = \begin{bmatrix} 1 & 1 \\ 1 & 2 \end{bmatrix}\begin{bmatrix} 8 & 6 \\ -3 & 5 \end{bmatrix} = \begin{bmatrix} 5 & 11 \\ 2 & 16 \end{bmatrix} = \begin{bmatrix} 7 & 10 \\ 9 & 13 \end{bmatrix} + \begin{bmatrix} -2 & 1 \\ -7 & 3 \end{bmatrix} = AB + AC.$

9. A, A, D, D. **11.** **(a)** I **(b)** I.

13. **(a)** $\begin{bmatrix} 3 & -4 \\ -5 & 7 \end{bmatrix}$ **(b)** $\begin{bmatrix} -3 & 5 \\ 5 & -8 \end{bmatrix}$ **(c)** $\begin{bmatrix} 2 & -\frac{7}{2} \\ -1 & 2 \end{bmatrix}$ **(d)** no inverse.

15. **(a)** $\begin{bmatrix} 6 & 7 & 3 \\ 6 & 2 & 0 \\ 13 & 6 & 1 \end{bmatrix}$ **(b)** $\begin{bmatrix} 5 & -2 & 6 \\ 2 & 0 & 2 \end{bmatrix}$ **17.** $l = m$; dimensions $k \times n$.

Problem Set 13.2 (Page 427)

1. $(AB)C = \begin{bmatrix} -7 & -2 \\ -11 & -3 \end{bmatrix}\begin{bmatrix} 2 & -1 \\ 1 & 4 \end{bmatrix} = \begin{bmatrix} -16 & -1 \\ -25 & -1 \end{bmatrix} = \begin{bmatrix} 1 & -2 \\ 2 & -3 \end{bmatrix}\begin{bmatrix} -2 & 1 \\ 7 & 1 \end{bmatrix}$
$= A(BC).$

3. $AB = \begin{bmatrix} -7 & -2 \\ -11 & -3 \end{bmatrix}$ $BA = \begin{bmatrix} -1 & 2 \\ 5 & -9 \end{bmatrix}$.

5. $A(B + C) = \begin{bmatrix} 1 & -2 \\ 2 & -3 \end{bmatrix}\begin{bmatrix} 1 & -1 \\ 4 & 5 \end{bmatrix} = \begin{bmatrix} -7 & -11 \\ -10 & -17 \end{bmatrix} = \begin{bmatrix} -7 & -2 \\ -11 & -3 \end{bmatrix} + \begin{bmatrix} 0 & -9 \\ 1 & -14 \end{bmatrix} = AB + AC.$

7. **(a)** $A^{-1} = \begin{bmatrix} -3 & 2 \\ -2 & 1 \end{bmatrix}$ **(b)** $C^{-1} = \begin{bmatrix} \frac{4}{9} & \frac{1}{9} \\ -\frac{1}{9} & \frac{2}{9} \end{bmatrix}$ **(c)** $(AC)^{-1} = \begin{bmatrix} -\frac{14}{9} & 1 \\ -\frac{1}{9} & 0 \end{bmatrix}$; $A^{-1}C^{-1} = \begin{bmatrix} -\frac{14}{9} & \frac{1}{9} \\ -1 & 0 \end{bmatrix}$.

9. $C^{-1}A^{-1} = \begin{bmatrix} -\frac{14}{9} & 1 \\ -\frac{1}{9} & 0 \end{bmatrix} = (AC)^{-1}$ $(CA)^{-1} = \begin{bmatrix} -\frac{14}{9} & \frac{1}{9} \\ -1 & 0 \end{bmatrix} = A^{-1}C^{-1}$.

11. $(\det A)(\det B) = 1(-1) = \det AB$. **17.** $\begin{bmatrix} 0 & 0 \\ 0 & 0 \end{bmatrix}$. **19.** No.

Problem Set 13.3 (Page 433)

1. (a) $x = 3, y = -2$ **(b)** $x = -2, y = \frac{5}{2}$ **(c)** $x = -1, y = 2$. **3. (a)** $x = -67, y = -17, z = 12$ **(b)** $x = -65, y = -18, z = 11$.

5. $\begin{bmatrix} 195 & 12 \\ 239 & 16 \\ 288 & 19 \end{bmatrix}\begin{bmatrix} 9 \\ 43 \end{bmatrix} = \begin{bmatrix} 2271 \\ 2839 \\ 3409 \end{bmatrix}$.

7.

	Dough	Cheese	Tomato Sauce	Sausage	Mushrooms
Cheese Spec.	1	3	1	0	0
Sausage	1	1	1	1	0
Super Sausage	1	1	1	$\frac{3}{2}$	1
Large Special	2	$\frac{5}{2}$	$\frac{5}{2}$	2	2

$$\cdot \begin{bmatrix} .35 & .28 \\ .35 & .28 \\ .12 & .09 \\ .79 & .63 \\ .18 & .15 \end{bmatrix} = \begin{bmatrix} 1.52 & 1.21 \\ 1.61 & 1.26 \\ 2.18 & 1.72 \\ 3.82 & 3.02 \end{bmatrix}.$$

Problem Set 14.1 (Page 442)

1. (a) No; $4 \div 2 \neq 2 \div 4$ **(b)** no; $(48 \div 12) \div 2 \neq 48 \div (12 \div 2)$. **3. (a)** -41 **(b)** -41 **(c)** yes **(d)** yes. **5.** Yes. **7. (a)** yes **(b)** yes **(c)** no; $2*3 = \frac{5}{2}$ is not an integer. **(d)** yes. **9. (a)** is determinative; **(c)** is questionable, since different-sized tires may be on the same car; (b) and (d) are matters of judgment. **11. (a)** R, S **(b)** S **(c)** S **(d)** R, T **(e)** R, S **(f)** R, S. **13.** "Separated from" is symmetric, not reflexive nor transitive. **15.** "Is at least as far north as."

Problem Set 14.2 (Page 450)

1. (a) $\bar{x} = 2$ **(b)** $\bar{y} = 2$ **(c)** $\overline{xy} = 1$ **(d)** $xy = 1$. **3. (a)** Doesn't exist **(b)** 1 **(c)** 3 **(d)** 2 **(e)** 4. **5. (a)** 9 **(b)** doesn't exist **(c)** 7 **(d)** 5 **(e)** 1 **(f)** 11. **7. (a)** $\begin{bmatrix} -1 & -2 \\ -1 & -3 \end{bmatrix}$ **(b)** $\begin{bmatrix} 3 & -2 \\ -1 & 1 \end{bmatrix}$ **(c)** $\begin{bmatrix} -5 & -5 \\ -1 & -1 \end{bmatrix}$ **(d)** doesn't exist. **9.** There is no additive identity, so it doesn't even make sense to ask if each element has an additive inverse. **11.** $(a + b)(a - b) = a^2 - ab + ba - b^2 = a^2 - b^2$ (if and only if $ab = ba$) The result depends on multiplication being commutative.

Problem Set 14.3 (Page 455)

1. $\overline{0} = 0, \overline{1} = 4, \overline{2} = 3, \overline{3} = 2, \overline{4} = 1; 1^{-1} = 1, 2^{-1} = 3, 3^{-1} = 2, 4^{-1} = 4.$ **3. (a)** $\overline{4} = 9$ **(b)** $4^{-1} = 10$ **(c)** $\overline{5} = 8$ **(d)** $5^{-1} = 8.$ **13. (a)** The universal set U acts as multiplicative identity. **(b)** The empty set Ø acts as the additive identity.

15. $\begin{bmatrix} a & -b \\ b & a \end{bmatrix}\begin{bmatrix} c & -d \\ d & c \end{bmatrix} = \begin{bmatrix} R & -S \\ S & R \end{bmatrix}$ where $\begin{matrix} R = ac - bd \\ S = ad + bc \end{matrix}$ **19.** Yes.

Problem Set 14.4 (Page 460)

1. To subtract a (positive) 3 from -5, write either

$$\text{subtract}\quad \begin{array}{r} -5 \\ 3 \\ \hline \end{array} \qquad \text{or} \qquad \begin{array}{r} -5 \\ -3 \\ \hline \end{array}$$

Either way, the answer is -8.

3. (a) $-\frac{5}{3}$ **(b)** $\frac{5}{2}$ **(c)** $-\frac{4}{9}$ **(d)** -1. **5.** 6 **7. (a)** $x_1 = -\frac{1}{3}$ $x_2 = 2$ **(b)** $\begin{bmatrix} \frac{1}{6} & 1 \\ \frac{3}{2} & 1 \end{bmatrix}$.

9. $\begin{bmatrix} 2 & 1 \\ 6 & 4 \end{bmatrix}^{-1} = \begin{bmatrix} 2 & 5 \\ 8 & 1 \end{bmatrix}$ $\begin{bmatrix} x_1 & x_2 \\ x_3 & x_4 \end{bmatrix} = \begin{bmatrix} 4 & 5 \\ 1 & 4 \end{bmatrix}$ **11. (a)** 7 **(b)** $x = 6 \bmod 26$.

Acknowledgments

PHOTO CREDITS

Unless otherwise acknowledged, all photographs are the property of ScottForesman. Page abbreviations are as follows: (L) left, (R) right.

3 AP/Wide World
16 Mark Antman/The Image Works
34L Graphic courtesy of the Boeing Corporation, Postcard from Joint Policy Board for Mathematics
34R Photo Courtesy of Pechiney Research Center, Postcard from Joint Policy Board for Mathematics
70 Photograph courtesy of James Hamilton
119 Paul Chesley/Tony Stone Images
132 Chris Mihalka/The Stock Market
138 Bob Daemmrich/The Image Works
172 Bruce Aynes/Tony Stone Images
182 Pat Watson/Tony Stone Images
268 Brownie Harris/The Stock Market
270 © 1995/Cordon Art—Baarn—Holland
281 California Institute of Technology
321 John Warden/Tony Stone Images
338 Bob Daemmrich/The Image Works
345 Bob Kalman/The Image Works
368 George Hunter/Tony Stone Images
379 Courtesy Prof. Gottfried E. Noether

LITERARY PERMISSIONS

26 From *Puzzles in Math and Logic* by Aaron J. Friedland. Copyright © 1970 by Aaron J. Friedland. Reprinted by permission of Dover Publications, Inc.
84 From *The Moscow Puzzles* by Boris A. Kordemsky, edited by Martin Gardner and translated by Albert Parry. Copyright © 1972 by Charles Scribner's Sons. Reprinted by permission.
121 From *The Participant,* November 1992. Reprinted by permission of Teachers Insurance and Annuity Association College Retirement Equities Fund.
263 *Kepler's Model* is reprinted courtesy of The Princeton University Press (Source: Hermann Weyl, Symmetry, p. 76, copyright 1952 by The Princeton University Press).
357 From *Collected Sonnets of Edna St. Vincent Millay.* Reprinted by permission.

Names and Faces Index

Boldface page numbers indicate that a picture appears on that page.

Subject Index

Italic entries indicate problem titles.